45.00

THE WONDERFUL WORLD OF STOCHASTICS
A TRIBUTE TO ELLIOTT W. MONTROLL

STUDIES IN STATISTICAL MECHANICS

VOLUME XII

SERIES EDITOR

J. L. LEBOWITZ
Rutgers University, New Brunswick, NJ

ADVISORY BOARD

E. G. D. Cohen
R. Dobrushin
G. Gallavotti
N. G. van Kampen
J. S. Langer
B. Widom

NORTH-HOLLAND
AMSTERDAM · OXFORD · NEW YORK · TOKYO

THE WONDERFUL WORLD OF STOCHASTICS

A TRIBUTE TO ELLIOTT W. MONTROLL

EDITORS

Michael F. SHLESINGER
Office of Naval Research, Arlington, VA

George H. WEISS
National Institute of Health, Bethesda, MD

1985

NORTH-HOLLAND
AMSTERDAM · OXFORD · NEW YORK · TOKYO

© Elsevier Science Publishers B.V., 1985

All rights reserved. No part of this publication may be reproduced, stored in a retrieval system, or transmitted, in any form or by any means, electronic, mechanical, photocopying, recording or otherwise, without the prior permission of the publisher, Elsevier Science Publishers B.V. (North-Holland Physics Publishing Division), P.O. Box 103, 1000 AC Amsterdam, The Netherlands.

Special regulations for readers in the USA: This publication has been registered with the Copyright Clearance Center Inc. (CCC), Salem, Massachusetts. Information can be obtained from the CCC about conditions under which photocopies of parts of this publication may be made in the USA. All other copyright questions, including photocopying outside of the USA, should be referred to the publisher.

ISBN: 0 444 86937 9

Published by:

North-Holland Physics Publishing

a division of
Elsevier Science Publishers B.V.
P.O. Box 103
1000 AC Amsterdam
The Netherlands

Sole distributors for the U.S.A. and Canada:

Elsevier Science Publishing Company, Inc.

52 Vanderbilt Avenue
New York, N.Y. 10017
U.S.A.

Library of Congress Cataloging-in-Publication Data
Main entry under title:

The Wonderful world of stochastics.

   (Studies in statistical mechanics; v. 12)
   Bibliography: p.
   Includes index.
   1. Random walks (Mathematics)--Addresses, essays, lectures.   2. Stochastic processes--Addresses, essays, lectures.   3. Statistical mechanics--Addresses, essays, lectures.   4. Montroll, E.W.   I. Montroll, E.W. II. Shlesinger, Michael F.   III. Weiss, George H. (George Herbert), 1930-   . IV. Series.
QC175.S77 vol. 12   530.1'3 s [530.1'5]   85-17210
[QC174.85.R37]
ISBN 0-444-86937-9

Printed in The Netherlands

PREFACE

There are gifted scientists and there are people endowed with the ability to bring pleasure to any gathering. Elliott Montroll belonged to that rare group who possess both. He had such an exceptional combination of desirable qualities: warmth, charm, intelligence, curiosity, creativity, humor and just plain good naturedness, and had all of them to such a high degree that he was beloved by many and liked by all. He is very sorely missed, not only by those contributing to this volume, but by many, many more. It would have been so much better if this volume were just a festschrift celebrating his birthday. We have a need for people like him as friends and as exemplars —rare individuals who show us how one can be both intelligent and happy.

He was a great friend, a great scientist, and a noble man and this volume is dedicated to his memory.

Joel L. Lebowitz

Elliott W. Montroll enjoying a meeting of the National Academy of Sciences, circa early 1980's.

ACKNOWLEDGEMENTS

E.W. Montroll, "Some Notes and Applications of the Characteristic Value Theory of Integral Equations", Abstract of a Doctor's Dissertation, University of Pittsburgh Bulletin 37 (1941) no. 3, Jan. 15, is reprinted by permission of the Faculty of Arts and Sciences, University of Pittsburgh.

The following articles are reprinted from Journal of Chemical Physics by permission of the publisher, American Institute of Physics:
E.W. Montroll and J.E. Mayer, "Statistical Mechanics of Imperfect Gases", J. Chem. Phys. 9 (1941) 626–637; E.W. Montroll, "Frequency Spectrum of Crystalline Solids", J. Chem. Phys. 10 (1942) 218–229; and E.W. Montroll and K.E. Shuler, "Studies in Nonequilibrium Rate Processes. I: The Relaxation of a System of Harmonic Oscillators", J. Chem. Phys. 26 (1957) 454–464.

The following articles are reprinted from Journal of Mathematical Physics by permission of the publisher, American Institute of Physics:
P. Mazur and E.W. Montroll, "Poincaré Cycles, Ergodicity, and Irreversibility in Assemblies of Coupled Harmonic Oscillators", J. Math. Phys. 1 (1960) 70–84; and E.W. Montroll and G.H. Weiss, "Random Walks on Lattices. II", J. Math. Phys. 6 (1965) 167–179.

E.W. Montroll, "On the Theory of Markoff Chains", Ann. Math. Stat. 28 (1947) 18–36, is reprinted by permission of the publisher, The Institute of Mathematical Statistics.

The following articles are reprinted from Physical Review by permission of the publisher, The American Physical Society:
E.W. Montroll and R.B. Potts, "Effects of Defects on Lattice Vibrations: Interaction of Defects and an Analogy with Meson Pair Theory", Phys. Rev. 102 (1956) 72–84; and H. Scher and E.W. Montroll, "Anomalous transit-time dispersion in amorphous solids", Phys. Rev. B12 (1975) 2455–2477.

V.M. Kenkre, E.W. Montroll and M.F. Shlesinger, "Generalized Master Equations for Continuous-Time Random Walks", J. Stat. Phys. 9 (1973) 45–50, is reprinted by permission of the publisher, Plenum Publishing Corporation.

R.E. Chandler, R. Herman and E.W. Montroll, "Traffic Dynamics: Studies in Car Following", Operations Research 6 (1958) 165–184, is reprinted by permission of the Operations Research Society of America.

E.W. Montroll, "Social Dynamics and the Quantifying of Social Forces", Proc. Nat. Acad. Sci. (USA) 75 (1978) 4633–4637, is reprinted by permission of the National Academy of Sciences.

CONTENTS

Preface	v
Acknowledgements	ix
Contents	xi

I. Elliott W. Montroll (May 4, 1916–December 3, 1983)
   M.F. Shlesinger and G.H. Weiss   1

II. List of Publications of Elliott W. Montroll   17

III. Invited Contributions   29

    Ch. 1. Dielectric relaxation via the Montroll–Weiss random walk of defects
        John T. Bendler and Michael F. Shlesinger   31

    Ch. 2. The fascination of old texts
        Cyril Domb   47

    Ch. 3. Some statistical and dynamical problems in quantum electronics
        F.T. Hioe   75

    Ch. 4. Theory of diffusion via an interstitial and vacancy mechanism
        Paul H.E. Meijer   99

    Ch. 5. Mathieu difference equations
        Renfrey B. Potts   111

    Ch. 6. Mayer–Montroll equations (and some variants) through history for fun and profit
        G. Stell   127

Ch. 7. Illumination in a random medium
N.G. van Kampen ... 157

Ch. 8. Random walks in crystallography
George H. Weiss and James E. Kiefer ... 169

Ch. 9. On the quantum Langevin equation: The linear oscillator
Bruce J. West and Katja Lindenberg ... 189

Ch. 10. Some inequalities for anisotropic rotators
J. Bricmont, J.L. Lebowitz and C.E. Pfister ... 205

IV. Reprints of Elliott W. Montroll ... 215

Some Notes and Applications of the Characteristic Value Theory of Integral Equations, Abstract of a Doctor's Dissertation
University of Pittsburgh Bulletin, Vol. 37 (1941) No. 3, Jan. 15. ... 217

With J.E. Mayer, Statistical Mechanics of Imperfect Gases
J. Chem. Phys. 9 (1941) 626–637. ... 223

On the Theory of Markoff Chains
Ann. Math. Stat. 28 (1947) 18–36. ... 235

Frequency Spectrum of Crystalline Solids
J. Chem. Phys. 10 (1942) 218–229. ... 218

With R.B. Potts, Effect of Defects on Lattice Vibrations: Interaction of Defects and an Analogy with Meson Pair Theory
Phys. Rev. 102 (1956) 72–84. ... 266

With P. Mazur, Poincaré Cycles, Ergodicity, and Irreversibility in Assemblies of Coupled Harmonic Oscillators
J. Math. Phys. 1 (1960) 70–84. ... 279

With K.E. Shuler, Studies in Nonequilibrium Rate Processes. I: The Relaxation of a System of Harmonic Oscillators
J. Chem. Phys. 26 (1957) 454–464. ... 294

With G.H. Weiss, Random Walks on Lattices. II
J. Math. Phys. 6 (1965) 167–181. ... 305

With V.M. Kenkre and M.F. Shlesinger, Generalized Master Equations for Continuous-Time Random Walks
J. Stat. Phys. 9 (1973) 45–50. ... 320

With H. Scher, Anomalous transit-time dispersion in amorphous solids
Phys. Rev. B12 (1975) 2455–2477. ... 326

With R.E. Chandler and R. Herman, Traffic Dynamics: Studies in Car Following
Operations Research 6 (1958) 165–184.   349

Social dynamics and the quantifying of social forces
Proc. Nat. Acad. Sci. (USA) 75 (1978) 4633–4637.   369

Contents of Volumes I–XII   375

Cumulative Index, Volumes VI–XII   379

I. Elliott Waters Montroll

(May 4, 1916–December 3, 1983)

M.F. SHLESINGER
Navy Department
ONR/412, 800 N. Quincy St
Arlington, VA 22217-5000
U.S.A.

G.H. WEISS
National Institutes of Health
Department of Health, Education and Welfare
Bethesda, MA 20205
U.S.A.

© Elsevier Science Publishers B.V. 1985

The Wonderful World of Stochastics
A Tribute to Elliott W. Montroll
Eds. M.F. Shlesinger and G.H. Weiss

Elliott W. Montroll at commencement exercises at the University of Rochester, circa 1980. (Photo by Andy Montroll)

Introduction

Elliott W. Montroll led a remarkably rich and varied life. He had a profound influence, beginning in the early 1940s, on mathematical and chemical physics through his important innovative ideas presented in his exceptionally lucid writings which blended common sense and beautiful mathematics. On 3–4 May 1984 a memorial symposium, sponsored by the Office of Naval Research and the Institute for Physical Science and Technology of the University of Maryland, was held at the National Academy of Sciences to honor his memory and his wide contributions to science. The speakers, session chairmen, and organizers, were John Toll, Robert Dorfman, Robert Rubin, Cyril Domb, Alex Maradudin, George Weiss, John Bendler, Harvey Scher, Joel Lebowitz, Kurt Shuler, Bruce West, Katja Lindenberg, Mel Lax, Michael Fisher, Leo Kadanoff, Robert Zwanzig, Irwin Oppenheim, George Stell, Michael Shlesinger, Howard Reiss, Nico Van Kampen, and Paul Meijer. This volume is a combination of some of the symposium lectures, invited contributions, and selected reprints from Elliott's classic works.

Elliott was an original with his own distinctive style—the elegant solution of mathematical models which captured the essential physics of a problem. One always learned from a Montroll paper. He never tried to impress the reader, as his goal was to explain with a penetrating clarity. The same held for his lectures. In person, Elliott was charismatic and he was continually sought after as a lecturer. In his lectures, he created a unique mixture of physical and mathematical insight and historical perspective.

As a person, Harvey Scher [1] called Elliott "an instinctively free man". Joel Lebowitz [2] wrote that Elliott "was one of a rare breed, a very intelligent, happy man". His steadfast warmth and good humor was another of his trademarks. He took joy from many sources, of which one of his most treasured was his warm acceptance by the Dutch statistical mechanics community. He was twice the Lorentz Professor at Leiden. He could also poke fun at himself. In a history of stochastic processes Elliott [3] wrote, "We are happy to see that Kolmogorov did not submit his paper on time to be included in the Proceedings ... since ... EWM has on occasion been in a similar predicament".

Elliott was a complex and highly sophisticated individual. He moved with ease through the top levels of universities, large corporate research centers, and government research institutions. He seemed to have traveled everywhere and to have met everyone. He lectured often, created new courses (such as Quantitative Aspects of Social Phenomena and the Physical Basis of Modern Technology) at the University of Rochester, helped create Physical Dynamics, Inc. and the La Jolla Institute, consulted often with major companies, and contributed to many important scientific committees. It appeared that he never wasted a single day of his life.

Elliott was also a modest man. He usually ate in the Student Union and not the Faculty Club, took buses instead of taxis, and did more than his share of the dirty work in joint publications. In summary one could simply say that Elliott Montroll was a wise man. When he turned sixty-five and became Einstein Professor Emeritus at the University of Rochester, he did not retire, but moved back to the University of Maryland as a Distinguished Professor. In the 1950s Elliott helped turn the Institute for Fluid Mechanics and Applied Mathematics of the University of Maryland into a center of excellence in the area of statistical mechanics. Despite several moves, Elliott, his wife Shirley, and their ten children considered the Washington area to be their home, where their hospitality and house in Chevy Chase were well known to many in the physics world. Elliott's loss is sorely felt by the physics community and more so by the many who knew him.

Elliott began his scientific education as an undergraduate chemistry major (B.S. 1937) at the University of Pittsburgh. He once remarked that he didn't like the smell of chemicals and so continued as a Ph.D. student in mathematics (Ph.D. 1940). How then did he become one of the most famous physicists of his time, and hold high-level positions in government, industry, and universities? His positions included Director of Physical Science at the Office of Naval Research (1953–1955), Vice president for Research at the Institute for Defense Analyses (1963–1966), Director of General Sciences at IBM (1960–1963), and Einstein Professor of Physics at the University of Rochester (1966–1981). To answer this question we outline, by topics, some of the areas in which he made major contributions.

Imperfect gases and cluster integrals

Elliott had the good fortune to learn of important outstanding physics problems (from Gregory Wannier who was then at the University of Pittsburgh) and the brilliance to make significant contributions towards their solution. Wannier asked Elliott to present a seminar on Joseph Mayer's recent work on the theory of imperfect gases based on a cluster integral representation. The cluster integrals involve integrals over the spatial distribution functions of interacting particles. All thermodynamic functions can formally be expressed in terms of these integrals. Elliott not only quickly learned Mayer's work, but realized that a certain class of cluster integrals (now called ring diagrams) could have their contribution summed exactly in Fourier space. This was the earliest of the diagrammatic resummation procedures that are now so prevalent in physics. Elliott next went to Columbia to postdoc with Mayer where they jointly published this work, which is our second reprinted article in this volume. The first is a reprint of the still interesting abstract of Elliott's Ph.D. thesis on cluster integrals. Ironically, Mayer, Montroll and Wannier all died within weeks of each other. The article by George Stell in this volume gives an excellent review of the Montroll-Mayer equations and their applications, especially to fluids. The extension of these ideas to quantum many-body interacting systems was achieved independently in the late 1950s by several groups, including Elliott and John Ward [4]. The analysis using propagator techniques with temperature treated as a complex time is reviewed by Elliott [5] in the book *The Theory of Neutral and Ionized Gases*. Also, Isihara's [6] superb text *Statistical Physics* covers these topics in detail.

Ising models

After the ring diagrams, Elliott's next major contribution to physics took place in 1941 while he was Lars Onsager's postdoc at Yale. This was the invention of the transfer matrix method for calculating the partition function of an Ising model [7] (another topic he got from Wannier). Elliott later discovered on a trip to Japan that Ryogo Kubo had arrived at the same method and result but it was never communicated to the West because of World War II. Elliott solved exactly the one-dimensional case, but could only find high- and low-temperature approximations in two dimensions. His method involved finding the largest eigenvalue of a linear operator equation. He concluded his 1941 paper [7] *Statistical Mechanics of Nearest*

Neighbor Systems II: General Theory and Application to Two-Dimensional Ferromagnets with "it seems apparent that the characteristic value method of treating nearest neighbor problems is adequate to give asymptotic formula for physical properties at high and low temperatures to locate the temperature of a phase transition if one exists, but at its present state of development the method cannot provide accurate values of physical quantities which are derivatives of the partition function, in the immediate neighborhood of a transition point." Elliott passed this problem on to Onsager [8] who solved the two-dimensional partition function exactly in 1944 using a transfer matrix approach. We have chosen to reprint a review of Elliott's *On the Theory of Markov Chains* which deals with his transfer matrix techniques. Never one to remain fixed with a technique, Elliott with Renfrey Potts and John Ward [9] in 1963 built upon Piet Kasteleyn's [10] exquisite Pfaffian approach, for calculating the partition function, to exactly solve for the Ising magnetic correlation function in two dimensions. This was published in the **Journal of Mathematical Physics** in a special issue dedicated to Lars Onsager. Elliott was the founding editor of that journal in 1960. In 1966 Elliott and Narendra Goel [11] modeled the melting of a double helix strand of DNA as an Ising model. A detailed account of this type of theory and experiments was given by Elliott and Roger Wartell [12] in *Advances in Chemical Physics* in 1972. In 1981 with Howard Reiss [13], Elliott again returned to the Ising model in the paper *Phase Transition Versus Disorder: A Criteria Derived from a Two Dimensional Ferromagnetic Model* to study its dynamics (rather than equilibrium properties) to model the formation of metastable states that can be frozen-in during rapid cooling. The spin-flip transition rates are temperature dependent and thus time dependent during the cooling process. Elliott [14] felt that significant progress could be made on this problem by deriving the hierarchy of spin moment equations and then projecting out an appropriate combination of variables whose equation of motion would represent a type of random walk. Elliott planned to also investigate random walk connections to the hierarchy of velocity correlations obtained from the Navier–Stokes equations of a fluid [14]. It will be interesting to see if these preliminary notions can be developed.

Lattice dynamics

We have gotten far afield, so let us return to 1942 as Elliott leaves Yale to work with John Kirkwood at Cornell where he also is strongly influenced by

Peter Debye and his work on lattice dynamics. Elliott stressed that while Debye's continuum theory gave better results for the optical and thermodynamical properties of crystals than Einstein's single-frequency approach, one really had to do the calculations for a lattice as first proposed by Born and von Karman. The prime quantity necessary for most calculations was the vibrational density of states $\rho(\omega)$. Debye's model has a single maximum while Elliott calculated, via two different methods [15,16], that the two-dimensional lattice should have a density of states with two maxima. Elliott [15] in his first approach in 1942 discovered the "method of moments" where the $2n$th moment of $\rho(\omega)$ is related to the trace of the nth power of the secular matrix. The density of states is then approximated as a linear combination of Legendre polynomials in which the coefficient of each polynomial is a linear combination of the moments. This work is our fourth reprint. Elliott [16] solved for $\rho(\omega)$ in the square lattice exactly in 1947. His solution can be written in terms of lattice Green's functions which are elliptic integrals with logarithmic singularities (nowadays known as Van Hove singularities due to Van Hove's later more general treatment of singularities in the density of states). Lattice dynamics became a major topic of Elliott's research. With Alex Maradudin and George Weiss, he co-authored the classic book [17] on this subject, *Lattice Dynamics in the Harmonic Approximation*. Elliott later used the lattice Green's functions with great success in the theory of lattice random walks.

Another major lattice dynamics paper, with Renfrey Potts, our fifth reprint, is a beautiful example of the power of complex analysis. The eigenfrequencies of the lattice are the roots of the characteristic equation $D(\omega) = 0$. A thermodynamic quantity S involving a function $g(\omega)$ summed over normal mode frequencies can be expressed as

$$S = \sum_j g(\omega_j) = \frac{1}{2\pi i} \int_C g(z) \frac{d}{dz} \log D(z) \, dz$$

when $g(z)$ is analytic inside of a closed contour C, and $D(z)$ has all of its zeroes and no poles inside the contour.

Elliott and Renfrey Potts were able to calculate the change ΔS in the quantity S due to the effect of defects in the lattice by subtracting out the infinite self-energy of the perfect lattice,

$$\Delta S = \frac{1}{2\pi i} \int g(z) \frac{d}{dz} \log[D(z)/D_0(z)] \, dz,$$

where D_0 represents the perfect lattice, and D the defective lattice. In the continuum limit this became equivalent to calculating the interaction between nucleons in scalar meson-pair theory.

Elliott and Peter Mazur employed the dynamics of harmonic-oscillator assemblies to shed light on the fundamental problem of the nature of irreversibility in physics arising from time-reversible equations. By calculating momentum autocorrelation functions they showed that a system, with a large number of degrees of freedom, will appear for most of the time to be in a "fluctuation" regime undergoing an irreversible decay. However, a return to an "improbable" state consistent with reversible behavior is not ruled out. It is just a very improbable unstable event. This work is our sixth reprinted article. On the topic of irreversibility Elliott [18] derived the generalized master equation, for the diagonal elements of the density matrix satisfying the Liouville equation, by separating interesting (diagonal) from uninteresting (non-diagonal) elements. He arrived at an infinite set of linear equations which he dealt with by applying Cramer's rule. While not as elegant as Zwanzig's operator techniques the spirit was very similar.

Stochastic processes

We still have not discussed the topic for which perhaps Elliott is best known, stochastic processes, particularly random walks. After leaving Cornell he taught at Princeton, where he met his future wife, Shirley, and then became the Head of the Mathematics Group of the Kellex Corporation in New York City. His work at the Kellex Corporation involved analyzing the separation performance of cascades used at the Oak Ridge uranium isotope separation plant.

Elliott [19] stressed that the cascade process is essentially never in a steady state when the quantity of material in the cascade is much greater than the daily output. Under this condition, the plant responds slowly to changes in operating procedures and the cascade adjusts slowly to changes in operating conditions and transients generated by regular plant maintenance. Although Elliott's work was classified at the time, part of it appeared later in 1952 in the joint paper with Gordon Newell [19] *Unsteady-State Separation Performance of Cascades*. A nonlinear partial differential equation was derived for the concentration of the separable materials in each stage of the cascade. This equation was first derived by Burgers (1940) in another context, as a model for one-dimensional turbulence [20], and linearized by Hopf [22] exactly to the diffusion equation using a transformation of the independent variable. Ever the good historian, Elliott traced this linearization procedure back to H. Thomas [21] (1944), E. Hopf [22] (1950), J. Cole [23] (1951), and C. Majumdar [24] (1951). These authors discovered this method independently and used it for quite different problems: ion exchange, turbulence, aerodynamics, and general relativity, respectively.

Elliott's [25] analysis of the behavior of an impurity in the cascade system led him to the equation of a random walk on a lattice, the lattice representing the levels in the cascade. This now much quoted paper was published as *A Note on Bessel Functions of Purely Imaginary Argument* in 1946 without a hint of its connection to the Manhattan Project. The work on random walks grew and was shown to be in many aspects equivalent to the lattice dynamics problem, and tight-binding models. Elliott's Green's function and generating function techniques were so simple that they could be readily applied to many outstanding problems. Elliott solved exactly Polya's problem of calculating the probability of return to the origin for very general lattices. The answers were given in terms of Green's functions for nearest-neighbor random walks on simple body-centered and face-centered cubic lattices. These could be expressed as the already known Watson integrals. In 1965 Elliott and George Weiss introduced a memory or waiting time density $\psi(t)$ into the random walk problem. This seminal paper is our eighth reprint. In our ninth reprinted article Elliott with Vasudev Kenkre and Michael Shlesinger rewrote the Montroll–Weiss random walk as a generalized master equation whose memory $\phi(t)$ is related to $\psi(t)$ in Laplace space as

$$\tilde{\phi}(s) = s\tilde{\psi}(s)/(1 - \tilde{\psi}(s))$$

This allowed the powerful Green's function techniques and the insight from random walk problems to be transformed into rate equations, and was quite useful for connecting multiple trapping and hopping models of charge transport in the Scher–Montroll model discussed below.

The novel idea that the mean waiting time between jumps of a random walker could be infinite was introduced by Elliott and Harvey Scher in 1973 and used in their treatment of charge transport in amorphous semiconductor films in their 1975 paper. This citation classic is our tenth reprint article. Since the mean time of the waiting time density is infinite, no characteristic time exists in which to gauge measurements. This has explained several unusual scaling laws discovered in time-of-flight experiments in amorphous semiconductor films. Today this waiting-time density would be said to generate a fractal-time stochastic process. The connection of these stochastic processes to the mathematics of Levy distributions (which govern random variables with infinite moments) is given in a paper by Elliott with Michael Shlesinger [26], *On the Wedding of Certain Dynamical Processes in Dis-*

ordered Complex Materials to the Theory of Stable (Levy) Distributions.

Random walks, on regular lattices, which have fractal (self-similar) trajectories were examined by Elliott with Barry Hughes and Michael Shlesinger [27]. For walks in one dimension, the random-walk jump distribution can be related to Weierstrass' famous continuous but non-differentiable function

$$\sum_{n=0}^{\infty} a^{-n} \cos(b^n k), \quad b > a.$$

For spherically symmetric walks in higher dimensions, Elliott and the above authors were able, as a by-product from the analysis, to generalize the Weierstrass function in the paper *Fractal Random Walks* [28]. Elliott's interest in Levy distributions and fractals came from Benoit Mandelbrot's lecture at the Montroll Sixtieth Birthday Celebration [29] at the University of Rochester.

Rayleigh's analysis of the separation of argon from atmospheric nitrogen involved a type of random walk similar to that discussed by Elliott and Gordon Newell. Rayleigh's last paper also discussed a random walk (or flight); the so-called Rayleigh–Pearson walk whose application to crystallography is described by George Weiss and James Kiefer in this volume. Elliott, as did Rayleigh whom he greatly admired, wrote his last paper on random walks [30]. With Michael Shlesinger in *On the Williams–Watts Function of Dielectric Relaxation*, Elliott again took up the concept of the fractal-time random walk that he and Harvey Scher had used in 1973 [30]. They describe the relaxation of electric dipoles in amorphous materials such as glasses and polymers, by postulating a diffusion reaction scheme where dipoles interact with mobile defects moving according to fractal-time dynamics. A fractional exponential decay (known to fit much data from the analysis of Williams and Watts in 1970) rather than a simple Debye exponential decay is derived for the relaxation function. This work is the subject of the article by John Bendler and Michael Shlesinger in this volume. Thus the Montroll–Scher ideas on transport in amorphous media governed by long-tailed waiting-time distributions appear to have uncovered a generic property of the amorphous state.

Elliott [31] had begun to consider diffusion-controlled reactions in 1946 with his *A Note on the Theory of Diffusion Controlled Reactions with Applications to the Quenching of Fluorescence*, where he stressed the importance of initial conditions and transient effects which can quench fluorescence before steady-state theory becomes applicable. He not only realized that the diffusion-regulated phase of the reaction can be modeled as a stochastic process, but that the entire process can be modeled as a first passage-time problem. Following the work of Robert Rubin and Kurt Shuler [32], Elliott with Kurt Shuler exactly solved a quantum-mechanical vibrational ladder climbing model for the dissociation of polyatomic molecules in our seventh reprinted article. Elliott being an omnivorous reader of the literature recognized that the obscure Gottlieb polynomials could be used to find a general solution to the Rubin–Shuler equation.

Social phenomena

Elliott did not let his interests be bound by strict academic discipline. He had a love of many subjects, including history, economics, social sciences, ecology, and technology. With Wade Badger [33], he wrote the delightful book *Quantitative Aspects of Social Phenomena* (Gordon and Breach, 1974) in which the reader is introduced to many complex problems such as pollution control, traffic flow, population dynamics, and the development of countries from an agricultural to an industrial society. Elliott felt that a physicist should be able to consult and contribute to these important problems of modern times. In fact, much of this book is directly related to his own consulting career.

Elliott [34] wrote that theorists should be able to "operate in the manner of the old family doctor, the general practitioner whose door is open to receive all clients with a collection of numbers, or other mathematically describable objects .... Such applied mathematicians must have a great (but not necessarily deep) storage of lore of old and new forms of mathematical science including some forgotten old relics of the 19th and 18th centuries; they should be experts in making rapid searches through the literature; they should be able to recall a wide variety of mathematical models that have been applied successfully and some unsuccessfully in a diversity of other sciences; and it is absolutely necessary that they have numerous learned friends who are easily available through the wonderful electronic data retriever, the telephone."

In our eleventh reprint, Elliott, Robert Chandler and Robert Herman study the laws of car-following and the propagation of a disturbance down a single flowing line of cars. Elliott [35] called this problem "the hydrogen atom of behavioral science because each vehicle can be characterized by the value of only one variable as a function of time, its acceleration". This model, later found to be oversimplified, nevertheless proved to be a great stimulant for further research in traffic research. Elliott assumed that a single lane of traffic moved according to a car-following law of the following form

$$\dot{v}_j(t) = \lambda \left[v_{j-1}(t - \Delta) - v_j(t - \Delta) \right],$$

where $v_j(t)$ is the speed of the jth car in line, Δ is the time for a given driver to react to a stimulus, and λ is a control parameter. Using this model which assumes that drivers react only to differences in speed, Elliott and his co-authors showed that if the reaction lag was sufficiently large, collisions would inevitably occur when the number of cars was sufficiently great. Such chain collisions are known to occur in foggy conditions which tend to artificially increase reaction lags. It is now known that the original equation is too simple, and car-following laws are nonlinear, containing terms that depend both on absolute speed and on relative distances. In later papers, Elliott added a noise term to the right-hand side of the linear car-following law, showing that instabilities due to large reaction lags also tended to amplify so-called traffic "noise". This and following work helped establish the theory of traffic as a new scientific discipline, and led to the awarding of the 1959 Lanchester Prize for Operations Research to R.E. Chandler, D.C. Gazis, R. Herman, E.W. Montroll, R.B. Potts, and R. Rothery.

In our last reprint Elliott is searching for the laws of social dynamics in analogy to Newton's laws for dynamics. Specifically, in this article Elliott studies how one process or device replaces another (e.g., steamships replacing clipper ships, etc.).

Another article began as an after-dinner bet that he could write a paper on any relevant table of numbers. He won the bet when he and Robert Herman co-authored *Some Statistical Observations from a 75 Year Run of Sears–Roebuck Catalogues* [36]. They found that while the mean prices varied each year the variance of the price distribution for goods was fixed and that the price distribution was Gaussian. The surprising conclusion was that Sears–Roebuck's price distribution maximized entropy, since the Gaussian distribution accomplishes this under the constraint that the variance remain fixed. Elliott [37] returned to the problems of traffic and prices in a study of sociotechnical systems governed by entropy considerations. Undoubtedly, he would have continued his endeavors in the field of quantitative social science.

Elliott's vast knowledge of the historical development of physical concepts never failed to elicit interest and enthusiasm. In his history of stochastic processes with Michael Shlesinger he traced the history of ideas in the theory of probability. He [38] began with "Since traveling was onerous (and expensive), and eating, hunting, and wenching generally did not fill the 17th century gentleman's day, two possibilities remained to occupy the empty hours, praying and gambling; many preferred the latter".

Elliott's own illness and the need for chemotherapy caused him to reflect on an old controversy from the 1760s between D'Alembert and Daniel Bernoulli. In the 1760s the practice of inoculating against smallpox was becoming widespread. However, the technique employed caused a number of toxic responses which became widely advertised. A vocal opposition arose to the practice. Since good statistics did not exist, strong arguments raged for and against inoculation. Bernoulli, the father of risk–benefit analysis, through a statistical study recommended inoculation. D'Alembert argued that one should not risk one's life today for only a chance of extending one's life in case of some possible future exposure to smallpox. Assume that inoculations against all diseases existed, but each carries a risk. The possibility of a toxic side effect rises with the number of inoculations. Would you risk death now against the expectation of leading a long disease-free life? D'Alembert [38,39] wrote, "To enjoy the present and not to trouble oneself about the future, is common logic; a logic half good, half bad". In a comment on the Bernoulli–D'Alembert controversy Elliott wrote [39] "We must learn to properly manage potentially risky situations rather than legislate blindly against them. No modern test for the characterization of a dangerous substance would allow gasoline to pass. Yet through proper management we sit close to gallons of this toxic explosive every day and dread the possibility of a future life without it."

As we have discussed, in just the years 1940–1942 Elliott assured his reputation and career with his classic work on imperfect gases, phase transitions, and lattice dynamics; and over the next forty-one years he fulfilled the promise of his youth. By 1946 he made important contributions to nonlinear diffusion problems, lattice random walks, and diffusion-controlled reactions; and his masterful applications of stochastic processes in physics were still to come. Elliott once said that he considered the paper *Markov Chains, Wiener Integrals, and Quantum Theory* to be his best work, in which he independently derived the Feynman path integral approach to quantum mechanics, although his paper appeared some time after Feynman's work. We have highlighted some of the directions in which his career matured, but as the reader may guess we have only mentioned a small fraction of his publications and activities. Some of these activities included positions as Chairman of the Applied Physics and Mathematics Section of the National Academy of Sciences, Chairman of the Commission on Sociotechnological Systems for the National Research Council, Founder of the Institute for Fundamental Studies at the University of Rochester, and Member of the EPA Science Advisory Board.

Other works we have not yet mentioned include the invention of the spherical model, the generation of $1/f$ noise from the log–normal distribution, models of interacting species with quadratic nonlinearities, free-radical statistics, trapping of excitons on photosynthetic units, nonlinear perturbation theory for anharmonic oscillators, and network theory of solids. In this short space we have not done justice to Elliott's work and can only urge the reader to follow his advice, that is, to go back and read the masters. We provide a list of his publications. Fortunately this list contains many excellent reviews.

The reputation of Elliott W. Montroll will continue to grow along with the sadness of his loss.

References

[1] H. Scher, eulogy for E.W. Montroll.
[2] J.L. Lebowitz, preface, in: From Stochastics to Hydrodynamics, Studies in Statistical Mechanics, Vol. 11, eds. J.L. Lebowitz and E.W. Montroll (North-Holland, Amsterdam, 1984). This contains the article *On the Wonderful World of Random Walks* by E.W. Montroll and M.F. Shlesinger.
[3] Reference [2], p. 45.
[4] E.W. Montroll and J.C. Ward, The Physics of Fluids 1 (1958) 55.
[5] E.W. Montroll, Statistical Mechanics of Interacting Particles, in: The Theory of Neutral and Ionized Gases, eds. C. DeWitt and J.F. Detoff (Wiley, New York, 1960).
[6] A. Isihara, Statistical Physics (Academic, New York, 1977).
[7] E.W. Montroll, J. Chem. Phys. 9 (1941) 706; 10 (1942) 61.
[8] L. Onsager, Phys. Rev. 65 (1944) 117.
[9] E.W. Montroll, R.B. Potts and J.C. Ward, J. Math. Phys. 4 (1963) 308.
[10] P.W. Kasteleyn, J. Math. Phys. 4 (1963) 287.
[11] E.W. Montroll and N.S. Goel, Biopolymers 4 (1966) 855.
[12] E.W. Montroll and R.M. Wartell, Adv. Chem. Phys. 22 (1972) 129.
[13] E.W. Montroll and H. Reiss, Proc. Nat. Acad. Sci. (USA) 78 (1981) 2659.
[14] Reference [2], p. 99.
[15] E.W. Montroll, J. Chem. Phys. 10 (1942) 218.
[16] E.W. Montroll, J. Chem. Phys. 15 (1947) 575.
[17] A.A. Maradudin, E.W. Montroll and G.H. Weiss, Lattice Dynamics in the Harmonic Approximation (Academic, New York, 1963; 2nd Ed. 1971).
[18] E.W. Montroll, Some Remarks on the Integral Equations of Statistical Mechanics, in: Fundamental Problems in Statistical Mechanics (North-Holland, Amsterdam, 1961).
[19] E.W. Montroll and G.F. Newell, J. Appl. Phys. 23 (1952) 184.

[20] J.M. Burgers, Proc. Acad. Sci. Amsterdam 43 (1940) 2; 53 (1950) 247.
[21] H. Thomas, J. Am. Chem. Soc. 66 (1944) 1664.
[22] E. Hopf, Comm. Pure and Applied Math. 3 (1950) 201.
[23] J.D. Cole, Q. Appl. Math. 9 (1951) 225.
[24] C.D. Majumdar, Phys. Rev. 81 (1951) 844.
[25] E.W. Montroll, J. Math. and Phys. 25 (1946) 37.
[26] E.W. Montroll and M.F. Shlesinger, On the Wedding of Certain Dynamical Processes in Disordered Complex Materials to the Theory of Stable (Levy) Distributions, in: The Mathematics and Physics of Disordered Media, eds. B.D. Hughes and B.W. Ninham, Lecture Notes in Mathematics 1035 (Springer, 1983).
[27] B.D. Hughes, M.F. Shlesinger and E.W. Montroll, Proc. Nat. Acad. Sci. (USA) 78 (1981) 3287.
[28] B.D. Hughes, E.W. Montroll and M.F. Shlesinger, J. Stat. Phys. 28 (1982) 111.
[29] U. Landman, ed., Statistical Mechanics and Statistical Methods in Theory and Application, a 60th Birthday Festschrift (Plenum, New York, 1977).
[30] M.F. Shlesinger and E.W. Montroll, Proc. Nat. Acad. Sci. (USA) 81 (1984) 1280.
[31] E.W. Montroll, J. Chem. Phys. 14 (1946) 202.
[32] R.J. Rubin and K.E. Shuler, J. Chem. Phys. 25 (1956) 59.
[33] E.W. Montroll and W.W. Badger, Quantitative Aspects of Social Phenomena (Gordon and Breach, London, 1974).
[34] Reference [25], p. 110.
[35] E.W. Montroll, in: Contemporary Physics, Vol. 1 (International Atomic Energy Agency, Vienna, 1969).
[36] E.W. Montroll and R. Herman, in: ref. [29].
[37] E.W. Montroll, Proc. Nat. Acad. Sci. (USA) 78 (1981) 7839.
[38] Reference [2], p. 7.
[39] Reference [2], p. 18.

Elliott W. Montroll at the International Conference on Statistical Mechanics, Florence, Italy. May 17–20, 1949.

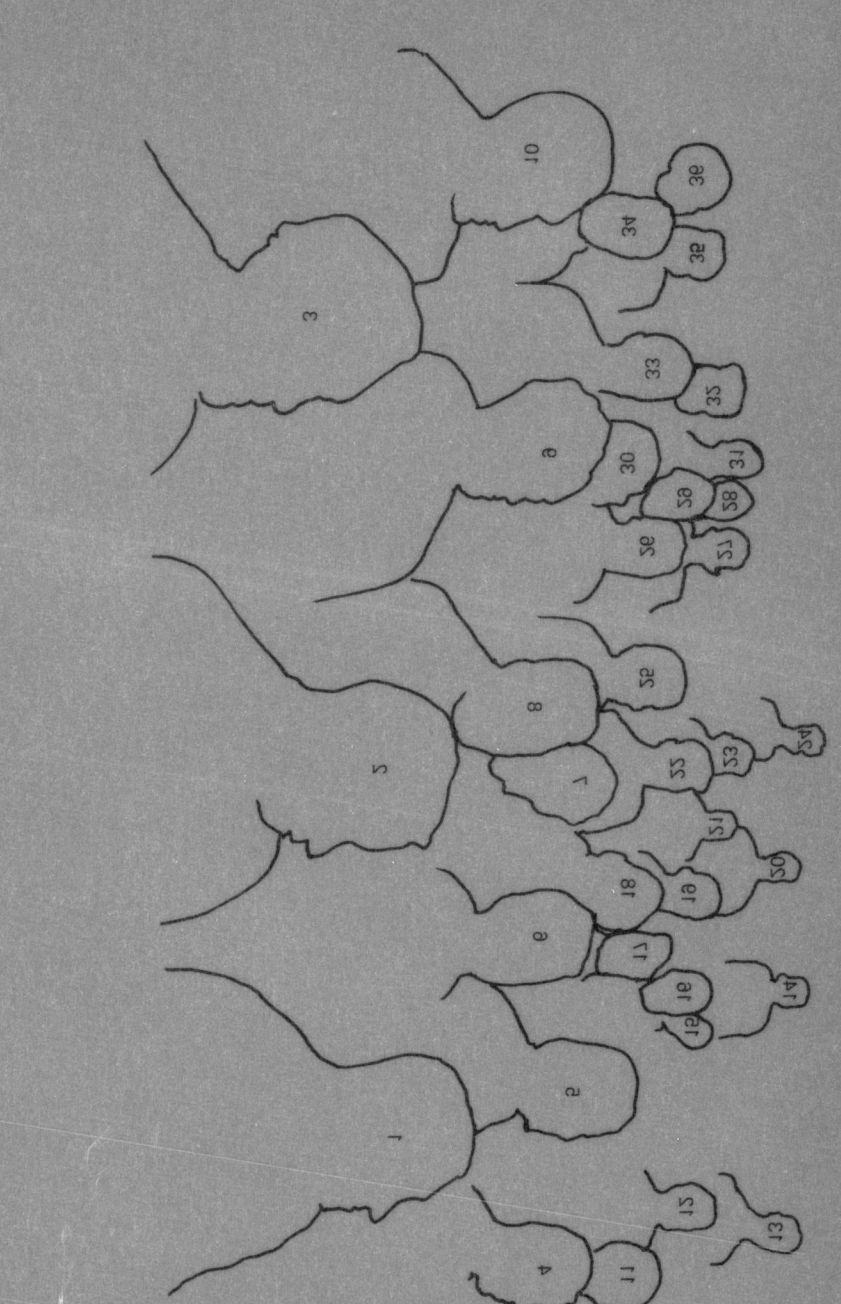

1. J. S. O. Richards, 2. 3. O. Cotter, 4. 14. C. J. Giles, 5. Sporleder, 6. L. O. Colter, 7. J. G. Piggins, 8. A. V. Guggenheim, 9. W. E. Monteith, 10. M. Boot, 11. Mayer, 12. J. T. Ortega, 13. 5. 33. C. O. Cotter, 25. 14. A. C. D. Copher, 22. 23. 24. C. O. invader, 20. C. Salvaer, 21. V. Krusemar, 19. 18. HBG Creant, 16. 17. limit, 15. Anderson, 29. 30. 28. E. Bauer, 32. 30. A. E. Baner, 31. 35. O. puper, 33. 36. Y. von, 31. 35. 28. 38. 37. 36.

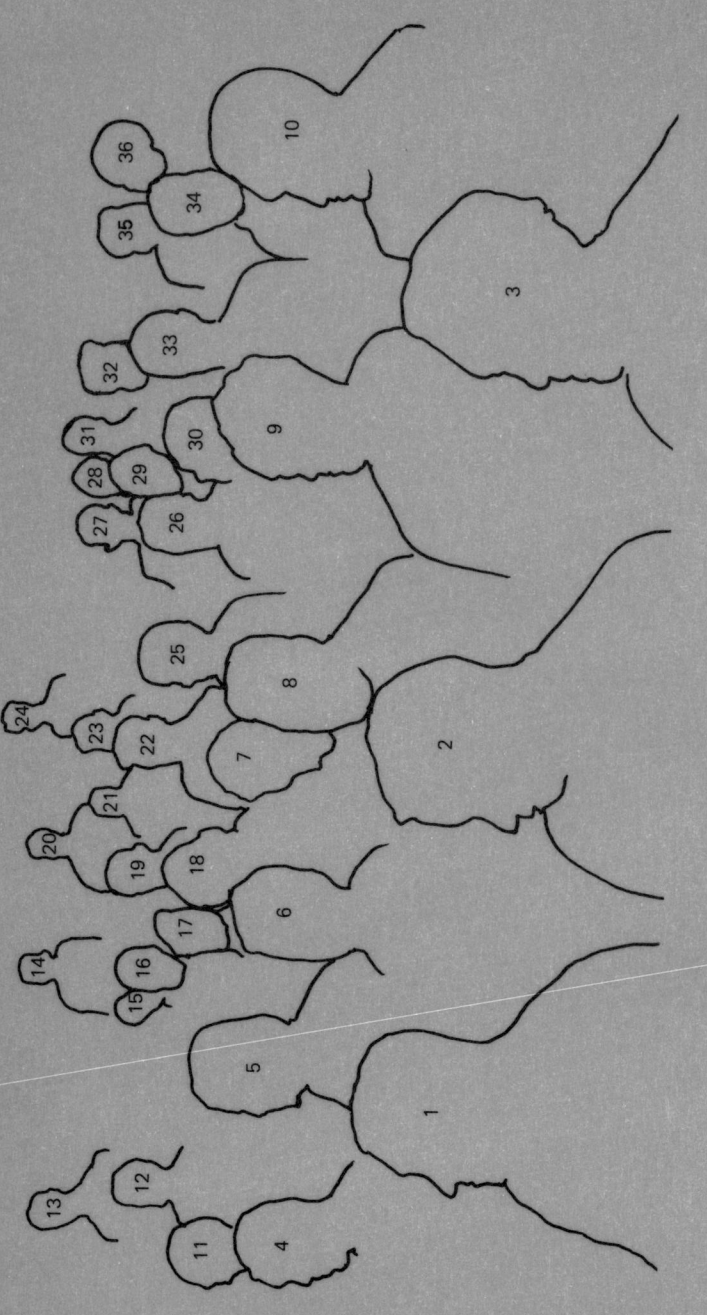

1, 2, 3? 4. G.S. Rushbrooke. 5. C.J. Gorter. 6. I. Prigogine. 7. E.A. Guggenheim. 8. E.W. Montroll. 9. M. Born. 10. J.E. Mayer. 11. L. Onsager. 12, 13, 14, 15? 16. P. Kasteleyn. 17? 18. H.B.G. Casimir. 19. J. v. Kranendonk. 20. C. Salvetti. 21? 22. E.G.D. Cohen. 23? 24. G. Toraldo. 25, 26, 27, 28, 29? 30. J. Yvon. 31? 32. G. Puppi. 33? 34. E. Bauer. 35? 36. S.R. de Groot.

II. Elliott W. Montroll
List of Publications

[1] With Robert Simha, Theory of Depolymerization of Long Chain Molecules
    J. Chem. Phys. 8 (9) (1940) 721–722.
[2] Some Notes and Applications of the Characteristic Value Theory of Integral Equations (an Abstract of a Doctor's Dissertation)
    Univ. of Pittsburgh Bull. 37 (3) (1941) 1–6.
[3] With Joseph E. Mayer, Molecular Distribution
    J. Chem. Phys. 9 (1) (1941) 2–16.
[4] Molecular Size Distributions and Depolymerization Reactions in Polydisperse Systems
    J. Amer. Chem. Soc. 63 (1941) 1215–1220.
[5] With Joseph E. Mayer, Statistical Mechanics of Imperfect Gases
    J. Chem. Phys. 9 (8) (1941) 626–637.
[6] Statistical Mechanics of Nearest Neighbor Systems
    J. Chem. Phys. 9 (9) (1941) 706–721.
[7] Statistical Mechanics of Nearest Neighbor Systems: II. General Theory and Application to Two-Dimensional Ferromagnets
    J. Chem. Phys. 10 (1) (1942) 61–77.
[8] Frequency Spectrum of Crystalline Solids
    J. Chem. Phys. 10 (4) (1942) 218–229.
[9] Frequency Spectrum of Crystalline Solids: II. General Theory and Applications to Simple Cubic Lattices
    J. Chem. Phys. 11 (10) (1943) 481–495.
[10] With David C. Peaslee, Frequency Spectrum of Crystalline Solids: III. Body-Centered Cubic Lattices
    J. Chem. Phys. 12 (3) (1944) 98–106.
[11] Phenomenological Theory of Emulsion Polymerization
    J. Chem. Phys. 13 (8) (1945) 337–348.
[12] A. Note on Bessel Functions of Purely Imaginary Argument
    J. Math. & Phys. XXV (1) (1946) 37–48.
[13] A. Note on the Theory of Diffusion Controlled Reactions With Application to the Quenching of Fluorescence
    J. Chem. Phys. 14 (3) (1946) 202–211.
[14] On the Theory of Markoff Chains
    Ann. of Math. Stat. XVIII (1) (1947) 18–36.
[15] Averages Over Normal Modes of Coupled Oscillators With Application to Theory of Specific Heats
    Qtrly. Appl. Math. 5 (2) (1947) 224–227.

[16] Dynamics of a Square Lattice: I. Frequency Spectrum
J. Chem. Phys. 15 (8) (1947) 575–591.
[17] Light-Scattering from Macromolecules
ONR Monthly Research Rpt. (1) (1949) 19–23.
[18] Continuum Models of Cooperative Phenomenon
Suppl. Nuovo Cimento VI (2), Ser. IX (1949) 1–14.
[19] Size Effect in Low Temperature Heat Capacities
J. Chem. Phys. 18 (2) (1950) 183–185.
[20] Markoff Chains and Excluded Volume Effect in Polymer Chains
J. Chem. Phys. 18 (5) (1950) 734–743.
[21] With T.H. Berlin, An Analytical Approach to the Ising Problem
Comm. Pure & Appl. Math IV (1) (1951) 23–30.
[22] With Robert W. Hart, On the Scattering of Plane Waves by Soft Obstacles: I. Spherical Obstacles
J. Appl. Phys. 22 (4) (1951) 376–386.
[23] With Robert W. Hart, Scattering of Plane Waves by Soft Obstacles: II. Scattering by Cylinders, Spheroids, and Disks
J. Appl. Phys. 22 (1951) 1278–1289.
[24] With T.H. Berlin, On the Free Energy of a Mixture of Ions: an Extension of Kramers' Theory
J. Chem. Phys. 20 (1) (1952) 75–84.
[25] With Gordon F. Newell, Unsteady-State Separation Performance of Cascades: I.
J. Appl. Phys. 23 (2) (1952) 184–194.
[26] With T.H. Berlin and R.W. Hart, Fonctions Delta et Intégrales Gaussiennes en Méchanique Statistique
Extrait des Comptes Rendus de la 2^2 Reunion de Chimie Physique, Paris, France (1952) 211–223.
[27] With J. Mayo Greenberg, Scattering of Plane Waves by Soft Obstacles: III. Scattering by Obstacles with Spherical and Circular Cylindrical Symmetry
Phys. Rev. 86 (6) (1952) 889–898.
[28] Markoff Chains, Wiener Integrals, and Quantum Theory
Comm. Pure & Appl. Math V (4) (1952) 415–453.
[29] With Gordon F. Newell, On the Theory of the Ising Model of Ferromagnetism
Rev. Mod. Phys. 25 (2) (1953) 353–389.
[30] With J. Mayo Greenberg, On the Theory of Scattering of Plane Waves by Soft Obstacles
In Proc. Symp. of Applied Mathematics, Wave Motion, and Vibration Theory, V (McGraw–Hill, New York, 1954) pp. 103–122.
[31] Frequency Spectrum of Vibrations of a Crystal Lattice
Amer. Math. Monthly LXI (7) (1954) 46–73.
[32] With M.S. Green, Statistical Mechanics of Transport and Nonequilibrium Processes
Ann. Rev. Phys. Chem. 5 (1954) 449–476.
[33] With Renfrey B. Potts, Effect of Defects on Lattice Vibrations
Phys. Rev. 100 (2) (1955) 525–543.
[34] With P. Mazur and R.B. Potts, Effect of Defects on Lattice Vibrations: II. Localized Vibration Modes in a Linear Diatomic Chain
J. Wash. Acad. Sci. 46 (1) (1956) 2–11.
[35] With Renfrey B. Potts, Effect of Defects on Lattice Vibrations: Interaction of Defects and an Analogy with Meson Pair Theory
Phys. Rev. 102 (1) (1956) 72–84.

[36] Theory of Vibration of Simple Cubic Lattices with Nearest Neighbor Interactions
in: Proc. 3rd Berkeley Symp. on Mathematical Statistics and Probability (Univ. of California Press, Berkeley, 1956) pp. 209–246.
[37] Random Walks in Multidimensional Spaces, Especially on Lattices
J. Soc. Industr. Appl. Math. 4 (4) (1956) 241–260.
[38] With Kurt E. Shuler, Studies in Nonequilibrium Rate Processes; I. The Relaxation of a System of Harmonic Oscillators
J. Chem. Phys. 26 (3) (1957) 454–464.
[39] With A.A. Maradudin, P. Mazur and G.H. Weiss, Remarks on the Vibrations of Diatomic Lattices
Rev. Mod. Phys. 30 (1) (1958) 175–196.
[40] With John C. Ward, Quantum Statistics of Interacting Particles: General Theory and Some Remarks on Properties of an Electron Gas
Phys. Fluids 1 (1) (1958) 55–72.
[41] With Robert E. Chandler and Robert Herman, Traffic Dynamics: Studies in Car Following
Oper. Research 6 (2) (1958) 165–184.
[42] With Julius L. Jackson, Free Radical Statistics
J. Chem. Phys. 28 (6) (1958) 1101–1109.
[43] With Kurt E. Shuler, The Application of the Theory of Stochastic processes to Chemical Kinetics
in: Advances in Chemical Physics I, ed. I. Prigogine (Interscience, New York, 1958) pp. 361–398.
[44] With N.W. Bazley, R.J. Rubin and K.E. Shuler, Studies in nonequilibrium Rate Processes. III The Vibrational Relaxation of a System of Anharmonic Oscillators
J. Chem. Phys. 28 (1958) 700–704.
[45] Theory of the Vibrational Relaxation of Diatomic Molecules
in: Proc. Int. Symp. on Transport Processes in Statistical Mechanics, part XII, Brussels, Belgium, August, 1956, ed. I. Prigogine (Interscience, New York, 1958), pp. 351–362.
[46] With S. Fujita and A. Isihara, On the Cluster Integral Development of Pair Distribution Functions
in: Bulletin de l'Academie Royale de Belgique (Classe des Sciences) Seance du 13 Decembre (1958), pp. 1018–1032.
[47] Principles of Statistical mechanics and Kinetic Theory of Gases
in: Handbook of Physics, Ch. 2 (McGraw–Hill, New York, 1958) pp. 5–11.
[48] Vibrations of Crystal Lattices and Thermodynamic Properties of Solids
in: Handbook of Physics, Ch. 10 (McGraw–Hill, New York, 1958) pp. 5–150.
[49] With Robert Herman, Renfrey B. Potts and Richard W. Rothery, Traffic Dynamics: Analysis of Stability in Car Following
Oper. Research 7 (1) (1959) 86–106.
[50] With John C. Ward, Quantum Statistics of Interacting Particles
Nuovo Cimento II (1959).
[51] With John C. Ward, Quantum Statistics of Interacting Particles: II. Cluster Integral Development of Transport Coefficients
Physica 25 (1959) 423–443.
[52] With C. Domb, A.A. Maradudin and G.H. Weiss, The Vibration Spectra of Disordered Lattices
J. Phys. Chem. Solids 8 (1959) 419–421.

[53] With C. Domb, A.A. Maradudin and G.H. Weiss, Vibration Frequency Spectra of Disordered Lattices: I. Moments of the Spectra for Disordered Linear Chains
Phys. Rev. 115 (1) (1959) 18–24.
[54] With C. Domb, A.A. Maradudin and G.H. Weiss, Vibration Frequency Spectra of Disordered Lattices: II Spectra of Disordered One-Dimensional Lattices
Phys. Rev. 115 (1) (1959) 24–36.
[55] On Statistical Mechanics of Transport Processes
in: Rendiconti della Scuola Internazionale di Fisica "Enrico Fermi", Suppl. Nuovo Cimento II, Corso X (1959) p. 217–261.
[56] Statistical Mechanics of Interacting Particles
in: La théorie des gas neutres et ionisés (Hermann, Paris, 1960) pp. 15–148.
[57] With P. Mazur, Poincaré Cycles, Ergodicity, and Irreversibility in Assemblies of Coupled Harmonic Oscillators
J. Math. Phys. 1 (1) (1960) 70–84.
[58] Sto Lat Mechaniki Statystycznes
Kosmos B, Rok vi, zeszyt 2 (22), (1960).
[59] With A.A. Maradudin, G.H. Weiss, R. Herman and H.W. Milnes, Green's Functions for Monatomic Simple Cubic Lattices
in: Memoires of the Belgian Acad. Sci. XIV, No. 1709 (1960).
[60] Nonequilibrium Statistical Mechanics
in: Lectures in Theoretical Physics III, eds. U.E. Brittin, B.W. Downs and J. Downs (Interscience, New York, 1961) pp. 221–325.
[61] Acceleration Noise and Clustering Tendency of Vehicular Traffic
in: Theory of Traffic Flow (Elsevier, Amsterdam, 1961) pp. 147–157.
[62] Some Remarks on the Integral Equations of Statistical Mechanics
in: Fundamental Problems in Statistical mechanics I (North-Holland, Amsterdam, 1961) pp. 230–249.
[63] With A.A. Maradudin and G.H. Weiss,
Theory of Lattice Dynamics in the Harmonic Approximation (Academic, New York, 1963; 2nd Ed. 1971).
[64] With R.B. Potts and J.C. Ward, Correlations and Spontaneous Magnetization of the Two-Dimensional Ising Model
J. Math. Phys. 4 (2) (1963) 308–322.
[65] With A.A. Maradudin and G.H. Weiss, Effect of Defects on the Vibration of Crystal Lattices
in: The Many-Body Problem, ed. J.K. Percus (Interscience, New York; Wiley, New York, 1963) Ch. XX, pp. 353–373.
[66] One Hundred Years of Statistical Mechanics
in: The Many-Body Problem, ed. J.K. Percus (Interscience, New York; Wiley, New York, 1963) Ch. XXXI, pp. 525–533.
[67] Theory and Observations of the Dynamics and Statistics of Traffic on an Open Road
in: Proc. 1st Symp. on Engineering Applications of Random Function Theory and Probability, eds. R. Bagdonoff and J. Kozin (Wiley, New York, 1963) pp. 231–269.
[68] Quantum Statistics of Interacting Particles
Appendix to Lectures in Statistical Mechanics, eds. George E. Uhlenbeck and George W. Ford (American Mathematical Society, Providence, RI, 1963) pp. 143–178.
[69] Lattice Statistics
in: Applied Combinatorial Mathematics (Wiley, New York, 1964) Ch. 4, pp. 96–143.

[70] With R.B. Potts, Car Following and Acceleration Noise
in: An Introduction to Traffic Flow Theory (Nas Nat. Res. Coun. Highway Research Board Pub. No. 1121, 1964) Ch. II.

[71] Random Walks on Lattices
in: Stochastic Processes in Mathematical Physics and Engineering, Proc. Symp. in Applied mathematics XVI (American Mathematical Society, Providence, RI 1964) pp. 193–220.

[72] Model Making in Biological and Behavioral Sciences
in: Proc. 2nd Techn. Meeting Soc. of Engineering Sci. Vol. I, ed. A.C. Eringen (Michigan State Univ., 1964) pp. 247–298.

[73] With G.H. Weiss, Random Walks on Lattices: II
J. Math. Phys. 6 (1965) 167–181.

[74] On the Theory of Coiling and Uncoiling of DNA Molecules
in: Proc. Symp. on Inelastic Scattering of Neutrons by Condensed Systems (Brookhaven Nat'l. Lab. Publ. BNL 940 (C-45), 1965) pp. 57–68.

[75] With N.S. Goel, Denaturation and Renaturation of DNA: I. Equilibrium Statistics of Copolymeric DNA
Biopolymers 4 (1966) 855–886.

[76] Lectures on the Ising Problem of Phase Transitions
in: Stat. Phys., Phase Transitions and Superfluidity, Vol. 11, Proc. Brandeis Univ. Summer Inst. in Theoretical Physics II (Gordon and Breach, New York, 1966) pp. 197–267.

[77] Stochastic Processes and Chemical Kinetics
Energet. Metallurg. Phenom. 3 (1967) 123–179.

[78] On Nonlinear Processes Involving Population Growth and Diffusion
J. Appl. Prob. 4 (1967) 281–290.

[79] The Science Curriculum (in Mathematics Education)
SIAM Rev. 9 (1967) 326–340. (1967).

[80] With N.S. Goel, Denaturation and Renaturation of DNA: II. Possible Use of Synthetic Periodic Copolymers to Establish Model and Parameters
Biopolymers 6 (1968) 731–765.

[81] Lectures on Nonlinear Rate Equations, Especially Those with Quadratic Nonlinearities
in: Lectures in Theoretical Physics (Boulder), X, (Gordon and Breach, New York, 1968) p. 531.

[82] With Lee-Po Yu, Analysis of Assemblies with Large Defect Concentrations with Special Application to the Theory of Denaturation of Copolymeric DNA
in: Localized Excitations in Solids, ed. R.F. Wallis (Plenum, New York, 1968) pp. 745–766.

[83] Random Walks on Lattices Containing Traps
in: Proc. Int. Conf. on Statistical Mechanics, 1968, Suppl. J. Phys. Soc. Japan 26 (1969) 6–10.

[84] Random Walks on Lattices. III. Calculation of First-Passage Times With Application to Exciton Trapping on Photosynthetic Units
J. Math. 10 (4) (1969) 753–765.

[85] Theoretical Basis of Techniques for the Investigation of Molecular Dynamics and Structure of Solids
in: Molecular Dynamics and Structure of Solids (National Bureau of Standards, Washington, 1969) pp. 3–56.

[86] Three Examples of One-Dimensional Systems
in: Contemporary Physics, Vol. I (IAEA, Vienna, 1969) pp. 177–193.

[87] Some Remarks on Turbulence
in: Contemporary Physics, Vol. I (IAEA, Vienna, 1969) pp. 274–294.

[88] Propagation of Waves in Discrete Media, Harmonic, Anharmonic, and Defective
in: Proc. Explosion Chain Reaction Seminar (Army Research Office, Durham, NC, 1969).

[89] Quantum Theory on a Network: I. A Solvable Model Whose Wave Functions Are Elementary Functions
J. Math. Phys. 11 (2) (1970) 636–648.

[90] With R.G.J. Mills, Quantum Theory on a Network: II A Solvable Model Which May Have Several Bound States per Node Point
J. Math. Phys. 11 (8) (1970) 2525–2538.

[91] With N.S. Goel and S.C. Maitra, On the Volterra and Other Nonlinear Models of Interacting Populations
Rev. Mod. Phys. 43 (1971) 216–231.

[92] With A. Isihara, A Note on the Ground State Energy of Interacting Electrons
Proc. Natl. Acad. Sci. 12 (1971) 3111–3115.

[93] With Robert Herman, A Manner of Characterizing the Development of Countries
Proc. Natl. Acad. Sci. 69 (10) (1972) 3019–3023.

[94] With Roger M. Wartell, Equilibrium Denaturation of Natural and of Periodic Synthetic DNA Molecules
Adv. Chem. Phys. 22 (1972) 129–203.

[95] Some Statistical Aspects of the Theory of Interacting Species
in: Some Mathematical Problems in Biology, IV (American Mathematical Society, Providence, RI, 1972) pp. 101–143.

[96] Nonlinear Rate Processes, Especially Those Involving Competitive Processes
in: Statistical Mechanics, New Concepts, New Problems, New Applications, eds. S.A. Rice, K.F. Freed and J.C. Light (Univ. of Chicago Press, IL, 1972) pp. 69–89.

[97] On Coupled Rate Equations with Quadratic Nonlinearities
Proc. Natl. Acad. Sci. 69 (1972) 2532–2536.

[98] With Bruce J. West, Models of Population Growth, Diffusion, Competition and Rearrangement
in: Synergetics, ed. H. Haken (Teubner, Stuttgart, 1973) pp. 143–56.

[99] With V.M. Kenkre and M.F. Shlesinger, Generalized Master Equations for Continuous-Time Random Walks
J. Stat. Phys. 9 (1973) 45–50.

[100] Surface Properties of a Network Model of Electrons in Solids
J. Phys. Chem. Solids 34 (1973) 567–610.

[101] With D.C. Gazis and T.E. Ryruker, Age-Specific, Deterministic Model of Predator–Prey Population: Application to Isle Royale
IBM Jour. of Res. & Devel. 17 (1973) 47–53.

[102] With F.T. Lee and Lee-Po Yu, Two-Component Ising Chain with Nearest-Neighbor Interaction
J. Stat. Phys. 8 (4) (1973) 309–333.

[103] With H. Scher, Random Walks on Lattices: IV. Continuous-Time Walks and Influence of Absorbing Boundaries
J. Stat. Phys. 9 (2) (1973) 101–135.

[104] Enzyme Cascades and Their Control in Blood Plasma
Adv. Chem. Phys. 26 (1974) 145–176.
[105] With R.H.G. Helleman, On a Nonlinear Perturbation Theory Without Secular Terms: I. Classical Coupled Anharmonic Oscillators
Physica 74 (1974) 22–74.
[106] On the Quantum Analogue of the Lévy Distribution
in: Physical Reality and Mathematical Description, eds. C. Enz and M. Mehra (Reidel, Dordrecht, The Netherlands, 1974) pp. 501–508.
[107] Propagation of Waves in Discrete Media, Harmonic, Anharmonic, and Defective
in: Proc. 3rd Advanced School for Statistical Mechanics and Thermodynamics, Austin, TX, 1971, Lecture Notes in Physics, Vol. 28 (Springer, Heidelberg, 1974) ch. 1, pp. 145–203.
[108] With C.H. Wu, A Network Model of Electronic States of Thin Films, Solid Interfaces, and Planar Defects
J. Nonmet. and Semicond. 2 (1975) 153–191.
[109] With Bruce J. West, Scattering of Waves by Irregularities in Periodic Discrete Lattice Spaces: I. Reduction of the Problem to Quadratures on a Discrete Model of the Schrödinger Equation
J. Stat. Phys. 13 (1) (1975) 17–42.
[110] With F.T. Hioe, Two-Level Radiative Systems and Perturbation Theory
J. Math. Phys. 16 (6) (1975) 1259–1270.
[111] With F.T. Hioe, Quantum Theory of Anharmonic Oscillators: I. Energy Levels of Oscillators with Positive Quartic Anharmonicity
J. Math. Phys. 16 (9) (1975) 1945–1955.
[112] With Harvey Scher, Anomalous Transit-Time Dispersion in Amorphous Solids
Phys. Rev. B12 (6) (1975) 2455–2477.
[113] With Robert H.G. Helleman, On a Nonlinear Perturbation Theory Without Secular Terms: ii. Carleman Embedding of Nonlinear Equations in an Infinite Set of Linear Ones
in: Topics in Statistical Mechanics and Biophysics, A Memorial to Julius L. Jackson, Wayne State Univ., ed. R.A. Piccirelli, AIP Conf. Proc. 27 (1976) 75–10.
[114] With Charles R. Eminhizer and Robert H.G. Helleman, On A Convergent Nonlinear Perturbation Theory Without Small Denominators or Secular Terms
J. Math. Phys. 17 (1) (1976) 121–140.
[115] With Uzi Landman, Adsorption in Heterogeneous Surfaces: I. Evaluation of the Energy Distribution Function Via the Wiener and Hopf Method
J. Chem. Phys. 64 (4) (1976) 1762–1767.
[116] With F.T. Hioe and Don MacMillen, Quantum Theory of Anharmonic Oscillators: II. Energy Levels of Oscillators with Anharmonicity
J. Math. Phys. 17 (7) (1976) 13120–1337.
[117] With U. Landman and M.F. Shlesinger, Motion of Clusters on Surfaces
Phys. Rev. Lett. 38 (6) (1977) 285–289.
[118] With U. Landman and M.F. Shlesinger, Random Walks and Generalized Master Equations with Internal Degrees of Freedom
Proc. Natl. Acad. Sci. 74 (2) (1977) 430–433.
[119] With Martin Schwarz Jr, On the Mapping of Chains of First Order Chemical Reactions on Random Walks
Biosystems 9 (1977) 175–186.
[120] With Robert Herman, Some Statistical Observations from a 75-Year Run of Sears Roebuck Catalogues

in: Statistical Mechanics and Statistical Methods in Theory and Application, ed. Uzi Landman (Plenum, New York, 1977) pp. 785–803.

[121] With Charles A. Ginsburg, Application of Novel Interpolative Perturbation Scheme to the Determination of Anharmonic Oscillator Wave Functions
J. Math. Phys. 19 (1978) 336–346.

[122] With F.T. Hioe and Don MacMillen, Quantum Theory of Anharmonic Oscillators: Energy Levels of a Single and a Pair of Coupled Oscillators with Quartic Coupling
Phys. Rep. 43 (7) (1978) 305–335.

[123] On Some Mathematical Models of Social Phenomena
in: Nonlinear Equations in Abstract Spaces, Proc. Int. Symp. on Nonlinear Equations in Abstract Spaces, Univ. of Texas at Arlington, TX, ed. V. Lakshmikantham (Academic, New York, 1978) pp. 161–216.

[124] On the Solution of Nonlinear Rate Equations by Matrix Inversion
in: Topics in Nonlinear Dynamics, A Tribute to Sir Edward Bullard, ed. S. Jorna, AIP Conf. Proc. 46 (1978) 337–359.

[125] Social Dynamics and the Quantifying Forces
Proc. Natl. Acad. Sci. 75 (10) (1978) 4633–4637.

[126] With Charles A. Ginsburg, On Wave Functions and Energy Levels of a Three Dimensional Anharmonic Oscillator with Quartic Anharmonicities
in: Science of Matter (a symposium in honor of Professor T.Y.Wu, SUNY, Buffalo, NY), ed. Shigeji Fujita (Gordon and Breach, New York, 1978) pp. 207–222.

[127] With Bruce J. West, On an Enriched Collection of Stochastic Processes
in: Fluctuation Phenomena, Studies in Statistical Mechanics VII, eds. E.W. Montroll and J.L. Lebowitz (North-Holland, Amsterdam, 1979) pp. 61–173.

[128] With K. Shuler, Dynamics of Technological Evolution: Random Walk Model for the Research Enterprise
Proc. Natl. Acad. Sci. 76 (12) (1979) 6030–6034.

[129] Some Historical Remarks on the Catalytic Process and on Stochastic Models of Chemical Kinetics
AIP Conf. Proc. 61 (1980) 1–37.

[130] With H. Scher and S. Alexander, Field-Induced Trapping as a Probe of Dimensionality in Molecular Crystals
Proc. Natl. Acad. Sci. 77 (7) (1980) 3758–3762.

[131] With F.T. Hioe and M. Yamawaki, On Higher Order WKB Approximation for the Calculation of Energy Levels
in: Perspectives in Statistical Physics, M.S. Green, Memorial Volume, Studies in Statistical Mechanics IX, ed. H.T. Raveché (North-Holland, Amsterdam, 1980) pp. 295–324.

[132] On the Dynamics of the Ising Model of Cooperative Phenomena
Proc. Natl. Acad. Sci. 78 (1) (1981) 36–40.

[133] With H. Reiss, Phase Transition Versus Disorder: Criterion Derived From a Two-Dimensional Dynamic Ferromagnetic Model
Proc. Natl. Acad. Sci. 78 (5) (1981) 2659–2663.

[134] With B.D. Hughes and M.F. Shlesinger, Random Walks with Self-Similar Clusters
Proc. Acad. Sci. 78 (6) (1981) 3287–3291.

[135] On the Entropy Function in Sociotechnical Systems
Proc. Natl. Acad. Sci. 78 (12) (1981) 7839–7843.

[136] With M.F. Shlesinger and B.D. Hughes, Fractal Random Walks
in: Proc. 6th Int. Symp. on Noise in Physical Systems, eds. P.H.E. Meijer, R. Mountain and R.J. Soulen Jr (Natl. Bureau of Standards Publ. 61, 1981) pp. 18–22.

[137] On the Dynamics of Technological Evolutions: Phase Transitions
in: Self-Organization and Dissipative Structures (Applications in the Physical and Social Sciences), eds. W.C. Schieve and P.M. Allen (Univ. of Texas Press, Austin, TX, 1982) ch. 3, pp. 63–90.

[138] With B.D. Hughes and M.F. Shlesinger, Fractal Random Walks
J. Stat. Phys. 28 (1) (1982) 111–126.

[139] With M.F. Shlesinger, On $1/f$ Noise and Other Distributions with Long Tails
Proc. Natl. Acad. Sci. 79 (5) (1982) 3380–3383.

[140] Some Introductory Philosophical Remarks on the Mathematical Modeling of Complex Systems
in: Proc. Conf. on Molecular Science, Carbondale, IL, ed. T.A. Burton (Pergamon Press, New York, 1982) pp. 179–207.

[141] With B.D. Hughes and M.F. Shlesinger, Fractal and Lacunary Stochastic Processes
J. Stat. Phys. 30 (2) (1983) 273–283.

[142] With C.H. Wu, Kinetics of Adsorption on Stepped Surfaces and the Determination of Surface Diffusion Constants
J. Stat. Phys. 30 (2) (1983) 537–547.

[143] With M.F. Shlesinger, Maximum Entropy Formalism, Fractals, Scaling Phenomena, and $1/f$ Noise: A Tale of Tails
J. Stat. Phys. 32 (2) (1983) 209–230.

[144] The Entropy Function in Complex Systems
in: The Study of Information: Interdisciplinary Messages, eds. F. Machlup and U. Mansfield (Wiley, New York, 1983) pp. 503–511.

[145] With M.F. Shlesinger, A Wonderful World of Random Walks
in: CCNY Physics Symposium: M. Lax's Sixtieth Birthday, ed. H. Falk (City College of New York Physics Department, New York, 1983) pp. 44–147.

[146] With M.F. Shlesinger, On the Wonderful World of Random Walks
in: Nonequilibrium Phenomena II. From Stochastics to Hydrodynamics, Studies in Statistical Mechanics XI, eds. J.L. Lebowitz and E.W. Montroll (North-Holland, Amsterdam, 1984) pp. 1–121.

[147] With J. Bendler, On the Lévy (or Stable) Distributions and The Williams–Watts model of Dielectric Relaxation: I.
J. Stat. Phys. 34 (1984) pp. 129–162.

[148] With M.F. Shlesinger, On the Wedding of Certain Dynamical Processes in Disordered Materials to the Theory of Stable (Lévy) Distribution Functions
in: The Mathematics and Physics of Disordered Media, eds. B.D. Hughes and B.W. Ninham, Lecture Notes in Mathematics, Vol. 1035 (Springer, Berlin, 1983) pp. 109–137.

[149] With M.F. Shlesinger, Fractal Stochastic Processes: Clusters and Intermittencies
in: [147], pp. 138–152.

[150] With M.F. Shlesinger, On the Williams–Watts Function of Dielectric Relaxation
Proc. Natl. Acad. Sci 81 (1984) 1280–1283.

[151] On the Vienna School of Statistical Thought
in: Random Walks and Their Applications to the Physical and Biological Sciences, eds. M.F. Shlesinger and B.J. West, AIP Conf. Proc. 109 (1984) 1–10.

III. Invited Contributions *

---

* Contributions to the Elliott W. Montroll Memorial Symposium on Mathematical Physics, held 3–4 May 1984 at the National Academy of Sciences, Washington, DC.

© Elsevier Science Publishers B.V. 1985

The Wonderful World of Stochastics
A Tribute to Elliott W. Montroll
Eds. M.F. Shlesinger and G.H. Weiss

CHAPTER 1

Dielectric Relaxation via the Montroll–Weiss Random Walk of Defects

John T. BENDLER

Polymer Physics and Engineering Branch
General Electric Corporate Research and Development
Schenectady, New York 12301
U.S.A.

Michael F. SHLESINGER

Physics Division
Office of Naval Research
800 North Quincy Street
Arlington, Virginia 22217
U.S.A.

© Elsevier Science Publishers B.V. 1985

The Wonderful World of Stochastics
A Tribute to Elliott W. Montroll
Eds. M.F. Shlesinger and G.H. Weiss

Contents

1. Introduction   33
2. Discrete-time random walks on a lattice   34
    2.1. Probabilities, generating functions, and first passage times   34
    2.2 Number of distinct sites visited   35
3. The Montroll–Weiss random walk   37
4. Dielectric relaxation: a defect diffusion model   39
    4.1. Introduction   39
    4.2. A frozen dipole amidst a swarm of mobile defects   41
    4.3. Temperature dependence of the Williams–Watts exponent   42
    4.4. Probability density $\rho_\alpha(\lambda)$ of relaxation times for $e^{-(t/\tau)^\alpha}$ 43
References   45

"The best advisors have simple models and courage. Genius consists in finding the models... that correspond closely to aspects of reality."
Paul A. Samuelson, as quoted in *The Economists*, by L. Silk (Avon Books, New York, 1978).

1. Introduction

While Samuelson was thinking in the context of economics and models such as "the market", we believe he gave an apt description of the scientific career of Elliott W. Montroll. In addition Montroll also had the ability to present his ideas in an extraordinarily interesting and clear fashion. Every new Montroll paper was a celebration.

Some of Montroll's most elegant work has been in the field of random walks. An important idea introduced by Montroll and Weiss [1] in 1965 is the concept of a waiting-time distribution governing the time interval between the jumps of a random walker. The case when the mean time between jumps is infinite is fascinating because then there does not exist a characteristic time in the process. This novel case was used by Montroll and Scher [2,3] in 1973 to give the first explanation of experiments measuring transient electrical currents in amorphous semiconductors. In Montroll's last paper [4], published in 1984 jointly with Shlesinger, dielectric relaxation in amorphous materials was considered to be initiated when mobile defects reached frozen-in dipoles. In this model the motion of the defects was governed by a waiting-time distribution with an infinite mean. It is this last problem which we present here, in this volume dedicated to the memory of Elliott Montroll, in addition to some new results on the temperature dependence of the relaxation.

We will first illustrate the power of Montroll's original analysis of discrete-time random walks [5–9] by showing, in some detail, how these concepts can be generalized to handle evermore complicated situations such as diffusion-reaction schemes in disordered materials as well as deep problems in mathematics [10–24]. In fact, Montroll's techniques have been the foundation upon which a whole class of random-walk problems problems in the physical and biological sciences have been approached [25].

2. Discrete-time random walks on a lattice

2.1. Probabilities, generating functions, and first passage times

Assume a random walker starts at the origin of a periodic lattice. The probability of a single jump going from the origin to site l is denoted by $p(l)$. The probability $P_{n+1}(l)$ of being at site l after $n + 1$ steps is

$$P_{n+1}(l) = \sum_{l'} p(l') P_n(l - l'), \tag{1}$$

i.e., one can reach an intermediate site $l - l'$ after n steps and then cover the remaining distance l' in one jump to bring the walker to site l. As eq. (1) is in the form of a convolution, it can be readily transformed to give

$$P_n(k) = \sum_l e^{ik \cdot l} P_n(l) = [p(k)]^n. \tag{2}$$

Inverse transforming gives, in D dimensions,

$$P_n(l) = \left(\frac{1}{2\pi}\right)^D \int_{-\pi}^{\pi} \cdots \int_{-\pi}^{\pi} e^{-ik \cdot l} [p(k)]^n \, d^D k. \tag{3}$$

If the mean square displacement per jump, denoted by

$$\overline{l^2} = \sum_l l^2 p(l),$$

is finite, then for small values of $|k|$:

$$p(k) \sim 1 - \tfrac{1}{2}\overline{l^2}|k|^2 \sim \exp\left(-\tfrac{1}{2}\overline{l^2}|k|^2\right). \tag{4}$$

Substituting the Gaussian form into eq. (3) gives $P_n(l)$ as a Gaussian,

$$P_n(l) \sim \frac{1}{(2\pi n \overline{l^2})^{D/2}} \exp\left(\frac{-|l|^2}{2n\overline{l^2}}\right). \tag{5}$$

This is a particular example of the Central Limit Theorem. While $P_n(l)$ contains all the information necessary to describe this stochastic process, this does not imply that every interesting statistical quantity is Gaussian or even that it can be easily expressed in terms of $P_n(l)$. One such quantity is the probability $F_n(l)$ for reaching site l for the first time at the nth step of the random walk. This quantity plays a crucial role in the theory of diffusion-controlled reactions. If two particles annihilate when they meet, this is described by $F_n(l)$ and not by $P_n(l)$. To calculate $F_n(l)$ consider all the ways a walker can get to site l after n steps,

$$P_n(l) = \sum_{m=0}^{n} F_{n-m}(l) \, P_m(0) + \delta_{n,0}\delta_{l,0}, \tag{6}$$

which states that a walker can reach site l after n steps if first it reaches l after $n-m$ steps and then leaves and returns to l (perhaps more than once) in the next m steps. At this point it is useful to introduce the generating functions

$$G(l, z) \equiv \sum_{n=0}^{\infty} P_n(l) \, z^n \tag{7}$$

and

$$F(l, z) \equiv \sum_{n=0}^{\infty} F_n(l) \, z^n. \tag{8}$$

In many cases it is easier to calculate a generating function rather than individual probabilities directly. Multiplying eq. (1) by z^{n+1} and summing over n gives an equation for the generating function

$$G(l, z) - z \sum_{l'} p(l') \, G(l-l') = \delta_{l,0}, \tag{9}$$

leading to the discovery that the generating function is the Green's function of eq. (9). Fourier-transforming eq. (9) gives

$$G(\mathbf{k}, z) = [1 - zp(\mathbf{k})]^{-1}, \tag{10}$$

and inverse transforming yields

$$G(l, z) = \left(\frac{1}{2\pi}\right)^D \int_{-\pi}^{\pi} \cdots \int_{-\pi}^{\pi} \frac{e^{i\mathbf{k}\cdot l}}{[1 - zp(\mathbf{k})]} \, d^D k. \tag{11}$$

The generating form of eq. (6) is

$$F(l, z) = \frac{[G(l, z) - \delta_{l,0}]}{G(l=0, z)}. \tag{12}$$

2.2. Number of distinct sites visited

The first passage time probabilities can be used to calculate the mean number of distinct sites S_n visited by an n-step random walk, an important quantity for calculating the survival of a random walker in a system of traps. To calculate S_n we must include the probability that the jth step visits a new site (for all j from 1 to n), i.e.

$$S_n = 1 + \sum_l [F_1(l) + \cdots + F_n(l)], \tag{13}$$

where the 1 on the rhs counts the origin as the first distinct site. Forming the generating function $S(z)$ of eq. (13) we have

$$S(z) = \sum_{n=0}^{\infty} S_n z^n = \frac{1}{1-z} + z\sum_l F_1(l) + z^2 \sum_l [F_1(l) + F_2(l)]$$

$$+ \cdots + z^n \sum_l [F_1(l) + \cdots + F_n(l)] + \cdots$$

$$= \frac{1}{1-z} \sum_l [zF_1(l) + \cdots + z^n F_n(l) + \cdots] + \frac{1}{1-z} \sum_l \delta_{l,0}$$

$$= \frac{1}{1-z} \sum_l F(l, z) = \frac{z}{(1-z)^2} \frac{1}{G(l=0, z)}, \tag{14}$$

where eq. (9) has been used together with

$$\sum_l G(l, z) = \frac{1}{1-z}.$$

In three dimensions, as $z \to 1$, $S(z)$ divergences as $[(1-z)^2 G(0, z)]^{-1}$, where $G(0, z=1)$ is a constant depending on the lattice structure. Using a Tauberian theorem this divergence as $z \to 1$ leads to the following divergence of S_n as $n \to \infty$:

$$S_n \sim \frac{n}{G(0, 1)}. \tag{15}$$

There are three ways to change this result: (i) consider a random-walk in continuous time with an infinite mean waiting-time [26] as shown in section 3; (ii) consider the number of discrete points visited in a special set rather than the whole plane [27]; or (iii) relax the restriction that the lattice is regular. In this last case:

$$S_n \sim n^{h/2}, \tag{16}$$

where h has been called the harmonic [28], spectral, and fracton dimension of the lattice. This dimension can also be calculated from the low-frequency density of states $\rho(\omega)$ of the lattice, i.e., $\rho(\omega) \sim \omega^h$. This connection between random walks and lattice vibrations was one of the early research achievements of Elliott Montroll [30]. These ideas have been applied by Alexander and Orbach [29] to discuss the low-frequency vibrational spectrum of glasses in terms of random walks on the glass lattice. Surprisingly, for randomly constructed fractal lattices h appears to be about 1.333 independent of the Euclidean dimension in which the lattice is embedded [29]. Note also that h is not necessarily equal to the fractal dimension of the lattice [31].

3. The Montroll–Weiss random walk

All the results from the previous section can be fundamentally changed if there is a waiting-time probability distribution governing the time interval between jumps. For example, if the mean time interval is infinite, then one need not arrive at the Gaussian distribution for the probability $P(l, t)$ of being at site l at time t. Montroll and Weiss [1] showed how to calculate $P(l, t)$ in terms of the Green's (generating) function for $P_n(l)$. For a walker to be at site l at time t, it can accomplish this by reaching site l at an earlier time $t - \tau$ and then not moving for at least a time τ, i.e.

$$P(l, t) = \int_0^t Q(l, t - \tau) R(\tau) \, d\tau, \tag{17}$$

where $Q(l, t)$ is the probability density to reach l exactly at time $t - \tau$, and $R(\tau)$ is the probability that no jump occurs for a time interval τ after the previous jump. Let $\psi(t) dt$ = the probability that a jump occurs in the time interval $(t, t + dt)$ given that the last jump was at $t = 0$. Both Q and R can be expressed in terms of ψ, i.e.

$$R(t) = 1 - \int_0^t \psi(\tau) \, d\tau \tag{18}$$

and

$$Q(l, t) = \sum_{l'} \int_0^t \psi(\tau) \, p(l') Q(l - l', t - \tau) \, d\tau + \delta_{l,0} \delta(t). \tag{19}$$

Laplace transforming, i.e.

$$Q(l, s) \equiv \int_0^\infty e^{-st} Q(l, t) \, dt,$$

we arrive at

$$Q(l, s) - \psi(s) \sum_{l'} p(l') \, Q(l - l', s) = \delta_{l, 0} \tag{20}$$

which is precisely eq. (9) with $z = \psi(s)$. Furthermore,
$$R(s) = [1 - \psi(s)]/s \tag{21}$$
so that
$$P(l, t) = L^{-1}\left[G(l, \psi(s))\frac{1 - \psi(s)}{s}\right], \tag{22}$$
where L^{-1} represents the inverse Laplace transform and G is the Green's function of eq. (9).

The first passage time $F_n(l)$ and the mean number of distinct sites visited can also be calculated as $F(l, t)$ and $S(t)$ in the Montroll–Weiss random walk:
$$F(l, t) = \sum_{n=0}^{\infty} F_n(l) \, \psi_n(t), \tag{23}$$
where
$$\psi_n(t) = \int_0^t \psi_{n-1}(t - \tau) \, \psi(\tau) \, d\tau \tag{24}$$
is the probability that the nth jump occurs at time t. In Laplace space
$$F(l, s) = \sum_{n=0}^{\infty} F_n(l) \, [\psi(s)]^n \tag{25}$$
which is precisely the generating function of eq. (8) with $z = \psi(s)$. The number of distinct sites visited by time t is
$$S(t) = \sum_{n=0}^{\infty} S_n \int_0^t \psi_n(t - \tau) \, R(\tau) \, d\tau. \tag{26}$$
In terms of an inverse Laplace transform
$$S(t) = L^{-1}\left[\sum_{n=0}^{\infty} S_n [\psi(s)]^n \frac{1 - \psi(s)}{s}\right]$$
$$= L^{-1}\left[S(z = \psi(s))\frac{1 - \psi(s)}{s}\right] \tag{27}$$
If the first moment of $\psi(t)$, denoted by $\bar{t}$, is finite, then
$$\psi(s) = \int_0^{\infty} e^{-st} \psi(t) \, dt \sim 1 - s\bar{t} + \cdots \quad \text{as} \quad s \to 0, \tag{28}$$
and all the results for discrete time random walks hold asymptotically at long times when n is replaced by $t/\bar{t}$. However, if
$$\psi(t) \sim At^{-1-\alpha} + \cdots \quad \text{as} \quad t \to \infty, \; 0 < \alpha < 1, \tag{29}$$

then $\bar{t}$ is infinite and [32]

$$\psi(s) \sim 1 - \frac{A}{\alpha}\Gamma(1-\alpha)s^\alpha. \tag{30}$$

In 3D this leads to

$$S(t) \sim [A\Gamma(1-\alpha)\Gamma(1+\alpha)G(0,1)]^{-1}\alpha t^\alpha = \frac{\sin \pi\alpha}{\pi A}\frac{t^\alpha}{G(0,1)}. \tag{31}$$

We now have all the ingredients to analyze a model of dielectric relaxation in disordered liquids and solids. One last remark is that a stochastic process with a $\psi(t)$ with $\bar{t}$ infinite has been called a fractal time process because it can be shown that since the jumps do not occur, on the average, at regular intervals they occur in self-similar bursts. These bursts have the structure of a Cantor set of fractal dimension α [33].

4. Dielectric relaxation: a defect diffusion model

4.1. Introduction

Consider a polarizable material with individual dipole moments $\mu(t)$ in an electric field. When the field is turned off at $t=0$ the decay of the autocorrelation function

$$\phi(t) = \frac{\langle \mu(t)\mu(0)\rangle}{\langle \mu^2(0)\rangle} \tag{32}$$

is related to the frequency-dependent dielectric constant $\varepsilon(\omega)$ of the material by

$$\frac{\varepsilon(\omega) - \varepsilon_\infty}{\varepsilon_0 - \varepsilon_\infty} = -\int_0^\infty e^{-i\omega t}\frac{d\phi(t)}{dt}dt \equiv \varepsilon'(\omega) - i\varepsilon''(\omega). \tag{33}$$

In 1913 Debye treated the relaxation of spherical molecules, of radius R, in a fluid of viscosity η and derived that

$$\phi_D(t) = e^{-t/\tau_D}, \tag{34}$$

with

$$\tau_D = \frac{4\pi R^3 \eta}{kT}, \tag{35}$$

and thus

$$\varepsilon'(\omega) = \frac{1}{1+\omega^2 \tau_D^2} \quad \text{and} \quad \varepsilon''(\omega) = \frac{\omega \tau_D}{1+\omega^2 \tau_D^2}. \tag{36}$$

Note that a plot (Cole–Cole plot) of ε' versus ε'' is in this case a circle of radius unity. A good example of "Debye relaxation" is i-butyl bromide at 25°C, which has $\tau_D = 5.3 \times 10^{-11}$ s. The relaxation comes about through the collisions of the fluid particles with the polar molecules. Not surprisingly, the Debye expression does not describe more complicated systems such as glasses and polymers. For these materials it was found by Williams and Watts [34] that a fractional exponential fits the data exceedingly well, i.e.

$$\phi(t) = e^{-(t/\tau)^\alpha}, \quad 0 < \alpha < 1. \tag{37}$$

Note that this decay is slower than exponential, and if one wishes, eq. (37) can be written as a continuous distribution of exponents. [Below we derive an expression for the inverse Laplace transform of $e^{-(t/\tau)^\alpha}$.] Why does eq. (37) hold for so many materials? The answer proposed by Shlesinger and Montroll [4] is that the Williams–Watts form is the limiting form for the distribution function of a stochastic process governing the relaxation mechanism which we now describe.

We assume that when the electric field is on, some of the dipoles have enough energy and time to reach new configurations which they are blocked, by the topology of the system, from relaxing back from when the field is turned off. However, it is also assumed that mobile defects exist in the material which are able to move to the trapped dipole and bring about its reorientation—in a sense, by transporting a packet of "free volume" to the frozen configuration. When a defect and a dipole meet, the dipole is assumed to relax instantaneously. This model was first proposed in 1960, by Glarum [35], who solved it for one fixed dipole undergoing a one-dimensional random walk. The Williams–Watts form was not found. Philips, Barlow and Lamb [36] treated Glarum's 1D model but with two defects, and Bordewijk [37] considered an infinite system with a concentration of mobile defects. Bordewijk concluded that the defect diffusion model will not give a Williams–Watts result except in 1D, where $\alpha = 1/2$ is obtained. The value $\alpha = 1$ is found in 3D. We now show that the missing ingredient is to treat the mobile defect motion by the Montroll–Weiss random walk with $\psi(t) \sim t^{-1-\alpha}$. This will lead to Williams–Watts eq. (34). We do not specify the precise nature of the defects, although candidates such as dangling bonds, grain boundaries, vacancies, and local conformational fluctuations are possibilities. In glassy polymers, the so-called "beta process" may be related to main-chain bond defect transport [45].

4.2. A frozen dipole amidst a swarm of mobile defects

Let our dipole of interest be immobile and situated at the origin of a lattice of V sites. Consider that N independent mobile defects are also on this lattice, but not initially at the origin. The probability that a given defect is initially at site l_0 is V^{-1}. The survival probability of the dipole orientation $\phi(t)$ is the probability that no defect has reached the origin by time t, and it is given by

$$\phi(t) = \left[1 - V^{-1} \sum_{l_0} \int_0^t F(l_0, \tau) \, d\tau \right]^N. \tag{38}$$

The term in brackets is one minus the probability that a defect which starts at l_0 reaches the origin (has its first passage to the origin) in the time interval $(0, t)$, i.e. the term in brackets is the probability that the defect does not reach the origin. Note we have averaed over all initial positions. The bracket is raised to the Nth power as this is the probability that none of the N walkers reaches the origin. In the thermodynamic limit $N, V \to \infty$ but with $N/V = \text{const.} = c$:

$$\phi(t) = \exp\left[-c \sum_{l_0} \int_0^t F(l_0, \tau) \, d\tau \right]. \tag{39}$$

In the finite system we calculated when the first defect would reach the origin, but this becomes the flux of walkers into the origin in the infinite system. Examining eq. (13) for S_n and eq. (27) for $S(t)$ we see that the survival probability can be rewritten as

$$\phi(t) = e^{-cS(t)} \tag{40}$$

since

$$\mathscr{L} \sum_l \int_0^t F(l, t) \, dt = \sum_l (1/s) F(l, s) = \frac{1}{s} \sum_l \frac{G(l, \psi(s)) - \delta_{l,0}}{G(l=0, \psi(s))}$$

$$= \frac{\psi(s)}{s(1 - \psi(s))} \frac{1}{G(0, \psi(s))} = \mathscr{L}^{-1} S(t).$$

For a random walk with $\bar{t} < \infty$ we find $\phi(t) = \exp[-\text{const.} t^{1/2}]$ in 1D, and

$\phi(t) = \exp[-\text{const.} t]$ in 3D, Bordewijk's results. However, if $\bar{t} = \infty$ as in eq. (29), then

$$\phi(t) = \exp\left[\frac{-c \sin \pi \alpha}{\pi A} t^\alpha\right] = e^{-(t/\tau)^\alpha}, \quad 0 < \alpha < 1, \tag{41}$$

the Williams–Watts form. Similar equations have been previously derived for problems in electron scavenging, by Hamill and Funabashi [38] and Tachiya [39].

4.3. Temperature dependence of the Williams–Watts exponent

The exact form of $\psi(t)$ will determine the constant A in eq. (41), as well as the α dependence of A in particular. Let us examine a model, due to Pfister and Scher [40], which will also provide us with the temperature dependence of α.

Assume a defect sees a distribution of activation barriers, described by the probability density $f(\Delta)$, over which it jumps. For a single barrier height Δ_0,

$$\psi(t) = \lambda_0 \exp(-\lambda_0 t), \tag{42}$$

where

$$\lambda_0 = \nu_0 \exp(-\Delta_0/kT). \tag{43}$$

For a distribution of barrier heights (but not prefactors).

$$\psi(t) = \int_0^\infty \lambda e^{-\lambda t} \rho(\lambda) \, d\lambda, \tag{44}$$

where $\rho(\lambda) = f(\Delta) \frac{d\Delta}{d\lambda}$ or $\lambda \rho(\lambda) = -f(\Delta) kT$. For

$$f(\Delta) = 0 \quad \text{for} \quad \Delta < \Delta_0 \quad \text{and} \quad f(\Delta) = q e^{-q(\Delta - \Delta_0)} \quad \text{for} \quad \Delta > \Delta_0, \tag{45}$$

where we can also write $q = (kT_0)^{-1}$, then

$$\rho(\lambda) = \frac{kTq}{\lambda} \left(\frac{\lambda}{\lambda_0}\right)^{kTq}, \quad 0 \leq \lambda \leq \lambda_0. \tag{46}$$

Call $kTq = \alpha$, and use the $\rho(\lambda)$ from eq. (46) in eq. (44) to derive

$$\psi(t) \sim \frac{\alpha \Gamma(\alpha + 1)}{\lambda_0^\alpha} t^{-1-\alpha} \quad \text{as} \quad t \to \infty, \tag{47}$$

the fractal-time form with $\bar{t}$ infinite. Finally for the barrier-hopping model using eqs. (29) and (31) we have the following relaxation law:

$$\phi(t) = \exp\left[\frac{-c \sin \pi\alpha (\lambda_0 t)^\alpha}{G(0, 1) \pi\alpha \Gamma(1+\alpha)}\right]. \tag{48}$$

Note that the model predicts that α will change linearly with absolute temperature. Such behavior has been reported in an analysis of dielectric relaxation data by Ngai [41]. Also note that if we rewrite eq. (48) as $\phi(t) = \exp(-t/\tau)^\alpha$ then τ will be a function of α.

4.4. Probability density $\rho_\alpha(\lambda)$ of relaxation times for $e^{-(t/\tau)^\alpha}$

Since a single relaxation time model such as $\phi_D(t)$ of eq. (34) plays a prominent theoretical and practical role in the interpretation of experimental data, it is natural to attempt to analyze non-exponential relaxation, such as described by $\exp[-(t/\tau)^\alpha]$ of eq. (37), in terms of a continuous distribution of exponential processes. Formally, this is accomplished by introducing the probability density $\rho_\alpha(\lambda)$;

$$e^{-(t/\tau)^\alpha} = \int_0^\infty \rho_\alpha(\lambda) e^{-\lambda t} \, dt. \tag{49}$$

By definition, then, the relaxation function $\exp[-(t/\tau)^\alpha]$ is the Laplace transform of the density $\rho_\alpha(\lambda)$ so that the latter is the inverse Laplace transform of $\exp[-(t/\tau)^\alpha]$. As a practical matter (e.g., for fitting NMR relaxation parameters) it is useful to be able to compute $\rho_\alpha(\lambda)$ for small α such as are found near or below the glass transition temperature. Lindsey and Patterson [42] employed special-purpose hardware and software to deal with this range, and Helfand [43] and Montroll and Bendler [44] examined asymptotic series. Helfand's results are disappointing for $\alpha = \frac{1}{3}$ and $\lambda > 1$ or so. Montroll and Bendler reported several new series expansions, but presented no numerical results to gauge convergence behavior. For very small α, $\rho_\alpha(\lambda)$ behaves as a log–normal density. Though of theoretical interest, this expansion (eq. 60b of ref. [44]) suffers from poor convergence and lack of normalization. Equation (59) of ref. [44] is superior in both respects. Rewriting it, using the present notation,

$$\rho_\alpha(\lambda) = \frac{\alpha}{\tau}\left(\frac{\tau}{\lambda}\right)^{\alpha-1} e^{-(\lambda/\tau)^\alpha}\left[1 - \alpha F_2 + \alpha^2 F_3 - \alpha^3 F_4 + \alpha^4 F_5 + \cdots\right], \tag{50}$$

Table 1
Comparison of the approximate relaxation density $\rho_\alpha(\lambda)$ of eq. (51) with the exact result of eq. (53) for $\alpha = \frac{1}{3}$. Relaxation times are scaled by setting $\tau = 1$.

λ	$\rho_{1/3}(\lambda)$		% error
	Exact (eq. 53)	Approximated (eq. 51)	
0.27×10^{-2}	11.803670	11.803660	-0.0001
0.27	0.404173	0.404176	-0.0007
3.3075	0.040636	0.040635	-0.004
15.1875	0.005994	0.005995	0.01
27	0.002379	0.002380	0.03
108	0.0001154	0.0001150	-0.3

where

$$F_2 = 0.577215665(1 - \mu^{-\alpha}), \tag{51a}$$

$$F_3 = -0.6558775(1 - 3\mu^{-\alpha} + \mu^{-2\alpha}), \tag{51b}$$

$$F_4 = -0.04200328(1 - 7\mu^{-\alpha} + 6\mu^{-2\alpha} - \mu^{-3\alpha}), \tag{51c}$$

$$F_5 = 0.16653857(1 - 15\mu^{-\alpha} + 25\mu^{-2\alpha} - 10\mu^{-3\alpha} + \mu^{-4\alpha}), \tag{51d}$$

where $\mu \equiv \tau/\lambda$. Higher-order terms are readily found. Closed-form expressions for $\rho_\alpha(\lambda)$ are known for $\alpha = \frac{1}{3}$ and $\frac{1}{2}$ [44]:

$$\rho_{1/3}(\lambda) = \frac{1}{3\pi\tau}\left(\frac{\tau}{\lambda}\right)^{1/2} K_{1/3}\left[\frac{2}{3^{1.5}}\left(\frac{\lambda}{\tau}\right)^{1/2}\right], \tag{52}$$

where $K_{1/3}$ is the modified Bessel function of the third kind; also

$$\rho_{1/2}(\lambda) = \tfrac{1}{2}\left(\frac{\tau}{\pi\lambda}\right)^{1/2} e^{-\tau/4\lambda}. \tag{53}$$

Table 2
Comparison of the approximate relaxation density $\rho_\alpha(\lambda)$ of eq. (51) with the exact result of eq. (54) for $\alpha = \frac{1}{2}$. Relaxation times are scaled by setting $\tau = 1$.

λ	$\rho_{1/2}(\lambda)$		% error
	Exact (eq. 54)	Approximated (eq. 51)	
1.0×10^{-3}	8.918391	8.918445	0.0006
1.0×10^{-1}	0.870037	0.870038	0.0002
1.0	0.2196956	0.2196399	-0.025
5.0	0.036144	0.0362828	0.38
1.0×10	0.007322	0.0074139	1.2

In tables 1 and 2 we compare the results of an eight order form of eq. (51) with the exact results of eqs. (53) and (54). From the tables we see that eq. (51) is an excellent approximation to the relaxation density $\rho_\alpha(\lambda)$ in the range $0 < \lambda \leq 0.5$ or so. It is both interesting and useful that eq. (51) is exactly normalized for any order.

References

[1] E.W. Montroll and G.H. Weiss, J. Math. Phys. 6 (1965) 167.
[2] E.W. Montroll and H. Scher, J. Stat. Phys. 9 (1973) 101.
[3] H. Scher and E.W. Montroll, Phys. Rev. B12 (1975) 2455.
[4] M.F. Shlesinger and E.W. Montroll, Proc. Nat. Acad. Sci (USA) 81 (1984) 1280.
[5] E.W. Montroll, J. Math. and Phys. 25 (1946) 37.
[6] E.W. Montroll, J. Soc. Ind. Appl. Math. 4 (1956) 241.
[7] E.W. Montroll, Am. Math. Soc. Sixteenth Symp. in Appl. Math. (1964) p. 193.
[8] E.W. Montroll, J. Phys. Soc. Japan, Suppl. 26 (1969) 6.
[9] E.W. Montroll, J. Math. Phys. 10 (1969) 753.
[10] V.M. Kenkre, E.W. Montroll and M.F. Shlesinger, J. Stat. Phys. 9 (1973) 45.
[11] U. Landman, E.W. Montroll and M.F. Shlesinger, Phys. Rev. Lett. 38 (1977) 285.
[12] U. Landman, E.W. Montroll and M.F. Shlesinger Proc. Nat. Acad. Sci. (USA) 74 (1977) 430.
[13] E.W. Montroll and M. Schwarz Jr., Biosystems 8 (1977) 175.
[14] E.W. Montroll and B.J. West, Fluctuation Phenomena (North-Holland, Amsterdam, 1979) pp. 134–175.
[15] H. Scher, S. Alexander and E.W. Montroll, Proc. Nat. Acad. Sci. (USA) 77 (1980) 3748.
[16] B.D. Hughes, M.F. Shlesinger and E.W. Montroll, Proc. Nat. Acad. Sci. (USA) 78 (1981) 2659.
[17] M.F. Shlesinger, B.D. Hughes and E.W. Montroll, Proc. 6th Int. Symp. on Noise in Physical Systems, eds. P.H.E. Meijer, R. Mountain and R.J. Soulen Jr. (National Bureau of Standards Publ. 61, 1981) pp. 18–22.
[18] B.D. Hughes, E.W. Montroll and M.F. Shlesinger, J. Stat. Phys. 28 (1982) 111.
[19] B.D. Hughes, E.W. Montroll and M.F. Shlesinger, J. Stat. Phys. 30 (1983) 293.
[20] C.H. Wu and E.W. Montroll, J. Stat. Phys. 30 (1983) 537.
[21] E.W. Montroll and M.F. Shlesinger, in: CCNY Physics Symposium: M. Lax's Sixtieth Birthday, ed., H. Falk (City College of New York Physics Dept.) pp. 44–147.
[22] E.W. Montroll and M.F. Shlesinger, in: Studies in Statistical Mechanics, vol. 11, Nonequilibrium Phenomena II: From stochastics to Hydrodynamics, eds. J.L. Lebowitz and E.W. Montroll (North-Holland, Amsterdam 1984) pp. 1–121.
[23] E.W. Montroll and M.F. Shlesinger, in: Lecture Notes in Math. Vol. 1035, The Mathematics and Physics of Disordered Solids, eds. B.D. Hughes and B.W. Ninham, Springer, Berlin, 1983) pp. 109–137.
[24] M.F. Shlesinger and E.W. Montroll, in: ref. [23], pp. 138–152.
[25] J. Stat. Phys. 30, No. 2 (1983), eds. G.H. Weiss and R.J. Rubin.
[26] M.F. Shlesinger, J. Stat. Phys. 10 (1974) 421.
[27] G.H. Weiss and M.F. Shlesinger, J. Stat. Phys. 27 (1982) 355.

[28] B.D. Hughes and M.F. Shlesinger, J. Math. Phys. 23 (1982) 1688 (see footnote 21).
[29] S. Alexander and R. Orbach, 1982, J. Physique Lett. 43 (1982) 1625.
[30] E.W. Montroll, in: Proc. 3rd Berkeley Symp. on Mathematical Statistics and Probability (Univ. of California, Berkeley, 1955).
[31] D. Dhar, J. Math. Phys. 18 (1977) 577.
[32] W. Feller, 1966, An Introduction to Probability Theory and Its Applications (Wiley, New York) vol. 2, p. 424.
[33] M.F. Shlesinger and B.D. Hughes, Physica A109 (1981) 597.
[34] G. Williams and D.C. Watts, Trans. Faraday Soc. 66 (1970) 80.
[35] S. Glarum, J. Chem. Phys. 33 (1960) 1371.
[36] M.C. Philips, A.J. Barlow and J. Lamb, Proc. Roy. Soc. London A329 (1972) 193.
[37] P. Bordewijk, Chem. Phys. Lett. 32 (1975) 592.
[38] W.H. Hamill and K. Funabashi, Phys. Rev. B16 (1977) 5523.
[39] M. Tachiya, Rad. Phys. Chem. 17 (1981) 447.
[40] G. Pfister and H. Scher, Adv. Phys. 27 (1978) 747.
[41] K.L. Ngai, Solid State Ionics 5 (1981) 27.
[42] C.P. Lindsey and G.D. Patterson, J. Chem. Phys. 73 (1980) 3348.
[43] H. Helfand, J. Chem. Phys. 78 (1983) 1931.
[44] E.W. Montroll and J.T. Bendler, J. Stat. Phys. 34 (1984) 129.
[45] A.A. Jones, Macromolecules, (in press).

CHAPTER 2

The Fascination of Old Texts

Cyril DOMB
Department of Physics
Bar-Ilan University
Ramat-Gan
Israel

© *Elsevier Science Publishers B.V. 1985*

The Wonderful World of Stochastics
A Tribute to Elliott W. Montroll
Eds. M.F. Shlesinger and G.H. Weiss

Contents

1. Introduction. Historical remarks — 49
2. Pierre Laplace (1749–1827) — 52
3. Logarithms: Napier: Briggs: Bonfils — 56
4. James Stirling (1692–1770) — 64
5. James Clerk Maxwell (1831–1872) — 68
6. Conclusion — 73
References — 74
Note added after submission — 74

1. Introduction. Historical Remarks

It is a privilege to be the opening speaker at this gathering of friends of Elliott Montroll to honour his memory. I am sure there are many here whose collaboration with Elliott has been more significant than mine. My major claim to distinction is the period of my association with Elliott which is, I believe, longer than that of anyone here today. I first met Elliott in the late 1940s during his period of service with ONR and at the time of Statphys 1, the Florence Conference on Statistical mechanics. I was not invited to attend the Conference myself, (being a graduate student), but Rushbrooke reported on some of my work. It was closely related to Elliott's 1941 paper in which he introduced the transfer matrix, and Elliott showed great interest in my results.

When we talked we soon found that we had interests in common other than the Ising model, and the statistical mechanics of interacting systems; among them were random walks, asymptotic expansions, Laplace transforms, and lattice dynamics. I soon learned of a number of features of Elliott's work. His mathematics possessed a characteristic elegance; he was never secretive and always tried to interest other people in problems on which he was working; and he always tried, and usually succeeded, in finding applications of one piece of mathematics in several different fields.

In the earliest published papers of Joe Mayer on the calculation of cluster integrals it is easy to identify Elliott's contribution in the neat application of Fourier transforms. It was Elliott who introduced Lars Onsager to the Ising problem. As Onsager recorded [1] in his autobiographical comments during the Conference to celebrate the 25th anniversary of his solution of the two-dimensional Ising model: "I got a letter from Joe Mayer recommending a young fellow, indeed recommending him very highly. The young fellow had worked with Wannier before and spent a year with Joe and Maria (Mayer). On the strength of that recommendation, I used the strongest pressure I had at my disposal to wedge an opening and Elliott (Montroll) came to occupy it. I was very glad I had made that effort. And incidentally, so is everybody else in the Chemistry Department at Yale. I think everybody found Elliott very interesting to talk to. He brought the news of this development on the Ising model that Wannier had brought home from Holland after working with Kramers." And Elliott soon found an application of his solution of the one-dimensional Ising chain to Markoff processes in probability [2].

I next met Elliott at Statphys 2 in Paris in 1952. Here he played an important role in laying the ghost of the exact solution of the three-dimensional Ising model. The story is as follows. John Maddox was a very able graduate student of Charles Coulson in London, who had subsequently taken up an appointment as an assistant lecturer at Manchester University. Whilst working with Coulson he had on his own derived an exact solution of the two-dimensional model, and he now claimed to have an exact solution of the three-dimensional model. Nothing had been published (this was before the preprint era!) but several visitors to Manchester returned convinced that Maddox indeed had a solution.

Prigogine and de Boer who organized Statphys 2 were unaware of Maddox's work, and did not invite him. Elliott prevailed on them soon after the start of the Conference to get in touch with Maddox by telephone, and to offer to cover his expenses if he would present his results during the Wednesday morning session. Maddox crossed the Channel on Tuesday night, and gave a lucid account in the morning session of the operator algebra he had used. Finally he wrote on the board a formula for the partition function which was a triple integral, the three-dimensional analogue of the Onsager two-dimensional double integral formula. He sat down in an atmosphere of great excitement. Elliott immediately jumped to his feet with the words "Please leave the formula on the board". He continued "In 1943 Onsager solved the two-dimensional Ising model. Ten minutes after he had arrived at the solution, he conjectured that the three-dimensional solution might be the formula on the board. He tested the conjecture by comparing it with exact series expansions for the partition function, and he found that the conjectured formula gave two terms correctly but failed at the third term. I cannot comment on the operator algebras discussed by Maddox, but I do know that the formula on the board is not the solution of the three-dimensional Ising model." Elliott sat down in an atmosphere of even greater excitement.

The Chairman of the session, Rushbrooke, suggested that all those with detailed knowledge should huddle together in a sub-session to iron out the situation, and Maddox discovered an error in his calculations.

He revealed exceptional talents as a writer on general scientific matters, and after a distinguished career as a scientific correspondent, he became editor of *Nature*, a post which he holds at the present day.

At this Conference Elliott invited me to spend a sabbatical with him at Maryland, and told me that he would be delighted to find post-doctoral positions for any of my bright post-graduate students. I soon sent him Ren Potts with whom he collaborated to produce an elegant treatment of the effect of impurities on the spectrum of lattice vibrations. (Some years later in the early 1960s I introduced him to Michael Fisher.) In 1958 I took up Elliott's invitation and arrived in Washington for a seven month stay at the University of Maryland.

I found there an international collection of bright young people, many of whom subsequently became leaders of research in different fields. It was an excellent place for learning of problems at the frontiers of current research in theoretical physics. Elliott was then very much concerned with lattice dynamics of disordered solids at about the time that Phil Anderson [3] was writing his "much quoted but little read" fundamental paper on localization in disordered systems. I learned there of the relevance of percolation processes to solid state physics (Hammersley had introduced such processes in statistics a year previously [4]).

Soon after my arrival I noticed that Elliott was scheduled to give a lecture at the Bureau of Standards on "101 Years of Statistical Mechanics". It seemed a strange title, but Elliott explained that last year he had delivered a centenary talk, and when he prepared a new talk he always planned to use it at least twice. From this talk I learned of his great interest in the history of science, and his fine judgement on matters connected with it. He considered that real statistical mechanics had started with the famous paper of Rudolf Clausius in 1857 "On the nature of the motion we call heat". There may have been activity before this by men of distinction like Newton, Boyle and Daniel Bernoulli, but activity is not synonymous with progress.

I subsequently found that Elliott's thesis had the support of Willard Gibbs, who in an obituary tribute to Clausius wrote [5]:

"The origin of the kinetic theory of gases is lost in remote antiquity and its completion the most sanguine cannot hope to see. But a single generation has seen it advance from the stage of vague surmises to an extensive and well established body of doctrine. This is mainly the work of three men, Clausius, Maxwell, and Boltzmann, of whom Clausius was the earliest in the field and has been called by Maxwell the principal founder of the science. We may regard his paper (1857), *Ueber die Art der Bewegung, welche wir Warme nennen*, as marking his definite entry into this field, although many points were incidentally discussed in earlier papers.

Elliott rarely delivered a public lecture without introducing some tit-bit of historical interest, and it is to his fascination with the history of science that I am devoting my lecture.

2. Pierre Laplace (1749–1827)

My association with Elliott provided a great deal of useful pragmatic information about the history of science; the importance of getting as close as possible to the source (in Abel's language "Read the masters not their pupils"); that only a relatively small number of fundamental papers or text-books have lasting value. Elliott managed to collect many of these in his library and he put them to good use - books like Gilbert's *de Magnete*, Maxwell's *Theory of Heat* and *Electricity and Magnetism*, Rayleigh's *Theory of Sound*, Hobson's *Spherical Harmonics*, Todhunter's *History of Probability*, McMahon's *Combinatorial Analysis*, Whitworth's *Choice and Chance*, Watson's *Bessel Functions*. He also had an eye for unusual books like Maxwell's editing of the papers of the Hon. Henry Cavendish, or Edwards' *Differential Calculus* and *Integral Calculus*.

I have already referred to the elegance which characterized Elliott's mathematical work; he took particular delight in clever mathematical devices like Laplace and Fourier transforms, generating functions, and asymptotic expansions. When I tried to identify the great mathematician in whose footsteps Elliott followed I was inevitably led to consider Pierre Laplace.

Let me make it clear that I should not wish to associate Elliott's character with that of Laplace, who did not maintain the highest ethical standards in either his personal or his scientific life. Laplace lived during one of the most turbulent periods of French history, during the revolution and rise and fall of Napoleon, yet he always managed to remain *persona grata* to those in authority. The first edition of his *Treatise on Probability* was dedicated to Napoleon; by the time the second edition appeared Napoleon was on Elba so the dedication was withdrawn. He was not always scrupulous about giving credit to the predecessors whose ideas he developed.

Laplace considered his major achievement to have been the *Mecanique Celeste* in which he applied Newtonian mechanics to the exploration of stability of the solar system. Nowadays this work does not command too much attention, but if one turns over the pages of his collected works, one finds many pioneering ideas which have proved extremely fruitful in the

20th century. At some points I must confess that I had the feeling that Elliott Montroll took over where Laplace left off. I should like to mention one or two of Laplace's elegant contributions to mathematics which must surely have delighted Elliott. In this connection the section on Laplace in Todhunter's *History of the Theory of Probability* provides a convenient summary of an important section of his work.

Firstly, in the evaluation of definite integrals Laplace was the first to derive the following:

$$\int_0^\infty \cos rx \exp - \alpha^2 x^2 \, dx = \frac{\pi}{2a} \exp - r^2/4\alpha^2, \tag{1}$$

$$\int_0^\infty \frac{\sin rx}{x} \, dx = \pi/2, \tag{2}$$

$$\int_0^\infty \frac{\cos ax}{1+x^2} \, dx = \frac{\pi}{2} e^{-a} = \int_0^\infty \frac{x \sin ax \, dx}{1+x^2}. \tag{3}$$

It must be remembered that the method of contour integration was not available (Cauchy was born only in 1789).

Secondly, Laplace discovered the most important single tool in combinatorial analysis, the generating function, which he used particularly for the solution of difference equations. Given an equation in one variable, $u(n)$,

$$u(n) = au(n-1) + bu(n-2) \tag{4}$$

he introduced the function

$$F(t) = \sum u(n) t^n \tag{5}$$

and it is easy to see that if one forms

$$(1 - at - bt^2) F(t) \tag{6}$$

all of the coefficients will vanish except for a few initial terms. Hence

$$(1 - at - bt^2) F(t) = h(t), \tag{7}$$

where $h(t)$ is determined by the initial terms $u(0)$ and $u(1)$. Hence

$$F(t) = \frac{h(t)}{1 - at - bt^2}, \tag{8}$$

from which $u(n)$ can readily be calculated.

The method can readily be extended to several variables. In his investigations of questions in probability, Laplace deals with the problem of points, which had been discussed by several of his predecessors in the theory. Two players A and B need respectively m and n points to win a set of games;

their chances of winning a single game are p and q ($p + q = 1$) and the first player to obtain his points wins the set. What is the probability in favour of each player?

Denoting A's probability by $\phi(m, n)$, Laplace derived the difference equation

$$\phi(m, n) = p\phi(m-1, n) + q\phi(m, n-1), \tag{9}$$

and proceeded to solve it by the technique of generating functions. Write

$$F(t, \tau) = \sum \phi(m, n) t^m \tau^n; \tag{10}$$

then if $F(t, \tau)$ is multiplied by $(1 - pt - q\tau)$ all terms except the initial ones vanish, and the general solution

$$F(t, \tau) = \frac{a(t) + b(\tau)}{1 - pt - q\tau} \tag{11}$$

is derived, where $a(t)$ and $b(\tau)$ depend on starting conditions. Taking these appropriately, the complete solution

$$F(t, \tau) = \frac{\tau(1 - q\tau)}{(1 - \tau)(1 - pt - q\tau)} \tag{12}$$

is obtained, from which $\phi(m, n)$ can readily be calculated explicitly.

Elliott devoted much effort to various aspects of the random-walk problem which was first formulated long after Laplace. It is interesting on turning the pages of Laplace's collected works to come across equations which are identical with those of the one-dimensional random walk,

$$y_{n,r} = \tfrac{1}{2}(y_{n-1, r+1} + y_{n-1, r-1}), \tag{13}$$

with a specific solution in the form of a definite integral

$$y_{n,r} = 1 - \frac{2}{\pi} \int_0^{\pi/2} \frac{(\cos \phi)^n \sin r\phi}{\sin \phi} d\phi. \tag{14}$$

Nowadays we use contour integrals as a method of calculating coefficients in series expansions, but before the discovery of contour integrals Laplace (ref. [6], p. 521) used the equally useful method of Fourier series. The coefficient independent of x in the series

$$(x^{-r} + x^{-(r-1)} + \cdots + x^{r-1} + x^r)^n \tag{15}$$

was derived by putting $x = e^{i\theta}$ and integrating from 0 to 2π. All other terms vanish and

$$a_0 = \int_0^{2\pi} \left[\frac{\sin(r+\tfrac{1}{2})\theta}{\sin\tfrac{1}{2}\theta} \right]^n d\theta. \tag{16}$$

The coefficient of x^m is obtained by multiplying by x^{-m} before integrating.

Another area of great interest to Elliott was that of asymptotic expansions, which have played an important part in 20th century mathematical physics. The method usually employed for the development of such expansions, the saddle-point and steepest-descent method, is associated with the name of Debye. But Dingle has noted [7] that the four different approaches to the evaluation of integrals as asymptotic series initiated by Laplace (1820), Riemann (1863), Stokes and Kelvin (1887) and Debye (1910) all yield the same asymptotic expansions. Laplace (ref. [6], p. 512) describes the evaluation of an integral of the form

$$\int y \, dx \tag{17}$$

in which y has a single maximum between the limits of integration. By transferring the origin to the maximum the integral is converted into one of the form

$$y_0 \int e^{-t^2} (b_1 + b_2 t + b_3 t^2 + \cdots) dt, \tag{18}$$

each term of which can readily be calculated. He generalizes the method to two dimensions, and describes its extension to n dimensions.

Finally I was intrigued by two numerical results in Laplace's collected works. On p. 318 of Vol. 10 in a probability calculation he quotes the following

$$\log 251\,527 = 5.400584610947. \tag{19}$$

Where did he find a logarithm to so many significant figures? And on p. 226 of Vol. 10 after expressing great admiration for Stirling he quotes the following result:

$$\sqrt{2\pi} \int \frac{du}{(1-u^4)^{1/2}} = 4\left(\int dt\, e^{-t^4}\right)^2, \tag{20}$$

and says that Stirling has evaluated the first integral as 1.31102877714605987. Stirling preceded Laplace by nearly 60 years; how had he evaluated the integral to such accuracy? The pursuit of these two questions led me to other interesting features of the history of mathematics.

3. Logarithms: Napier: Briggs: Bonfils

There are at least two interesting queries to be made in relation to the discovery of logarithms:

(i) The first tables of logarithms were published in 1614, well over 50 years before the development of the calculus (Newton was born in 1642). How were they calculated without Taylor series?

(ii) John Napier (1550–1617), who first conceived the idea, developed natural logarithms to base e. (Common logarithms to base 10 came later.) How was he led to such a sophisticated concept?

In introducing the Napier tercentenary celebration in 1914 Lord Moulton said the following [8]:
"The invention of logarithms came on the world as a bolt from the blue. No previous work had led up to it, nothing had foreshadowed it or heralded its arrival. It stands isolated, breaking in upon human thought abruptly without borrowing from the work of other intellects or following known lines of mathematical thought."

In fact the addition of exponents corresponding to the multiplication of numbers had been noted nearly 200 years previously, but surprisingly no one thought of making practical use of this result. Instead Napier concentrated on establishing a correspondence between an arithmetic and a geometric progression, and he did this by comparing the motion of a point A on a line whose velocity decreases and is proportional to the distance to be covered with a point B on a second line whose velocity is constant. He could not deal with the continuum problem before the advent of the calculus, and instead took intervals of size 10^{-7} which seemed sufficiently small to give the accuracy he needed. In fig. 1,

$$A_0 Z = 10^7,$$
$$A_1 Z = 10^7(1 - 10^{-7}),$$
$$A_2 Z = 10^7 - 1 - (1 - 10^{-7})$$
$$\quad = 10^7(1 - 10^{-7})^2,$$
$$A_3 Z = 10^7(1 - 10^{-7})^2 - (1 - 10^{-7})^2$$
$$\quad = 10^7(1 - 10^{-7})^3,$$
$$\vdots$$
$$A_r Z = 10^7(1 - 10^{-7})^r.$$

$$B_0 Z' = 10^7,$$
$$B_1 Z' = 10^7(1 - 10^{-7}),$$
$$B_2 Z' = 10^7(1 - 2 \times 10^{-7}),$$
$$B_3 Z' = 10^7(1 - 3 \times 10^{-7}),$$
$$\vdots$$
$$B_r Z' = 10^7(1 - r \times 10^{-7}).$$

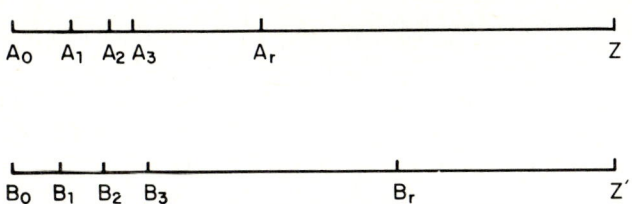

Fig. 1 (Following Napier). Point A_r has velocity proportional to A_rZ. Point B_r moves uniformly.

The function which Napier was constructing with points A_rZ is thus

$$N(1 - N^{-1})^{xN} \qquad (N = 10^7, \; r/N = x) \tag{21}$$

and this tends to Ne^{-x} with error of order N^{-2}. Hence the distances along the B line are the natural logarithms of those along the A line.

Napier spent 20 years calculating his tables of natural logarithms by the tedious straightforward method. His tables were not of great practical use since the base e is not convenient for computation (although from the theoretical point of view they provide a remarkable anticipation of the ideas of the calculus). However their publication made a deep impression on Henry Briggs (1561–1631) who was then Professor of Mathematics at Gresham College, London. (He later became second Savilian Professor of Geometry at Oxford.) Briggs realized that a considerable simplification could be achieved by changing the base, and decided to journey up to Scotland to consult personally with Napier. In the summer of 1615 he spent a month with Napier at Merchiston Castle (which Napier had inherited from his father). It seems that Napier had independently arrived at a similar conclusion, and Briggs left with Napier's blessing to construct new logarithm tables to base 10. In 1617 he published tables of common logarithms of all numbers from 1 to 1000 to 14 decimal figures, and he planned to extend the tables to all numbers from 1 to 100 000. In 1624 he published the "Arithmetica Logarithmica" (cover page, see fig. 2) containing tables to 14 figures of all numbers from 1 to 20 000 and from 90 000 to 100 000. The logarithms from 20 000 to 90 000 were filled in by A. Vlacq, a Dutchman, and the Briggs–Vlacq tables published in 1628 (to 10 figures) were not superseded for nearly 300 years.

The method by which Briggs calculated his tables was ingenious and we shall describe it briefly. He calculated square roots successively 54 times to 27 figures starting with 10 (see fig. 3) and thus provided a binary scale from

ARITHMETICA
LOGARITHMICA
SIVE

LOGARITHMORVM
CHILIADES TRIGINTA, PRO
numeris naturali serie crescentibus ab vnitate ad
20,000 : et a 90,000 ad 100,000. Quorum ope multa
perficiuntur Arithmetica problemata
et Geometrica.

HOS NVMEROS PRIMVS
INVENIT CLARISSIMVS VIR IOHANNES
NEPERVS Baro Merchistonij : eos autem ex eiusdem sententia
mutavit, eorumque ortum et vsum illustravit HENRICVS BRIGGIVS,
in celeberrima Academia Oxoniensi Geometriæ
professor SAVILIANVS.

DEVS NOBIS VSVRAM VITÆ DEDIT
ET INGENII, TANQVAM PECVNIÆ,
NVLLA PRÆSTITVTA DIE.

LONDINI,
Excudebat GVLIELMVS
IONES. 1624.

Fig. 2. Cover page of Briggs' 1624 logarithm tables to 14 decimal places using base 10.

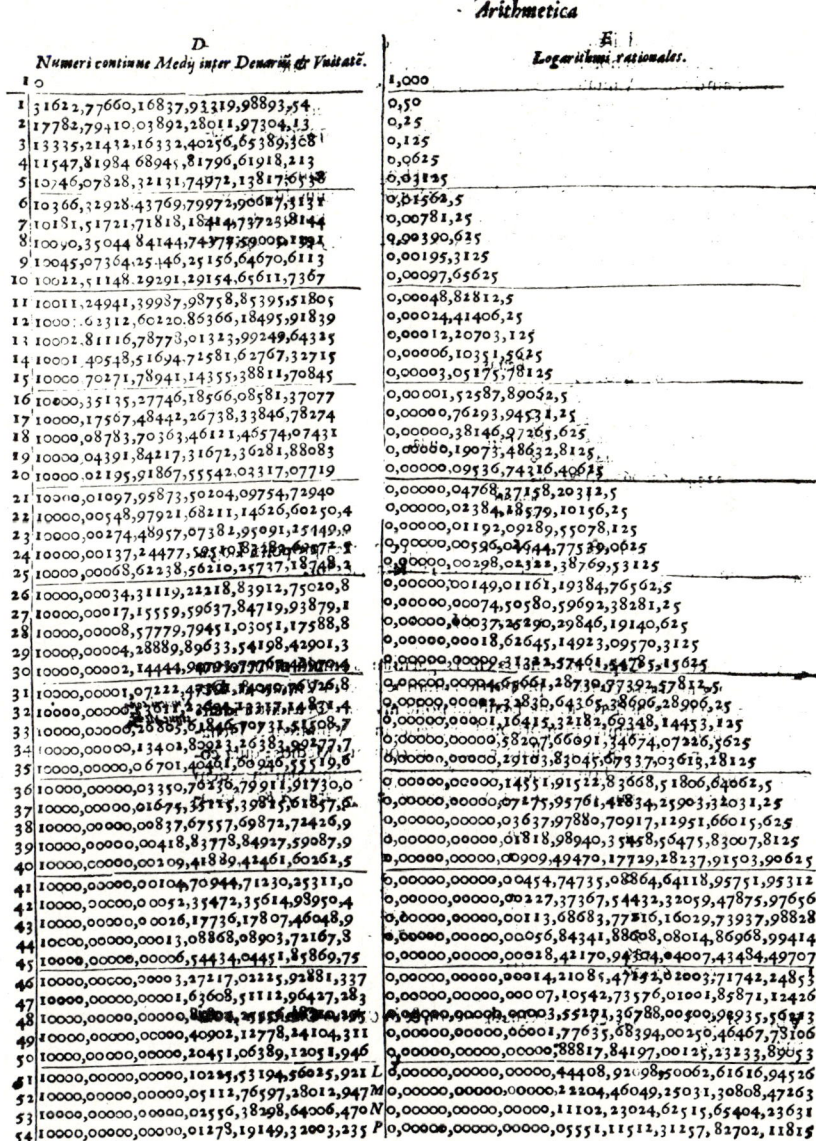

Fig. 3. Briggs calculation of 54 successive square roots of 10.

which the logarithm of any number can be calculated. This is illustrated in table 1 for the calculation of log 2 keeping 10 figures. We look down the scale to find the number closest to and just below 2; this is 1.778279410. Divide 2 by this number to obtain 1.124682650; down the scale to find the number closest to and just below this; it is 1.074607828; divide by this to obtain 1.046598229; look down the scale to 1.036632928 and so on. Log 2 is found by adding the logarithms of all the sub-product numbers from the right-hand side of the table. Using a pocket calculator we find for log 2 the value 0.3010299955 which has only a unit error in the last place.

Table 1
Calculation of log 2 keeping 10 figures.

	10.00000000		1.00000000
1	3.16227766		0.50000000
2	1.778279410	2.00000000	0.25000000
3	1.333521432		0.12500000
4	1.154781985		0.06250000
5	1.074607829	1.124682650	0.03125000
6	1.036632928	1.046598239	0.01562500
7	1.018151722		0.00781250
8	1.009035045	1.009613144	0.00390625
9	1.004507364		0.001953125
10	1.002251148		0.0009765625
11	1.001124941		0.0004882812
12	1.000562313	1.000572922	0.0002441406
13	1.000281117		0.0001220703
14	1.000140549		0.0000610352
15	1.000070272		0.0000305176
16	1.000035135		0.0000252588
17	1.000017567		0.000076294
18	1.000008784	1.000010603	0.0000038147
19	1.000004392		0.0000019073
20	1.000002196		0.0000009537
21	1.000001098	1.000001819	0.0000004768
22	1.000000549	1.000000721	0.0000002384
23	1.000000274		0.0000001192
24	1.000000137	1.000000172	0.0000000596
25	1.000000069		0.0000000298
26	1.000000034	1.000000035	0.0000000149
27	1.000000019		0.0000000075
28	1.000000009		0.0000000037
29	1.000000004		0.0000000018
30	1.000000002		0.0000000009
31	1.000000001	1.0000000001	0.0000000005

The above method is very suitable for the logarithms of specific numbers. But Briggs wished to construct a table of logarithms and he therefore devised appropriate methods of interpolating and assessing the errors arising *. Figure 4 shows a typical page of these tables.

For the sake of historical completeness we should add that J. Burgi, a Swiss mathematician, hit on the idea of logarithms independently at about the same time and constructed tables to base 1.0001 (these were published in 1620). However these tables made little impact in view of the superiority of the Briggs and Briggs–Vlacq tables.

A question which naturally suggests itself is why Briggs made the effort to keep so many significant figures. The answer to this may lie in the very important arithmetical tool which had just become available to him. Cajori [9] considers that the three most important mathematical advances in the mediaeval period were the introduction of arabic numerals, the discovery of decimal fractions, and the invention of logarithms. Arabic numerals had become well established long before Briggs, but decimal fractions had only recently come into practical use. In the literature of the mid-16th century one finds [10] vulgar fractions like

$$31\tfrac{2227971490}{6847124257} \quad \text{and} \quad 2\tfrac{1724912242}{3377560879}$$

appearing. In 1585 Stevin published a book on decimal fractions in Flemish with a French translation. His notation was clumsy, but the book had a wide impact, and influenced Briggs, who may well have been the first in England to use the modern decimal point notation. (This notation first appeared in the second edition of Napier's tables in 1619, edited by Briggs after Napier's death.)

One can well imagine the excitement which this discovery engendered, and the new apparatus which it provided for multiplying, dividing, and extracting square roots. There would clearly be a desire to push the new tool to the limit of its capacity.

Briggs' tabulation to 14 figures (fig. 4) was so good that when nearly 100 years later Edmund Halley [11] (one of his successors in the Savilian Chair) wished to demonstrate the superiority of Newton's power series method, he felt obliged to take calculations to 60 figures. Halley used two combinations of the Taylor series for $\ln(1+z)$:

(i) $\quad \tfrac{1}{2}\ln\dfrac{z+x}{z-x} = \dfrac{x}{z} + \dfrac{x^3}{3z^3} + \dfrac{x^5}{5z^5} + \dfrac{x^7}{7z^7} + \cdots, \qquad x = \dfrac{a-b}{2}, \quad z = \dfrac{a+b}{2},$

(22)

---

* The actual method used by Briggs is described in ref. [17].

Chilias quarta.

Num. absolu.	Logarithmi.	Num. absolu.	Logarithmi.	Num. absolu.	Logarithmi.
3401	3,53160,66319,3273 12,76773,4383	3434	3,53580,02908,2489 12,64505,7068	3467	3,53995,38416,5640 12,52471,4778
3402	3,53173,43092,7656 12,76398,1926	3435	3,53592,67413,9557 12,64137,6364	3468	3,54007,90888,0418 12,52110,3789
3403	3,53186,19490,9582 12,76023,1674	3436	3,53605,31551,5921 12,63769,7802	3469	3,54020,42998,4207 12,51749,4881
3404	3,53198,95514,1256 12,75648,3626	3437	3,53617,95321,3723 12,63402,1380	3470	3,54032,94747,9088 12,51388,8054
3405	3,53211,71162,4882 12,75273,7778	3438	3,53630,58723,5103 12,63034,7098	3471	3,54045,46136,7142 12,51028,3304
3406	3,53224,46436,2660 12,74899,4129	3439	3,53643,21758,2201 12,62667,4952	3472	3,54057,97165,0446 12,50668,0630
3407	3,53237,21335,6789 12,74525,2678	3440	3,53655,84425,7153 12,62300,4940	3473	3,54070,47833,1076 12,50308,0031
3408	3,53249,95860,9467 12,74151,3423	3441	3,53668,46726,2093 12,61933,7061	3474	3,54082,98141,1107 12,49948,1506
3409	3,53262,70012,2890 12,73777,6361	3442	3,53681,08659,9154 12,61567,1313	3475	3,54095,48089,2613 12,49588,5050
3410	3,53275,43789,9250 12,73404,1490	3443	3,53693,70227,0467 12,61200,7695	3476	3,54107,97677,7663 12,49229,0663
3411	3,53288,17194,0740 12,73030,8808	3444	3,53706,31427,8162 12,60834,6203	3477	3,54120,46906,8326 12,48869,8344
3412	3,53300,90224,9548 12,72657,8315	3445	3,53718,92262,4365 12,60468,6836	3478	3,54132,95776,6670 12,48510,8089
3413	3,53313,62882,7863 12,72285,0008	3446	3,53731,52731,1201 12,60102,9594	3479	3,54145,44287,4759 12,48151,9899
3414	3,53326,35167,7871 12,71912,3884	3447	3,53744,12834,0795 12,59737,4472	3480	3,54157,92439,4658 12,47793,3770
3415	3,53339,07080,1755 12,71539,9942	3448	3,53756,72571,5267 12,59372,1471	3481	3,54170,40232,8428 12,47434,9702
3416	3,53351,78620,1697 12,71167,8179	3449	3,53769,31943,6738 12,59007,0589	3482	3,54182,87667,8130 12,47076,7693
3417	3,53364,49787,9876 12,70795,8595	3450	3,53781,90950,7327 12,58642,1822	3483	3,54195,34744,5823 12,46718,7739
3418	3,53377,20583,8471 12,70424,1188	3451	3,53794,49592,9149 12,58277,5168	3484	3,54207,81463,3562 12,46360,9840
3419	3,53389,91007,9659 12,70052,5954	3452	3,53807,07870,4317 12,57913,0627	3485	3,54220,27824,3402 12,46003,3995
3420	3,53402,61060,5613 12,69681,2893	3453	3,53819,65783,4944 12,57548,8199	3486	3,54232,73827,7397 12,45646,0201
3421	3,53415,30741,8506 12,69310,2002	3454	3,53832,23332,3143 12,57184,7879	3487	3,54245,19473,7598 12,45288,8455
3422	3,53428,00052,0508 12,68939,3280	3455	3,53844,80517,1022 12,56820,9664	3488	3,54257,64762,6053 12,44931,8758
3423	3,53440,68991,3788 12,68568,6723	3456	3,53857,37338,0686 12,56457,3555	3489	3,54270,09694,4811 12,44575,1108
3424	3,53453,37560,0511 12,68198,2333	3457	3,53869,93795,4241 12,56093,9549	3490	3,54282,54269,5919 12,44218,5500
3425	3,53466,05758,2844 12,67828,0105	3458	3,53882,49889,3790 12,55730,7646	3491	3,54294,98488,1419 12,43862,1935
3426	3,53478,73586,2949 12,67458,0038	3459	3,53895,05620,1436 12,55367,7842	3492	3,54307,42350,3354 12,43506,0411
3427	3,53491,41044,2987 12,67088,2129	3460	3,53907,60987,9278 12,55005,0135	3493	3,54319,85856,3765 12,43150,0927
3428	3,53504,08132,5116 12,66718,6378	3461	3,53920,15992,9413 12,54642,4524	3494	3,54332,29006,4692 12,42794,3479
3429	3,53516,74851,1494 12,66349,2783	3462	3,53932,70635,3937 12,54280,1009	3495	3,54344,71800,8171 12,42438,8067
3430	3,53529,41200,4277 12,65980,1341	3463	3,53945,24915,4946 12,53917,9585	3496	3,54357,14239,6238 12,42083,4688
3431	3,53542,07180,5618 12,65611,2049	3464	3,53957,78833,4531 12,53556,0252	3497	3,54369,56323,0926 12,41728,3341
3432	3,53554,72791,7667 12,65242,4907	3465	3,53970,32389,4783 12,53194,3007	3498	3,54381,98051,4267 12,41373,4025
3433	3,53567,38034,2574 12.64873,9915	3466	3,53982,85583,7790 12,52832,7850	3499	3,54394,39424,8292 12,41018,6736
3434	3,53580,02908,2489 12,64505,7068	3467	3,53995,38416,5640 12,52471,4778	3500	3,54406,80443,5028 12,40664,1475

3501

Fig. 4. Typical page of Briggs' 1624 tables.

(ii) $\quad \ln\dfrac{ab}{z^2} = \dfrac{x^2}{2z^2} + \dfrac{x^4}{4z^4} + \dfrac{x^6}{6z^6} + \cdots.$ (23)

These could be used to calculate the logarithm of a prime number between two composite numbers. For example if $a = 22 = 11 \times 2$, $b = 24 = 2^3 \times 3$, the series (22) and (23) could be used to calculate ln 23 very rapidly since $(x/z)^2 = \frac{1}{529}$. Halley published the following for log 2 to 60 places

0.30102 99956 63981 19521 37388 94724 49302 67681 89881 46210 85413 10427, (24)

and asked for volunteers to calculate the logarithms of all prime numbers to 25 or 30 figures! None seems to have been forthcoming.

Before leaving this topic I should like to mention that the first person to formulate clearly the background theory of decimal fractions was a Jewish Rabbi [12], Immanuel ben Jacob Bonfils of Tarascon (c.1350). Bonfils was a mathematician and astronomer who had published astronomical tables, *Kanfei Nesharim*, wings of eagles (also called *Shesh Kenafayim*, six wings, because of its division into six parts). Their main purpose was to help in determining the Jewish calendar. He had also written a brief note entitled *Derech Chilluk* (a method of division), a manuscript of which was to be found in the French National Library in Paris. This manuscript had been catalogued, but the assumption had been that it contained no material of great novelty of interest.

In 1935 the eminent historian of science George Sarton decided to have a look at the work personally, sent for a copy, and gave it to Solomon Gandz to translate [13]. Gandz found that it contained a short but thorough discussion of decimal fractions, based on a unified description of exponents which included fractions as well as integers. As a result of this discovery, Sarton places Bonfils among the great mathematicians of the 14th century.

Bonfils did not introduce a notation, or undertake any demonstrative calculations by numerical examples. In his manuscript he describes the operation of decimals and states step by step rules for multiplying and dividing. His work unfortunately does not seem to have exercised a significant influence on his contemporaries or successors. The first person to establish a decimal symbolism was Nicholas Choquet a hundred years later (1484), and there seems to have been no interaction between Bonfils and Stevin who first successfully launched decimal fractions, and to whom we have referred above. But Bonfils' work has now received its proper recognition.

4. James Stirling (1692–1770)

James Stirling was a younger contemporary of Newton. He is best remembered for the asymptotic approximation which he derived for $n!$, but he also discovered Maclaurin's theorem (this was first published in a book by Maclaurin entitled *Treatise on Fluxions*).

Stirling's own most famous publication was *Methodus Differentialis* (1730) in which he developed a number of theorems of relevance to the calculus of finite differences, interpolation, and summation of series. His methods have a characteristic elegance, and he devoted a great deal of attention to the transformation of slowly convergent series. One of the ideas he exploited was the use of auxiliary series whose general term r_n could be expressed as a difference of successive terms

$$r_n = t_n - t_{n-1}. \tag{25}$$

The series r_n can be easily summed since successive terms cancel.

For example in trying to find a formula for

$$\sum_{n_0}^{\infty} \frac{1}{n^2}, \tag{26}$$

Stirling made use of auxiliary series whose nth terms are

$$\frac{1}{n(n+1)}, \quad \frac{1}{n(n+1)(n+2)}, \quad \frac{1}{n(n+1)(n+2)(n+3)}, \quad \ldots,$$

all of which satisfy a relation of the form (25). Thus

$$\frac{1}{n(n+1)} = \frac{1}{n} - \frac{1}{(n+1)},$$

$$\frac{1}{n(n+1)(n+2)} = \tfrac{1}{2}\left(\frac{1}{n(n+1)} - \frac{1}{(n+1)(n+2)}\right),$$

$$\frac{1}{n(n+1)(n+2)(n+3)} = \tfrac{1}{3}\left(\frac{1}{n(n+1)(n+2)} - \frac{1}{(n+1)(n+2)(n+3)}\right). \tag{27}$$

Hence,

$$\sum_{n_0}^{\infty} \frac{1}{n(n+1)} = \frac{1}{n_0}, \quad \sum_{n_0}^{\infty} \frac{1}{n(n+1)(n+2)} = \frac{1}{2n_0(n_0+1)},$$

$$\sum_{n_0}^{\infty} \frac{1}{n(n+1)(n+2)(n+3)} = \frac{1}{3n_0(n_0+1)(n_0+2)}. \tag{28}$$

Using the identity

$$\frac{1}{n^2} = \frac{1}{n(n+1)} + \frac{1!}{n(n+1)(n+2)} + \frac{2!}{n(n+1)(n+2)(n+3)} + \cdots,$$
(29)

Stirling derived the formula

$$\sum_{n_0}^{\infty} \frac{1}{n^2} = \frac{1}{n_0} + \frac{1!}{2n_0(n_0+1)} + \frac{2!}{3n_0(n_0+1)(n_0+2)} + \cdots.$$
(30)

This formula has the advantage that it is convergent, not asymptotic.

To calculate the sum $\Sigma(1/n^2)$ (the exact value of $\pi^2/6$ was given by Euler some years later) Stirling evaluated the first 12 terms directly, and calculated the remainder by (30). He obtained

$$\sum_{1}^{12} \frac{1}{n^2} = 1.564976\,638, \qquad \sum_{13}^{\infty} \frac{1}{n^2} = 0.079957427,$$

$$\sum_{1}^{\infty} \frac{1}{n^2} = 1.644934065,$$
(31)

which is in error by two units in the last place.

More generally for a series whose nth term is given by

$$r_n = x^{n+r}\left(\frac{a}{n} + \frac{b}{n(n+1)} + \cdots\right),$$
(32)

Stirling derived the formula

$$\sum_{n_0}^{\infty} r_n = x^{n_0+r}\left(\frac{a}{n_0(1-x)} + \frac{b - Ax}{n_0(n_0+1)(1-x)} + \frac{c - 2Bx}{n_0(n_0+1)(n_0+2)(1-x)}\right),$$
(33)

where

$$A = \frac{a}{1-x}, \quad B = \frac{b - Ax}{1-x}, \quad \ldots.$$

This could be used to evaluate $\ln(1 + x)$ or $\tan^{-1} x$. Taking $x = -1$ Stirling obtained the approximation

$$\pi/4 = 0.785398\,163\,4.$$
(34)

The Beta and Gamma functions were introduced by Euler, but Stirling was thinking independently how to interpolate the factorial function; to evaluate $(11\frac{1}{2})!$ he used the formula which Newton had introduced in 1711:

$$f(a + xw) = f(a) + x\Delta f(a) + \tfrac{1}{2}x(x - a)\Delta^2 f(a - w)$$
$$+ \tfrac{1}{6}(x + 1)x(x - 1)\Delta^3 f(a - w) + \cdots. \tag{35}$$

Dividing by $11\frac{1}{2} \times 10\frac{1}{2} \times 9\frac{1}{2} \times \cdots \times \frac{1}{2}$ he then obtained an approximation to $\Gamma(\frac{1}{2})$

$$\Gamma(\tfrac{1}{2}) = 1.7724538502. \tag{36}$$

(The first ten digits are correct, but the last digit should read 9 instead of 2.)

We now turn to the integral (20) quoted by Laplace, which can easily be related to a Beta function integral:

$$\int_0^1 v^{-3/4}(1 - v)^{-1/2}\, dv = B(\tfrac{1}{4}, \tfrac{1}{2}). \tag{37}$$

Stirling had been concerned more generally with the integral

$$\int_0^z z^{m-1}(1 - z)^{n-1}\, dz, \tag{38}$$

and using the binomial theorem introduced by Newton he showed that (38) could be expanded as a series

$$\frac{z^m}{n} F(m, 1 - n, m + 1, z), \tag{39}$$

with the modern notation $F(a, b, c, z)$ for the hypergeometric function,

$$F(a, b, c, z) = 1 + \frac{ab}{c \cdot 1} z + \frac{a(a + 1)b(b + 1)}{c(c + 1)2!} z^2 + \cdots. \tag{40}$$

Hence from (39)

$$B(m, n) = \frac{1}{m} F(m, 1 - n, m + 1, 1). \tag{41}$$

Stirling derived an asymptotic formula for the sum of a series

$$S(p_0) = \sum_{p_0}^{\infty} a_p \tag{42}$$

whose successive terms satisfied the relation

$$a_{p+1} = \frac{p - r}{p} \frac{p - s}{p - s + 1} a_p \tag{43}$$

58 *Summatio Serierum.*

$T\ \ =.0050.1271.8406.8834.46$
$T2=\ \text{- - -}\ 1833.9213.6837.20$
$T3=\ \text{- - - -}\ 15.5605.4494.38$
$T4=\ \text{- - - - -}\ 2425.8219.16$
$T5=\ \text{- - - - - -}\ 56.8742.44$
$T6=\ \text{- - - - - - -}\ 1.7922.51$
$T7=\ \text{- - - - - - - -}\ 707.95$
$T8=\ \text{- - - - - - - -}\ 33.45$
$T9=\ \text{- - - - - - - - -}\ 1.83$

$.0952.4164.9730.7854.7$
$8069.2540.2083.7$
$43.2237.3595.5$
$5224.8472.0$
$103.7118.6$
$2.9017.4$
1047.8
46.1
2.3

$S=.0953.2277.9839.9238.1$

Adjiciatur jam S aggregato initialium, & obtinebis 1.3110.2877.7146.0598.7 pro valore Seriei, id eft, pro longitudine Curvæ Elafticæ, modo in lineam rectam extenfa foret. Hunc autem numerum determinavit *Bernoullius* confiftere inter limites 1.308 & 1.315. Quod fi longitudini Elafticæ adjiciatur fua Ordinata, habebitur numerus 1.9100.9889.4513.8559.8 qui eft femiperiferia Ellipfeos habentis 1 & $\sqrt{2}$ pro Axibus. Et hæc Exempla fufficiant; haud enim immoror Seriebus quæ per hanc Propofitionem fummari poffint accurate.

SCHOLION.

Hoc Theorema nullo fere negotio exhibet Areas Curvarum binomialium quarum Ordinatæ continentur fub forma $x^{\prime}\times\overline{e+fx^{\eta}}^{\lambda}$, in unico tamen cafu, fcilicet ubi eft $e+fx^{\eta}=0$, five $x^{\eta}=-\frac{e}{f}$; hoc eft in eo cafu in quo Series pro Area convergit lentiffime. Sed ubi Areæ non producendæ funt ultra octo aut novem figuras, fufficit quærere Summam quatuor Terminorum initialium, nam S dabit eam reliquorum exiguo labore; imo fi nulli Termini initiales colligantur, fed inchoetur tranfmutatio ad primum Terminum, valor ipfius S approximabit ad valorem totius Seriei fat celeriter. Series autem in Theoremate, redditur generalior, atque extenditur ad cafus qui ad Quadraturas non attinent, ut fequitur. Sit Æquatio ad Seriem $T'=\frac{zz+m}{zz+nz+r}T$, & ponatur

Fig. 5. Page from Stirling's *Methodus Differentialis* giving the calculation quoted by Laplace.

[which is the case for (41) with $r = 1 - m$, $s = n$]. He showed that

$$S(p_0) = \frac{(r+1)p - 1 \cdot s}{r(r+1)} a_{p_0} + \frac{(r+3)(p_0 + 2) - 2(s+1)}{(r+2)(r+3)} T_2$$
$$+ \frac{(r+5)(p_0 + 4) - 3(r+2)}{(r+4)(r+5)} T_3 + \cdots, \quad (44)$$

where

$$T_2 = \frac{1}{r} \frac{r}{r+1} \frac{s}{p_0} \frac{r-s+1}{p_0 - s + 1} a_{p_0},$$

$$T_3 = \frac{2}{r+2} \frac{r+1}{r+3} \frac{s+1}{p_0 + 1} \frac{r-s+2}{p_0 - s + 2} T_2, \quad \cdots. \quad (45)$$

He summed the first nine terms of $F(\frac{1}{4}, \frac{1}{2}, \frac{5}{4}, 1)$ to give 1.21570 59973 06136 06; and used (44) and (45) up to T_9 to give 0.09532 27798 39923 81; hence he found (fig. 5) for $F(\frac{1}{4}, \frac{1}{2}, \frac{5}{4}, 1)$ the value 1.31102 87771 46059 87, which is the figure quoted in (20).

5. James Clerk Maxwell (1831–1879)

There is a general consensus among physicists nowadays that James Clerk Maxwell was the greatest physicist of the 19th century. But Maxwell was not rated so highly in his own day, and William Thomson (later Lord Kelvin) was given far more respect and attention. Excellent biographies are available of both of these outstanding scientists, which were written shortly after their deaths and contain a variety of interesting information. Elliott had personal copies of the *Life of Lord Kelvin* by Silvanus P. Thompson, and of the famous *Baltimore Lectures* which Kelvin undertook in 1884. He often referred to the latter in his public lectures.

Kelvin was invited by Johns Hopkins University to spend a month with them giving lectures on some topic of his choice which they hoped would be attended by graduate students and by leading applied mathematicians of the USA. It was a highly innovative idea, a precursor of what has become a widespread fruitful practice in the 20th century.

Silvanus Thompson tells of the negotiations with Kelvin for a suitably payment for his services. Rather diffidently Johns Hopkins offered the sum of 1000 dollars plus travelling expenses. In a footnote we are told that the eventual sum agreed on was 400 pounds sterling. Elliott drew attention to the significance of these sums; firstly in those days the rate of exchange was about 4.87 dollars to the pound [14] so that Kelvin settled for nearly twice the sum initially offered. Secondly it would not be unreasonable to suggest a factor of 15 for inflation during the past 100 years (since the end of World War II the inflation factor has been at least 10). Hence the modern equivalent would be about 29 250 dollars!

The value of the payment can be assessed alternatively as follows. Sir Brian Pippard has informed me that when Maxwell was appointed to the Cavendish Chair at Cambridge in 1871 his annual salary was £800 (presumably this covered himself and the laboratory). Thus Kelvin was paid the equivalent of 6 months of a top professorial salary for one month of lecturing.

Incidentally it is interesting to note in the preface to the publication of the lectures, Kelvin's statement that "the so-called electromagnetic theory of light has not helped us hitherto". Although 20 years had elapsed since the publication of Maxwell's classic papers, Kelvin did not yet appreciate their significance.

But it is largely to the biography of Maxwell by Campbell and Garnett that I wish to devote this section. I have the privilege to claim a personal connection with Maxwell since it was during his stay at King's College, London (where I served as professor of theoretical physics from 1954 to 1981) that Maxwell's classic papers on the electromagnetic field were published. A few years ago in 1979 the College organized a celebration of the centenary of Maxwell's death, and I looked up a number of original letters and documents relating to his biography. I followed the advice which Elliott had given me of going to original sources wherever possible, and checking on the assessments of professional historians.

Maxwell came to King's College in 1860 after he had been made redundant at Aberdeen. He stayed until 1865 when he resigned his appointment to return to the family estate in Glenlair in Southern Scotland. The period from 1860 to 1865 was the most creative in his carrier. Why did he resign? The natural explanation was that he wished to write a treatise on electromagnetic theory based on the concepts of Faraday and himself; that as a man of independent means he did not need the College salary; that teaching students full time in those days involved a great deal of hard labour. This is borne out by letters quoted by his biographer.

However *The Centenary History of King's College, London* by F.J.C. Hearnshaw offered a different reason—that he was asked to resign because he could not keep order at lectures. The authority for this statement was a certain Canon Abbay, who had written to Hearnshaw in response to a letter in the press soliciting information of relevance to the history of the College.

Fortunately the Abbay letters had been preserved in the College archives, and a study of them revealed the following:

(1) Abbay had spent a total of one year as a junior lecturer at King's College from 1868–9 three years after Maxwell had left.

(2) He had not known Maxwell personally.

(3) His sole authority for the statement was Prof. Clifton at Oxford who had never been at King's College, and whose only association with Maxwell was that both had been Smith's Prizemen at Cambridge (at periods separated by several years) and Fellows of the Royal Society.

(4) Abbay was 83 at the time when he wrote the letters, and he was writing from memory about events which had taken place more than 60 years earlier. He himself ends the first of his letters with the apologetic sentence "I am afraid this is only a gossipy letter and of no use to you".

(5) In his letters Abbay makes two significant errors in relation to Maxwell's life and scientific career.

There is not a shred of evidence in any official records or in the biography to support Abbay's contention. Maxwell was scrupulously honest, and it is most unlikely that he would have tried to conceal any facts of this kind. Finally I managed to lay hands on the official College Minute quoting Maxwell's letter of resignation in which he expresses his readiness to continue his work until the appointment of a successor. This is hardly the reaction to be expected of one who has been asked to resign because he is unable to keep order!

The readiness of the official College historian to swallow a story having such a flimsy basis shows how great is the temptation towards anything a little colourful or sensational, and re-emphasizes the importance of checking sources.

Continuing the story after Maxwell's resignation we shall find that even biographers have their temptations. In November 1868 Maxwell's former teacher I.D. Forbes resigned from the Principalship of United College, St. Andrews. Lewis Campbell, his life-long friend and biographer, who was then at United College writes "an effort was made by several of the professors to induce Maxwell to stand for the vacant post, which was in the gift of the Crown, and had been held by Brewster and Forbes successively. He was touched by the kindness, and travelled a whole day from Galloway to confer with us, but on mature consideration relinquished the idea". Campbell added a footnote on the word "professors". "It is right that I should add the suggestion did not proceed from me".

The following letter from Maxwell to Lewis Campbell is reproduced to confirm this account.

Glenlair, Dalbeattie,
3rd November 1868

I have given considerable thought to the subject of the candidature, and have come to the decision not to stand. The warm interest which you and the other professors have taken in the matter has gratified me very much, and the idea of following Principal Forbes had also a great effect on my feelings, as well as the prospect of residing among friends; but I still feel that my proper path does not lie in that direction – Your Afft. friend,

J. CLERK MAXWELL

Whilst looking around letters from Maxwell in various archives, I was surprised to find unequivocal evidence that Campbell was not telling the whole truth. The following letter to W.R. Grove is preserved in the archives of the Royal Institution.

Glenlair, Dalbeattie
7 November, 1868

Dear Sir,

I have received an invitation from two thirds of the Professors of the United College of St Salvator and St Leonard, St Andrews, to come forward as a candidate for the office of Principal of that College, of which Dr. J D Forbes has given in his resignation. They wish a scientific man to succeed Brewster and Forbes and have done me the honour to think me qualified.

I have therefore become a candidate and as far as I know there is no other professedly scientific man in the field.

If you are of opinion that I am qualified for the situation and could bring my claims in any way before the Home Secretary or Lord Advocate, I should esteem it a great favour.

The vacancy occurs on the 11th November and Government will probably lose no time in making the appointment.

I have paid so little attention to the political sympathies of scientific men that I do not know which of the scientific men I am acquainted with have the ear of the Government. If you can inform me, it would be of service to me.

I remain
Yours truly
J Clerk Maxwell

W.R. (later Sir William) Grove was a distinguished scientist, inventor of the Grove cell, and Vice-President of the Royal Institution; he was also a barrister who later became a Judge, and was a member of the Royal Commission on Patent Law in 1864. He may well have wielded considerable influence in government circles.

This evidence is supported by a letter to William Thomson written two days later (in the archives of Glasgow University Library)

Glenlair, Dalbeattie
1868 Nov 9

Dear Thomson

When I last wrote I had not been at St Andrews. I went last week and have gone in for the Principalship. If you can certify my having been industrious etc. since 1856, or if you can tell me what scientific men are conservative, or still better if you can use any influence yourself in my favour pray do so. 6 Professors out of 9 have memorialized the Ld Adv. & Home Sec. for me together with Principal Tulloch the V. Chancellor. Of the other 3, one Prof Shairp, is a candidate and one, Prof Bell does not approve of memorials at all and is neutral, I have written to Sabine, Airey, Stokes and Grove.

Yours truly
J Clerk Maxwell

From this it is clear that Maxwell changed his mind after the November 3rd letter to Campbell. In fact the previous letter to Thomson referred to above is also preserved in the archives of Glasgow University Library:

Glenlair, Dalbeattie
1868 Oct 30

Dear Thomson

I got your letter about St Andrew's. Swan and Campbell have also written. One great objection is the East Wind which I believe is severe in those parts. Another is that my proper line is in working not in governing, still less in reigning and letting others govern...

Yours truly
J Clerk Maxwell

It shows that Campbell was one of the initiators of the suggestion in contrast to the denial in the footnote, and contains the marvelously succinct phrase "reigning and letting others govern".

Putting the letters in chronological order, it seems that between 3 and 7 November Maxwell visited St Andrews and changed his mind about his candidature; his biographer has apparently mistimed his visit to St Andrews. The sequence of events would then probably be as follows: Friday 30 October Maxwell writes negatively to Thomson, Tuesday 3 November likewise to Campbell. Some communication on 4 or 5 November persuades him to visit St Andrews. The visit probably takes place on Thursday 5 November (it was nearly a day's journey), he confers with the professors and agrees to become a candidate, returns to Glenlair, writes to Grove,

Sabine, Airy and Stokes on Saturday 7 November, and to Thomson on Monday 9 November. It is difficult to find any other picture consistent with Maxwell's statement in his second letter to Thomson, 'I went (to St Andrews) last week'.

If this is the correct picture two obvious questions spring to mind. Why did Campbell who was at St Andrews and knew all that was going on deliberately suppress this information? And how did it come about that a man of Maxwell's calibre with the backing of the Vice Chancellor, nearly all the professors, the President of the Royal Society (Sabine) and other eminent scientists of the day was not appointed? The appointment went to J.C. Shairp a professor of Latin who had been acting as assistant to Forbes during the previous few years.

To suggest an answer to the first question one must remember that when the biography was published in 1882 Campbell was still at St Andrews and Shairp was still Principal. One can reasonably speculate that Campbell acted in consideration of Shairp's feelings (and perhaps Maxwell's memory). There was nothing about the candidature in official records and therefore a good chance that the true story would not come to light. The footnote may also have been inserted to maintain good relations with Shairp.

Regarding the second question my former assistant Mrs. E. Fisk has made a very plausible suggestion. A great political upheaval was taking place at this time. Parliament was dissolved on 11 November 1868. At subsequent elections the Liberals under Gladstone obtained a majority of 115 and Disraeli resigned. Maxwell's attempt to cultivate Conservative scientists seems to have been counterproductive. On 4 December Gladstone was summoned to Windsor and asked to form his ministry; by 9 December his government was complete. Shairp who had good Liberal connections was appointed Principal on 12 December.

6. Conclusion

I have tried to put before you some topics from the history of science which would have interested Elliott Montroll. I hope that I have convinced you of how fascinating and informative it can be to go oneself to primary sources rather than relying completely on secondary material.

Acknowledgement

I am indebted to Mr. N.H. Robinson, the Librarian of the Royal Society, and his staff for their great help in tracing books and documents.

References

[1] L. Onsager, in: Critical Phenomena in Alloys, Magnets and Superconductors, eds. R.E. Mills, E. Ascher and R.I. Jaffee (McGraw-Hill, New York, 1971) p. xxi.
[2] E.W. Montroll, Ann. Math. Stat. 18 (1947) 18.
[3] P.W. Anderson, Phys. Rev. 109 (1958) 1492.
[4] J.M. Hammersley, Proc. Camb. Phil. Soc. 53 (1957) 642.
[5] J.W. Gibbs, Collected Works (Longman's Green, 1906; reprinted Dover, New York, 1961) vol. 2, p. 261.
[6] I. Todhunter, A History of the Mathematical Theory of Probability (Cambridge, 1865; reprinted Chelsea, London 1965) ch. 20.
[7] R.B. Dingle, Asymptotic Expansions (Academic, New York, 1973) p. 134.
[8] Lord Moulton, Inaugural Address, The Invention of Logarithms, Napier Tercentenary Volume (London, 1915) p. 1.
[9] F. Cajori, History of Mathematics (MacMillan, 1894; reprinted Chelsea, London, 1960) p. 161.
[10] D.E. Smith, History of Mathematics, Vol. 2 (1925; reprinted Dover, New York, 1958) p. 235.
[11] Edmund Halley, Phil. Trans. Roy. Soc. No. 216 (1695) p. 13.
[12] I am deeply grateful to Dr. Shimon Bolag of the Jerusalem College of Technology for telling me about Immanuel Bonfils.
[13] G. Sarton, Isis 25 (1936) 16.
[14] The gold standard exchange rate in operation from 1870–1914. See Encyclopedia Brittanica, 11th Ed. (1911), article on "Exchange".
[15] Lewis Campbell and William Garnett, The Life of James Clerk Maxwell (London, 1882; reprinted Johnson, New York, 1969) pp. 327–328.
[16] Ref. [15], p. 345–346.
[17] H.H. Goldstine, A History of Numerical Analysis (Springer, Berlin, 1977).

Note added after submission

When this MS had been completed I received from Dr. Michael Shlesinger a reprint of the chapter "On the Wonderful World of Random Walks" which Elliott and he had written for *Nonequilibrium Phenomena II - From Stochastics to Hydrodynamics*, vol. XI of the present series. This had been published after Elliots death. It is there pointed out that generating functions had been mentioned before Laplace by Jacob Bernoulli, De Moivre, and Stirling.

CHAPTER 3

*Some Statistical and Dynamical Problems
in Quantum Electronics*

F.T. HIOE
*St. John Fisher College
Rochester, New York 14618
U.S.A.*

© *Elsevier Science Publishers B.V. 1985*

*The Wonderful World of Stochastics
A Tribute to Elliott W. Montroll
Eds. M.F. Shlesinger and G.H. Weiss*

Contents

Introduction	77
1. Bistability in two-mode lasers	77
1.1. Basic equations	78
1.2 Minima of the potential	82
1.2.1. Potential minimum at (0)	83
1.2.2. Potential minimum at (1)	84
1.2.3. Potential minimum at (2)	84
1.2.4. Potential minimum at (3)	85
1.3. Coupling modes	86
1.3.1. Weak coupling ($0 \leq \xi < 1$)	86
1.3.2. Strong coupling ($\xi > 1$)	87
1.4. Concluding note	89
2. Bloch equations involving amplitude and frequency modulations	89
2.1. Special cases	94
2.1.1. The case $\beta = 0$	94
2.1.2. The case $\beta_0 = 0$	95
2.1.3. The case $\beta = \beta_0 = 0$	96
2.2. Summary	97
References	98

Introduction

Elliott Montroll's outstanding contributions to the fields of statistical mechanics and nonlinear dynamics are well known to everyone in the fields. He was also a great teacher not only in the clarity of his lectures and writings, but also in setting a style which greatly influenced the people who had worked with him.

In this article dedicated to his memory, I shall discuss some statistical and dynamical problems involving competing and coherent effects in laser–atom interactions.

1. Bistability in two-mode lasers

The dynamics of the laser field amplitudes, according to the theory originally developed by Lamb [1], is described in terms of a set of coupled nonlinear equations. The theory has been very successfully applied to one-mode lasers. More remarkably, it has given even more interesting results for two-mode lasers because of the effects of mode competition for the atomic population. Examples of the two-mode lasers include the so-called ring lasers involving two counter-propagating lasers used for optical gyroscopes, and certain dye lasers involving, under certain conditions, two dominant modes. The results on the two-mode lasers [2,3] which we shall discuss demonstrate an interesting analogy between the bistability phenomenon in two-mode lasers and the thermodynamic phase transition phenomenon in statistical mechanics. Mathematically the theory also resembles that of interacting biological species studied by Montroll [4].

A novel feature of a two-mode laser is the display of bistable behavior under certain conditions. When this occurs, the laser system exhibits two quasistable states, switching from one to the other at random time intervals. One observes experimentally that the modes appear and disappear continually in an anticorrelated fashion. The conditions under which the bistable behavior occurs can be expressed in terms of the physical parameters a_1, a_2, the pump parameters for the two modes. Mathematically, we show that the onset of bistability is associated with the appearance of two minima of a potential associated with the Fokker–Planck equation, which give rise to two almost degenerate ground-state eigenvalues. The presence of noise makes it possible for the system to go from one metastable state corresponding to one of the minima, to the other metastable state corresponding to the other minimum of the potential. We show that when the bistability conditions are not satisfied, even though the potential has several local minima, only one contains the stable steady state. All this is made more precise by the analytic expressions we present.

1.1. Basic equations

We consider a dynamical system composed of two-laser modes which are coupled via their intensities. The equations of motion for the slowly varying complex field amplitudes E_1 and E_2 are:

$$\frac{dE_1}{dt} = E_1\left(a_1 - |E_1|^2 - \xi|E_2|^2\right) + q_1(t), \tag{1.1a}$$

$$\frac{dE_2}{dt} = E_2\left(a_2 - |E_2|^2 - \xi|E_1|^2\right) + q_2(t), \tag{1.1b}$$

where a's are the pump parameters for the two-laser modes and ξ is the mode coupling constant, $q_1(t)$ and $q_2(t)$ are Langevin noise terms which will be taken to be delta-correlated Gaussian random processes with a zero mean and

$$\langle q_m^*(t)\, q_n(t')\rangle = 4\delta_{mn}\delta(t - t'), \quad m, n = 1, 2. \tag{1.2}$$

If we express the complex field amplitudes E_1 and E_2 in terms of real and imaginary parts

$$E_n = x_n + iy_n, \quad n = 1, 2, \tag{1.3}$$

the vector $x \equiv (x_1, y_1, x_2, y_2)$ represents the state of the laser, and its components obey a set of coupled Langevin equations of motion which are easily derived from the set of equations (1.1). The Fokker–Planck equation for the probability density $p(x, t)$ associated with these Langevin equations is given by

$$\frac{\partial}{\partial t} p(x, t) = \sum_{i=1}^{2}\left[-\frac{\partial}{\partial x_i}A_i^{(x)}p - \frac{\partial}{\partial y_i}A_i^{(y)}p + \left(\frac{\partial^2}{\partial x_i^2} + \frac{\partial^2}{\partial y_i^2}\right)p\right], \tag{1.4}$$

where
$$A_i^{(\eta)} = \left[a_i - (x_i^2 + y_i^2) - \xi(x_j^2 + y_j^2)\right]\eta_i,$$
$$i, j = 1, 2, \quad i \neq j, \quad \eta = x, y. \tag{1.5}$$

Equation (1.4) describes the evolution of the probability density in a four-dimensional coordinate space spanned by the components of the vector x.

The stationary state probability density $p_s(x)$, which corresponds to $\partial p(x, t)/\partial t = 0$, is obtained easily in this case, since the drift coefficients A satisfy the potential conditions [5]:

$$A_i^{(\eta)}(x) = -\frac{\partial U(x)}{\partial \eta_i}, \tag{1.6a}$$

$$\frac{\partial A_i^{(\eta)}}{\partial \zeta_j} = \frac{\partial A_j^{(\zeta)}}{\partial \eta_i}, \quad \zeta, \eta = x, y, \tag{1.6b}$$

where the potential $U(x)$ is given by
$$U(x) = -\tfrac{1}{2}a_1(x_1^2 + y_1^2) - \tfrac{1}{2}a_2(x_2^2 + y_2^2)$$
$$+ \tfrac{1}{4}\left[(x_1^2 + y_1^2)^2 + (x_2^2 + y_2^2)^2 + 2\xi(x_1^2 + y_1^2)(x_2^2 + y_2^2)\right]. \tag{1.7}$$

The steady-state probability density $p_s(x)$ is then given by
$$p_s(x) = B^{-1} \exp[-U(x)], \tag{1.8}$$

where B is the normalization constant. This distribution depends on three parameters a_1, a_2, and ξ. The statistical properties of the optical field that follow from eq. (1.8) may exhibit entirely different fluctuation properties for $\xi < 1$ and $\xi > 1$.

To discuss the time-dependent solution of $p(x, t)$ we put
$$p(x, t) = \exp[-\tfrac{1}{2}U(x)]\, g(x)\, T(t), \tag{1.9}$$

where $U(x)$ is given by eq. (1.7). A substitution of eq. (1.9) in (1.4) yields the following eigenvalue equation

$$-\frac{T'}{T} = \frac{\mathscr{L}g}{g} = \lambda, \tag{1.10}$$

where
$$\mathscr{L} = \sum_{i \neq j = 1}^{2} \left\{ -\frac{\partial^2}{\partial x_i^2} - \frac{\partial^2}{\partial y_i^2} + \tfrac{1}{4}(x_i^2 + y_i^2)\left[a_i - (x_i^2 + y_i^2) - \xi(x_j^2 + y_j^2)\right]\right.$$
$$\left. + a_i - 2(x_i^2 + y_i^2) - \xi(x_j^2 + y_j^2)\right\}. \tag{1.11}$$

Notice that the operator $\mathscr{L}$ is a Sturm–Liouville self-adjoint operator in four dimensions. It follows that the eigenvalues and the eigenfunctions will be labeled by four indices. The general solution of the Fokker–Planck equation (1.4) can then be written as

$$p(x, t) = \exp[-\tfrac{1}{2}U(x)]$$
$$\times \sum_{n_1, n_2, m_1, m_2} C_{n_1 n_2 m_1 m_2} g_{n_1 n_2 m_1 m_2}(x) \exp(-\lambda_{n_1 n_2 m_1 m_2} t), \quad t \geq 0, \tag{1.12}$$

where the coefficients $C_{n_1 n_2 m_1 m_2}$ are determined from the boundary conditions. Since the steady-state solution has been shown to exist, the lowest eigenvalue must be zero and the corresponding eigenfunction $g_{0000}(x)$ will be

$$g_{0000}(x) = [p_s(x)]^{1/2} = B^{-1/2} \exp[-\tfrac{1}{2}U(x)]. \tag{1.13}$$

We shall assume that $g_{n_1 n_2 m_1 m_2}$ satisfy the orthonormality and completeness conditions.

To calculate correlations we need the second-order joint probability density $p_2(x, t; x', t')$ for the field at two different times. To this end we note that the conditional probability density $G(x, t | x', t')$ for the field to be characterized by x at time t if it was characterized by x' at time t' is also the Green's function for the problem, which is a solution of the Fokker–Planck equation with the initial condition

$$G(x, t | x', t') = \delta(x - x') \delta(t - t'). \tag{1.14}$$

$G(x, t | x', t')$ can be expressed in terms of the orthogonal set $\{g_{n_1 n_2 m_1 m_2}\}$ as

$$G(x, t | x', t') = \sum_{n_1 n_2 m_1 m_2} g^*_{n_1 n_2 m_1 m_2}(x) g_{n_1 n_2 m_1 m_2}(x')$$
$$\times \exp[-\lambda_{n_1 n_2 m_1 m_2}(t - t')], \quad t \geq t'. \tag{1.15}$$

The joint probability density $p_2(x, t; x', t')$ for the field at two different times is independent of the origin of time in the stationary state and can be written as

$$p_2(x, t + \tau; x', t)$$
$$= G(x, \tau | x', 0) p_s(x') = [p_s(x) p_s(x')]^{1/2}$$
$$\times \sum_{n_1 n_2 m_1 m_2} g^*_{n_1 n_2 m_1 m_2}(x) g_{n_1 n_2 m_1 m_2}(x') \exp(-\lambda_{n_1 n_2 m_1 m_2} \tau), \quad \tau \geq 0. \tag{1.16}$$

Higher-order joint probabilities can be expressed similarly. However, since we are dealing with a Markov process all the higher-order probabilities can be expressed in terms of p_2, and eqs. (1.12)–(1.16) solve the problem in principle. We shall find it convenient to use polar coordinates,

$$x_i = r_i \cos \varphi_i, \qquad y_i = r_i \sin \varphi_i,$$
$$i = 1, 2, \qquad r_i \geq 0, \qquad 0 \leq \varphi_i \leq 2\pi. \tag{1.17}$$

The volume element in these variables is

$$d^4x = r_1 dr_1 \, d\varphi_1 \, r_2 dr_2 \, d\varphi_2. \tag{1.18}$$

The operator $\mathscr{L}$ given by eq. (1.11) becomes

$$\mathscr{L} = -\sum_{i=1}^{2} \left(\frac{\partial^2}{\partial x_i^2} + \frac{\partial^2}{\partial y_i^2} \right) + V, \tag{1.19}$$

where

$$V = (a_1 + a_2) - (2 + \xi)(r_1^2 + r_2^2)$$
$$+ \tfrac{1}{4} r_1^2 (a_1 - r_1^2 - \xi r_2^2) + \tfrac{1}{4} r_2^2 (a_2 - r_2^2 - \xi r_1^2), \tag{1.20}$$

and

$$-\sum_{i=1}^{2} \left(\frac{\partial^2}{\partial x_i^2} + \frac{\partial^2}{\partial y_i^2} \right) = -\sum_{i=1}^{2} \left(\frac{1}{r_i} \frac{\partial}{\partial r_i} r_i \frac{\partial}{\partial r_i} + \frac{1}{r_i^2} \frac{\partial^2}{\partial \varphi_i^2} \right). \tag{1.21}$$

The steady-state potential U and the steady-state wave function may be rewritten as

$$U = -\tfrac{1}{2}(a_1 r_1^2 + a_2 r_2^2) + \tfrac{1}{4}(r_1^4 + r_2^4 + 2\xi r_1^2 r_2^2) \tag{1.22}$$

and

$$\chi_{0000} = C \exp\left[\tfrac{1}{4}(a_1 r_1^2 + a_2 r_2^2) - \tfrac{1}{8}(r_1^4 + r_2^4 + 2\xi r_1^2 r_2^2)\right], \tag{1.23}$$

where C is the appropriate normalization factor and χ is the function g expressed in terms of the new variables. If we write the eigenfunction $g(x)$ of $\mathscr{L}$ in the form

$$g_{n_1 n_2 m_1 m_2}(x) = \chi_{n_1 n_2 m_1 m_2}(r_1, r_2) \frac{1}{\sqrt{2\pi}} e^{im_1\varphi_1} \frac{1}{\sqrt{2\pi}} e^{im_2\varphi_2}, \tag{1.24}$$

the radial part $\chi_{n_1 n_2 m_1 m_2}$ satisfies the equation

$$\left[\sum_{i=1}^{2} \left(-\frac{1}{r_i} \frac{\partial}{\partial r_i} r_i \frac{\partial}{\partial r_i} + \frac{m_i^2}{r_i^2} \right) + V(r_1, r_2) \right] \chi_{n_1 n_2 m_1 m_2} = \lambda_{n_1 n_2 m_1 m_2} \chi_{n_1 n_2 m_1 m_2}. \tag{1.25}$$

The crucial point to note is that the potential $V(r_1, r_2)$ can have one or several minima depending on the values a_1, a_2 and ξ, and that for large magnitudes of a_1 and a_2 every potential well can be well approximated by a harmonic potential well. We should mention at this point that for the most general problem a_1 and a_2 should be independent variables. However, in practice a_1 and a_2 are close and differ only by a small amount, e.g., in lasers as excitation is increased, both a_1 and a_2 increase in such a way that $a_1 - a_2$ remains approximately constant. It is, of course, possible to introduce asymmetries in the system but we are mostly concerned with the cases of small differences in the pump parameters. Situations that involve independent variations of a_1 and a_2 are easily handled by a straightforward application of the methods discussed below.

As the pump parameters a_1 and a_2 are increased, mode competition effects become important and the dynamical system under consideration becomes capable of exhibiting entirely new phenomena, like growth of one mode at the expense of the other, switching of field excitation from one mode to the other, etc. Depending on the value of ξ, three regions of distinct fluctuation properties appear, and the potential V may have several minima.

The potential V in eq. (1.20) may be expressed in terms of the stationary-state potential U given by eq. (1.22) as

$$V(r_1, r_2) = \tfrac{1}{4}\left[\left(\frac{\partial U}{\partial r_1}\right)^2 + \left(\frac{\partial U}{\partial r_2}\right)^2\right] - \left(\frac{\partial^2 U}{\partial r_1^2} + \frac{\partial^2 U}{\partial r_2^2}\right) + (r_1^2 + r_2^2). \quad (1.26)$$

1.2. Minima of the potential

The extrema of V are determined by setting the first derivative of V with respect to r_1 and r_2 equal to zero. From eqs. (1.22) and (1.26) it follows that for large values of $|a_1|$ and $|a_2|$ we can neglect derivatives of U higher than the second and the $r_1^2 + r_2^2$ term in eq. (1.26). This approximation is valid not only for small deviations from the stationary state but also for arbitrarily large deviations from it. This can be seen clearly from eq. (1.22). The neglected terms always turn out to be smaller by a factor on the order of a_i^{-1} compared to the terms that have been retained. Then, setting $\partial V/\partial r_1 = 0 = \partial V/\partial r_2$ is equivalent to setting $\partial U/\partial r_1$ and $\partial U/\partial r_2$ equal to zero. This leads to the following equations for the extrema

$$\frac{\partial U}{\partial r_1} = -r_1(a_1 - r_1^2 - \xi r_2^2) = 0, \quad (1.27a)$$

$$\frac{\partial U}{\partial r_2} = -r_2(a_2 - r_2^2 - \xi r_1^2) = 0. \quad (1.27b)$$

The set of equations (1.27) admits the following four solutions which we label as 0, 1, 2, and 3:

0: $r_1^2 = 0$, $\quad r_2^2 = 0$, $\hfill (1.28)$

1: $r_1^2 = a_1$, $\quad r_2^2 = 0$, $\hfill (1.29)$

2: $r_1^2 = 0$, $\quad r_2^2 = a_2$, $\hfill (1.30)$

3: $r_1^2 = \dfrac{a_1 - \xi a_2}{1 - \xi^2} \equiv \rho_1^2$, $\quad r_2^2 = \dfrac{a_2 - \xi a_1}{1 - \xi^2} \equiv \rho_2^2$. $\hfill (1.31)$

Solutions 0, 1, and 2 always correspond to a minimum solution. Solution 3 may or may not correspond to a minimum depending on the values of a_1, a_2, and ξ.

1.2.1. Potential minimum at (0)

We first consider the potential V near its minimum given by eq. (1.28), which occurs at r_1, r_2 of order $a_1^{-1/2}$, $a_2^{-1/2}$. It is easy to show that near this minimum V can be approximated as

$$V \simeq a_1 + a_2 + \tfrac{1}{4}a_1^2 r_1^2 + \tfrac{1}{4}a_2^2 r_2^2. \quad (1.32)$$

We find that this potential is for the problem of two uncoupled harmonic oscillators. Thus, writing the eigenfunctions of $\mathscr{L}$ as

$$\chi_{n_1 n_2 m_1 m_2}(r_1, r_2) = (r_1 r_2)^{-1/2} \psi_{n_1 m_1}^{(1)}(r_1) \psi_{n_2 m_2}^{(2)}(r_2), \quad (1.33)$$

we obtain the following eigenvalue equation:

$$\left(\sum_{i=1}^{2} -\frac{\partial^2}{\partial r_i^2} + a_i + \omega_i r_i^2 + \frac{m_i^2 - \tfrac{1}{4}}{r_i^2} \right) \psi_{n_1 m_1}^{(1)} \psi_{n_2 m_2}^{(2)} = \lambda_{n_1 n_2 m_1 m_2} \psi_{n_1 m_1}^{(1)} \psi_{n_2 m_2}^{(2)}, \quad (1.34)$$

where

$$\omega_i = a_i/2, \quad i = 1, 2. \quad (1.35)$$

We find that eq. (1.34) has the eigenvalues

$$\lambda_{n_1 n_2 m_1 m_2} = a_1 + a_2 + (2n_1 + |m_1| + 1)a_1 + (2n_2 + |m_2| + 1)a_2, \quad (1.36)$$

$$n_1, n_2, |m_1|, |m_2| = 0, 1, \ldots. \quad (1.37)$$

Note that the lowest eigenvalue is not zero and therefore the steady state is not contained in this well.

1.2.2. Potential minimum at (1)

Next we consider the minimum of V given by eq. (1.29). This minimum occurs at $r_1^2 = a_1$ and $r_2^2 = 0$. The potential V near this minimum can be expanded as

$$V \simeq (a_2 - \xi a_1) - a_1 + a_1^2 \left(r_1 - \sqrt{a_1}\right)^2 + \tfrac{1}{4}(a_2 - \xi a_1)^2 r_2^2. \tag{1.38}$$

If we write the eigenfunctions

$$\chi_{n_1 n_2 m_1 m_2}(r_1, r_2) = \psi^{(1)}_{n_1 m_1}(r_1) \, r_2^{-1/2} \, \psi^{(2)}_{n_2 m_2}(r_2), \tag{1.39}$$

we obtain the following eigenvalue problem:

$$\left(-\frac{\partial^2}{\partial r_1^2} - \frac{\partial^2}{\partial r_2^2} + (a_2 - \xi a_1) - a_1 + a_1^2 \left(r_1 - \sqrt{a_1}\right)^2 + \frac{m_1^2}{a_1} \right.$$
$$\left. + \tfrac{1}{4}(a_2 - \xi a_1)^2 r_2^2 + \frac{m_2^2 - \tfrac{1}{4}}{r_2^2} \right) \psi^{(1)}_{n_1 m_1} \psi^{(2)}_{n_2 m_2} = \lambda_{n_1 n_2 m_1 m_2} \psi^{(1)}_{n_1 m_1} \psi^{(2)}_{n_2 m_2}. \tag{1.40}$$

Equation (1.40) again turns out to describe the problem of two uncoupled harmonic oscillators. The eigenvalues may readily be shown to be

$$\lambda_{n_1 n_2 m_1 m_2} = 2n_1 a_1 + m_1^2/a_1 + (2n_2 + |m_2| + 1)|a_2 - \xi a_1| + (a_2 - \xi a_1),$$

$$n_1, n_2, |m_1|, |m_2| = 0, 1, 2, \ldots. \tag{1.41}$$

1.2.3. Potential minimum at (2)

The well located at $r_1^2 = 0$ and $r_2^2 = a_2$ can be treated in a similar manner. It is easy to see that the eigenvalues can be obtained from eq. (1.41) simply by interchanging the subscripts 1 and 2; they are given by

$$\lambda_{n_1 n_2 m_1 m_2} = 2n_2 a_2 + m_2^2/a_2 + (2n_1 + |m_1| + 1)|a_1 - \xi a_2| + (a_1 - \xi a_2),$$

$$n_1, n_2, |m_1|, |m_2| = 0, 1, 2, \ldots. \tag{1.42}$$

1.2.4. Potential minimum at (3)

Finally we consider the solution of the problem near the minimum given by eq. (1.31) which is located at

$$r_1^2 = (a_1 - \xi a_2)/(1 - \xi^2) \equiv \rho_1^2, \qquad r_2^2 = (a_2 - \xi a_1)/(1 - \xi^2) \equiv \rho_2^2.$$

The potential V near this minimum can be approximated as

$$V \simeq \frac{a_1 + a_2}{1 + \xi} - \left[A(r_1 - \rho_1)^2 + 2B(r_1 - \rho_1)(r_2 - \rho_2) + C(r_2 - \rho_2)^2 \right],$$
(1.43)

where

$$A = \tfrac{1}{4}(\alpha^2 + \beta^2), \qquad B = \tfrac{1}{4}\beta(\alpha + \gamma), \qquad C = \tfrac{1}{4}(\beta^2 + \gamma^2), \qquad (1.44)$$

and

$$\alpha = \frac{2(a_1 - \xi a_2)}{1 - \xi^2}, \qquad \beta = \frac{2\xi}{1 - \xi^2}\left[(a_1 - \xi a_2)(a_2 - \xi a_1)\right]^{1/2},$$

$$\gamma = \frac{2(a_2 - \xi a_1)}{1 - \xi^2}. \qquad (1.45)$$

Again replacing the terms $(1/r_i)\partial/\partial r_i$ and the terms m_i^2/r_i^2 by m_i^2/ρ_i^2 in eq. (1.25) we obtain the following eigenvalue equation:

$$\left(-\frac{\partial^2}{\partial r_1^2} - \frac{\partial^2}{\partial r_2^2} + \frac{m_1^2}{\rho_1^2} + \frac{m_2^2}{\rho_2^2} - \frac{a_1 + a_2}{1 + \xi} \right.$$

$$\left. + A(r_1 - \rho_1)^2 + 2B(r_1 - \rho_1)(r_2 - \rho_2) + C(r_2 - \rho_2)^2 \right) \chi_{n_1 n_2 m_1 m_2}$$

$$= \lambda_{n_1 n_2 m_1 m_2} \chi_{n_1 n_2 m_1 m_2}. \qquad (1.46)$$

Equation (1.46) describes two coupled harmonic oscillators, and it can again be solved exactly. By a suitable coordinate transformation we can convert this equation into an equation for two uncoupled harmonic oscillators. The eigenvalues of eq. (1.46) are found to be

$$\lambda_{n_1 n_2 m_1 m_2} = (2n_1 + 1)\mu^{1/2} + (2n_2 + 1)\nu^{1/2} + \frac{m_1^2}{\rho_1^2} + \frac{m_2^2}{\rho_2^2} - \frac{a_1 + a_2}{1 + \xi},$$

$$n_1, n_2, |m_1|, |m_2| = 0, 1, 2, \ldots, \qquad (1.47)$$

where μ, ν are given by

$$\mu = \tfrac{1}{2}\{A + C + [(A-C)^2 + 4B^2]^{1/2}\},$$
$$\nu = \tfrac{1}{2}\{A + C - [(A-C)^2 + 4B^2]^{1/2}\}. \tag{1.48}$$

As mentioned earlier not all the extrema given by eqs. (1.28)–(1.31) correspond to minima of the potential V. Depending on the values of a_1, a_2 and ξ, we have to select the appropriate minima and the corresponding solutions. Moreover, as we shall see, not all the true minima contain the ground-state zero eigenvalue.

1.3. Coupling modes

1.3.1. Weak coupling ($0 \leq \xi -1$)

If the mode coupling constant ξ is less than unity, two modes are said to be weakly coupled. With increasing excitation, mean intensities of both modes grow and the relative intensity fluctuations of both modes tend to zero. In other words, mode competition effects are not too severe, as the two modes draw their gain from two partially overlapping sets of atoms. Examples of such systems are provided by an inhomogeneously broadened bidirectional ring laser away from line center, some two-mode standing-wave lasers, and some Zeeman lasers.

We assume for the sake of definiteness that $a_1 > a_2 > 0$. Then it can be shown that the extrema (0), (1) and (2) given by eqs. (1.28)–(1.30) describe three minima of V. Equation (31) describes a minimum only when

$$a_1 - \xi a_2 > 0, \quad a_2 - \xi a_1 > 0. \tag{1.49}$$

The first of these inequalities is already satisfied because $a_1 > a_2$ and $\xi < 1$. The condition for both inequalities to be satisfied is

$$\xi a_1 < a_2 < a_1/\xi, \quad a_1, a_2 > 0. \tag{1.50}$$

It follows from this equation that for a range of values of a_1, viz.,

$$0 < a_1 < (a_1 - a_2)/(1 - \xi), \tag{1.51}$$

we will have $a_1 - \xi a_2 > 0$ but $a_2 - \xi a_1 < 0$. In this case the potential $V(r_1, r_2)$ has only three minima given by eqs. (1.28)–(1.30). The eigenvalues for the minima located at (0), (1), and (2) are

$$\lambda^{(0)}_{n_1 n_2 m_1 m_2} = (2n_1 + |m_1| + 2)a_1 + (2n_2 + |m_2| + 2)a_2, \tag{1.52a}$$

$$\lambda^{(1)}_{n_1 n_2 m_1 m_2} = 2n_1 a_1 + m_1^2/a_1 + (2n_2 + |m_2|)|a_2 - \xi a_1|, \quad (1.52b)$$

$$\lambda^{(2)}_{n_1 n_2 m_1 m_2} = 2n_2 a_2 + m_2^2/a_2 + (2n_1 + |m_1| + 2)|a_1 - \xi a_2|. \quad (1.52c)$$

Here we have used the facts that a_1, a_2, and $a_1 - \xi a_2$ are all positive but $a_2 - \xi a_1$ is negative. The superscripts in eq. (1.52) label the wells to which these eigenvalues belong. The eigenvalue spectrum $\{\lambda_{n_1 n_2 m_1 m_2}\}$ for the full problem comprises all the eigenvalues given by eqs. (1.52) arranged in ascending order. The three sets of eigenvalues do not add simply to give the eigenvalue spectrum $\{\lambda_{n_1 n_2 m_1 m_2}\}$. In case of degeneracies, the degeneracy is lifted because of tunneling between various wells under the influence of noise. The splitting of eigenvalues is usually very small compared to the magnitude of the eigenvalues themselves and except for the case of degenerate zero eigenvalues, this splitting may be ignored. Thus if the ground-state eigenvalue zero is nondegenerate, the eigenvalue spectrum of the full problem $\{\lambda_{n_1 n_2 m_1 m_2}\}$ may be taken to be a superposition of three spectra given by eq. (1.52). In the present case the zero eigenvalue is contained only in well (1) and as expected this is the only stable solution of the deterministic equations given by eq. (1.29).

1.3.2. Strong coupling (ξ –1)

We have seen in our discussion that with increasing coupling constant, mode competition effects tend to become more pronounced. The system of two strongly coupled modes, e.g., is characterized by strong anticorrelations, large relative intensity fluctuations, and spontaneous switching of radiation from one mode to the other. Examples of such systems are provided by some Zeeman lasers, some two-mode lasers, some two-mode lasers with a homogeneously broadened gain medium, and some solid state two-mode lasers.

First, let us consider the case $a_1 \neq a_2$, and for definiteness, let $a_1 > a_2$. Then noting that $1 - \xi^2 < 0$ we find that if the conditions

$$a_1 - \xi a_2 > 0, \quad a_2 - \xi a_1 < 0 \quad (1.53)$$

are satisfied, the potential V has only three minima given by eqs. (1.28)–(1.30). In this case the eigenvalues in various minima are found from eqs. (1.36), (1.41) and (1.42) to be

$$\lambda^{(0)}_{n_1 n_2 m_1 m_2} = (2n_1 + |m_1| + 2) a_1 + (2n_2 + |m_2| + 2) a_2, \quad (1.54a)$$

$$\lambda^{(1)}_{n_1 n_2 m_1 m_2} = 2n_1 a_1 + m_1^2/a_1 + (2n_2 + |m_2|)|a_2 - \xi a_1|, \quad (1.54b)$$

$$\lambda^{(2)}_{n_1 n_2 m_1 m_2} = 2n_2 a_2 + m_2^2/a_2 + (2n_1 + |m_1| + 2)|a_1 - \xi a_2|. \quad (1.54c)$$

From eqs. (1.54) it follows that only the well at $r_1^2 = a_1$ and $r_2^2 = 0$ has the zero eigenvalue.

Let us now assume that the conditions

$$a_1 - \xi a_2 < 0, \qquad a_2 - \xi a_1 < 0 \tag{1.55}$$

are satisfied. Then it follows from eqs. (1.28)–(1.31) that the potential V now has four minima with eigenvalues

$$\lambda^{(0)}_{n_1 n_2 m_1 m_2} = (2n_1 + |m_1| + 2) a_1 + (2n_2 + |m_2| + 2) a_2, \tag{1.56a}$$

$$\lambda^{(1)}_{n_1 n_2 m_1 m_2} = 2n_1 a_1 + m_1^2/a_1 + (2n_2 + |m_2|)|a_2 - \xi a_1|, \tag{1.56b}$$

$$\lambda^{(2)}_{n_1 n_2 m_1 m_2} = 2n_2 a_2 + m_2^2/a_2 + (2n_1 + |m_1|)|a_1 - \xi a_2|, \tag{1.56c}$$

$$\lambda^{(3)}_{n_1 n_2 m_1 m_2} = (2n_1 + 1)\mu^{1/2} + (2n_2 + 1)\nu^{1/2} + \frac{m_1^2}{\rho_1^2} + \frac{m_2^2}{\rho_2^2} - \frac{a_1 + a_2}{1 + \xi}, \tag{1.56d}$$

where μ, ν are given by eq. (1.48). It can be shown that $\lambda^{(3)}_{0000} \neq 0$ when $\xi > 1$. Then eqs. (1.56) imply that $\lambda^{(1)}_{0000} = 0 = \lambda^{(2)}_{0000}$. Thus, both the minima located at (1.29) and at (1.30) contain the zero eigenvalue. This implies that, depending on the initial condition, the system may end up in one minimum or the other and will stay there. However, noise can push the system out of one well into the other. Thus, the system will switch back and forth spontaneously between the two minima characterized by $r_1^2 = a_1$, $r_2^2 = 0$ [eq. (1.29)], and $r_1^2 = 0$, $r_2^2 = a_2$ [eq. (1.30)] and the stationary states of our description turn out to be quasistationary at best. For large pump parameters, when the two minima are deep and well separated, the system will spend a long time in a minimum before it receives a sufficiently large noise excitation to switch over to the other minimum. With increasing pumping the depth of the minima increases, and switching from one minimum to the other becomes a rare event. In this limit we can treat the two wells as being independent of each other. Our results describe the dynamics well in this region. However, for moderate excitations the system switches frequently from one minimum to the other, and the assumption of two independent wells becomes invalid. In this operating range there are two aspects to the system dynamics: one that is related to the relaxation within a well, and another that is related to the switching of the system. The time scales for the two types of relaxations may be quite different. Near threshold they may be similar. High above threshold, the switching aspect becomes unimportant because of its rarity. But in the intermediate range, both local and global aspects are important and can be distinguished by their time scales. It is clear that so long as the system remains in one minimum or the other, eqs. (1.56b) or (1.56c) and the associated eigenfunctions describe the local transient properties, and the corresponding joint probability density can be used to describe local correlations. It is also clear that a description of global transients is missing from our approach so far. By a suitable modification of our approach, it is possible to give a reasonable description of the global dynamics and the switching behavior [2].

If we call the change from a "normal" to a "bistable" behavior a phase transition, then the conditions under which a phase transition can occur can be best described by the following three parameters: $a_1 - \xi a_2$, $a_2 - \xi a_1$, and ξ, and can be stated as follows (a_1, a_2, and ξ are all non-negative here).

 (i) If $a_1 - \xi a_2$ and $a_2 - \xi a_1$ are of opposite signs, i.e., if $a_1 > \max(\xi a_2, \xi^{-1} a_2)$ or $a_1 < \min(\xi a_2, \xi^{-1} a_2)$, then the system is normal for all values of ξ.

 (ii) If $a_1 - \xi a_2$ and $a_2 - \xi a_1$ are of the same signs, i.e., if a_1 lies between ξa_2 and $\xi^{-1} a_2$, then the system, considered as a function of ξ, undergoes a phase transition at $\xi = 1$. The system is normal when $\xi < 1$ and is bistable when $\xi > 1$.

1.4. Concluding note

Experiments on the He–Ne gas ring laser ($\xi < 1$) and certain dye laser ($\xi > 1$) have given rather accurate confirmation of our theoretical results. The recent observation of the sudden disappearance of mode correlations in a free running broad band, standing-wave cw dye laser [7] as the laser power is increased above a critical point may suggest a confirmation of our predicted "phase transition" as the laser-power dependent ξ value of the laser changes continuously from below 1 to above 1.

2. Bloch equations involving amplitude and frequency modulations

A fundamental set of equations in magnetic resonance [8] and optical resonance [9] problems is the set of so-called Bloch equations given by

$$\begin{pmatrix} \dot{u} \\ \dot{v} \\ \dot{w} \end{pmatrix} = \begin{pmatrix} 0 & \dot{B} & 0 \\ -\dot{B} & 0 & \dot{A} \\ 0 & -\dot{A} & 0 \end{pmatrix} \begin{pmatrix} u \\ v \\ w \end{pmatrix}, \qquad (2.1)$$

where, in the optical resonance problems [10], u and v are the components of the atomic dipole moment in-phase and in quadrature with the incident laser field, and w is the population inversion for the atom. The matrix elements $\dot{A}$ and $\dot{B}$ have the dimension of angular frequency and are related to the generally time-dependent amplitude and detuning of the incident laser pulse by

$$\dot{A} = \kappa \mathscr{E}(z, t) \equiv \Omega(z, t), \qquad (2.2a)$$

and

$$\dot{B} = \omega_0 - \omega(z, t) \equiv \Delta(z, t), \qquad (2.2b)$$

where $\mathscr{E}(z, t)$ and $\omega(z, t)$ are the amplitude and frequency of the incident laser pulse, ω_0 the frequency difference between the two levels of the atom and $\kappa = 2d/\hbar$, d being the atomic dipole moment.

A beautiful solution of eq. (2.1) for the case of constant detuning Δ is the famous hyperbolic secant pulse of McCall and Hahn [11]:

$$\dot{A} = \kappa \mathscr{E}(z, t) = \frac{2}{\tau} \operatorname{sech} \frac{t - t_0}{\tau}, \qquad (2.3a)$$

$$\dot{B} = \Delta(z, t) = \Delta, \qquad (2.3b)$$

where τ is an arbitrary pulse length. The important concept of envelope area of the pulse was also introduced by McCall and Hahn as

$$\theta(z, t) = \kappa \int_{-\infty}^{t} \mathscr{E}(z, t') \, dt', \qquad (2.4)$$

which is also identified as the dipole turning angle. The area $\theta(z, \infty)$ of the pulse (2.3a) is 2π, and it is this special 2π property which makes the pulse stable. As the pulse propagates through the atomic medium, it excites but returns the atoms to their initial state as it emerges from the medium completely unattenuated. McCall and Hahn named the phenomenon "self-induced transparency".

An interestingly related solution of eq. (2.1) for the case involving both amplitude and frequency modulations is the non-adiabatic excitation solution of Allen and Eberly [12]:

$$\dot{A} = \kappa \mathscr{E}(z, t) = \frac{1}{\tau} \sqrt{1 + \delta^2 \tau^2} \operatorname{sech} \frac{t - t_0}{\tau}, \qquad (2.5a)$$

$$\dot{B} = \Delta(z, t) = -\delta \tanh \frac{t - t_0}{\tau}, \qquad (2.5b)$$

where 2δ is the magnitude of the frequency sweep. Starting with an atom initially in the lower state (at $t = -\infty$), the pulse completely inverts the atom and leaves it in the upper state at $t = +\infty$. If the frequency modulation is zero, then $\mathscr{E}$ becomes a standard π pulse and it would naturally be expected to invert an atom. On the other hand, if the frequency modulation is substantial enough so that, for example, $\delta^2 \tau^2 = 3$, then we have a 2π pulse which again inverts the atoms, and does not return them to their ground states. Clearly in the presence of frequency modulation, the identification of the pulse area to dipole turning angle is no longer true. The appropriate characterization of pulse area has never been made in that case.

We present here a more general analytic solution of the Bloch equations (2.1) which includes the solution of McCall and Hahn, eq. (2.3), and the solution of Allen and Eberly, eq. (2.5), as special cases. We shall introduce a characterization of pulse area and show that the pulses given by eqs. (2.3) and (2.5) correspond to the special cases of 2π and π pulses respectively. Our solution provides an exact analytic result for a specific type of amplitude- and frequency-modulated pulse of *any* area.

We shall derive our solution of eq. (2.1) through solving the following set of equations:

$$\begin{pmatrix} \dot{C}_1 \\ \dot{C}_2 \end{pmatrix} = \begin{pmatrix} 0 & \tfrac{1}{2}i A e^{-iB} \\ \tfrac{1}{2}i A e^{iB} & 0_1 \end{pmatrix} \begin{pmatrix} C_1 \\ C_2 \end{pmatrix}. \tag{2.6}$$

It can be verified that given eq. (2.6), eq. (2.1) follows if u, v, w are related to C_1 and C_2 by

$$u = e^{-iB} C_1^* C_2 + e^{iB} C_2^* C_1, \tag{2.7a}$$

$$v = -i\left(e^{-iB} C_1^* C_2 - e^{iB} C_2^* C_1\right), \tag{2.7b}$$

$$w = C_1^* C_1 - C_2^* C_2 = |C_1|^2 - |C_2|^2. \tag{2.7c}$$

When the solutions of $|C_1|$ and $|C_2|$ are determined from eq. (2.6), w can be determined from eq. (2.7c), and v and u can be determined from eq. (2.1) as

$$v = -\dot{w}/\dot{A}, \tag{2.8}$$

and

$$u = (-\dot{v} + \dot{A}w)/\dot{B}. \tag{2.9}$$

The steps (2.7c), (2.8) and (2.9) make the determination of the phases of C_1 and C_2 unnecessary for our solution of eq. (2.1).

Equations (2.6) are an extension of the set of equations studied by Rosen and Zener [13] in connection with the double Stern–Gerlach experiment in which they had $B = \omega t$. We shall follow the method used by them, but our extension of B to the form which we shall specify is crucial in incorporating the frequency modulation in eq. (2.1).

Elimination of C_2 from eq. (2.6) leads to the second-order differential equation

$$\ddot{C}_1 + (i\dot{B} - \ddot{A}/\dot{A})\dot{C}_1 + (\tfrac{1}{2}\dot{A})^2 C_1 = 0. \tag{2.10}$$

We now make the choice

$$\dot{A} = \frac{\alpha}{\pi \tau} \operatorname{sech} \frac{t - t_0}{\tau}, \tag{2.11}$$

$$\dot{B} = \frac{1}{\pi \tau}\left(\beta_0 - \beta \tanh \frac{t - t_0}{\tau}\right), \tag{2.12}$$

where α, β_0, β are arbitrary constants which may be considered to have the dimension of angles in radians for our problem, and τ is an arbitrary pulse length. These functions, (2.11) and (2.12), together with the transformation

$$z = \tfrac{1}{2}\left(1 + \tanh \frac{t - t_0}{\tau}\right) \tag{2.13}$$

reduce eq. (2.10) to the hypergeometric equation

$$z(1-z)\frac{\mathrm{d}^2 C_1}{\mathrm{d} z^2} + \{c - (a + b + 1)z\}\frac{\mathrm{d} C_1}{\mathrm{d} z} - ab C_1 = 0, \tag{2.14}$$

where

$$a = \frac{1}{2\pi}\left(\sqrt{\alpha^2 - \beta^2} + i\beta\right), \tag{2.15a}$$

$$b = \frac{1}{2\pi}\left(-\sqrt{\alpha^2 - \beta^2} + i\beta\right), \tag{2.15b}$$

$$c = \tfrac{1}{2}\left(1 + i\frac{\beta_0 + \beta}{\pi}\right). \tag{2.15c}$$

As t goes from $-\infty$ to ∞, z goes from 0 to 1. The general solution of eq. (2.14) which is defined in this range is

$$C_1 = a_1 F(a, b, c, z) + a_2 z^{1-c} F(a + 1 - c, b + 1 - c, 2 - c, z). \tag{2.16}$$

Consider the boundary conditions

$$C_1(-\infty) = 0, \tag{2.17a}$$

$$|C_2(-\infty)| = 1, \tag{2.17b}$$

so that $w(-\infty) = -1$ in eq. (2.1). To satisfy the boundary condition (2.17a), we must set $a_1 = 0$. Using the first of eqs. (2.6), we find that the boundary condition (2.17b) is satisfied when

$$a_2 = e^{i\varphi_1} \frac{\alpha}{2\pi |c|}, \tag{2.18}$$

where φ_1 is an arbitrary (real) phase factor. Thus we find

$$C_1 = e^{i\varphi_1} \frac{\alpha}{2\pi |c|} z^{1-c} F(a+1-c, b+1-c, 2-c, z), \tag{2.19}$$

and similarly, we find

$$C_2 = e^{i\varphi_2} F(a^*, b^*, c^*, z), \tag{2.20}$$

where φ_2 is another arbitrary phase factor, and we note that

$$|C_1|^2 + |C_2|^2 = 1. \tag{2.21}$$

By using the relations

$$F(a, b, c, 1) = \frac{\Gamma(c)\,\Gamma(c-a-b)}{\Gamma(c-a)\,\Gamma(c-b)}, \tag{2.22}$$

and

$$\Gamma(z)\Gamma(-z) = -\frac{\pi}{z \sin \pi z}, \tag{2.23}$$

we find

$$|C_1(+\infty)|^2 = \operatorname{sech}\tfrac{1}{2}(\beta_0 + \beta)\operatorname{sech}\tfrac{1}{2}(\beta_0 - \beta)$$
$$\times \left(\sin^2 \tfrac{1}{2}\Phi \cosh^2 \tfrac{1}{2}\beta + \cos^2 \tfrac{1}{2}\Phi \sinh^2 \tfrac{1}{2}\beta\right) \tag{2.24}$$

and

$$|C_2(+\infty)|^2 = 1 - |C_1(+\infty)|^2, \tag{2.25}$$

where

$$\Phi = \sqrt{\alpha^2 - \beta^2} \tag{2.26}$$

turns out to be an important parameter which we can use to characterize the "area" of the pulse given by eqs. (2.11) and (2.12) under various conditions which we shall discuss.

Equations (2.24) and (2.25) give us the fraction of electron population inversion for the atoms at $t = +\infty$ of eq. (2.1) to be

$$w(+\infty) = 2 \operatorname{sech}\tfrac{1}{2}(\beta_0 + \beta)\operatorname{sech}\tfrac{1}{2}(\beta_0 - \beta)$$
$$\times \left(\sin^2 \tfrac{1}{2}\Phi \cosh^2 \tfrac{1}{2}\beta + \cos^2 \tfrac{1}{2}\Phi \sinh^2 \tfrac{1}{2}\beta\right) - 1, \tag{2.27}$$

when the atoms, starting with $w(-\infty) = -1$, are subjected to an incident laser pulse which has a modulated amplitude and frequency given by eqs. (2.11) and (2.12). The special cases discussed below when one or both of the parameters β_0 and β equals zero give us further insight into the nature and wide applicability of the solution which we have obtained.

2.1. Special cases

2.1.1. The case $\beta = 0$

We have from eqs. (2.26) and (2.27),

$$\Phi = \alpha, \tag{2.28}$$

and

$$w(+\infty) = 2\sin^2\tfrac{1}{2}\Phi \operatorname{sech}^2\tfrac{1}{2}\beta_0 - 1. \tag{2.29}$$

If

$$\Phi = 2n\pi, \quad n = 1, 2, 3, \ldots, \tag{2.30}$$

then the atoms are always returned to their initial state. The self-induced transparency solution of McCall and Hahn corresponds to the *particular* case of eq. (2.30) when $n = 1$ for which a simple analytic solution of C_1 is available for all t:

$$C_1 = e^{i\varphi_1} \frac{1}{(1+\beta_0^2/\pi^2)^{1/2}} z^{(1+i\beta_0/\pi)/2}(1-z)^{(1-i\beta_0/\pi)/2}, \tag{2.31}$$

which yields

$$u = -\frac{2\beta_0/\pi}{1+\beta_0^2/\pi^2} \operatorname{sech}\frac{t-t_0}{\tau}, \tag{2.32a}$$

$$v = \frac{2}{1+\beta_0^2/\pi^2} \operatorname{sech}\frac{t-t_0}{\tau} \tanh\frac{t-t_0}{\tau}, \tag{2.32b}$$

$$w = -1 + \frac{2}{1+\beta_0^2/\pi^2} \operatorname{sech}^2\frac{t-t_0}{\tau}. \tag{2.32c}$$

Equations (2.32) can be written in the more familiar forms when we identify β_0/π to be $\Delta\tau$ from eqs. (2.12) and (2.3b).

That a pulse with $\Phi = 2n\pi$ always returns the atoms to their initial state was conjectured and numerically confirmed by McCall and Hahn. Equation (2.29), derived much earlier by Rosen and Zener, was in fact a more precise statement stating that effect.

It will be noted that a pulse with $\Phi = (2n-1)\pi$, $n = 1, 2, \ldots$, does *not* completely invert the atoms to their upper states except when $\beta_0 = 0$.

2.1.2. The case $\beta_0 = 0$

We have from eqs. (2.26) and (2.27).

$$\Phi = \sqrt{\alpha^2 - \beta^2} \tag{2.33}$$

and

$$w(+\infty) = 1 - 2\cos^2\tfrac{1}{2}\Phi \operatorname{sech}^2\tfrac{1}{2}\beta. \tag{2.34}$$

If

$$\Phi = (2n-1)\pi, \qquad n = 1, 2, 3, \ldots, \tag{2.35}$$

then the atoms are always *fully* excited to their upper states. The non-adiabatic excitation solution of Allen and Eberly corresponds to the particular case of eq. (2.35) when $n = 1$ for which again simple analytic expressions of C_1 and C_2 are available for all t:

$$C_1 = e^{i\varphi_1} z^{(1-i\beta/\pi)/2}, \tag{2.36a}$$

$$C_2 = e^{i\varphi_2}(1-z)^{(1+i\beta/\pi)/2}, \tag{2.36b}$$

which yield

$$u = \frac{\beta}{\alpha} \operatorname{sech} \frac{t-t_0}{\tau}, \tag{2.37a}$$

$$v = -\frac{\pi}{\alpha} \operatorname{sech} \frac{t-t_0}{\tau}, \tag{2.37b}$$

$$w = \tanh \frac{t-t_0}{\tau}. \tag{2.37c}$$

It will be noted that in the presence of modulated detuning, the area of the pulse is no longer characterized by α but by $\Phi = \sqrt{\alpha^2 - \beta^2}$, and that for the pulse considered by Allen and Eberly, eqs. (2.5), $\alpha = \pi\sqrt{1+\delta^2\tau^2}$, $\beta = \pi\delta\tau$ from eqs. (2.11), (2.12), and (2.5), and hence

$$\sqrt{\alpha^2 - \beta^2} = \{\pi^2(1+\delta^2\tau^2) - \pi^2\delta^2\tau^2\}^{1/2} = \pi \tag{2.38}$$

no matter what δ is. Our equations (2.34) and (2.35) express a more general result not previously stated, that a pulse with $\Phi = (2n-1)\pi$ in the case $\beta_0 = 0$ always inverts the atoms *completely*, but that a pulse with $\Phi = 2n\pi$ does not always return an atom to its initial state except when $\beta = 0$.

2.1.3. The case $\beta_0 = \beta = 0$

This case is of separate interest because the atomic evolutions can be expressed in terms of elementary functions for all t for an incident pulse of modulated amplitude of any area.

From eq. (2.20),

$$C_2 = e^{i\varphi_2} F(a, -a, \tfrac{1}{2}, z) \tag{2.39}$$

where

$$a = \alpha/2\pi. \tag{2.40}$$

Since

$$F(a, -a, \tfrac{1}{2}, \sin^2 x) = \cos(2ax), \tag{2.41}$$

we find

$$|C_1|^2 = \sin^2(\tfrac{1}{2}a\theta), \tag{2.42a}$$

$$|C_2|^2 = \cos^2(\tfrac{1}{2}a\theta), \tag{2.42b}$$

where

$$\sin\tfrac{1}{2}\theta = \operatorname{sech}\frac{t-t_0}{\tau}, \tag{2.43a}$$

$$\cos\tfrac{1}{2}\theta = -\tanh\frac{t-t_0}{\tau}. \tag{2.43b}$$

Thus we have

$$u = 0, \tag{2.44a}$$

$$v = -\sin\left(\frac{\alpha}{2\pi}\theta\right), \tag{2.44b}$$

$$w = -\cos\left(\frac{\alpha}{2\pi}\theta\right), \tag{2.44c}$$

where $\dot\theta$ can be verified to be related to $\dot A$ of eq. (2.11) by

$$\dot\theta = \frac{2\pi}{\alpha}\dot A. \tag{2.45}$$

Equations (2.44), together with eqs. (2.43), express the atomic evolutions when the atoms are subjected to an incident resonant pulse given by eq. (2.11) of any area α. A $(2n-1)\pi$ pulse always fully inverts the atoms and a $2n\pi$ pulse always returns the atoms to their initial state. The atomic evolutions when α is an integral multiple of π can be written down more explicitly by using the multiple angle expansions of the relevant trigonometric functions. In particular,

(a) $\alpha = \pi$,

$$v = -\text{sech}\frac{t-t_0}{\tau}, \tag{2.46}$$

$$w = \tanh\frac{t-t_0}{\tau},$$

(b) $\alpha = 2\pi$,

$$v = 2\,\text{sech}\frac{t-t_0}{\tau}\tanh\frac{t-t_0}{\tau}, \tag{2.47}$$

$$w = -1 + 2\,\text{sech}^2\frac{t-t_0}{\tau},$$

(c) $\alpha = 3\pi$,

$$v = -3\,\text{sech}\frac{t-t_0}{\tau} + 4\,\text{sech}^3\frac{t-t_0}{\tau}, \tag{2.48}$$

$$w = -3\tanh\frac{t-t_0}{\tau} + 4\tanh^3\frac{t-t_0}{\tau},$$

(d) $\alpha = 4\pi$,

$$v = 8\tanh^3\frac{t-t_0}{\tau}\,\text{sech}\frac{t-t_0}{\tau} - 4\tanh\frac{t-t_0}{\tau}\,\text{sech}\frac{t-t_0}{\tau}, \tag{2.49}$$

$$w = -1 + 8\tanh^2\frac{t-t_0}{\tau}\,\text{sech}^2\frac{t-t_0}{\tau},$$

etc.

2.2. Summary

To summarize, we have found the population inversion of electrons in atoms, eq. (2.27), when subjected to an incident laser field involving both amplitude and frequency modulation of the forms given by eqs. (2.11) and (2.12). The special results, eqs. (2.29) and (2.34) for the cases when one of the parameters β and β_0 equals zero, suggest the usefulness of the parameter Φ, eq. (2.26), for characterizing the pulse area, which also unifies the self-induced transparency solution of McCall and Hahn and the non-adiabatic excitation solution of Allen and Eberly which are shown to be two special cases ($\Phi = 2\pi$ and π respectively) of our special solutions. Explicit expressions for the atomic evolutions at all times, eqs. (2.44) and (2.43) for the case $\beta = \beta_0 = 0$ are also presented.

These results are immediately applicable to the multiple soliton and multilevel excitation problems recently studied [14–16] where specific dynamic symmetries were shown to produce dynamical subspaces of the form of eq. (2.1).

Acknowledgement

This research is supported in part by the U.S. Department of Energy, under Grant number DE-FG02-84ER 13243.

References

[1] W.E. Lamb Jr., Phys. Rev. 134 (1964) A1429.
[2] F.T. Hioe and S. Singh, Phys. Rev. A24 (1981) 2050.
[3] R. Roy and L. Mandel, Opt. Commun. 34 (1980) 133;
R. Roy, J. Durnin, R. Short and L. Mandel, Phys. Rev. Lett. 45 (1980) 1486.
[4] E.W. Montroll, Some Statistical Aspects of the Theory of Interacting Species, in: Some Mathematical Problems in Biology, Vol. 4 (Amer. Math. Soc., 1972) pp. 101–143.
[5] R.L. Stratonovich, Topics in Theory of Random Noise, Vol. I (Gordon and Breach, New York, 1963).
[6] M. Sargent III, M.O. Scully and W.E. Lamb Jr., Laser Physics (Addison–Wesley, Reading, MA, 1974).
[7] L.A. Westling, M.G. Raymer, M.G. Sceats and D.F. Coker, Opt. Commun. 47 (1983) 212.
[8] F. Bloch, Phys. Rev. 70 (1946) 460.
[9] R.P. Feynman, F.L. Vernon and R.W. Hellwarth, J. Appl. Phys. 28 (1957) 49.
[10] See, e.g., L. Allen and J.H. Eberly, Optical Resonance and Two-Level Atoms (Wiley, New York, 1975).
[11] S.L. McCall and E.L. Hahn, Phys. Rev. 183 (1969) 457.
[12] Ref. [10], Sec. 4.6.
[13] N. Rosen and C. Zener, Phys. Rev. 40 (1932) 502.
[14] F.T. Hioe, Phys. Rev. A26 (1982) 1466; A28 (1983) 879; A29 (1984) 3434; Phys. Lett. A99 (1983) 150.
[15] F.T. Hioe, Theory of Atomic Excitation by Multiple Laser Pulses, in: Coherence and Quantum Optics V, eds. L. Mandel and E. Wolf (Plenum Press, New York, 1984) p. 965.
F.T. Hioe and J.H. Eberly, Phys. Rev. A29 (1984) 1164.
[16] F.T. Hioe, Phys. Rev. A30 (1984) 3097.

Theory of Diffusion via an Interstitial and Vacancy Mechanism

Paul H.E. MEIJER

Physics Department
Catholic University of America
Washington, DC 20064
U.S.A.

National Bureau of Standards
Gaithersburg, MD 20899
U.S.A.

© Elsevier Science Publishers B.V. 1985

The Wonderful World of Stochastics
A Tribute to Elliott W. Montroll
Eds. M.F. Shlesinger and G.H. Weiss

Contents

1. Introduction — 101
2. Continuous description — 102
3. General description — 103
4. Major and minor Fourier transforms — 105
5. Comparison with the continuous case — 107
6. Comparison with other work — 108
References — 109

1. Introduction

In many applications of solid-state physics the diffusion of foreign atoms in semiconductors is of major practical importance [1].

The model which is used here is based on the diffusion in those lattices where the atom moves by an interstitial as well as by a vacancy mechanism [2]. In this chapter we describe a diffusion mechanism which uses jumps from vacancy to vacancy at one hand, and jumps from interstitial to interstitial at the other hand. Moreover there may be transitions between the vacancy sites and the interstitial sites and vice versa. This type of transition of the atom from a vacancy to an interstitial usually does not have the same waiting time as the transition in the other direction. For the sake of simplicity let us call the vacancies black sites and the interstitials white sites.

Let us assume that there are about n interstitial (white) sites for every vacancy (black) site. If the vacancies are more or less homogeneously distributed through the crystal, one can imagine cells, which contain each one vacancy site and n interstitial sites. This is shown in fig. 1, for a one dimensional crystal, in a schematic way. The waiting-time function will be described by a simple exponential, as used by Landman and Shlesinger [3]. For other choices than the exponential see Kehr and Haus [4]. Using the phenomenological equations of Svob and Marfaing [2] we obtain an integral

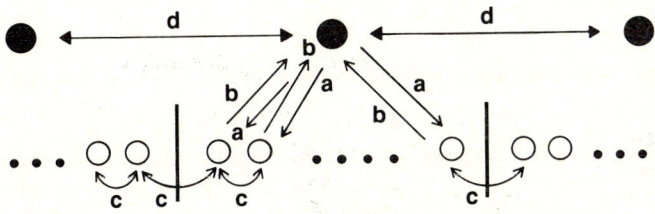

Fig. 1.

for the diffusion profile of the interstitials and the vacancy distribution.

Montroll and Weiss [5] showed how the random-walk problem can be solved by using a generating function. In a number of papers [6–8] the model was generalized for the case that the walker may attain several internal states. In this case we deal with $n + 1$ internal states, n of which are equivalent. Since the interstitials are on a periodic lattice, a Fourier transform is useful.

In order to describe the transition behaviour inside the unit cell, a matrix is used indicating the possible transitions from site to site. The submatrix referring to the interstitials is cyclic, except for the elements describing the jumps to the neighboring cells. Hence in order to solve the main equations a double Fourier transform is used. One major transform is enacted on the cells, and a minor transform is done on the inside of the cell, in order to diagonalize the interstitial transitions. The wave number of the minor transform is the branch-label of the folded Brillouin zone of the crystal without vacancies.

The result is compared with the solution for the continuous case.

2. Continuous description

This chapter describes the behavior of a random walker who has two modes of mobility: the vacancy-to-vacancy motion and the motion via interstitials. The walker can switch from one mode to the other at any given moment. This behavior is described in a phenomenological way by the following two diffusion equations. Let C_i be the concentration of interstitial atoms and C_v the concentration of foreign atoms that are placed in vacancies. The behavior is governed by a set of diffusion equations, coupled by terms which describe the disappearance of one type with the simultaneous reappearance as the other type of atom, as follows:

$$\frac{\partial C_i}{\partial t} = D_i \nabla^2 C_i + \alpha C_v - \gamma C_i, \tag{1a}$$

$$\frac{\partial C_v}{\partial t} = D_v \nabla^2 C_v + \gamma C_i - \alpha C_v. \tag{1b}$$

The second derivative in the diffusion equation describes the fact that the random walker makes in the average no progress. The coefficients α and γ describe the switch mode. The sum of the two concentrations is described by a diffusion equation without extra terms.

The picture is similar, but not quite the same, as the motion of a random walker on a lattice with a base, a non-Bravais lattice. This case was considered by Ishioka and Koiwa [10], who treated the random walk on a diamond lattice. In their case the two sublattices are equivalent, in our case they are not. Moreover one of our subsystems is multiple degenerated since there are in general $n > 1$ interstitial sites for every vacancy. Furthermore the transition and also the waiting times for interstitial-to-vacancy and for vacancy-to-interstitial transitions are different as is clear from the physical picture.

In order to solve the phenomenological equation, one starts with a Laplace and Fourier transform of eq. (1) which leads to two coupled algebraic equations that are solved by

$$uC_i - C_i(t=0) = -k^2 D_i C_i - \alpha C_i + \gamma C_v, \tag{2a}$$

$$uC_v - C_v(t=0) = -k^2 D_v C_v + \alpha C_i - \gamma C_v. \tag{2b}$$

In order to obtain the inverse Laplace transform the denominator D is factorized,

$$D = (u - u_1)(u - u_2), \tag{3}$$

with

$$2u_{1,2} = -\left(D_i k^2 + \alpha + D_2 k^2 + \gamma\right) \\ \pm \sqrt{\left(D_i k^2 + \alpha - D_v k^2 - \gamma\right)^2 + 4\alpha\gamma}.$$

The result is

$$C_i(t) = \left[\left\{(u_1 + D_2 k^2 + \gamma)C_1(0) + \gamma C_2(0)\right\} \exp(u_1 t) \right. \\ \left. - \left\{(u_2 + D_2 k^2 + \gamma)C_1(0) + \gamma C_2(0)\right\} \exp(u_2 t)\right]/(u_1 - u_2),$$

$$C_v(t) = \left[\left\{(u_1 + D_1 k^2 + \alpha)C_2(0) + \alpha C_1(0)\right\} \exp(u_1 t) \right. \\ \left. - \left\{(u_2 + D_1 k^2 + \alpha)C_2(0) + \alpha C_1(0)\right\} \exp(u_2 t)\right]/(u_1 - u_2).$$

3. General description

We subdivide the solid in cells of such size that they contain each *one* vacancy. Although the vacancies may not be homogeneously dispersed through the solid, it is of great help in the calculation to assume that they are, and it seems unlikely that this will influence the result. Since each cell contains many interstitial sites, say n, which are potential stopovers for the diffusing atom, a label j is introduced to indicate in which (inner) state of the cell the atom may reside. This inner state is either one of the n interstitial states or the vacancy state. The quantity of concern is the probability $P_m(j, s; j', s'; t)$. This is the probability that the atom, originally inserted in cell s' in state j' at time $t = 0$, will be found in cell s in state j at time t, after m steps. Let the value $j = 0$ be assigned to the vacancy and the remaining labels to the interstitials. The indices j, and j' can attain $n + 1$ values. Subsequently we will assume that both the cells and the internal states lie on a periodic lattice. It is convenient to treat the labels j and j' as rows and columns of an $n + 1$ by $n + 1$ matrix. Each element represents the time-dependent probability for the atom to jump:

(1) from one interstitial site to another, either inside the cell or if the site is close to the boundary to the neighboring cell;
(2) from an interstitial into a vacancy or vice versa;
(3) from a vacancy in the cell to the vacancy in the neighboring cell.

This leads to four different transition probabilities. Let us assume, and this is not necessary for the calculation, that the time dependence is exponential. (For choices other than exponential see Kehr and Haus [4].) Hence the transition matrix $\psi(s)$ will have the following form:

$$\begin{pmatrix} d(\delta_1 + \delta_{-1}) & b\delta_0 & b\delta_0 & b\delta_0 & \cdots & b\delta_0 \\ a\delta_0 & 0 & c\delta_0 & 0 & \cdots & c\delta_1 \\ a\delta_0 & c\delta_0 & 0 & c\delta_0 & \cdots & 0 \\ a\delta_0 & 0 & c\delta_0 & 0 & \cdots & 0 \\ \vdots & & & & & \vdots \\ a\delta_0 & c\delta_{-1} & 0 & 0 & \cdots & 0 \end{pmatrix}, \quad (4)$$

where $\delta_0 = \delta_{s,s'}$; $\delta_{\pm 1} = \delta_{s,s'+1}$; and where s is the label of the cells. In this matrix, a, b, c and d are short for $a(t)$, $b(t), \ldots$; their interpretation is as follows: $a(t)$ is the waiting time distribution for an atom in the vacancy, i.e. a substitutional atom, to jump into an interstitial site; $b(t)$ is the reverse process; $c(t)$ represents the jump from an interstitial to an interstitial; and $d(t)$ the jump from a vacancy to a vacancy. The time functions are given by

$$\begin{aligned} a(t) &\equiv ae^{-at}, & b(t) &\equiv be^{-bt}, \\ c(t) &\equiv ce^{-ct}, & d(t) &\equiv de^{-dt}, \end{aligned} \quad (5)$$

with $2d + na = 1$ and $b + 2c = 1$. This expresses the fact that the walker, whether he is on a white or on a black site, has to make a move.

Let us label the rows and columns of the matrix with labels $0, 1, \ldots, n$.

The submatrix with rows and columns 1 to n is cyclic when considered as part of the total system, since it contains two sets of elements $c\delta_0$ above and below the diagonal and elements $c\delta_{\pm 1}$ in the upper right and lower left corner. These elements describe the boundary effect: the jump from an interstitial in one cell to its nearest neighbor in the next cell.

4. Major and minor Fourier transforms

Let the probability that the random walker, i.e. the diffusing atom, is in cell s after n steps be $P_m(s, j)$. The label j indicates in which internal state of the cell number s the atom is located. This may be a vacancy ($j = 0$) or one of the n interstitials. It is convenient to introduce the random-walk generating function used by Montroll and Weiss [5] given by:

$$P(s, j) \equiv P(s, j; z) = \sum_{m=0}^{\infty} z^m P_m(s, j). \tag{6}$$

From now on P without the subscript m refers to the generating function. To solve the problem we need the Fourier transform of the quantity

$$P(K, j) = \sum_{s=1}^{N} P(s, j) e^{iKs}, \tag{7}$$

where N is the total number of cells. The periodic boundary condition requires that:

$$K \equiv K_\nu = \frac{2\pi}{N} \nu \quad (\nu = 0, \ldots, N-1), \tag{8}$$

and we will refer to this as the major Fourier transform.

Since it was assumed that the interstitials are on a periodic lattice, it is possible to diagonalize the cyclic submatrix using a minor Fourier transform of the type

$$P(K, k) = \sum_{j=0}^{n} P(K, j) e^{ikj}, \tag{9}$$

with

$$k \equiv k_\mu = \frac{2\pi}{n} \mu \quad (\mu = 0, 1, \ldots, n-1).$$

The minor transform could also have been exercised on the $P(s, j)$, leading to a partial Fourier transform $P(s, k)$ of the main spacial function.

In the language of lattice vibrations, the Brillouin zone for the case of interstitials only is given by $k_B = \pi$, if the interstitials are one length-unit apart. If one introduces one vacancy per n interstitials, the periodicity is n, in the same length units, and the Brillouin zone is given by $K_B = \pi/n$. However in the last case there are n branches, labelled by μ, so that the number of possible Fourier components is again nN, the total number of lattice points.

With this notation it is easy to establish the eigenvalues of the n by n submatrix of ψ for fixed K, i.e. after taking the major Fourier transform. These are given by

$$\lambda_\mu = 2c \cos\left(\frac{K}{n} + k_\mu\right) \quad (\mu = 1, \ldots, n), \tag{10}$$

where $nk_\mu = 0 \pmod{2\pi}$.

As a result of this transformation the matrix can be written as

$$\psi' = \psi_I + \psi_V + \psi_C, \tag{11}$$

where I means the interstitial, V the vacancy and C the coupling. The first matrix has diagonal elements $0, \lambda_1, \lambda_2, \ldots, \lambda_n$, the second matrix has diagonal elements $1, 0, \ldots, 0$ and the coupling consists of a row

$$b' = \langle 0|\psi_C|j\rangle = \frac{1}{\sqrt{n}} b(1 - \alpha_j) \quad (j = 1, \ldots, n) \tag{12}$$

and a column

$$a' = \langle j|\psi_C|0\rangle = \frac{1}{\sqrt{n}} a(1 - \alpha_j^{-1}) \quad (j = 1, \ldots, n), \tag{13}$$

with $\alpha_j = \exp i(K/n + k_j)$.

If we now use the standard random-walk solution procedure as used by Montroll and Weiss [5] and by Landman and Shlesinger [3], the solution is given by

$$P(K, j) = \sum_i \left(\frac{1}{1 - z\psi'}\right)_{ji} e^{iKs_0} P_i(0), \tag{14}$$

where $P(K, j)$ is the Laplace and Fourier transform of the generating function. $P_i(0)$ is the probability of finding the walker in state i when he started out at $t = 0$ in cell s_0. After taking the inverse transforms the generating function can be expanded in powers of z and the coefficient of the mth power in z is the probability that the random walker finds himself in cell s in state j, provided he started out in cell s_0 in state i after the mth step. In many cases one is interested in his arrival in cell s independent of the number of steps needed; this is found by taking $z = 1$ in eq. (14).

The inverse matrix $V = (1 - \psi')^{-1}$ can be obtained by straightforward algebra. The non-zero matrix elements are:

$$V_{00} = \left(1 - 2d \cos K - \sum_{i=1}^{n} \frac{b'_i a'_i}{1 - \lambda_i}\right)^{-1},$$

$$V_{0\mu} = -V_{00} b'_\mu (1 - \lambda_\mu)^{-1},$$

$$V_{\mu 0} = -V_{00} a'_\mu (1 - \lambda_\mu)^{-1},$$

$$V_{\mu\mu} = -(1 - \lambda_\mu)^{-1}, \tag{14a}$$

with $\mu = 1, \ldots, n$, and with λ_μ given by eq. (10). This gives a complete solution of the problem.

5. Comparison with the continuous case

In order to obtain the relations between the random-walk results and the phenomenological equations (1), a transformation will be made to replace the random-walk equations by the equivalent master equation. This transformation is described by Kenkre et al. [9]. The time functions $\psi_a(t) = a \exp(-at)$, etc. in the transition matrix become delta functions: $\phi_a(t) = 2a\delta(t)$, etc. in the master-equation description. Hence the integrals over time in the generalized master equation are trivial. The equation used in ref. [9], extended to a multiple-state system, has now the form

$$\frac{\mathrm{d}P(s, i)}{\mathrm{d}t} = \int \mathrm{d}t' \sum_{s'j} \left[\phi_{ij}(t - t') p_{ij}(s - s') P(s, j; t')\right.$$

$$\left. - \phi_{ji}(t - t') p_j(s' - s) P(s, i; t')\right]. \tag{15}$$

Since

$$\sum_{s'j} \phi_{ji} p_{ji}(s' - s) = 1 \quad (\text{all } i), \tag{16}$$

and $\phi_{ij} = \alpha_{ij} \delta(t)$, one has

$$\frac{\mathrm{d}P(s, i)}{\mathrm{d}t} = \sum_{s'j} \alpha_{ij} p_{ij}(s - s') P(s', j) - P(s, i). \tag{17}$$

Since the condition (16) limits the number of parameters α_{ij}, we introduce

$$d \equiv \alpha_{00} = \tfrac{1}{2}(1-\epsilon), \qquad b \equiv \alpha_{0\mu} = \gamma,$$
$$a \equiv \alpha_{\mu 0} = \epsilon/n, \qquad c \equiv \alpha_{\mu\mu} = \tfrac{1}{2}(1-\gamma),$$
$$\mu = 1, \ldots, n.$$

It is possible to rewrite eq. (17) as follows:

$$\frac{dP(s, 0)}{dt} = \tfrac{1}{2}(1-\epsilon)[P(s+1, 0) + P(s-1, 0) - 2P(s, 0)]$$

$$+ \gamma \sum_{\mu=1}^{n} P(s, \mu) - \epsilon P(s, 0), \tag{17a}$$

which leads to $D_v = \tfrac{1}{2}(1-\epsilon)$ using $\Delta s = 1$, after the bracketted term is replaced by the second derivative. A similar treatment for the total interstitial probability gives:

$$\frac{d\Sigma_\mu P(s, \mu)}{dt} = D_i \frac{d^2 P}{dx^2} + \epsilon P(s, 0) - \gamma \sum_\mu P(s, \mu), \tag{17b}$$

with $D_i = \tfrac{1}{2}(1-\gamma)n^2$ since $\Delta v = 1/n$, using the cell size as length unit. Hence the coefficient α in eq. (1) is the same as ϵ used here.

If the random walker is a "self" atom, it disappears in the vacancy [11], and the problem needs additional attention in its description.

6. Comparison with other work

The idea of coupled diffusion equations was used in the work of Cann, Kirkwood and Brown [12,13] on electrophoresis. In this problem the particles are proteins that exist in two interconvertible states. In addition to the exchange term, where the walker goes from one state to another, there is also a drift term. The last term is similar to the field term used by Svob and Marfaing [2]. If the mobilities in the drift term in each of the equations are different, the solution shows an interesting feature. For certain values of the parameters the diffusion current may go against the gradient instead of with the gradient. This happens in a limited region and for certain boundary conditions. This is sometimes called "uphill" diffusion and Cann et al. call it a bimodular distribution in the gradient.

Acknowledgement

The author is grateful for the Night Vision and Electro-Optics Laboratory at Fort Belvoir for support of this work. I would like to thank M. Shlesinger for several discussions on this subject and G. Weiss for bringing my attention to the paper of Cann, Kirkwood and Brown. Part of this work was done during a stay at the University of Nancy, an invitation which is hereby gratefully acknowledged.

References

[1] B. Tuck, Introduction to Diffusion in Semiconductors. I.E.E. Monograph Series 16 (P. Peregrino, London, 1974).
[2] L. Svob and Y. Marfaing, J. Crystal Growth 59 (1982) 276.
[3] U. Landman and M.F. Shlesinger, Phys. Rev. 19 (1979) 6207.
[4] K.W. Kehr and J.W. Haus, Physica 93A (1978) 412.
[5] E.W. Montroll and G. Weiss, J. Math. Phys. 6 (1965) 167.
[6] U. Landman, E.W. Montroll and M.F. Shlesinger, Proc. Nat. Acad. Sci. U.S.A. 74 (1979) 430.
[7] U. Landman and M.F. Shlesinger, Phys. Rev. B16 (1977) 3389.
[8] E.W. Montroll and H. Scher, J. Stat. Phys. 9 (1973) 101.
[9] V.M. Kenkre, E.W. Montroll and M.F. Shlesinger, J. Stat. Phys. 9 (1973) 45. (Preceded by ref. [8].)
[10] S. Ishioka and M. Koiwa, Phil. Mag. 37 (1978) 517.
[11] H.F. Schaake, J.H. Tregilgas, D. Beck and M.A. Kinch, Solid State Commun. 50 (1984) 133.
[12] J.R. Cann, J.G. Kirkwood and R.A. Brown, Arch. Biochem. and Biophys. 72 (1975) 37.
[13] G.H. Weiss and R.J. Rubin, Adv. Chem. Phys. 52 (1983) 363; see sec. IIIB: Multistate random Walk.

CHAPTER 5

Mathieu's Difference Equation *

Renfrey B. POTTS
Applied Mathematics Department
University of Adelaide
South Australia, 5001

* Research supported by an award under the Australian Research Grant Scheme.

© *Elsevier Science Publishers B.V. 1985*

The Wonderful World of Stochastics
A Tribute to Elliott W. Montroll
Eds. M.F. Shlesinger and G.H. Weiss

Contents

Dedication 113
1. Introduction 113
2. Discrete Floquet theory 115
3. Hill's difference equation 118
4. Evaluation of the basic determinants 122
5. Mathieu's difference equation 123
6. Choice of H 123
7. Discussion 124
References 125

Dedication

It is an honour for me to be asked to contribute a paper to the proceedings of the Montroll Symposium. As a Fulbright Scholar from Australia, I spent 1955 with Elliott Montroll at the University of Maryland, and what a year it was. The research work on lattice vibrations, initiated and guided by Elliott, was exciting and productive, and my wife and I were overwhelmed by the open house friendship of Elliott and Shirley who included us into their family—an extra two were hardly noticed. Although I settled in Adelaide, Australia, about as far from the East Coast of America that one can get and still be on dry land, I was able to arrange to meet with Elliott on many different occasions and in many different places and without fail he was always able to trigger off some fruitful research. Work with Elliott was always great fun. I remember on one occasion, when we were playing with the Ising model at IBM, Yorktown Heights, Elliott remarking 'What would T.J.W. say if he knew what we were doing!'

The present paper is one of a series that has developed from a suggestion by Elliott that I should investigate difference equations, with the aim of throwing more light on the problem of scattering by lattices. The matrices occurring in the paper are just the sort that Elliott enjoyed and handled with great skill.

It is with humility that I dedicate this paper to the memory of a great scientist, a great person, a great friend.

1. Introduction

Mathieu's differential equation
$$\ddot{x}(t) + (a - 2q\cos 2t)\, x(t) = 0, \tag{1.1}$$
where a and q are real parameters, occurs in the modelling of an extensive

and diverse class of applied problems and for more than a century the theory of Mathieu functions has gradually developed into a considerable mathematical edifice [3].

The standard approach in the numerical analysis of a differential equation (DE) is to replace it by a difference equation (ΔE) which is carefully chosen so that its numerical solution can be obtained in a stable manner and without trouble from round-off errors. In this approach, a derivative term, such as $\ddot{x}(t)$ in eq. (1.1), is approximated by a difference, such as $(x_{n+1} - 2x_n + x_{n-1})/h^2$, where h is a constant step-size.

In a series of recent papers [4–6] it has been shown that it can be of considerable advantage to allow instead of denominators such as h^2 certain functions of h which can be chosen to greatly improve the approximation of the solution of the ΔE to the solution of the DE. In the simplest case of a linear DE with constant coefficients it has been shown [4] that the ΔE can be chosen so that the agreement of the solutions is in fact exact. For example, the DE

$$\ddot{x}(t) + ax(t) = 0 \tag{1.2}$$

and the ΔE

$$(x_{n+1} - 2x_n + x_{n-1})/H^2 + ax_n = 0 \tag{1.3}$$

with

$$H^2 = 4a^{-1} \sin^2 \tfrac{1}{2} h\sqrt{a} \tag{1.4}$$

have the same solution in the sense that if $x(t)$ is the solution of the DE, then $x_n = x(nh)$ is the solution of the ΔE. Explicitly, the solution of (1.2) is

$$x(t) = A \cos\sqrt{a}\, t + B \sin\sqrt{a}\, t, \tag{1.5}$$

while that of (1.3) with (1.4) is

$$x_n = A \cos\sqrt{a}\, nh + B \sin\sqrt{a}\, nh. \tag{1.6}$$

Geometrically the solution points x_n of the ΔE lie precisely on the solution curve $x(t)$ of the DE, independent of the sign and magnitude of h.

This approach has been exploited in constructing ΔE's for approximating non-linear DE's with constant coefficients, such as Duffing's equation [5] and the van der Pol equation [6]; precise agreement between the two solutions was obtained in the former case and while this was not possible for the latter case, good agreement between the qualitative features was achieved.

It is the purpose of the present paper to follow this approach for the analysis of ΔE's with non-constant coefficients—in particular periodic coefficients [1]. Of these DE's the Mathieu DE is the most important example. A discrete form of the Floquet theory will be developed and then applied to the analysis of a ΔE for the Hill's equation. This will finally be specialised to the case of the Mathieu equation and it will be shown that indeed the characteristic features of the solutions of a suitably chosen Mathieu ΔE are very similar to those for the DE.

2. Discrete Floquet theory

Consider the second-order linear homogeneous DE

$$\ddot{x}(t) + J(t)\, x(t) = 0, \tag{2.1}$$

where the coefficient $J(t)$ is periodic. Without loss of generality, the period can be taken as π so that

$$J(t+\pi) = J(t). \tag{2.2}$$

A corresponding ΔE can be obtained by choosing

$$h = \pi/p, \quad p \text{ +ve integer} \tag{2.3}$$

and approximating (2.1) by

$$(x_{n+1} - 2x_n + x_{n-1})/H^2 + J_n x_n = 0 \tag{2.4}$$

with

$$H^2 = h^2 + O(h^3), \tag{2.5}$$

$$J_n = J_{n+p}. \tag{2.6}$$

In the limit $h \to 0$, the ΔE $\to$ the DE but the precise form of the function H is as yet left undecided.

The Floquet theory [1] of the DE (2.1) establishes conditions for *basically-periodic solutions*, of period π or 2π. Analogous results for the ΔE (2.4) can be established by the following lemma and theorem.

Lemma 2.1. *If the solution of the ΔE (2.4) satisfies*

$$x_p = sx_0, \quad s \text{ non-zero constant} \tag{2.7}$$

$$x_{p+1} = sx_1, \tag{2.8}$$

then

$$x_{n+p} = sx_n. \tag{2.9}$$

Proof. It is an immediate consequence of (2.4) that, for $n = p + 1$,
$$x_{p+2} - 2x_{p+1} + x_p + H^2 J_{p+1} x_{p+1} = 0, \tag{2.10}$$
so that, using (2.6), (2.7) and (2.8),
$$s^{-1} x_{p+2} - 2x_1 + x_0 + H^2 J_1 x_1 = 0. \tag{2.11}$$
But (2.4) with $n = 1$ yields
$$x_2 - 2x_1 + x_0 + H^2 J_1 x_1 = 0 \tag{2.12}$$
which, compared with (2.11), forces
$$x_{p+2} = s x_2. \tag{2.13}$$
Proceeding with (2.4) for $n = p + 2$ and $n = 2$ forces $x_{p+3} = s x_3$ and so on until (2.9) is finally proved.

Theorem 2.1. *The ΔE (2.4) has a solution of the form*
$$x_n = \exp(i\nu n h) P_n(\nu), \tag{2.14}$$
where ν is a constant and $P_n(\nu)$ is periodic:
$$P_{n+p}(\nu) = P_n(\nu). \tag{2.15}$$

Proof. The ΔE (2.4) taken with the conditions (2.7), (2.8) can be represented by the matrix equation
$$G(s) x = 0, \tag{2.16}$$
where $G(s)$ is the $p \times p$ matrix

$$G(s) = \begin{pmatrix} g_0 & -1 & 0 & . & 0 & 0 & -s^{-1} \\ -1 & g_1 & -1 & . & 0 & 0 & 0 \\ 0 & -1 & g_2 & . & 0 & 0 & 0 \\ . & . & . & . & . & . & . \\ 0 & 0 & 0 & . & g_{p-3} & -1 & 0 \\ 0 & 0 & 0 & . & -1 & g_{p-2} & -1 \\ -s & 0 & 0 & . & 0 & -1 & g_{p-1} \end{pmatrix}, \tag{2.17}$$

with
$$g_n = 2 - H^2 J_n \quad n = 0, 1, \ldots, p-1, \tag{2.18}$$
$$x = (x_0, x_1, \ldots, x_{p-2}, x_{p-1})^T. \tag{2.19}$$

A non-trivial solution of (2.16) requires $\det G(s) = 0$. The Laplace expan-

sion of the determinant using the first and last rows yields
$$\det \mathbf{G}(s) = \varphi(\mathbf{g}) - s - s^{-1}, \tag{2.20}$$
where $\varphi(\mathbf{g})$ is a real function of $g_0, g_1, \ldots, g_{p-1}$ but independent of s. The *characteristic equation* equivalent to $\det \mathbf{G}(s) = 0$ is the quadratic
$$s^2 - \varphi(\mathbf{g})s + 1 = 0, \tag{2.21}$$
which has at least one root which may be written as
$$s = \exp(i\nu\pi). \tag{2.22}$$
The constant ν, in general complex, is called a *characteristic exponent*, as in the standard Floquet theory. For this value of s, the corresponding solution of (2.16) is written as
$$x_n = \exp(i\nu nh) \, P_n(\nu). \tag{2.23}$$
By the preceding lemma,
$$x_{n+p} = \exp(i\nu h) \, x_n \tag{2.24}$$
so that
$$\begin{aligned} P_{n+p}(\nu) &= \exp[-i\nu(n+p)h] x_{n+p} \\ &= \exp(-i\nu nh) x_n \\ &= P_n(\nu) \end{aligned} \tag{2.25}$$
which completes the proof of the theorem.

It will be noted that there is some ambiguity in the definition (2.22) for ν, because ν could be replaced by $\nu + 2k$ where k is an arbitrary integer. It will also be noted that the product of the roots of the characteristic equation (2.21) is unity, a result familiar in the Floquet theory.

As for the homogeneous DE (2.1), any solution of the ΔE (2.4) or equivalently of (2.16) is defined to within an arbitrary multiplicative constant. To conform with the most usual convention for Mathieu functions, the normalisation
$$x'x = p/2 \tag{2.26}$$
is adopted. This is analogous to forcing for the Mathieu functions a mean square value of $\frac{1}{2}$ over the interval $(0, 2\pi)$.

The consequences of theorem 2.1 for the ΔE (2.4) are similar to the Floquet theory of the DE (2.1). For example, the unstable and stable solutions of the ΔE are separated by $\varphi(\mathbf{g}) = \pm 2$ or, as is evident from

(2.20), by

$$\det \mathbf{G}(\pm 1) = 0. \tag{2.27}$$

For $|\varphi| < 2$, the imaginary part of ν is non-zero and $x_n \to \infty$ either as $n \to \infty$ or as $n \to -\infty$, the solution being *unstable*. On the other hand, if $|\varphi| > 2$, ν is real, x_n remains finite as $n \to \pm \infty$, and the solution is *stable*.

For $\varphi \neq \pm 2$, or $\det \mathbf{G}(\pm 1) \neq 0$, the characteristic equation has two distinct roots s and s^{-1} and hence yields two linearly independent solutions. But for $\varphi = 2$, or $\det \mathbf{G}(1) = 0$, only one root $s = 1$ is obtained and the corresponding solution has period $ph = \pi$ since

$$x_{n+p} = x_n. \tag{2.28}$$

And for $\varphi = -2$, or $\det \mathbf{G}(-1) = 0$, there is only one root $s = -1$ and the corresponding solution has period $2ph = 2\pi$ since

$$x_{n+p} = -x_n. \tag{2.29}$$

As in the Floquet theory [1], these two periodic solutions, one of period π and the other of period 2π, will be referred to as *basically-periodic solutions*.

The analogy between the continuous and discrete Floquet theories has been discussed from a somewhat different point of view by Hochstadt [2].

3. Hill's difference equation

For Hill's DE the function $J(t)$ in (2.1) not only has period π but is also an even function. In the corresponding ΔE (2.4) we take $J_{-n} = J_n$ or, equivalently,

$$J_n = J_{p-n}. \tag{3.1}$$

Without serious loss of generality and to simplify the situation, p is now taken to be an *even* integer, and the integer r defined by

$$r = p/2. \tag{3.2}$$

Because $g_n = g_{n-p}$ follows from (3.1), the matrices $\mathbf{G}(\pm 1)$ corresponding to the basically-periodic solutions display a symmetry which can be exploited to reduce them to direct sums. For $\mathbf{G}(1)$ we introduce the symmetric orthogonal transformation

$$x_0 = u_0, \tag{3.3}$$

$$x_n = 2^{-1/2}(u_n + u_{p-n}) \quad n = 1, \ldots, r-1, \tag{3.4}$$

$$x_r = u_r, \tag{3.5}$$
$$x_n = 2^{-1/2}(u_n - u_{p-n}) \quad n = r+1, \ldots, p-1, \tag{3.6}$$

and for $G(-1)$ the symmetric orthogonal transformation

$$x_0 = v_0, \tag{3.7}$$
$$x_n = 2^{-1/2}(v_n - v_{p-n}) \quad n = 1, \ldots, r-1, \tag{3.8}$$
$$x_r = v_r, \tag{3.9}$$
$$x_n = -2^{-1/2}(v_n + v_{p-n}) \quad n = r+1, \ldots, p-1. \tag{3.10}$$

The consequent similarity transformations reduce the matrices to the following direct sums:

$$G(1) \sim C_2 + S_2, \tag{3.11}$$
$$G(-1) \sim C_1 + S_1, \tag{3.12}$$

where

$$C_2 = \begin{pmatrix} g_0 & -\sqrt{2} & 0 & \cdot & 0 & 0 & 0 \\ -\sqrt{2} & g_1 & -1 & \cdot & 0 & 0 & 0 \\ 0 & -1 & g_2 & \cdot & 0 & 0 & 0 \\ \cdot & \cdot & \cdot & \cdot & \cdot & \cdot & \cdot \\ 0 & 0 & 0 & \cdot & g_{r-2} & -1 & 0 \\ 0 & 0 & 0 & \cdot & -1 & g_{r-1} & -\sqrt{2} \\ 0 & 0 & 0 & \cdot & 0 & -\sqrt{2} & g_r \end{pmatrix}, \tag{3.13}$$

$$S_2 = \begin{pmatrix} g_{r-1} & -1 & 0 & \cdot & 0 & 0 & 0 \\ -1 & g_{r-2} & -1 & \cdot & 0 & 0 & 0 \\ 0 & -1 & g_{r-3} & \cdot & 0 & 0 & 0 \\ \cdot & \cdot & \cdot & \cdot & \cdot & \cdot & \cdot \\ 0 & 0 & 0 & \cdot & g_3 & -1 & 0 \\ 0 & 0 & 0 & \cdot & -1 & g_2 & -1 \\ 0 & 0 & 0 & \cdot & 0 & -1 & g_1 \end{pmatrix}, \tag{3.14}$$

$$C_1 = \begin{pmatrix} g_0 & -\sqrt{2} & 0 & \cdot & 0 & 0 & 0 \\ -\sqrt{2} & g_1 & -1 & \cdot & 0 & 0 & 0 \\ 0 & -1 & g_2 & \cdot & 0 & 0 & 0 \\ \cdot & \cdot & \cdot & \cdot & \cdot & \cdot & \cdot \\ 0 & 0 & 0 & \cdot & g_{r-3} & -1 & 0 \\ 0 & 0 & 0 & \cdot & -1 & g_{r-2} & -1 \\ 0 & 0 & 0 & \cdot & 0 & -1 & g_{r-1} \end{pmatrix}, \tag{3.15}$$

Table 1
Distinguishing properties of basic matrices and solution vectors.

Basic matrix	Solution vector	Period	Symmetry
C_2	ce even	π	even, symmetric about $\pi/2$
S_2	se even	π	odd, antisymmetric about $\pi/2$
C_1	ce odd	2π	even, antisymmetric about $\pi/2$
S_1	se odd	2π	odd, symmetric about $\pi/2$

$$S_1 = \begin{pmatrix} g_r & \sqrt{2} & 0 & . & 0 & 0 & 0 \\ \sqrt{2} & g_{r-1} & -1 & . & 0 & 0 & 0 \\ 0 & -1 & g_{r-2} & . & 0 & 0 & 0 \\ . & . & . & . & . & . & . \\ 0 & 0 & 0 & . & g_3 & -1 & 0 \\ 0 & 0 & 0 & . & -1 & g_2 & -1 \\ 0 & 0 & 0 & . & 0 & -1 & g_1 \end{pmatrix} \qquad (3.16)$$

These four matrices will be called *basic* matrices and it is to be noted that C_2 is of order $r+1$, S_2 of order $r-1$, and C_1 and S_1 each of order r.

Corresponding to these four basic matrices are four types of periodic solution vectors which, in anticipation of the notation used in the theory of Mathieu functions, will be denoted by *ce* and *se*; their distinguishing properties are listed in table 1. The classification of the solutions is the same as that for the Hill's DE. The listed properties of the solutions for the ΔE are simple to derive. Thus for det $C_2 = 0$ or det $S_2 = 0$, det $G(1) = 0$ and the solutions have period π. For det $C_1 = 0$ or det $S_1 = 0$, det $G(-1) = 0$ so that solutions have period 2π.

For det $C_2 = 0$, the non-trivial solution satisfying $C_2 u_1 = 0$ may be denoted

$$u_1 = (u_0, u_1, \ldots u_{r-1}, u_r)^T. \qquad (3.17)$$

Then the corresponding basically-periodic solution of the ΔE is obtained from the transformation (3.3)–(3.6):

$$x = (u_0, 2^{-1/2}u_1, \ldots, 2^{-1/2}u_{r-1}, u_r, 2^{-1/2}u_{r-1}, \ldots, 2^{-1/2}u_1)^T, \qquad (3.18)$$

and since it is of period $ph = \pi$, it is an even function and symmetric about $rh = \pi/2$. It is classified as being of type *ce* even.

For det $S_2 = 0$, if the non-trivial solution satisfying $S_2 u_2 = 0$ is written as

$$u_2 = (u_{r+1}, u_{r+2}, \ldots, u_{p-1})^T, \tag{3.19}$$

then the corresponding *se* even solution, period π, of the ΔE is

$$x = \left(0, 2^{-1/2} u_{p-1}, \ldots, 2^{-1/2} u_{r+1}, 0, -2^{-1/2} u_{r+1}, \ldots, -2^{-1/2} u_{p-1}\right)^T, \tag{3.20}$$

which is an odd function and antisymmetric about $\pi/2$.

For det $C_1 = 0$, if the non-trivial solution satisfying $C_1 v_1 = 0$ is

$$v_1 = (v_0, v_1, \ldots, v_{r-2}, v_{r-1})^T \tag{3.21}$$

then the corresponding *ce* odd solution, period 2π, of the ΔE is

$$x = \left(v_0, 2^{-1/2} v_1, \ldots, 2^{-1/2} v_{r-1}, 0, -2^{-1/2} v_{r-1}, \ldots, -2^{-1/2} v_1\right)^T \tag{3.22}$$

which is an even function antisymmetric about $\pi/2$.

Finally, if for det $S_1 = 0$ the non-trivial solution satisfying $S_1 v_2 = 0$ is

$$v_2 = (v_r, v_{r+1}, \ldots, v_{p-2}, v_{p-1})^T \tag{3.23}$$

then the corresponding *se* odd solution, period 2π, of the ΔE is

$$x = \left(0, -2^{-1/2} v_{p-1}, \ldots, -2^{-1/2} v_{r+1}, v_r, \right.$$
$$\left. -2^{-1/2} v_{r+1}, \ldots, -2^{-1/2} v_{p-1}\right)^T. \tag{3.24}$$

That the solutions representing the basic types are mutually orthogonal over the interval $2ph = 2\pi$ is evident from the fact that over this interval the solution vectors are respectively

$$(x^T, x^T)^T, \quad (x^T, x^T)^T, \quad (x^T, -x^T)^T, \quad (x^T, -x^T)^T \tag{3.25}$$

where x is given by (3.18), (3.20), (3.22) and (3.24) respectively. The correct normalisation (2.26) is achieved by taking

$$u_1' u_1 = u_2' u_2 = v_1' v_1 = v_2' v_2 = p/2 = r. \tag{3.26}$$

Thus it is seen that the general features of the basically-periodic solutions of the Hill's ΔE are precisely the same as for the Hill's DE.

4. Evaluation of the basic determinants

Essential to the derivation of the basically-periodic solutions of the Hill's ΔE is the evaluation of the determinants of the basic matrices C_2, S_2, C_1, S_1 given by (3.13)–(3.16); equating these determinants to zero gives the conditions required for the special solutions. The basic determinants, as they will be called, can all be expressed in terms of

$$D_n(g_1, g_2, \ldots, g_{n-1})$$

$$= \det \begin{pmatrix} g_{n-1} & -1 & 0 & . & 0 & 0 & 0 \\ -1 & g_{n-2} & -1 & . & 0 & 0 & 0 \\ 0 & -1 & g_{n-3} & . & 0 & 0 & 0 \\ . & . & . & . & . & . & . \\ 0 & 0 & 0 & . & g_3 & -1 & 0 \\ 0 & 0 & 0 & . & -1 & g_2 & -1 \\ 0 & 0 & 0 & . & 0 & -1 & g_1 \end{pmatrix} \quad (4.1)$$

defined for $n \geq 2$ and with $D_1 = 1$.

The expressions for the basic determinants, valid for $r \geq 3$ or $p \geq 6$ are:

$$\det C_2 = g_0 g_r D_r(g_1, \ldots, g_{r-1}) - 2g_0 D_{r-1}(g_1, \ldots, g_{r-2})$$
$$- 2g_r D_{r-1}(g_2, \ldots, g_{r-1}) + 4D_{r-2}(g_2, \ldots, g_{r-2}), \quad (4.2)$$

$$\det S_2 = D_r(g_1, \ldots, g_{r-1}), \quad (4.3)$$

$$\det C_1 = g_0 D_r(g_1, \ldots, g_{r-1}) - 2D_{r-1}(g_2, \ldots, g_{r-1}), \quad (4.4)$$

$$\det S_1 = g_r D_r(g_1, \ldots, g_{r-1}) - 2D_{r-1}(g_1, \ldots, g_{r-2}). \quad (4.5)$$

It is interesting to note that the determinant D_n is itself a solution of the Hill's ΔE. Thus for $n \geq 2$

$$D_{n+1}(g_1, \ldots, g_n) = g_n D_n(g_1, \ldots, g_{n-1}) - D_{n-1}(g_1, \ldots, g_{n-2}). \quad (4.6)$$

Since $D_2(g_1) = g_1$ and $D_1 = 1$, this recurrence relation is valid for $n = 1$ if it is assumed that $D_0 = 0$. From (2.18) follows the difference equation in the form

$$(D_{n+1} - 2D_n + D_{n-1})/H^2 + J_n D_n = 0 \quad n \geq 1, \quad (4.7)$$

with $D_0 = 0$ and $D_1 = 1$.

An immediate and interesting consequence is that for the *se* even functions, $\det S_2 = D_r = 0$. Thus D_n is a solution of the Hill's ΔE (4.7) such that $D_0 = D_r = 0$—and hence is itself a multiple of an *se* even function.

5. Mathieu's difference equation

We may now turn to a consideration of Mathieu's DE (1.1), which is the most important example of the Hill's DE; the function $J(t) = a - 2q \cos 2t$ is indeed an even function with period π.

For the corresponding ΔE (2.4) we take

$$(x_{n+1} - 2x_n + x_{n-1})/H^2 + (a - 2q \cos 2nh)x_n = 0 \tag{5.1}$$

and so by (2.18)

$$g_n = 2 - H^2(a - 2q \cos 2nh). \tag{5.2}$$

The discrete Floquet theory outlined above applies to the ΔE (5.1) and in particular the basically-periodic solutions are of the four types designated *ce* even, *se* even, *ce* odd, *se* odd (table 1). For the respective solutions it is necessary that the basic determinants should be zero.

A central problem in the theory of Mathieu functions is the determination of the relations between the parameters a and q which characterise the existence of the basically-periodic solutions; these in turn demarcate the stability and instability regions in the q, a plane [3]. For the ΔE (5.1) these relations are expressed as det $C_2 = 0$, for example, and are complicated expressions involving the parameters a and q, the step size h or equivalently $p = \pi/h$, and also the function H.

6. Choice of H

So far the only restriction on H is that expressed by (2.5) which simply ensures that in the limit $h \to 0$, the ΔE (5.1) $\to$ the DE (1.1). We now choose for H the function given by (1.4) so that when $q = 0$ the DE and ΔE have the same solution in the sense described in the Introduction. Indeed the solution of the ΔE

$$(x_{n+1} - 2x_n + x_{n+1})/(4a^{-1} \sin^{-2} \tfrac{1}{2} h\sqrt{a}) + ax_n = 0 \tag{6.1}$$

is

$$x_n = A \cos\sqrt{a}\, nh + B \sin\sqrt{a}\, nh.$$

For period-π solutions, $\pi\sqrt{a} = 0, 2\pi, 4\pi, \ldots$, so that

$$a = 4m^2 \quad m = 0, 1, 2, \ldots, \tag{6.2}$$

while for period-2π solutions

$$a = (2m+1)^2 \quad m = 0, 1, 2, \ldots.$$

These are the same results as for the DE (1.1) with $q = 0$ and would not be realised if the standard choice $H^2 = h^2$ were made for the ΔE.

The final chosen form for the Mathieu ΔE is accordingly

$$(x_{n+1} - 2x_n + x_{n-1})/(4a^{-1} \sin^2 \tfrac{1}{2}h\sqrt{a}) + (a - 2q \cos 2nh)x_n = 0 \tag{6.3}$$

resulting in

$$g_n = 2 \cos h\sqrt{a} + 8qa^{-1} \sin^{-2} \tfrac{1}{2}h\sqrt{a} \cos 2nh \tag{6.4}$$

with $h = \pi/p = \pi/(2r)$.

In the special case $q = 0$,

$$g_n = g = 2 \cos h\sqrt{a}, \tag{6.5}$$

which is independent of n and the basic determinants can be evaluated explicitly. Indeed D_n given by (4.1) is a well-known determinant with value

$$D_n = \sin nh\sqrt{a} \operatorname{cosec} h\sqrt{a} \quad n \geqslant 0. \tag{6.6}$$

It follows from (4.2)–(4.5) that

$$\det C_2 = -4 \sin \pi\sqrt{a}/2 \sin h\sqrt{a}, \tag{6.7}$$

$$\det S_2 = \sin \pi\sqrt{a}/2 \operatorname{cosec} h\sqrt{a}, \tag{6.8}$$

$$\det C_1 = \det S_1 = 2 \cos \pi\sqrt{a}/2. \tag{6.9}$$

It is interesting to check that D_n given by (6.6) does satisfy the ΔE (6.1), and the boundary conditions $D_0 = D_r = 0$ when $a = 4m^2$.

7. Discussion

This paper has been concerned with difference approximations to ordinary linear homogeneous second-order differential equations with periodic coefficients. For the simple differential equation (2.1) it has been shown that a discrete theory based on a difference equation approximation can be developed which has features quite similar to the standard Floquet theory. In particular, conditions for basically-periodic solutions of periods π and 2π are established, as was their role in separating the parameter regions for stable and unstable solutions.

The specialisation to the Hill's differential equation has yielded a classification of the solutions of the corresponding difference equation which exhibits the same symmetry characteristics.

The further specialisation to the Mathieu's differential equation has yielded a discrete theory of a carefully chosen difference equation which mimics closely the theory of Mathieu functions. Although closed form results for the solutions of the proposed Mathieu ΔE cannot be expected, it has been shown that the well-known subdivision of the (q, a) parameter space into stable and unstable regions for solutions of the Mathieu DE [3] is qualitatively the same for the ΔE. For $q = 0$, agreement is exact because of the particular choice of H. For small q, a typical result for the ΔE is the expression for the characteristic number a_0 (in the standard notation):

$$a_0 = \tfrac{1}{2}(h^2 \operatorname{cosec}^2 h^2) q^2 + \mathrm{O}(q^3). \tag{7.1}$$

In the limit of $h \to 0$ this gives

$$a_0 = -\tfrac{1}{2} q^2 + \mathrm{O}(q^3), \tag{7.2}$$

which is the correct result [3] for the DE.

References

[1] F.M. Arscott, Periodic Differential Equations (Pergamon Press, Oxford, 1964).
[2] H. Hochstadt, On the theory of Hill's matrices and related inverse scattering problems, Lin. Alg. & Appl. 11 (1975) 41–52.
[3] N.W. McLachlan, Theory and Application of Mathieu Functions (Clarendon Press, Oxford, 1947).
[4] R.B. Potts, Differential and difference equations, Am. Math. Monthly 89 (1982) 402–407.
[5] R.B. Potts, Best difference equation approximation to Duffing's equation, J. Aust. Math. Soc. B23 (1982) 349–356.
[6] R.B. Potts, Van der Pol difference equation, Nonlin. Anal. 7 (1983) 801–812.

CHAPTER 6

*Mayer–Montroll Equations (and Some Variants)
Through History for Fun and Profit*

G. STELL

*Departments of Mechanical Engineering and Chemistry
State University of New York
Stony Brook, New York 11794
U.S.A.*

© *Elsevier Science Publishers B.V. 1985*

*The Wonderful World of Stochastics
A Tribute to Elliott W. Montroll
Eds. M.F. Shlesinger and G.H. Weiss*

Contents

1. Introduction   129
2. Boltzmann's program   129
3. Some immediate applications and implications of Boltzmann's program and connection with other work   139
   - 3.1. Scaled particle theory   140
   - 3.2. Bounds   140
   - 3.3. Zero-separation theorems   141
   - 3.4. Potential distribution theory   142
   - 3.5. Representations for dispersions and composite media; polydispersivity   143
   - 3.6. Connectivity, the continuum Potts model, and the Widom–Rowlinson model   146
4. Generalizations and final remarks   147
   - 4.1. Generalization to arbitrary potentials   147
   - 4.2. Generalization to nonequilibrium ensembles   150
   - 4.3. Further generalizations and closing remarks   151

References   154

1. Introduction

In 1941 a landmark paper [1] by Joseph Mayer and Elliott Montroll appeared in which functions that describe the distribution of molecules were precisely characterized in two ways. One characterization was in terms of fugacity and density expansions, greatly extending earlier work of Mayer and Harrison [2] on such expansions for the pressure of a system. The other characterization was in terms of a set of integral equations that have come to be called the Mayer–Montroll equations. In this lecture my purpose will be to show how these equations and their close variants, especially the Kirkwood–Salsburg equations [3], represent a development that was initiated by Boltzmann and runs through statistical mechanics in a way that is not generally recognized. Here I draw very heavily on a 1966 report of mine [4]. I shall then note how extensions and new applications of such equations continue to yield interesting results, and promise to play an important role in future work.

2. Boltzmann's program

A natural place to begin in order to gain a physical understanding of the significance of the Mayer–Montroll [1] and the Kirkwood–Salsburg [3,5] hierarchies and their relation to other approaches is with Boltzmann's representations for a hard-sphere system of the two-particle probability distribution function $g_2(r_1, r_2)$ and of the quantity that Boltzmann calls the available space, A, which for a hard-sphere system at equilibrium is $V\rho/z$, where V is the volume of the container of the system, ρ is its number density, and z its thermodynamic activity, normalized so that $\rho/z \to 1$ as $\rho \to 0$.

Boltzmann considers in some detail three different forms of the equation of state of a three-dimensional hard-sphere fluid (*Gastheorie* [6] *II*: Sections

2–5, Sections 49–52, Section 61), all of which involve this notion of available space. His primary interest in $g_2(r_1, r_2)$ *per se* lies in its "contact value" $G(\sigma)$ which appears in the equation that follows from the virial theorem of Clausius:

$$\beta P/\rho = 1 + b\rho G(\sigma). \tag{2.1}$$

Here $\beta = 1/kT$ where k is Boltzmann's constant, and T the temperature. The number density ρ is $\lim_{N \to \infty} = N/V$ where N is the number of particles in a vessel whose volume V must go to infinity as N does, in order for ρ to remain finite. Boltzmann is almost always interested in systems in this limit. If the particles have diameter σ, then b, which is four times the hard-sphere volume, is given by $[2\pi/3]\sigma^3$. $G(\sigma)$ is the value of the two-particle ("radial") distribution function $g_2(r_1, r_2)$ for $r_{12} = \sigma$ where $r_{12} = |r_1 - r_2|$. Boltzmann defines $g_2(r_1, r_2)$ just as we do today. In his *Gastheorie II* (Section 51) he considers it for the special case of $r_{12} = \sigma$, but in a subsequent 1899 paper [7] he touches on the more general problem of finding $g_2(r_1, r_2)$ for any r_{12}, also written $r_{1,2}$ as appropriate.

For a hard-sphere system, Boltzmann has a beautiful way of representing both the available space A (i.e., N/z) and $g_2(r_1, r_2)$. Before discussing Boltzmann's representations it is convenient to first introduce the following notation. Let $m(r; \sigma)$ be the step function

$$m(r; \sigma) = \begin{cases} 1 & \text{for} \quad r < \sigma, \\ 0 & \text{for} \quad r > \sigma. \end{cases} \tag{2.2}$$

This is just $-f(r)$, where $f(r)$ is the Mayer f-function. The volume of the interaction sphere or covering sphere that is unavailable to the center of a specified particle because of the impenetrability of any other particle is then just

$$\int m(r; \sigma) \mathrm{d}r = \tfrac{4}{3}\pi\sigma^3, \tag{2.3}$$

which is 8 times the volume of a single particle of diameter σ and twice the value of b, introduced in eq. (2.1). We note that

$$\frac{\partial m(r; \sigma)}{\partial r} = -\delta(r - \sigma), \tag{2.4}$$

$$\frac{\partial m(r; \sigma)}{\partial \sigma} = +\delta(\sigma - r). \tag{2.5}$$

It is also useful to have symbols for the volume of intersection (the "overlap volume") of n covering spheres and the surface area of this volume. We shall call them O_n and S_n respectively. For n spheres centered at $r_2, r_3, \ldots, r_{n+1}$, O_n is a function of σ and the r_i and can be written as a volume integral

$$O_n(\sigma; r_2, \ldots, r_{n+1}) = \int dr_1 \, m(r_{1,2}; \sigma) \, m(r_{1,3}; \sigma) \cdots m(r_{1,n+1}; \sigma). \tag{2.6}$$

The surface area S_n is related to O_n by the equation

$$S_n = \partial O_n / \partial \sigma \tag{2.7}$$

and we have

$$S_n(\sigma; r_2, \ldots, r_{n+1}) = \sum_{j=2}^{n+1} S_{n,j}, \tag{2.8}$$

where $S_{n,j}$ represents the piece of the overlap surface S_n that is also part of the surface of the jth covering sphere (all δ's are three-dimensional)

$$S_{n,j} = \int dr_1 m(r_{1,2}; \sigma) \cdots m(r_{1,j-1}; \sigma) \, \delta(r_{1,j} = \sigma)$$
$$\times m(r_{1,j+1}; \sigma) \cdots m(r_{1,n+1}; \sigma). \tag{2.9}$$

The above expressions (2.3) to (2.9) give us a precise way of analytically expressing equations involving simple geometric notions associated with overlapping spheres. (By and large, Boltzmann himself did not bother writing such equations analytically. Instead he gave them in words, and typically only wrote down in symbols such expressions through the order of approximation he actually evaluated.) We shall also need symbols for certain probabilities associated with subsets of particles in a system of N particles. We shall define $P_n(r_1, r_2, \ldots, r_n)$ by the statement that $P_n \, dr_1 \cdots dr_n$ is the probability that particle 1 is in dr_1, particle 2 is in dr_2, etc. If we let

$$\rho_n(r_1, \ldots, r_n) dr_1 \cdots dr_n = \rho^n g_n(r_1, \ldots, r_n) dr_1 \cdots dr_n \tag{2.10}$$

be the probability that exactly one (unspecified) particle is in dr_1, exactly one other (unspecified) particle is in dr_2, etc., then it is not hard to see that

$$\rho_n(r_1, \ldots, r_n) = \rho^n g_n = N(N-1) \cdots (N-n+1) P_n. \tag{2.11}$$

We are now in a notational position to transcribe Boltzmann's representation of $g_2(r_1, r_2)$ into precise analytic terms. He starts by considering (*Gastheorie II*, Sections 51, 59) the "space available for the center of a specified molecule". Let us call this space A and suppose that the specified molecule is the $(N+1)$st molecule that we contemplate putting into a vessel in which N molecules are already present. Neglecting the boundary effects at the walls of the vessel, A will be V minus at most the volume taken up by the covering spheres of the other N molecules, which is

$$N\int m(r)dr = \tfrac{4}{3}\pi\sigma^3 N. \tag{2.12}$$

[In passing we note that A is V minus at *least* the volume taken up by the N hard spheres themselves, which is

$$\tfrac{4}{3}\pi(\tfrac{1}{2}\sigma)^3 N = \tfrac{1}{6}\pi\sigma^3 N. \tag{2.13}$$

Again we are neglecting wall effects.] However, there is expected overlapping of spheres that we must consider. That is, we must subtract the expected overlap volume between all (unordered) pairs of covering spheres [there are $\tfrac{1}{2}(N(N-1))$ of them]:

$$\tfrac{1}{2}N(N-1)\int O_2(\sigma; r_2, r_3)P_2(r_2, r_3)dr_2\, dr_3. \tag{2.14}$$

But now we have overestimated this overlapping because we have overcounted the overlap whenever three or more covering spheres happen to simultaneously overlap. We can continue this line of reasoning, and if we are only interested in the limit $N \to \infty$, $N/V = \rho$ finite, we can continue to neglect wall effects. [In this limit, $g_1(r) = 1$ if there are no external fields present. Boltzmann always restricts his attention to such a uniform system and we shall do likewise in this section.] We then arrive at the expression for A:

$$A = V - A_1 + A_2 - \cdots, \tag{2.15a}$$

where

$$A_n = \frac{\rho^n}{n!}\int O_n(\sigma; r_2,\ldots,r_{n+1})g_n(r_2,\ldots,r_{n+1})dr_2\cdots dr_{n+1}, \tag{2.15b}$$

with O_n given by

$$O_n = \int dr_1\, m(r_{1,2}; \sigma)\cdots m(r_{1,n+1}; \sigma). \tag{2.15c}$$

This is Boltzmann's representation for the average space available to the $(N+1)$st sphere, which we shall denote more precisely as $A^{(N+1)}$ for the purposes of the discussion that immediately follows. (By "average" Boltzmann has in mind the canonical-ensemble average, as he spells out in detail in *Gastheorie II*, at the beginning of Section 51.) We turn now to the thermodynamic meaning of $A^{(N+1)}$. This follows from Boltzmann's identification (*Gastheorie II*, Section 61) of $A^{(N+1)}$ as the ratio of the configuration integrals for $N+1$ and N particles, respectively:

$$A^{(N+1)} = Q_{N+1}/Q_N. \tag{2.16}$$

Thus we can denote $A^{(N)}$ as N/z or $V\rho/z$. We note that the identification of z/ρ and V/A brings out in a particularly satisfactory way the meaning of z as an "effective density" for the hard-sphere system. To obtain thermodynamic quantities from (2.16), Boltzmann rewrote it as

$$\ln Q_N = \sum_{\nu=0}^{N-1} \ln A^{(\nu+1)}. \tag{2.17}$$

For his hard-sphere system, he identified $\ln A_N$ with the configurational entropy and used as his bridge between the $A^{(\nu+1)}$ and thermodynamics the equation

$$S_{(N)}^{\text{CONF}} = \sum_{\nu=0}^{N-1} \ln A^{(\nu+1)} \tag{2.18}$$

from which he was able to obtain the exact values of the first four virial coefficients of a hard-sphere fluid. In order to do this, Boltzmann had to compute the $g_2(r_1, r_2)$ appearing in (2.1b) through first order in ρ and the $g_3(r_1, r_2, r_3)$ through zeroth order. His treatment of $g_2(r_1, r_2)$ was a straightforward extension of his somewhat simpler treatment of $G(\sigma)$, which we shall give first.

In order to obtain $G(\sigma)$ Boltzmann points out that $\rho G(\boldsymbol{\sigma}) d\boldsymbol{\sigma}$ is a ratio of two volumes (where we use bold $\boldsymbol{\sigma}$ in a volume element or vector)

$$\rho G(\boldsymbol{\sigma}) d\boldsymbol{\sigma} = F/A, \tag{2.19}$$

where F, which Boltzmann calls the favorable space, is defined (*Gastheorie II*, Section 51) as the "space in which the center (of the specified molecule) must be found in order that its distance from the center of one of the remaining molecules may lie between σ and $\sigma + d\sigma$." $F/d\sigma$ is thus the expected surface area associated with the expected volume A. Boltzmann derives an expression for F by means of the same sort of argument he uses to obtain A. The only difference is that instead of looking at the O_n themselves, that is, at the intersection volume of covering spheres involving expected encounters between the specified molecule and single molecules, pairs of molecules, etc., we must now look at the functions $d\sigma (dO_n/d\sigma) =$

dσ S_n, which represent shells of thickness dσ associated with the surfaces of the O_n. Thus, neglecting wall terms of order $1/N$, the favorable space is at most

$$N \, d\sigma \frac{d}{d\sigma}\left[\int m(\mathbf{r};\sigma)d\mathbf{r}\right] = N \, d\sigma \frac{dO_1}{d\sigma} = N \, d\sigma \, S_1 = 4\pi\sigma^2 \, d\sigma \, N = d\sigma \, N. \tag{2.20}$$

However, we have ignored here the expected overlapping of covering spheres and therefore we must subtract the expected surface shell of the intersection volume O_2. This expected surface shell is obtained by multiplying the actual surface shell dσ $S_2(\sigma; \mathbf{r}_2, \mathbf{r}_3)$ of a given overlap configuration by the probability $P_2(\mathbf{r}_2, \mathbf{r}_3)$ associated with this configuration and integrating over all configurations. Thus we must subtract from the $4\pi\sigma^2$ dσ N of eq. (2.20) the expression

$$\tfrac{1}{2}N(N-1)\int S_2(\sigma; \mathbf{r}_2, \mathbf{r}_3) \, P_2(\mathbf{r}_2, \mathbf{r}_3) \, d\mathbf{r}_2 \, d\mathbf{r}_3 \, d\sigma. \tag{2.21}$$

Using (2.8), (2.9), and (2.11) this can be written as one half the sum of two identical terms that differ notationally only in the names of the dummy variables of integration. One of them is

$$d\sigma \, \rho^2 \int \delta(r_{1,2}=\sigma) \, g_2(\mathbf{r}_2, \mathbf{r}_3) \, m(\mathbf{r}_{1,3}; \sigma) \, d\mathbf{r}_1 \, d\mathbf{r}_2 \, d\mathbf{r}_3, \tag{2.22}$$

or

$$4\pi\sigma^2 \, d\sigma \, \rho^2 V \int m(\mathbf{r}_{1,3}; \sigma) \, g_2(\mathbf{r}_2, \mathbf{r}_3) d\mathbf{r}_3 \big|_{r_{1,2}=\sigma}. \tag{2.23}$$

We have subtracted too much, however, since we must always expect to find some overlapping of 3, 4, or more covering spheres, which we have not considered. We are thus led to the expression

$$F = F_1 - F_2 + F_3 - \cdots, \tag{2.24}$$

where

$$F_n = \frac{\rho^n}{n!} \int S_n(\sigma; \mathbf{r}_2, \ldots, \mathbf{r}_{n+1}) \, g_n(\mathbf{r}_2, \ldots, \mathbf{r}_{n+1}) \, d\mathbf{r}_2 \ldots d\mathbf{r}_{n+1} d\sigma \tag{2.25}$$

or

$$F_n = 4\pi\sigma^2 \, d\sigma \frac{N\rho^{n-1}}{(n-1)!} \int m(\mathbf{r}_{1,3}; \sigma) \cdots m(\mathbf{r}_{1,n+1}; \sigma)$$
$$\times g_n(\mathbf{r}_2, \ldots, \mathbf{r}_{n+1}) d\mathbf{r}_3 \ldots d\mathbf{r}_{n+1}\big|_{r_{1,2}=\sigma}. \tag{2.26}$$

To obtain his more general result for $g_2(r_1, r_2)$, Boltzmann noted [7] that $\rho d\mathbf{r}_{1,2} g_2(r_1, r_2)$ would be a ratio of two volumes, $F(r_{1,2})/A$, just as it is in the special case $r_{1,2} = \sigma$, and he obtained $g_2(r_1, r_2)$ by finding $F(r_{1,2})$ and A both through the first order in ρ. For $r_{1,2} \leq \sigma$, $g_2(r_1, r_2)$ is clearly zero; for $r_{1,2} \geq \sigma$, Boltzmann finds

$$\rho d\mathbf{r}_{1,2} g_2(r_1, r_2) = \frac{4\pi N r_{1,2}^2 dr_{1,2}\left[1 - 2\rho b + \rho O_2(\sigma, r_{1,2}) + \cdots\right]}{V[1 - 2\rho b + \cdots]}. \tag{2.27}$$

Boltzmann does not go on to discuss the higher order terms in $F(r_{1,2})$ but there is no essential difference between finding them and finding the expansion for F. Again we can write a series (denoting the argument of F as x for convenience)

$$F(x) = F_1(x) - F_2(x) + F_3(x) - \cdots, \tag{2.28}$$

but in evaluating $F_n(x)$ we must consider not n covering shells of radius σ but rather $n - 1$ shells of radius σ and one shell of radius x, corresponding to the fact that we are interested in knowing about the expectation of finding particles of diameter σ a distance x away from a specified particle (labeled 1 here) rather than a distance σ away.

Thus instead of (2.8) and (2.9) we deal with

$$S_n(x; \mathbf{r}_2, \ldots, \mathbf{r}_{n+1}) = \sum_{j=2}^{n+1} S_{n,j}(x), \tag{2.29}$$

where

$$S_{n,j}(x) = \frac{\partial O_{n,j}(x; \mathbf{r}_2, \ldots, \mathbf{r}_{n+1})}{\partial x} \tag{2.30}$$

and

$$O_{n,j}(x; \mathbf{r}_2, \ldots, \mathbf{r}_{n+1}) = \int d\mathbf{r}_1\, m(r_{1,2}; \sigma) \cdots m(r_{1,j-1}; \sigma)$$
$$\times m(r_{1,j}; x)\, m(r_{1,j+1}; \sigma) \cdots m(r_{1,n+1}; \sigma),$$

or

$$S_{n,j}(x) = \int d\mathbf{r}_1 m(r_{1,2}; \sigma) \cdots m(r_{1,j-1}; \sigma)\, \delta(r_{1,j} = x)$$
$$\times m(r_{1,j+1}; \sigma) \cdots m(r_{1,n+1}; \sigma). \tag{2.31}$$

Finally instead of (2.24), (2.25), and (2.26) we have (2.28) with

$$F_n(x) = \frac{\rho^n}{n!} \int S_n(x, r_2, \ldots, r_{n+1}) \, g_n(r_2, \ldots, r_{n+1}) \, dr_2 \cdots dr_{n+1} \, dx \tag{2.32}$$

or

$$F_n(r_{1,2}) = \frac{dr_{1,2} N \rho^{n-1}}{(n-1)!} \int m(r_{1,3}; \sigma) \cdots m(r_{1,n+1}; \sigma)$$
$$\times g_n(r_2, \ldots, r_{n+1}) dr_3 \cdots dr_{n+1}. \tag{2.33}$$

Now clearly $\rho g_2(r_1, r_2) \, dr_{1,2} = 0$ for $r_{1,2} < \sigma$ from its definition, and only for $r_{1,2} > \sigma$ will it equal $F(r_{1,2})/A$, which can be easily shown to be nonzero for $r_{1,2} < \sigma$ according to our prescription. Physically, however, our prescription for $F(r_{1,2})/A$ is still of interest for $r_{1,2} < \sigma$, since it defines a cavity-sphere or equivalently a cavity-cavity distribution function [which we shall denote as $y_2(r_1, r_2)$, or $y_2(r_{1,2})$ as convenience dictates], rather than the sphere-sphere function $g_2(r_1, r_2)$. The equation

$$\rho A y_2(r_1, r_2) \, dr_{1,2} = F(r_{1,2}), \tag{2.34}$$

with $F(r_{1,2})$ given by (2.28) and (2.33), proves to be the second of the Kirkwood–Salsburg [3,5] equations [eq. (2.15) is the first] for the special case of a hard-sphere fluid. For such a system, (2.15) is also the first of the Mayer–Montroll [1] equations.

To make contact with the full set of Kirkwood–Salsburg (KS) or Mayer–Montroll (MM) equations for a hard-sphere system using Boltzmann's approach we must generalize his machinery somewhat. To get the KS equations we consider $s - 1$ shells of radii $r_{1,2}, r_{1,3}, \ldots, r_{1,s}$ instead of just one shell. Extending the function $F(x)$ and its treatment in (2.28)–(2.33) to this s-shell case we arrive at

$$F_s(r_{1,2}, \ldots, r_{1,s}) = \sum_{n \geq 1} (-1)^{n-1} F_s^{(n)}(r_{1,2}, \ldots, r_{1,s}), \tag{2.35a}$$

where

$$F_s^{(n)}(r_{1,2}, \ldots, r_{1,s}) = \frac{dr_{1,2} \ldots dr_{1,s} N \rho^{n-1}}{(n-1)!}$$
$$\times \int g_{s-2+n}(r_2, \ldots, r_{s+n-1})$$
$$\times \prod_{j=s+1}^{s+n-1} \left[m(r_{1,j}; \sigma) dr_j \right]. \tag{2.35b}$$

Here F_s/A is the expected local number density $\rho y_s(r_{1,j}; r_2, \ldots, r_s)$ about a cavity of diameter σ centered at r_1 given $s-1$ hard spheres of diameter σ centered at $r_2, \ldots, r_s$, respectively. This can related to the corresponding g_{s+1} for a hard sphere at r_1 in the presence of the other spheres by the relation

$$\frac{F_s(r_{1,2},\ldots,r_{1,s})}{A} = \rho y_s(r_1; r_2, \ldots, r_s) = \frac{\rho g_s(r_1,\ldots,r_s)}{\prod_{j=2}^{s} e(r_{1,j})}, \quad (2.36a)$$

where

$$e(r_{i,j}) = 1 - m(r_{i,j}; \sigma). \quad (2.36b)$$

After some change of index notation, (2.35) becomes

$$\frac{\rho g_s(1,\ldots,s)}{z \prod_{2 \leq j \leq s} e(1, j)}$$

$$= \sum_{n \geq 0} \frac{(-1)^n \rho^n}{n!} \int g_{s-1+n}(2, \ldots, s+n) \prod_{j=s+1}^{s+n} [m(1, j; \sigma) dj], \quad (2.37)$$

where we represent arguments by their indices here and below for notational simplicity whenever appropriate. In (2.37) $g_0 \equiv 1$ and $g_1(i) = 1$. Equation (2.37) is the full Kirkwood–Salsburg hierarchy for a hard-sphere system. The KS equations for a system with interparticle potential energy of pairwise additive form, $\sum_{1 \leq i < j \leq N} \phi_2(i, j)$, but $\phi_2(i, j)$ arbitrary, have the same form as (2.37), but with $m(i, j)$ replaced everywhere by the negative of the Mayer f-function,

$$f(i, j) = e_2(i, j) - 1, \quad e_2(i, j) = \exp - \beta \phi_2(i, j). \quad (2.38a)$$

Thus in (2.37)

$$m(i, j; \sigma) \to -f(i, j), \quad e(i, j) \to e_2(i, j). \quad (2.38b)$$

In a variety of contexts (several of which we shall encounter below) it is useful to divide both sides of the KS equations given above by $g_{s-1}(2, \ldots, s)$. On the left-hand side this gives rise to the replacement of $g_s(1, \ldots, s)$ by the conditional function

$$g_s(1/2, \ldots, s) \equiv g_s(1, \ldots, s)/g_{s-1}(2, \ldots, s)$$

and on the right-hand side $g_{s-1+n}(2, \ldots, s+n)$ is replaced by the condi-

tional function
$$g_{s-1+n}(s+1,\ldots,s+n/2,\ldots,s)$$
$$= g_{s-1+n}(2,\ldots,s+n)/g_{s-1}(2,\ldots,s).$$

This gives a transparent and general meaning to the KS hierarchy as a sequence of equations for a conditional one-particle probability density given the location of an increasingly larger set of particles. This interpretation does not rest upon Boltzmann's surface argument and is applicable to an arbitrary pair potential.

To obtain the MM equations for hard spheres using Boltzmann's approach we must generalize his notion of the expected volume available to the center of a single additional particle A to that of expected volume available to n such additional particles. From the definition of A it follows that A/V (i.e., ρ/z) is the probability of finding that the centers of no particles are in a region defined by the volume of a spherical cavity of radius σ. Suppose we introduce the more general quantity $p_0[\Omega]$ where

$$p_0[\Omega] = \text{the probability of finding that the centers of no particles are in a region defined by the volume } \Omega. \qquad (2.39)$$

Now consider in particular the union volume $U_s(1,\ldots,s;\sigma)$ of s spherical cavities of radius σ centered at $r_1, r_2, \ldots, r_s$, respectively. We denote the volume simply as $U_s(\sigma)$ when no confusion can arise by doing so. The representation for A/V given by (2.15) generalizes to the representation for $p_0[U_s(0)]$ if one replaces everywhere the indicator function $m(r_{i,j};\sigma)$ for a single sphere of radius σ centered at r_1 by the indicator function that is 1 inside $U_s(\sigma)$ and 0 outside $U_s(\sigma)$. Denoting this function as $m_s(1,\ldots,s;j,\sigma)$ we have

$$p_0[U_s(\sigma)] = 1 + \sum_{n \geq 1} \frac{(-\rho)^n}{n!} \int g_n(s+1,\ldots,s+n)$$
$$\times \prod_{j=s+1}^{s+n} [m_s(1,\ldots,s;j,\sigma)\mathrm{d}j]. \qquad (2.40a)$$

It can be readily seen that m_s can be represented as

$$m_s(1,\ldots,s;j,\sigma) = 1 - \prod_{i=1}^{s}[1 - m(i,j;\sigma)]. \qquad (2.40b)$$

Finally we note from its definition that $p_0[U_s(\sigma)]$ is proportional to an s-cavity distribution function $y_s(1,\ldots,s)$, which in a hard-sphere system at equilibrium is simply related to the s-sphere distribution function g_s:

$$y_s(1,\ldots,s) = g_s(1,\ldots,s)/\prod_{1 \leq i < j \leq s} e(i,j). \qquad (2.41)$$

The proportionality constant relating $p_0[U_s(\sigma)]$ and y_s is determined by the $r_{i,j} \to \infty$ conditions $p_0[U_s(\sigma)] \to (\rho/z)^s$ and $y_s(1,\ldots,s) \to 1$ so that

$$p_0[U_s(\sigma)] = (\rho/z)^s g_s(1,\ldots,s) / \prod_{1 \leq i < j \leq s} e(i,j). \tag{2.42}$$

Equations (2.40) and (2.41) yield the MM equations for a hard-sphere system at equilibrium,

$$(\rho/z)^s g_s(1,\ldots,s) / \prod_{1 \leq i < j \leq s} e(i,j)$$

$$= 1 + \sum \frac{(-\rho)^n}{n!} \int g_n(s+1,\ldots,s+n)$$

$$\times \prod_{j=s+1}^{s+n} [m_s(1,\ldots,s; j, \sigma) \mathrm{d}j]. \tag{2.43}$$

As in the case of the KS equations one can generalize the equations to a pair potential of arbitrary form by making the replacement given in (2.38), obtaining the equations considered by Mayer and Montroll [1].

It is worth indicating here for future reference, however, that for an arbitrary pair potential, (2.40) *also* continues to hold exactly as it stands, with $p_0[U_s(\sigma)]$ well defined by (2.39) and the m_s also well defined as the indicator function of $U_s(\sigma)$. However, (2.40) in this context, although of MM structure, no longer represents a formally closed set of equations for the equilibrium g_s. Instead it yields the $p_0[U_s(\sigma)]$ induced by a particular set of g_s associated with a system at density ρ, whether the system is in equilibrium or not. In the case of the KS hierarchy similar remarks hold. The geometric surface arguments leading to (2.35) continue to apply for particles with arbitrary potentials if they are regarded as arguments applied to cavity surfaces. In this context, however, closure is lost, since the left-hand side of (2.35) then no longer can be related to g_s as in (2.36). However, the left-hand side then has a meaning, even in a nonequilibrium situation, which can be translated into the language of the matrix functions which we shall discuss in section 3.5.

3. Some immediate applications and implications of Boltzmann's program and connection with other work

In this section we make a sequence of observations that enable one to further utilize Boltzmann's picture and also facilitate contact with more recent work.

3.1. Scaled particle theory

We can generalize the notion of available space A given by (2.15) by supposing that the $(N+1)$st particle labeled 1 has radius σ_1 while the N spheres already in the system have radius σ_0. Then the interaction sphere involving 1 and any other particle will have radius $\sigma_1 + \sigma_0$. If we let

$$\sigma_1 + \sigma_0 = \sigma \tag{3.1}$$

then the space available to the center of particle 1 continues to be given by (2.15), except that the g_n are now independent of σ (depending instead upon σ_0) and A and $G(\sigma)$ depend upon σ only through particle 1.

In terms of the $p_0[\Omega]$ introduced in (2.39) this generalized A continues to be given by the same relation

$$A/V = p_0[U_1], \quad U_1 = \text{the volume of a sphere or cavity of radius } \sigma. \tag{3.2}$$

Upon comparing (2.15b) and (2.25) we find that (2.7) now implies

$$-F = d\sigma \frac{\partial A}{\partial \sigma} \quad \text{or} \quad -F/A = d\sigma \frac{\partial \ln(A/V)}{\partial \sigma}. \tag{3.3}$$

Thus from (2.19) we have an interesting equation that elegantly relates $G(\sigma)$ and A:

$$4\pi\rho\sigma^2 G(\sigma) = -\frac{\partial \ln(A/V)}{\partial \sigma} = -\frac{\partial \ln p_0[U_1]}{\partial \sigma}. \tag{3.4}$$

The above generalization of A/V was made in a 1959 paper by Reiss, Frisch and Lebowitz [8], who rediscovered Boltzmann's physical picture of ρ/z in a hard-sphere fluid and went on to use (3.2) and (3.4) as the starting points of what has proved to be an extremely fruitful and influential development of an approximate thermodynamic theory of a hard-sphere fluid, the "scaled-particle" theory.

We see from its derivation that (3.3) remains valid whether or not the N particles of the fluid are regarded as mutually impenetrable, as long as they are impenetrable spheres with respect to their interaction with the "test particle" labeled 1.

3.2. Bounds

It is geometrically clear that the partial sums of eqs. (2.15), (2.24), (2.28) and (2,35a) represent sequences of alternating upper and lower bounds which we can use to get families of bounds on ρ/z, $G(\sigma)$ [as well as pressure through (2.1)] and the $F_s(r_{1,2}, \ldots, r_{1,s})$. In expressing these results it is convenient to first rewrite (2.15a) as

$$A/V = \rho/z = 1 - R_1 + R_2 + \cdots, \quad R_i = A_i/V. \tag{3.5}$$

We find then, for example, from (2.19) and (2.1) and the bounding properties of our partial sums,

$$G(\sigma) \leq (1 - R_1)^{-1}, \quad \beta p/\rho \leq (1 - \rho b)(1 - 2\rho b)^{-1},$$
$$G(\sigma) \geq 1 - (F_2/F_1) \geq (1 - \tfrac{27}{16} R_1)(1 - R_1)^{-1},$$
$$\beta p/\rho \geq (1 - \rho b - \tfrac{27}{16} \rho^2 b^2)(1 - 2\rho b)^{-1}, \quad \text{etc.} \tag{3.6}$$

Similarly

$$y_2(x) \leq (1 - R_1)^{-1},$$
$$y_2(x) \geq [1 - 2R_1 + \rho O_2(x)](1 - R_1)^{-1}, \quad \text{etc.} \tag{3.7}$$

The obvious bound $\rho/z \geq 1 - \rho v_c$, where v_c is the average volume per particle at close packing, helps us to improve some of these results a bit. For example, it implies

$$G(\sigma) \geq \frac{1 - (F_2/F_1)}{1 - \rho v_c} \geq \frac{1 - \tfrac{27}{16} R_1}{(1 - R_1)(1 - \rho v_c)}. \tag{3.8}$$

Bounds similar to (but generally weaker than) (3.6) and (3.7) can also be derived from (2.43).

The bounding properties of the MM and KS partial sums for hard-sphere systems (and more generally for non-negative pair potentials) appear to have been first noticed and systematically exploited in the early 1960s, when some of the above bounds were first derived by various authors. We refer to the work of Lebowitz and Percus [9] for extensive references and further closely related bounds. For a more complete listing of various hard-sphere bounds that follow from the KS equations, see ref. [4].

3.3. Zero-separation theorems

For hard spheres, the terms in (2.37) for $n \geq 12$ are identically zero because $m(r) = 0$ for $r > \sigma$ while $g_n(r_2, \ldots, r_{n+1}) = 0$ for $r_{i,j} < \sigma$, $2 \leq i < j \leq n + 1$. (Hence there are no convergence problems.) For $r_{i,j} = 0$, only the first term, which is $g_{s-1}(r_2, \ldots, r_s)$, is nonzero. Thus in terms of the $y_s(r_1, r_2, \ldots, r_s)$ of

eq. (2.36a):
$$y_2(r_1; r_1) = y_2(r_1, r_1) = z/\rho, \tag{3.9}$$
$$(\rho/z) y_3(r_1; r_2, r_3)|_{r_{1,2}=r_{1,3}=0} = g_2(r_2, r_3) \tag{3.10}$$
etc.

If (2.37) is divided by $\prod_{2 \leq i < j \leq s} e(r_{i,j})$ then one has, comparing (2.36a) and (2.41), a KS equation for the function $y_s(r_1, \ldots, r_s)$, the full s-cavity function. Evaluating this equation for $r_{i,j} = 0$, $1 \leq i < j \leq s$, yields (by recursion) perhaps the most striking "zero-separation" result for hard spheres
$$y_s(r_1, \ldots, r_1) = (z/\rho)^{s-1} \tag{3.11a}$$
as well as results such as
$$y_3(r_1, r_1, r_3) = y_2(r_1, r_3). \tag{3.11b}$$
These were both first given by the author [10]. Results of this type were also subsequently rederived by Meeron and Siegert and by Henderson and Grundke [11]. For $s = 1$, (3.11a) reduces to (3.9) which was first obtained using a different method by Hoover and Poirier [12]. Various generalizations of such zero-separation results have subsequently been considered by a number of authors [13]. They represent a class of exact expressions that can be used to test various assumptions and approximations. (For example the Kirkwood superposition approximation
$$g_3(r_1, r_2, r_3) \approx g_2(r_{1,2}) g_2(r_{1,3}) g_2(r_{2,3})$$
is associated with the cavity function result
$$y_3(r_1, r_1, r_1) \approx y_2(r_1, r_1)^3,$$
yielding via (3.9) $y_3(r_1, r_1, r_1) \approx (z/\rho)^3$ instead of the exact $(z/\rho)^2$. The use of such exact relations between y_3 and y_2 to test and improve the superposition approximation is discussed in detail in Section 12 of ref. [10].)

3.4. Potential distribution theory

In 1963, Widom [14] noted that in a uniform system the quantity ρ/z can be characterized as the expected value $\langle \exp - \beta \psi \rangle$, where $\psi(1/2, \ldots, N+1)$ is the difference between the potential energy $W_{N+1}(1, \ldots, N+1)$ of the system in which particle 1 has been inserted and the potential energy $W_N(2, \ldots, N+1)$ of the system before such insertion. Similarly he observed that $(\rho/z)^2 g_2(1, 2)/e_2(1, 2)$ [$e_2(1, 2)$ given by (2.38a)] can be characterized as the expected value $\langle \exp - \beta[\psi(1/3,\ldots,N+2)+\psi(2/3,\ldots,N+2)]\rangle$, where $\psi(1/3,\ldots,N+2)+\psi(2/3,\ldots,N+2)$ is the additional potential energy introduced into the system of particles $3,\ldots,N+2$ upon insertion of particles 1 and 2. These characterizations are the natural extensions of Boltzmann's geometric characterization of ρ/z when one is dealing with particles that are not necessarily hard spheres. Conversely the first two MM equations can be regarded as giving a precise and powerful representation of these two expected values, which form the basis of an approach to liquid-state and lattice-gas theory developed by Widom that has come to be called potential distribution theory [14,15]. The theory can be immediately extended to encompass the full set of functions described by the MM hierarchy by observing that

$$(\rho/z)^s y_s(1,\ldots,s) = \langle \exp - \beta \sum_{i=1}^{s} \psi(i/s+1,\ldots,s+n)\rangle. \quad (3.12)$$

Here $y_s(1,\ldots,s)$ is $g_s(1,\ldots,s)$ divided by the Boltzmann factor for all pair potential terms $\phi_2(i,j)$ as well as intrinsic higher order terms $\phi_3(i,j,k)$, etc., if they appear. Thus the expectation value of (3.12) is exactly the left-hand side of the MM hierarchy [except for external-field Boltzmann factors, if they appear, which are included in the right-hand side of (3.12), but are more naturally pulled into the denominator of the left-hand side of the MM hierarchy, as in (4.1) below].

3.5. Representations for dispersions and composite media; polydispersivity

A wide variety of interesting systems can each be described as an included phase of one material distributed throughout a matrix (fluid, solid, or void) of another material. Many systems, both "dry" composites (alloys, rocks, porous beds) and "wet" dispersions (suspensions, micro-emulsions, etc.) as well as molecular systems, can be so described. In characterizing the microstructure of such a system, it is often convenient to use so-called matrix functions [16,17] $S_s(1,\ldots,s)$, where

$$S_s(1,\ldots,s) = \text{the probability that the } s \text{ points}$$
$$r_1,\ldots,r_s \text{ are all in the matrix phase, i.e.,}$$
$$\text{none are interior to the included phase.} \quad (3.13)$$

Knowing the S_i for successively higher i permits one to obtain successively better bounds and estimates for various material and transport properties of these systems, such as elastic moduli, conductivities, diffusivities, etc. [17]. Even for the simple model of an included phase built up of randomly centered spherical inclusions [for which the $g_n(1,\ldots,n) \equiv 1$] the problem of calculating the S_s and the resulting properties of the composite medium so defined is an extremely interesting and highly nontrivial one.

Some time ago I realized that the $S_s(1,\ldots,s)$ must satisfy MM and KS type equations and with Sal Torquato investigated the derivation and implications of these equations for the S_s [17,18]. To establish the MM equations it is useful to think of inserting s hypothetical "solute" particles that are simply points into the system of N actual inclusions (the "solvent" particles). To facilitate contact with our earlier derivation of the MM equations in terms of the function $p_0[\Omega]$ we shall consider here inclusions or solvent particles to be spheres of diameter σ. The definition of the S_s immediately implies that if Ω is the union volume of s spheres of radius $\sigma/2$ centered at $r_1,\ldots,r_s$, which we shall denote as $U_s(\sigma/2)$, then [17]

$$S_s(1,\ldots,s) = p_0[U_s(\sigma/2)]$$

= the probability of finding that no inclusion
centers are in $U_s(\sigma/2)$. (3.14)

But $p_0[U_s(\sigma/2)]$ clearly has the same MM representation as $p_0[U_s(\sigma)]$ does, except with $m(i,j;\sigma)$ replaced by $m(i,j;\sigma/2)$ and $m_s(1,\ldots,s;j,\sigma)$ replaced by $m_s(1,\ldots,s;j,\sigma/2)$. Thus eq. (2.40) immediately yields a MM representation of $S_s(1,\ldots,s)$ via (3.14).

To obtain the KS representation [18] of S_s, one instead considers a system in which $s-1$ "solute" points labeled $2,\ldots,s$, respectively, have already been inserted along with the "solvent" inclusions. When divided by $g_{s-1}(r_2,\ldots,r_s)dr_{1,2}\ldots dr_{1,s}$ the lhs of (2.35)—again with $m(r_{i,j};\sigma)$ replaced by $m(r_{i,j};\sigma/2)$—represents the conditional probability

$$S_s(1/2,\ldots,s) = S_s(1,\ldots,s)/S_{s-1}(2,\ldots,s). \tag{3.15}$$

On the rhs of the equation, the division by the solute g_{s-1} means that the g_{s-1+N} of (2.37) is replaced by conditional g's for the n solvent particles, given the presence of $s-1$ solute points. Denoting these conditional functions as $g_{s-1n+n}(s+1,\ldots,s+n/2,\ldots,s)$ we have finally

$$S_s(1/2,\ldots,s) = \sum_{n \geq 0} \frac{(-1)^n \rho^n}{n!} \int g_{s-1+n}(s+1,\ldots,s+n/2,\ldots,s)$$

$$\times \prod_{j=s+1}^{s+n} [m(1,j;\sigma/2)dj]. \tag{3.16}$$

In the spatially homogeneous case, $S_1(1)$ is simply the volume fraction of matrix material; usually denoted ϕ. In the nonhomogeneous case, the results of this section all continue to hold but the one-particle inclusion distribution function $g_1(i)$ is no longer unity. The results continue to go through when the g_i's of (2.40) and (3.16) are associated with nonequilibrium ensembles, since in obtaining them we have made no equilibrium assumptions that close the equations.

We note in passing that (3.16) is of considerable interest even in the case in which only particle 1 is a point, and particles $2, \ldots, s$ are inclusions, identical to all the other inclusions. We are back to the special simple case of a single point solute particle considered in section 3.1 (except that our meaning of σ is a bit different here) and $S_s(1/2, \ldots, s)\Pi_{2 \leqslant j \leqslant s} e(1, j; \sigma/2)$ now has the meaning of a conditional probability of finding a point in the matrix phase, given inclusions at $r_2, \ldots, r_s$.

We note also that the form of our results is independent of dimension so they hold for circular inclusions in two dimensions (and hence for parallel cylindrical inclusions in three dimensions).

The MM representation of S_1 and S_2 for the special cases of randomly centered inclusions and hard-sphere inclusions reduce to known representations. The general representations are new and have subsequently proved especially useful in obtaining new results for partially penetrable [19,20] and polydisperse inclusions [21] (e.g., spherical and cylindrical inclusions with a continuous distribution of diameters). Although such polydispersivity has long been a basic aspect of composite-media models as well as polymer and other macromolecular models, a general formalism for its treatment has heretofore been lacking.

If $F(\sigma)$ represents the probability density function of diameters (no relation to Boltzmann's favorable space, but we are running out of alternative notations), the KS equations for the polydisperse case are obtained with the replacement of

$$\int \prod_{j=s+1}^{s+n} \{m(1, j; \sigma) \mathrm{d}j\} \quad \text{by}$$

$$\int \prod_{j=s+1}^{s+n} \{m(1, j; \sigma_j) \, \mathrm{d}j \, F(\sigma_j) \, \mathrm{d}\sigma_j\} \tag{3.17}$$

while the MM equations require the replacement of

$$\int \prod_{j=s+1}^{s+n} \{m_n(1, \ldots, s; j, \sigma) \mathrm{d}j\} \quad \text{by}$$

$$\int \prod_{j=s+1}^{s+n} \{m_s(1, \ldots, s; j, \sigma_j) \, \mathrm{d}j \, F(\sigma_j) \, \mathrm{d}\sigma_j\}. \tag{3.18}$$

This holds whether the indicator functions m and m_s refer to particle–particle interactions, as in (2.37) and (2.43), or point–particle interactions, as in the representations of the S_s. The MM representation of the S_s in the randomly-centered sphere case reduces to a remarkably simple form. It is

$$S_s = \exp - \rho \int F(\sigma) \, U_s(\sigma/2) \, d\sigma, \tag{3.19}$$

where $U_s(\sigma/2)$ is the volume of the union of s spheres of radius $\sigma/2$.

As observed at the end of section 3.1, eq. (3.3) remains true for inclusions that are mutually penetrable to an arbitrary degree; in this case it becomes a nontrivial and fundamental relation in composite media theory. (To our knowledge, however, it has not heretofore appeared in the context of that theory.) Using σ_0, σ_1, and σ in the sense of eq. (3.1), the A and $F/d\sigma$ of (3.3), evaluated at $\sigma_1 = 0$ (i.e., for $\sigma_0 = \sigma$) are the matrix-phase volume and the expected inclusion surface area, respectively. For arbitrary σ_1 the A and $F/d\sigma$ continue to have direct physical meaning. They represent the matrix volume and surface area available to a particle of radius σ_1 that is being inserted into the medium containing inclusions of radius σ_0. Writing $s(\sigma) = F(\sigma)/d\sigma \, V$ and $\phi(\sigma) = A(\sigma)/V$ to exhibit the explicit dependence upon σ we thus have the relation, valid for arbitrary σ,

$$s(\sigma) = - \frac{\partial \phi(\sigma)}{\partial \sigma}, \tag{3.20}$$

that holds between the specific surface area available to particle 1 and the probability $\phi(\sigma)$ that the same particle lies entirely within the matrix.

3.6. Connectivity, the continuum Potts model, and the Widom–Rowlinson model

Although some questions concerning the properties of composite media and related systems are most readily answered in terms of the S_s, others are more directly considered in terms of the s-particle connectivity functions $g_s^{\ddagger}(1, \ldots, s)$ which give the probability that s particles or inclusions are part of the same connected cluster. In particular, questions concerning the percolation point, at which the mean cluster size becomes infinite, can be conveniently considered in terms of $g_2^{\ddagger}(1, 2)$ [22]. Bill Klein and I have begun investigating the $g_s^{\ddagger}$ in terms of the KS and MM equations for the continuous g state Potts model that Klein recently introduced to study continuum percolation problems [23]. In Klein's model the pair potential between a particle in state i at r_1 and a particle in state j at r_2 is given by

$$\phi_{i,j}(1,2) = \phi(r_{i,j})\delta_{i,j}, \qquad (3.21)$$

with $\delta_{i,j}$ the Kronecker delta and $\phi(r_{i,j})$ a hard-sphere interaction. The s-particle distribution functions for particles in the same state can be shown to yield connectivity functions $g_s^\ddagger$ for the randomly-centered sphere model upon differentiation with q and evaluation at $q=1$. Moreover we can show Klein's model to be isomorphic to the q-component Widom–Rowlinson model [24], which also has a pair potential given by (3.21) and has been already extensively studied in a variety of contexts [25]. This establishes an important identity between the critical exponents of the two models.

4. Generalizations and final remarks

4.1. Generalization to arbitrary potentials

It is clear that the MM and KS equations can be generalized to a potential energy that is not simply the sum of pair terms, although explicit results in this direction seem to be lacking in the literature. They are not only useful in themselves, but form an important bridge to applications to nonequilibrium ensembles.

With inclusion of an external field term $\phi_1(i)$ we have
MM

$$\frac{\rho_s(1,\ldots,s)}{\prod_{1\leqslant i\leqslant s} z(i) \prod_{1\leqslant i<j\leqslant s} e_2(i,j)} = \sum_{n=0}^{\infty} \frac{1}{n!} \int \rho_n(s+1,\ldots,s+n)$$

$$\times \prod_{j=s+1}^{s+n} \{f(1,\ldots,s;j)\mathrm{d}j\}, \qquad (4.1)$$

KS

$$\frac{\rho_s(1,\ldots,s)}{z(1)\prod_{2\leqslant i\leqslant s} e_2(1,i)} = \sum_{n=0}^{\infty} \frac{1}{n!} \int \rho_{s-1+n}(2,\ldots,s+n)$$

$$\times \prod_{j=s+1}^{s+n} \{f(1,j)\mathrm{d}j\}, \qquad (4.2)$$

where the external field appears in the quantity

$$z(i) = z\exp{-\beta\phi_1(i)}. \qquad (4.3)$$

As before, we have

$$f(i, j) = e_2(i, j) - 1, \tag{4.4}$$

$$f_s(1, \ldots, s; j) = \left[\prod_{i=1}^{s} e_2(i, j) \right] - 1. \tag{4.5}$$

The first MM and KS equation now describes a function $\rho(1)/z(1)$ that in general will be spatially varying. From Boltzmann's integrals we see that in the hard-sphere case his average available space is still spatially independent, however, since it can be identified as

$$A = \int [\rho(1)/z(1)] \, d1. \tag{4.6}$$

For potentials consisting only of pair terms the MM and KS equations have already been extended to mixtures [26] and the KS equations to particles that are not spherically symmetric [27]. In the latter case the potential terms depend upon arguments $i, j, \ldots$, that involve orientation ω_i, $\omega_j, \ldots$, as well as translational coordinates $r_i, r_j, \ldots$. When the KS and MM equations are written in the form given by (4.1) and (4.2), no change of notation is necessary to accommodate this generalization if $\int dj$ is interpreted to include orientational as well as volume integration. In fact, the same notation can be used in the mixture case if $\int dj$ is further generalized to include summation over species. Our polydisperse equations (3.17) and (3.18) are examples of this generalization.

For arbitrary potentials in which intrinsic n-body terms $\phi_n(1, \ldots, n)$, $n \geq s$, appear as well as ϕ_1 and ϕ_2, (4.1) and (4.2) must be generalized to include these terms. Introducing

$$e_s(1, \ldots, s) = \exp -\beta \phi_s(1, \ldots, s), \tag{4.7a}$$

where again

$$z(i) = z \exp -\beta \phi_1(i) = z e_1(i), \tag{4.7b}$$

we shall write
MM

$$\frac{\rho_s(1, \ldots, s)}{\prod_{1 \leq i \leq s} z(i) \prod_{1 \leq i < j \leq s} e_2(i, j) \prod_{1 \leq i < j < k \leq s} e_3(i, j, k) \cdots e_s(1, \ldots, s)}$$

$$= \sum_{n=0}^{\infty} R_{s, \text{MM}}^{(n)}, \tag{4.8}$$

and
KS
$$\frac{\rho_s(1,\ldots,s)}{\prod_{1\leqslant i\leqslant s}[z(i)e_2(1,i)]\prod_{1\leqslant i<j\leqslant s}e_3(1,i,j)\cdots e_s(1,i,\ldots,s)}$$
$$=\sum_{n=0}^{\infty}R_{s,\mathrm{KS}}^{(n)}. \tag{4.9}$$

It is convenient to express the $R_{s,\mathrm{MM}}^{(n)}$ and $R_{s,\mathrm{KS}}^{(n)}$ in graphical notation that has become more-or-less standard [28]

$R_{s,\mathrm{MM}}^{(n)}$ = sum of all distinct graphs consisting of a ρ_n-face (spanning n black 1-circles and s white 1-circles labeled 1 through s, resp.) and one or more f-faces such that each f-face spans one of more white circles and one or more black circles. (4.10)

$R_{s,\mathrm{KS}}^{(n)}$ = sum of all distinct connected graphs consisting of a ρ_{s-1+n} face (spanning n black 1-circles and $s-1$ white 1-circles labeled 2 through s, resp.) and one or more f-faces. Each f-face spans the white circle labeled 1, one or more black circles and some or no white circles labeled 2 through s. (4.11)

Here the f-faces spanning s circles represent the function
$$f_s(i,\ldots,i+s)=e_s(i,\ldots,i+s)-1, \tag{4.12}$$
with e_s given by (4.7). "Spanning" means "directly attached to". If one divides both sides of the KS equation (4.10) by $R_{s,\mathrm{KS}}^{(0)}$, which is just $\rho_{s-1}(2,\ldots,s)$, the $\rho_s(1,\ldots,s)$ on the left-hand side is replaced by the conditional density function $\rho_s(1/2,\ldots,s)$ and the ρ_{s-1+n} faces in (4.11) representing $\rho_{s-1+n}(2,\ldots,s+n)$ are replaced by faces—call them ρ_{s-1+n}^c faces—that represent the conditional densities $\rho_{s-1+n}(s+1,\ldots,n/2,\ldots,s)$. Thus
KS
$$\frac{\rho_s(1/2,\ldots,s)}{\prod_{1\leqslant i\leqslant s}[z(i)e_2(1,i)]\prod_{1\leqslant i<j\leqslant s}e_3(1,i,j)\cdots e_s(1,i,\ldots,s)}$$
$$=\sum_{n=0}^{\infty}R_{s,\mathrm{KS}}^{(n)c}, \tag{4.12}$$

where

$R_{KS}^{(n)c}$ = sum of all distinct connected graphs consisting of a ρ_{s-1+n}^c face (spanning n black 1-circles and $s-1$ white 1-circles labeled 2 through s, resp.) and one or more f-faces. Each f-face spans the white circle labeled 1, one or more black circles and some or no white circles labeled 2 through s. (4.13)

4.2. Generalization to nonequilibrium ensembles

We have already touched upon nonequilibrium at the end of section 2 and in section 3.5 in noting that equations such as (2.40) and (3.16) continue to hold for nonequilibrium ensembles since they involve no closure assumptions that force the probability densities ρ_n (i.e., $\rho^n g_n$) appearing on the right-hand sides of the equations to be equilibrium densities. On the other hand, in writing (2.37) and (2.43) for a hard-sphere system or (4.1), (4.2), (4.8)–(4.10) for more general potentials we *have* made such closure assumptions. Equation (2.42), for example, represents such a closure for a hard-sphere system. (To avoid misunderstanding we point out that such closure typically involves the full infinite set of MM or KS equations rather than some finite subset. Such closure readily yields recursion relations that enable one to construct the ρ_n term by term in successively higher powers of ρ or z, but the resulting series are rigorously known to converge and uniquely represent the equilibrium ρ_n only for sufficiently small ρ or z.)

Here we wish to point out that a formal closure scheme can also be implemented in the nonequilibrium case. It involves using the structure given by eqs. (4.8)–(4.10), appropriately generalizing the meaning of the $z(i)$ and the e_s and f_s found there. At equilibrium the denominator of the left-hand side of (4.8) is proportional to the probability density associated with an isolated set of s particles where the $e_i(1,\ldots,i)$ represent a decomposition of that density into intrinsic i-body terms, $0 \leq i \leq s$, and z is so defined as to yield the appropriate proportionality constant. For a system not in equilibrium let us similarly consider the probability density $\rho_s(1,\ldots,s)$ but let the argument i represent a vector that describes the momentum as well as the spatial coordinates of the ith particle. Here too the probability density associated with an isolated set of s particles can be used to define a set $e_i(1,\ldots,i)$ of intrinsic i-body terms $0 \leq i \leq s$ and an appropriate normalizing factor z. Perhaps the most satisfactory way of systematically defining these functions is through a sequence of approximating nonequilibrium ensembles that starts with the structure of (4.1) and (4.2) and approaches the full structure of (4.8) and (4.9) as a limiting case. Each ensemble in the sequence can be defined by the maximization of entropy subject to constraints that guarantee $e_j \equiv 1$ for j greater than a prescribed integer J and yield a unique set z, $e_1, e_2, \ldots, e_J$. This maximization-of-entropy program goes back to Mayer [29] who considered it for an open system. (McLennan and Harris [30] showed how to use it to recover the z, $e_1, e_2, \ldots, e_J$ for a closed system.) The program was rediscovered and generalized by Lewis [31] and extended by Karkheck and myself [32] through the introduction of auxiliary constraints not considered by the above workers. We are continuing to develop it in collaboration with van Beijeren [33] and de Schepper [34]. Independent of this program, van Beijeren and Ernst proposed a revised Enskog theory for hard spheres [35], which also yields a z, e_1, and e_2 and $\rho_s(1,\ldots,s)$ connected by exactly the structure of (4.1) and (4.2), and which has been shown to follow from the maximization of entropy [32].

The maximization-of-entropy program enables us to effect closure in a sequence of approximations, each of which is described by a special case of (4.8)–(4.10). The closure holds for arbitrary time but tells us nothing about the time evolution of the ρ_s or the z and e_j (except that $e_j \equiv 1$ for j greater than some fixed J). To find the time evolution of the ρ_s, one must turn to other equations such as one of the BBGKY hierarchy. The MM and KS hierarchies instead hold the possibility of bounds and other rigorous results for the sequence of approximations we have discussed.

4.3. Further generalizations and closing remarks

In section 3.5 we arrived at a MM and KS hierarchy of the form

$$S_s(1,\ldots,s) = \sum_{n \geq 0} \frac{(-1)^n \rho^n}{n!} \int g_n(s+1,\ldots,s+n)$$
$$\times \prod_{j=s+1}^{s+n} [m(1,\ldots,s;j)\,\mathrm{d}j], \qquad (4.14)$$

$$m_s(1,\ldots,s;\ j) = 1 - \prod_{i=1}^{s}[1-m(i,\ j)],$$

$$S_s(1/2,\ldots,s) = \sum_{n\geqslant 0} \frac{(-1)^n \rho^n}{n!} \int g_{s-1+n}(s+1,\ldots,s+n/2,\ldots,s)$$

$$\times \prod_{j=s+1}^{s+n}[m(i,\ j)\,\mathrm{d}j], \qquad (4.15)$$

respectively where $m(i,\ j)$ is the indicator function for a volume that defines the inclusion boundaries via a point–inclusion interaction that delineates the inclusion boundary as traced out by a point that ranges over the inclusion surface. Our first generalization of the discussion of section 3.5 comes from the observation that the inclusions need not be spheres in order for these equations to hold; they can be mixtures of particles of far more general size and shape. The sizes and shapes will in all cases be defined by the function $m(i,\ j)$, where the arguments associated with the ith solute point and the jth solvent particle in the general case will describe the orientation of the solvent particles as well as the locations of their volume centers.

We note that the KS equations of Barboy and Gelbart [13] for the g_s of hard nonspherical particles can immediately be utilized to obtain equations for the S_s of the same system. We also note that for hard needles (a model of nematic systems) the analogs of the inequalities (3.6)–(3.8) are apt to be of considerably more importance than they have been for spheres. Such inequalities have proved to be sharp only on the second virial coefficient level, roughly speaking, but for sufficiently long needles, this level already describes much of the thermodynamics and correlation of interest.

Our second generalization comes from the observation that we can consider solute particles that are not points but, like the solvent inclusions, have nonzero volume. The simplest case to consider is the one in which there is only a single such solute particle, labeled as particle one. For a spherical solute particle in a hard-sphere solvent we are back to exactly the scaled-particle geometry of section 3.1. $S_1(1)$ now gives the probability that the solute sphere lies wholly in the matrix phase. For a uniform system this will no longer simply reduce to the matrix volume fraction ϕ. Instead it will be the expected volume available to a solute sphere that one inserts into the system, according to eq. (3.2). Similarly, as remarked at the end of section 3.5, the $F/\mathrm{d}\sigma$ of eq. (3.3) is the expected inclusion surface area available to a solute particle coming into the medium. The generalization to the case of s non-point solute particles is immediate.

Throughout all of these considerations it should be kept in mind that the solvent–solvent (inclusion–inclusion) interaction can be regarded as arbitrary even in those cases in which the solute–solvent interaction is a point–inclusion interaction that is used to sharply define an inclusion shape and volume. For example, in some applications it is very natural to regard the inclusions as having a degree of surface adhesion with respect to each other. In others it is useful to introduce a degree of interpenetrability among the inclusions, as we have already discussed.

We are now in a position to summarize what we have learned concerning the various forms of the MM and KS equations. For arbitrary potentials we can identify three forms that are in general nonequivalent:

(i) There are the closed equations for the equilibrium g_n given in section 4.1, in which the e_s and f_s are directly related to potential terms.

(ii) There are the (nonclosed) equations for the $p_0[U_n]$ discussed in section 2 in which the functions $f_2(i, j)$ are replaced by $-m(i, j)$ where the $m(i, j)$ are indicator functions for cavities. These in general hold for arbitrary interparticle interactions (which do not explicitly appear in the equations but enter through the g_s) and hold for nonequilibrium as well as equilibrium. These equations can also be written as equations for matrix functions S_s through the introduction of "solute" points as discussed in section 3.5 and the present section; the $m(i, j)$ are then interpreted as indicator functions that delineate particle boundaries via point–particle interaction. At equilibrium, for a hard-sphere system there is a degeneracy in the sense that the $p_0[U_s(\sigma)]$ of equation (2.40) can be written in terms of the hard-sphere g_s via (2.42) and the resulting equations are exactly the MM equations of section 4.1 written for a hard-sphere system. At equilibrium for hard-sphere inclusions, however, the equations for the S_s do not become a closed set of equations for the g_s, even though the S_s can be re-expressed in terms of the solute g_s, because the S_s equations represent not the full set of MM (or KS) equations for the solvent–solute mixture, but only the subset of equations that give the solute g_s in terms of the solvent g_n (in the MM case) or in terms of mixed solute–solvent g_{s-1+n} (in the KS case).

(iii) Finally there are the nonequilibrium equations of section 4.2 that can be formally closed for a sequence of increasingly refined approximations in the sense we have discussed there. At equilibrium they will reduce to closed equations of section 4.2.

In closing, we have these additional remarks:

Although we have endeavored to give a good cross section here of the literature that is directly related to Boltzmann's program and the MM and KS hierarchies, we should emphasize that we have just scratched the surface

—dealing with all appropriate references is far beyond our grasp. There is one particular question raised by Boltzmann's eq. (2.19) that has been pursued in the recent literature that we have not yet cited and which is worth pointing out, since its further consideration may prove useful in computer simulation of the g_n and related functions. The question is: What is the relationship between the surface area and volume available to an $N + 1$st particle one wishes to insert into a system of N particles and the surface area and volume available to an $N + 1$st particle that *already is in* the system? Remarkably enough, the ratio of the expected values of the first pair (which is what Boltzmann considered) is equal to the expected value of the ratio of the second pair, which has been discussed by Hoover et al. [36]. In one [37] of an interesting series of papers [38] in which Boltzmann's program is rediscovered and used in obtaining new approximations, R.J. Speedy investigates the connections among the expectation values associated with these volumes and surfaces. Independently M.J. de Olivera [15] and K.S. Shing [39] each derived a very closely related result that can be expressed in the notation of section 3.4 as

$$\rho/z = \langle \exp - \beta\psi \rangle = \langle \exp + \beta\psi \rangle_{N+1}, \qquad (4.16)$$

where the second expectation value is taken for the system that already includes the $(N + 1)$st particle labeled 1. These relations prove to be a very useful [40] basis for the Monte Carlo evaluation of ρ/z. Since the KS representation of $\rho g_2(1, 2)/ze_2(1, 2)$, $\rho g_2(1; 2, 3)/ze_2(1, 2)e_2(1, 3)$, etc. can be regarded as that of ρ/z in the presence of successively more particles that already are in the system but are fixed (and hence not integrated over), the use of (4.16) in the presence of such particles promises to be a useful means of generating g_2 and possibly g_3 in Monte Carlo computations.

Acknowledgements

The author gratefully acknowledges the National Science Foundation for their long-term general support and the Office of Basic Energy Sciences, U.S. Department of Energy for support of his research concerning dispersions and nonequilibrium systems.

References

[1] J.E. Mayer and E. Montroll, J. Chem. Phys. 9 (1941) 2.
[2] J.E. Mayer and S.F. Harrison, J. Chem. Phys. 6 (1938) 87.

[3] J. Kirkwood and Z. Salsburg, Disc. Farad. Soc. 15 (1953) 28.
[4] G. Stell, Boltzmann's Method of Evaluating and Using Molecular Distribution Functions, Polytechnic Institute of Brooklyn Report (Sept. 1966). This report is available from the author upon request.
[5] The Kirkwood–Salsburg equations are really a special case of eq. (42) and eq. (54′) appearing in a paper by Mayer [J. Chem. Phys. 15 (1947) 187], but these very general equations of Mayer's are written in a form that necessitates further manipulation before they yield the equations that Kirkwood and Salsburg and Boltzmann derived and discussed. Mayer went through this manipulation for the particular case that reduces to the first Kirkwood–Salsburg equation, and subsequently Sarolea and Mayer [Phys. Rev. 101 (1956) 1627] examined in greater detail other of Mayer's results. Their eqs. (6.3) and (6.4) are the first two Kirkwood–Salsburg equations. Kirkwood and Salsburg also singled out these two equations [their eq. (19a) and eq. (19b)] for special attention. All of these authors (except Boltzmann) remark upon various superior properties of these equations compared to the usual cluster expansions and the integral equations of the Yvon–Kirkwood–Born–Green–Bogoliubov type.
[6] L. Boltzmann, Vorl. über Gastheorie II (1898). The last edition of this volume, bound with Gastheorie I, is the third (Leipzig, 1920). Both volumes have been beautifully translated into English by Stephen Brush and published as the single volume, Lectures on Gas Theory (University of California Press, Berkeley, 1964).
[7] L. Boltzmann, Kon. Akad. Wet. Amsterdam, Wis-en natuurk. afdeling, VII (1898/99) 477.
[8] H. Reiss, H.L. Frisch and J.L. Lebowitz, J. Chem. Phys. 31 (1959) 369.
[9] J.L. Lebowitz and J.K. Percus, J. Math. Phys. 4 (1963) 1495.
[10] G. Stell, in: The Equilibrium Theory of Classical Fluids, eds. H.L. Frisch and J.L. Lebowitz (Benjamin, New York, 1964). For (3.11) see pp. II–246, II–251. The results are given in terms of ln y_s, denoted in [4] as W_s.
[11] E. Meeron and A.J.F. Siegert, J. Chem. Phys. 48 (1968) 3139.
D. Henderson and E.W. Grundke, Mol. PHys. 24 (1972) 669.
In the first of these papers, a significant piece of the Boltzmann progress is rediscovered and certain interesting extensions are pursued.
[12] WG. Hoover, and J.C. Poirier, J. Chem. Phys. 37 (1962) 1041.
[13] See:
E.W. Grundke and D. Henderson, Mol. Phys. 24 (1972) 269.
B. Barboy and R. Tenne, Mol. Phys. 31 (1976) 1749; 33 (1977) 331.
B. Barboy and W.M. Gelbart, J. Stat. Mech. 22 (1980) 685. See also the papers of ref. [11].
[14] B. Widom, J. Chem. Phys. 39 (1963) 2808; 41 (1964) 74. See also J.L. Jackson and L.S. Klein, Phys. Fluids 7 (1964) 228.
[15] B. Widom, J. Phys. Chem. 86 (1982) 869 and references therein.
[16] See, e.g.,
G. Stell, in: The Mathematics and Physics of Disordered Media, eds. B.D. Hughes and B.W. Ninham (Springer, Berlin, 1983) p. 347, and references therein.
[17] S. Torquato and G. Stell, J. Chem. Phys. 77 (1982) 2071.
[18] S. Torquato and G. Stell, J. Chem. Phys. 78 (1983) 3262; 79 (1983) 1505.
[19] S. Torquato and G. Stell, J. Chem. Phys. 80 (1984) 878.
[20] P.A. Rikvold and G. Stell, to be published.
[21] G. Stell, R. Korlipara, K. Chelliah and C. Joslin, work in progress.
[22] A. Coniglio, V. De Angelis, A. Forlani and G. Lauro, J. Phys. A10 (1977) 219.
[23] W. Klein, Phys. Rev. B26 (1982) 2677.

[24] B. Widom and J.S. Rowlinson, J. Chem. Phys. 52 (1970) 1670.
[25] See, e.g.,
D. Ruelle, Phys. Rev. Lett. 27 (1971) 1040.
J. Karkheck and G. Stell, J. Chem. Phys. 71 (1979) 3620.
C.A. Lang, J.S. Rowlinson and S.M. Thompson, Proc. R. Soc. London A352 (1976) 1.
[26] S. Baer and J.L. Lebowitz, J. Chem. Phys. 40 (1964) 3474.
[27] B. Barboy and W.M. Gelbart, J. Stat. Phys. 22 (1980) 685.
[28] See ref. [10], for example. For a reference that explicitly deals with representing s-particle functions by faces, see
G. Stell, in: Phase Transitions and Critical Phenomena, Vol. 5B, eds. C. Domb and M.S. Green (Academic, London, 1976) p. 206.
[29] J.E. Mayer, J. Chem. Phys. 33 (1960); 34 (1961) 1207.
[30] A.D. McLennan and R.A. Harris, J. Chem. Phys. 34 (1961) 1451.
[31] R.M. Lewis, J. Math. Phys. 8 (1967) 1448.
[32] J. Karkheck and G. Stell, Phys. Rev. A25 (1982) 3328.
[33] G. Stell, J. Karkheck and H. van Beijeren, J. Chem. Phys. 79 (1983) 3199.
[34] Work in progress.
[35] H. van Beijeren and M.H. Ernst, Physica 68 (1973) 437; 70 (1973) 225.
[36] W.G. Hoover, W.T. Ashurst and R. Grover, J. Chem. Phys. 57 (1972) 1259.
[37] R.J. Speedy, J. Chem. Soc. F2 77 (1981) 329.
[38] R.J. Speedy, J. Chem. Soc. F2 73 (1977) 714; 75 (1979) 1643; 76 (1980) 693.
[39] K.S. Shing and K.E. Gubbins, Mol. Phys. 46 (1982) 1109.
[40] K.E. Gubbins, K.S. Shing and W.B. Streett, J. Phys. Chem. 87 (1983) 4573.

CHAPTER 7

Illumination in a Random Medium

N.G. VAN KAMPEN

Instituut voor Theoretische Fysica
Rijksuniversiteit Utrecht
Nederland

© *Elsevier Science Publishers B.V. 1985*

The Wonderful World of Stochastics
A Tribute to Elliott W. Montroll
Eds. M.F. Shlesinger and G.H. Weiss

Contents

1. Geometrical optics in inhomogeneous isotropic media — 159
2. Fokker–Planck equation for random medium — 161
3. The illumination of a surface element — 163
4. Alternative formulation as a first-passage problem — 164
References — 167

1. Geometrical optics in inhomogeneous isotropic media

Geometrical optics in a medium with refractive index $n(r)$ can be based on Fermat's principle [1]. It states that the rays are those curves in space for which

$$\int n \, ds \tag{1}$$

is stationary with respect to small variations with fixed end points r_1, r_2. As the path length s itself varies it is preferable to specify the curves by a vector function $r(\tau)$ of some parameter τ with fixed values τ_1, τ_2 at the end points r_1, r_2. Then (1) can be written as a more familiar type of variational problem:

$$\delta \int_{\tau_1}^{\tau_2} n(r(\tau)) \sqrt{\left(\frac{dr}{d\tau}\right)^2} \, d\tau = 0.$$

This is the Lagrange formulation; the associated Euler equations are second-order differential equations determining $r(\tau)$.

The corresponding Hamilton formulation, however, cannot be found in the way one has learned (and taught) in classical mechanics. If the canonical momenta are defined in the usual way by

$$p = \frac{\partial L}{\partial \dot{r}} = n(r) \frac{\dot{r}}{\sqrt{\dot{r}^2}}, \tag{2}$$

it turns out that the Hamilton function vanishes identically:

$$H = p \cdot \dot{r} - L = 0.$$

This difficulty seems to have been overlooked by Whittaker [2]. In most texts on geometrical optics [3] it is avoided by taking one of the components of r as independent variable rather than the parameter τ, so that only two unknowns remain. Such a description is suitable for lens systems, but awkward for our purpose. We shall therefore give a Hamiltonian formulation which does not spoil the spatial symmetry.

First note that it is not possible to solve eq. (2) for $\dot{r}$ so as to express $\dot{r}$ in p and r. The three equations are not independent but linked by one identity not containing $\dot{r}$:

$$p^2 = [n(r)]^2.$$

It is easy to see in general, that if the canonical momenta turn out to be linked by an identity $\Omega(r, p) = 0$ the Hamilton equations must be modified and become [4]

$$\frac{dr}{dt} = \frac{\partial H}{\partial p} + \lambda \frac{\partial \Omega}{\partial p}, \quad \frac{dp}{dt} = -\frac{\partial H}{\partial r} - \lambda \frac{\partial \Omega}{\partial r}.$$

The additional variable λ is determined by the additional equation $\Omega = 0$. In the present case

$$H = 0, \quad \Omega = \tfrac{1}{2}(p^2 - n^2), \quad t = \tau. \qquad (3)$$

Hence the rays are determined by the pseudo-Hamilton equations

$$\frac{dr}{d\tau} = \lambda p, \quad \frac{dp}{d\tau} = \lambda \nabla(\tfrac{1}{2}n^2), \quad p^2 = n^2. \qquad (4)$$

The value of λ depends on the choice of τ. The following three choices are of particular interest.

(a) τ is the arc length s, $\lambda = n^{-1}$, and

$$\frac{dr}{ds} = \frac{p}{n}, \quad \frac{dp}{ds} = \frac{\partial n}{\partial r}. \qquad (5)$$

(b) τ is the optical depth t, $\lambda = n^{-2}$,

$$\frac{dr}{dt} = \frac{p}{n^2}, \quad \frac{dp}{dt} = \frac{\partial \log n}{\partial r}. \qquad (6)$$

Here t is also the time during which a light pulse (or photon) has travelled.

(c) τ is taken to be $\theta = \int ds/n$, $\lambda = 1$,

$$\frac{dr}{d\theta} = p, \quad \frac{dp}{d\theta} = \frac{\partial}{\partial r}(\tfrac{1}{2}n^2). \qquad (7)$$

These equations [5] are truly Hamiltonian and describe the propagation of the light pulse as the motion of a classical particle in a potential $-\tfrac{1}{2}[n(r)]^2$. Note, however, that in this analogy only the solutions with zero energy are relevant, and that θ is not the real time. Apart from this freedom in the choice of the parameter τ one is also free to multiply the function Ω in (3) with any nonvanishing factor $g(r, p)$. This has the effect that λ is multiplied with the reciprocal,

$$\Omega^* = \Omega g, \qquad \lambda^* = \lambda/g.$$

This makes it possible to write eqs. (4) as true Hamilton equations by taking $\Omega^* = \Omega\lambda$. In this way eqs. (5) can be interpreted as the Hamilton equations belonging to the Hamilton function $\Omega^* = \frac{1}{2}(p^2/n - n)$, or alternatively [6] $\Omega^* = |p| - n$; while eqs. (6) have the Hamilton function $\Omega^* = \frac{1}{2}(p^2/n^2 - 1)$.

Finally it is sometimes advantageous to replace the variable p by the unit vector $e = p/n$. The equations (5), (6), (7) then become respectively

$$\text{(a')} \quad \frac{dr_i}{ds} = e_i, \qquad \frac{de_i}{ds} = (\delta_{ij} - e_i e_j)\frac{\partial \log n}{\partial r_j}, \tag{8}$$

$$\text{(b')} \quad \frac{dr_i}{dt} = \frac{e_i}{n}, \qquad \frac{de_i}{dt} = \frac{1}{n^2}(\delta_{ij} - e_i e_j)\frac{\partial n}{\partial r_j}, \tag{9}$$

$$\text{(c')} \quad \frac{dr_i}{d\theta} = ne_i, \qquad \frac{de_i}{d\theta} = (\delta_{ij} - e_i e_j)\frac{\partial n}{\partial r_j}. \tag{10}$$

2. Fokker–Planck equation for random medium

Now suppose that the refractive index involves a small random term [7]:

$$n(r) = n_0(r) + \alpha n_1(r), \qquad \langle n_1(r)\rangle = 0. \tag{11}$$

The dimensionless constant α serves as a scale factor. We have supposed that the fluctuations in n do not depend on time since light traverses the atmosphere so fast that the density fluctuations appear frozen. We also suppose that n_1 has a finite correlation length r_c such that $n(r_1)$ and $n(r_2)$ are statistically independent when $|r_1 - r_2| > r_c$.

Each of the equations in section 2 has now been turned into a nonlinear stochastic differential equation for six variables and one supplementary condition. It is then possible [8,9] by expanding in α to derive a Fokker–Planck equation for the probability $P(r, p, \tau)$ or $P(r, e, \tau)$. These give the probability density of position and direction of the light pulse *at a given value of the parameter* τ, whichever one has chosen. At first it might seem that in this connection only the use of time as a parameter makes physical sense, but as we are going to derive properties of the entire ray we are free to employ any parameter that is convenient.

The choice between **p** and **e** is decided by the initial condition. If the light source produces a ray in a given direction the initial value of **e** is presented. The initial value of **p** is not given, rather it is a random quantity, and even correlated with the random coefficient $n(r)$ in the equation itself. Although such a situation can be handled [10], it is obviously more natural to use **e** as variable rather than **p**. In this case the equations (8), (9), (10) should be used.

When the light source is located outside the turbulent medium, there is no preference. This is true in the case of stars, but also for any source that is separated from the turbulent medium by a lens or glass pane, provided one formulates the initial condition at that source or at least outside the medium. If, however, the initial point is taken on the *outer* surface of the pane, inside the medium, then the initial value of **p** is fixed, while that of **e** is random, as can be seen from Snell's law. For this reason we shall here use the variable **p**—with the added advantage that these equations are simpler.

Among the three options (5), (6), (7) the last one has two important simplifying features. The first three of its six equations have no random element, and the volume in phase space is conserved. We therefore opt for eqs. (7), but first introduce some notation.

We set instead of eq. (11)

$$\tfrac{1}{2} n^2 = \mu(r) + \alpha \kappa(r),$$

where $\mu = \tfrac{1}{2} n_0^2 + \tfrac{1}{2} \alpha^2 \langle n_1^2 \rangle$ is sure and κ is random with $\langle \kappa \rangle = 0$. Partial derivatives of μ and κ with respect to r_j are indicated by a subscript j. Then our equation is

$$\frac{dr_j}{d\theta} = p_j, \qquad \frac{dp_j}{d\theta} = \mu_j + \alpha \kappa_j.$$

We write this temporarily as an equation for the six-component vector $u_\nu = \{r_j, p_j\}$,

$$\frac{du_\nu}{d\theta} = F_\nu(u) = F_\nu^{(0)}(u) + \alpha F_\nu^{(1)}(u),$$

$$F_\nu^{(0)} = \{p_j, \mu_j\}, \qquad F_\nu^{(1)} = \{0; \kappa_j\}.$$

This set of six coupled nonlinear stochastic differential equations is of the type considered in ref. [9]. The probability density $P(u, \theta)$ obeys to second order in α a Fokker–Planck equation, which works out to be

$$\frac{\partial P(r, p, \theta)}{\partial \theta} = -p_i \frac{\partial P}{\partial r_i} - \frac{\partial}{\partial p_i} A_i P + \alpha^2 \frac{\partial^2}{\partial p_i \partial p_j} B_{ij} P, \qquad (12)$$

$$A_i(\boldsymbol{r}, \boldsymbol{p}) = \mu_i(\boldsymbol{r}) + \alpha^2 \int_0^\infty \theta' \langle \kappa_{ij}(\boldsymbol{r}) \kappa_j(\boldsymbol{r} - \theta'\boldsymbol{p}) \rangle \mathrm{d}\theta',$$

$$B_{ij}(\boldsymbol{r}, \boldsymbol{p}) = \int_0^\infty \langle \kappa_i(\boldsymbol{r}) \kappa_j(\boldsymbol{r} - \theta'\boldsymbol{p}) \rangle \mathrm{d}\theta.$$

In this equation not only higher orders of α have been omitted, but also terms of order $\alpha^2 r_c / l$, where l is a typical distance on which n_0 varies.

3. The illumination of a surface element

Let the solution of eq. (12) with initial value $\delta(\boldsymbol{r} - \boldsymbol{r}')\delta(\boldsymbol{p} - \boldsymbol{p}')$ be denoted by

$$P(\boldsymbol{r}, \boldsymbol{p}, \theta | \boldsymbol{r}', \boldsymbol{p}'). \tag{13}$$

It has the following meaning. I have an ensemble of media with the same $\mu(\boldsymbol{r})$ but each with its own $\kappa(\boldsymbol{r})$. In each of them I emit a light pulse at $\boldsymbol{r}'$ whose direction $\boldsymbol{e}'$ is given by $\boldsymbol{p}' = n(\boldsymbol{r}') \boldsymbol{e}'$. On each of these rays I mark the point that corresponds with a fixed value of θ. The ensemble of points $\boldsymbol{r}$, $\boldsymbol{p}$ obtained in this way has the distribution (13). Physically this is not a meaningful quantity: what one wants to know is the amount of light that passes through a given surface element that is located at a given point and has a given orientation. We shall now show how this information can be obtained from P.

Let the surface element be ΔS with normal unit vector $\boldsymbol{d}$. Extend it into a volume element of thickness Δh (fig. 1). The path length of the light ray inside this volume element is

$$\Delta s = \frac{\Delta h}{\boldsymbol{d} \cdot \boldsymbol{e}} = \frac{n}{\boldsymbol{d} \cdot \boldsymbol{p}} \Delta h,$$

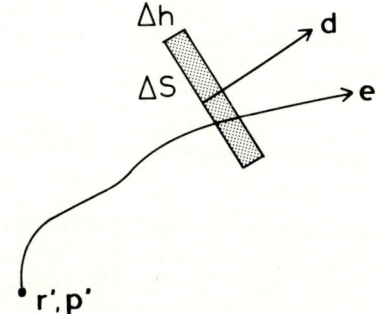

Fig. 1. Light ray traversing the surface element ΔS.

which corresponds to
$$\Delta\theta = \frac{\Delta s}{n} = \frac{\Delta h}{\boldsymbol{d}\cdot\boldsymbol{p}}.$$

The probability to find a light pulse, travelling along the ray, in the volume element $\Delta S \Delta h$ is
$$P\Delta S\Delta h = P(\boldsymbol{d}\cdot\boldsymbol{p})\,\Delta\theta\,\Delta S.$$

The probability that it passes through ΔS at any time is
$$\Delta S \int (\boldsymbol{d}\cdot\boldsymbol{p})\mathrm{d}^3 p \int_0^\infty P(\boldsymbol{r},\boldsymbol{p},\theta)\mathrm{d}\theta \equiv \Delta S \boldsymbol{d}\cdot\boldsymbol{K}(\boldsymbol{r}),$$
$$\boldsymbol{K}(\boldsymbol{r}) = \int \boldsymbol{p}\,\mathrm{d}^3 p \int_0^\infty P(\boldsymbol{r},\boldsymbol{p},\theta)\mathrm{d}\theta. \tag{14}$$

This defines the averaged intensity $\boldsymbol{K}(\boldsymbol{r})$ as a vector field in space. The essential point is that to find $\boldsymbol{K}$ one does not need the entire $P(\boldsymbol{r},\boldsymbol{p},\theta)$, but merely its integral over θ, that is,
$$R(\boldsymbol{r},\boldsymbol{p}) = \int_0^\infty P(\boldsymbol{r},\boldsymbol{p},\theta)\mathrm{d}\theta;$$

and that this R obeys an equation by itself. In fact, by integrating eq. (12) over θ one obtains
$$-p_i\frac{\partial R}{\partial r_i} - \frac{\partial}{\partial p_i}A_i R + \alpha^2\frac{\partial^2}{\partial p_i \partial p_j}B_{ij}R = -\delta(\boldsymbol{r}-\boldsymbol{r}')\delta(\boldsymbol{p}-\boldsymbol{p}'). \tag{15}$$

Once this equation is solved one also knows the value of
$$\boldsymbol{K}(\boldsymbol{r}) = \int \boldsymbol{p} R(\boldsymbol{r}\cdot\boldsymbol{p})\mathrm{d}^3 p.$$

Incidentally, one finds immediately
$$\operatorname{div} \boldsymbol{K} = \delta(\boldsymbol{r}-\boldsymbol{r}'). \tag{16}$$

The quantity $\boldsymbol{K}$, which utilizes only part of the information contained in R, is not always sufficient. Suppose I look at a star. Then ΔS is the pupil of my eye and the twinkling is described by the probability distribution of the directions of incident rays. Assume that the interior of my eye does not participate in the density fluctuations of the air, then the direction I see depends on the vector $\boldsymbol{p}$ on my cornea. Its distribution is given by $R(\boldsymbol{r},\boldsymbol{p})$, where $\boldsymbol{r}$ is taken at the position of the cornea. Hence in this case knowledge of $\boldsymbol{K}$ is not sufficient, but the solution of eq. (15) is.

4. Alternative formulation as a first-passage problem

The problem of finding the average illumination of ΔS can also be formulated as a first-passage problem, about which Elliott Montroll has done so much work [11]. In fact, eq. (12) describes a diffusion problem in the six-dimensional (r, p)-space. Let Σ be a hypersurface surrounding the initial point r', p', and let σ be a part of Σ. We ask for the probability $Q(r', p')$ that the point at which a light pulse, emitted at r', p', reaches Σ for the first time lies in σ. The answer is well known [12].

First one observes that

$$Q(r', p') = 1 \quad \text{for} \quad r', p' \in \sigma,$$
$$Q(r', p') = 0 \quad \text{for} \quad r', p' \in \Sigma - \sigma. \tag{17}$$

Next one can show [12] that Q obeys

$$p'_i \frac{\partial Q}{\partial r'_j} + A'_i \frac{\partial Q}{\partial p'_i} + \alpha^2 B'_{ij} \frac{\partial^2 Q}{\partial p'_i \partial p'_j} = 0. \tag{18}$$

The operator is the adjoint of the one appearing in (12); A'_i, B'_{ij} are the same functions as in eq. (12) but with arguments r', p'. Thus one has to solve eq. (18) in the region enclosed by Σ with the boundary condition (17).

Suppose the problem (17), (18) can be solved; then in order to find the illumination of ΔS one first extends ΔS into a surface S surrounding the source. The physical meaning of S is that all rays that reach S are absorbed; if I look at a star, ΔS is my pupil and S covers the remaining surface of the earth. Next take for the five-dimensional hypersurface Σ the cartesian product of S with the entire three-dimensional space of p. Similarly σ is the product of ΔS with p space. In this geometry solve eq. (18) with boundary condition (17). The quantity $Q(r', p')$ found is the average illumination of ΔS by a source at r' emitting in the direction determined by p'.

We shall now show that this result is equivalent with that in section 2. Multiply eq. (15) with $Q(r, p)$ and integrate over the interior of Σ

$$\int Q \left(-p_i \frac{\partial R}{\partial r_j} - \frac{\partial}{\partial p_i} A_i R + \alpha^2 \frac{\partial^2}{\partial p_i \partial p_j} B_{ij} R \right) d^3r \, d^3p = -Q(r', p'). \tag{19}$$

The interior of Σ consists of the product of the three-dimensional interior of S with the entire p-space. Partial integration gives for the left-hand side zero, plus boundary terms. The derivatives with respect or p generate no boundary terms because clearly both R and Q vanish for $|p| \to \infty$. The first term gives (d is again the normal to the surface element dS)

$$- \int d^3p \int Q (p \cdot d) R \, dS = -\Delta S \, d \cdot K.$$

According to eq. (19) this is equal to $-Q(r', p')$ so that both methods yield the same value for the illumination of ΔS.

Actually this equivalence is somewhat surprising for the following reason. In deriving the expression (14) for the intensity K we made an error. Any ray whose meandering path happens to cross ΔS twice contributes twice to (14) (fig. 2). Also any ray that crosses ΔS from the wrong side is counted as a negative contribution to the illumination. (Of course for light rays in the atmosphere this error is exceedingly small, provided one chooses the orientation of ΔS in a sensible way.) On the other hand, the result based on Q is exact, but required the choice of an absorbing surface S. Such an absorbing surface prevents the rays from crossing ΔS twice or in the wrong direction; we therefore conclude that actually such an absorbing surface was tacitly assumed in the derivation of K as well.

Yet this is still not the entire story, because such a surface would also suppress the contribution of those rays that happen to cross it on their way towards ΔS, although these contributions are included in K. Thus we would be forced to take the surface very far away, which is not required in the calculation of Q. The redeeming factor, however, is eq. (16). It states that K does not depend on the actual choice of S, as long as it serves its purpose of cutting out the rays that cross ΔS twice or from the wrong side. Our conclusion is that the derivation in section 3 is better than it deserved to be, and that it is only really justified by the alternative derivation of section 4.

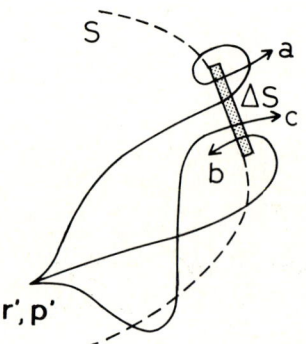

Fig. 2. The ray a is counted twice and the ray b gives a negative distribution. This is remedied by the absorbing surface S, but at the same time the ray c is erroneously cut off.

Acknowledgement

I am indebted to Dr. S. Golynski for many discussions and for acquainting me with the extensive research in this field done in the USSR.

References

[1] See, e.g.,
M. Kline and I.W. Kay, Electromagnetic Theory and Geometrical Optics (Interscience, New York, 1965).
[2] E.T. Whittaker, A Treatise on the Analytical Dynamics of Particles and Rigid Bodies, 4th Ed. (University Press, Cambridge, 1937) p. 292.
[3] M. Herzberger, Strahlenoptik (Springer, Berlin, 1931); Modern Geometrical Optics (Interscience, New York, 1958).
C. Carathéodory, Geometrische Optik (Springer, Berlin, 1937).
H.A. Buchdahl, An Introduction to Hamiltonian Optics (University Press, Cambridge, 1970).
[4] This can readily be derived by a slight extension of the usual derivation of Hamilton's equation.
P.A.M. Dirac, Can. J. Math. 2 (1950) 129.
E.C.G. Sudarshan and N. Mukunda, Classical Dynamics: A Modern Perspective (Wiley, New York, 1974) p. 81 ff.
[5] J.L. Synge, Geometrical Optics (Cambridge University Press, London, 1937).
[6] Yu. A. Kravtsov and Yu. I. Orlov, Geometrical Optics of Inhomogeneous Media (Nauka, Moscow, 1980; in Russian) p. 16.
[7] See also
J.B. Keller, in: Hydrodynamic Instability, Proc. Symp. in Applied Mathematics 13 (American Mathematical Society, Providence, RI, 1962) p. 227.
U. Frisch, in: Probabilistic Methods in Applied Mathematics I, ed. A.T. Bharucha-Reid (Academic, New York, 1968).
[8] B.J. West, Phys. Rev. A18 (1978) 1646.
S. Golynski and N.G. van Kampen, Phys. Lett. 102A (1984) 220.
S. Golynski, Phys. Lett. 103A (1984) 104.
[9] N.G. van Kampen, Phys. Rep. 24C (1976) 171, equation (19.9).
[10] J.B.T.M. Roerdink, Physica 109A (1981) 23.
[11] E.W. Montroll and K.E. Shuler, in: Advances in Chemical Physics 1, ed. I. Prigogine (Interscience, New York, 1958).
E.W. Montroll, in: Stochastic Processes in Mathematical Physics and Engineering, Proc. Symp. in Applied Mathematics 16 (American Mathematical Society, Providence, RI, 1964).
E.W. Montroll and G.H. Weiss, J. Math. Phys. 6, (1965) 167.
E.W. Montroll and H. Scher, J. Stat. Phys. 9 (1973) 101.
E.W. Montroll and B.J. West, in: Fluctuation Phenomena, Studies in Statistical Mechanics VII, eds. E.W. Montroll and J.L. Lebowitz (North-Holland, Amsterdam, 1979) p. 61.
[12] E.B. Dynkin and A.A. Yushkevich, Markov Processes, Theorems and Applications (Plenum, New York, 1969) p. 47.
Z. Schuss, Theory and Applications of Stochastic Differential Equations (Wiley, New York, 1980) p. 120.

CHAPTER 8

Random Walks in Crystallography

George H. WEISS and James E. KIEFER

National Institutes of Health
Bethesda, MD 20205
U.S.A.

The Wonderful World of Stochastics
A Tribute to Elliott W. Montroll
Eds. M.F. Shlesinger and G.H. Weiss

© *Elsevier Science Publishers B.V. 1985*

Contents

1. Introduction 171
2. Outline of the problem 172
3. The difficulty and its resolution 174
4. Some numerical results 183
References 187

1. Introduction

Elliott Montroll left his imprint on several areas in statistical mechanics, solid-state physics, and applied mathematics. The mathematical ideas in his papers were generally simple, direct, and elegant. Elliott had the gift of choosing extremely simple physical models whose analysis shed an extraordinary amount of light on a given subject area, and whose deeper analysis generally occupied the attention of many investigators thereafter. No subject was more closely linked to Elliott in recent years than that of the random walk [1]. His polished classical analysis of random walks still provides a vocabulary to many physical scientists making use of extensions of his work. Elliott's interests in random walks had diverse roots. They were first motivated by problems in chemical physics [2], then by problems in solid state physics [3–8], and finally by the beauty of the mathematical problems that arise in the analysis [9].

One area to which he never contributed, but which has been vigorously pursued for over forty years, is that of the role of random walks in the interpretation of crystallographic data [10]. This role is central to most of modern crystallography. The principal method now in use for approximating to the probability density functions describing heterogeneous crystallographic random walks involves the Edgeworth expansion, which is essentially an expansion in terms of polynomials orthogonal with respect to a Gaussian weight function, i.e., Hermite polynomials. The analogue of this technique with a different weight function was applied very successfully by Elliott and his collaborators to the approximation of the frequency spectrum of harmonic lattices [12,15]. Notwithstanding successful applications of this method there are cases in crystallography involving heterogeneity in which it gives relatively poor results. In this paper we report on some recent work in this field by us and other collaborators [16,17], in which approximate solutions of the random walk problems provided by the central limit theorem are replaced by exact solutions. More than mathematical elegance will be seen to be involved, since when the unit cell of a crystal has atoms with a considerable variation in atomic weight the Gaussian approximation and the Edgeworth expansions and related expansions in terms of orthogonal polynomials can give qualitatively incorrect results as will be demonstrated below.

2. Outline of the problem

To define our problem let us consider a unit cell containing n atoms, the jth of which has coordinates $r_j = (x_j, y_j, z_j)$ together with the diffraction amplitude, or structure factor, F_h, associated with radiation whose direction is described by Miller indices $h = (h, k, l)$,

$$F_h = \sum_j f_j \exp(2\pi i h \cdot r_j). \tag{1}$$

This expression is written in terms of the scattering factors, the f_j, which are known once the chemical composition of the unit cell is known. Since F_h is a complex quantity both an amplitude and a phase are required for its specification. However, only the amplitude or intensity is measurable so that the determination of crystal structure requires considerable further analysis, both theoretical and experimental, to establish the unknown phases. One technique widely used by crystallographers for this purpose is that of the study of intensity statistics, also known as Wilson statistics. This technique was initially proposed and analyzed by A.J.C. Wilson [18,19], and requires the analysis of the statistical properties of detected intensities when the crystal is bombarded by X-rays at many uniformly distributed values of h. Let us clarify the sorts of assumptions made to start such an analysis, and discuss the problems that must be tackled before successfully completing it.

The principal assumption used in the analysis is that since the r_j are initially unknown, the quantities $\theta_j = 2\pi h \cdot r_j$ are independent and uniformly distributed in the interval $(0, 2\pi)$. An equivalent formulation is that the atomic positions are fixed and not random, but that the crystal is irradiated uniformly. The ultimate justification for such an assumption must be experimental, and indeed such experiments are in good agreement with it. Let us follow the usual crystallographic custom of dealing not with the F_h but rather with

$$E_h = F_h / \langle |F_h|^2 \rangle^{1/2} \tag{2}$$

so that if the F's are regarded as random variables the E's have unit variance. We can rewrite E_h as

$$E_h = \sum_{j=1}^{n} \frac{f_j e^{i\theta_j}}{\left(\sum_{j=1}^{n} f_j^2\right)^{1/2}} = \sum g_j e^{i\theta_j} = A_h + i B_h, \qquad (3)$$

where

$$A_h = \sum_j g_j \cos \theta_j, \quad B_h = \sum g_j \sin \theta_j. \qquad (4)$$

It is readily seen from this representation that E_h can be regarded as a vector coordinate of an n-step random walk in two dimensions, in which the step sizes (the g_j) are known but the angles between successive steps are uniformly distributed in $(0, 2\pi)$. This is precisely the random walk mentioned by Pearson in his note of 1905 [20], for which Kluyver shortly thereafter provided an exact expression for the pdf (probability density function) of $|E_h|$ [21]. There are several ways of proving that this pdf is

$$f(|E|) = |E| \int_0^\infty J_0(\omega|E|) \prod_{j=1}^{n} J_0(\omega g_j) d\omega \qquad (5)$$

and the pdf of a single projection, A, is

$$g(A) = \frac{1}{2\pi} \int_{-\infty}^{\infty} \cos \omega A \prod_{j=1}^{n} J_0(\omega g_j) d\omega. \qquad (6)$$

These formulae would appear to suffice in resolving any problems related to statistical properties of A and $|E|$. From the practical point of view they do not. One can trivially observe that the formulae in eqs. (5) and (6) do not obviously show that both $f(|E|)$ and $g(A)$ are identically equal to zero when

$$|E|, |A| > \sum_{j=1}^{n} f_j. \qquad (7)$$

Furthermore, numerical integration of the highly oscillatory integrands also leads to difficulties.

Before we discuss some of the mathematical problems raised by this simple random walk, it is useful to indicate how the statistical properties of $|E|$ can be used to determine structural information. Let us consider the simplest possible distinction resolvable through the use of intensity statistics, that of the difference between centrosymmetric and non-centrosymmetric crystals. A centrosymmetric crystal is one in which for every atom located at r there is also one located at $-r$, while the noncentrosymmetric crystal is one that does not have this property. The random-walk picture indicates centrosymmetry by having $B \equiv 0$ in eq. (3) since for every angle θ appearing in eq. (4), a term with the same f and $-\theta$ also occurs. Thus, scattering from a centrosymmetric crystal can be modelled as a one-dimensional random walk, while the scattering from a noncentrosymmetric crystal is modelled as a two-dimensional random walk. With this observation we can describe Wilson's technique for distinguishing between P$\bar{1}$ (the crystallographers nomenclature for the simplest centrosymmetric space group) and P1 (noncentrosymmetry). The assumption is made that there are at least ten atoms per unit cell allowing the use of the central limit theorem to approximate to the density function $f(|E|)$:

$$f(|E|) = \sqrt{\frac{2}{\pi}} \exp(-E^2/2). \tag{8}$$

Since the noncentrosymmetric crystals can be represented as a two-dimensional random walk so that $\langle |E|^2 \rangle = \langle A^2 \rangle + \langle B^2 \rangle$, the pdf is approximated by that for the sum of squares of two Gaussian random variables. This is known to be a gamma density, so that in the absence of centrosymmetry one has

$$f(|E|) = |E| e^{-E^2/2}. \tag{9}$$

A best fit to the intensity data is then compared to the two theoretical forms in eqs. (8) and (9) (usually the cumulative distributions $N(E) = \int_0^E f(u) \, du$ rather than the pdf's are compared) and the appropriate choice between P1 and P$\bar{1}$ is made.

3. The difficulty and its resolution

Although the method just outlined is quite straightforward and works satisfactorily in many cases, there is nevertheless a difficulty which calls into question its use in others. The difficulty lies in the use of central limit theorem. A strategy often used by crystallographers to solve the phase problem is that of inserting a heavy atom, say uranium, in the unit cell [22]. When this is feasible one can still use the random-walk picture of eq. (3) to study intensities but naive application of the Gaussian approximation can lead to qualitative errors. This is because one or more of the g_j in eq. (4)

will then be much larger than the remaining scattering factors. To see how this would affect the shape of the pdf consider first the pdf corresponding to the projection of a one-step random walk, the bond being randomly oriented with respect to the A axis. The pdf is easily calculated to be

$$g(A) = \frac{1}{\pi} \frac{1}{\sqrt{L^2 - A^2}}, \quad A^2 < L^2,$$
$$= 0, \quad L^2 > A^2, \qquad (10)$$

where L is the length of the step. If one plots $g(A)$ against A, the resulting curve is U-shaped with a singularity at the two ends, $A = \pm L$. When a few additional steps are added to the random walk with values of the g_j very much smaller than L, one expects that the singularities at the end points will be smoothed out. This is indeed the case. It is evident that for such a pdf a Gaussian approximation will be far off the mark. We will discuss some of the ways that can be used to bypass this difficulty.

The most frequently used technique for taking heterogeneity into account is to use an expansion in terms of polynomials orthogonal to the approximation furnished by the central limit theorem [11]. Thus for space group P$\bar{1}$ the expansion is in terms of Hermite polynomials and for P1 it is given in terms of Laguerre polynomials. The coefficients of the nth polynomial are a linear combination of the first n moments of the normalized structure factor. That is, if $\rho(R)$ is the appropriate pdf and the $\phi_j(R)$ are orthogonal with respect to $\rho(R)$ then

$$f_N(R) = \rho(R) \sum_{j=1}^{N} \langle \phi_j(R) \rangle \phi_j(R) \qquad (11)$$

yields a refined approximation to the exact pdf. Calculation of the moments may be complicated for some space groups but some of the algebraic difficulties can be alleviated by using symbolic manipulation routines on a computer as has been done by Shmueli and Kaldor [23]. A second improvement to the straightforward expansion in terms of orthogonal polynomials has been exploited by Karle and his collaborators [24,25]. For P$\bar{1}$, for example, in which $\rho(E) = (2/\pi)^{1/2} \exp(-E^2/2)$, the orthogonal polynomials are $\phi_j(E) = (2^j j!)^{-1/2} H_j(E/\sqrt{2})$, Karle writes

$$f_N(E) = \sqrt{\frac{2}{\pi}} \exp(-E^2/2) \sum_{j=1}^{N} b_j H_j(E/\sqrt{2})$$
$$\approx \sqrt{\frac{2}{\pi}} \exp\left(-E^2/2 - \sum_{j=1}^{N} c_j H_j(E/\sqrt{2})\right), \qquad (12)$$

where the c_j's can be found in terms of the b_j's. Quicker convergence is claimed for the exponential form of the approximation than for the straightforward expansion. Other approximations are possible that generalize this approach. As an example, one might try replacing the polynomial terms in eq. (12) by a Padé approximation rather than modifying the exponent.

Another technique that has been explored for approximating the pdf of the projection A, whose exact expression is given in eq. (6), is that of using the method of steepest descents [16]. Let the characteristic function for a given space group be $\prod_{j=1}^{n} C(\omega g_j)$ so that

$$g(A) = \frac{1}{2\pi} \int_{-\infty}^{\infty} \exp(-i\omega A) \prod_{j=1}^{n} C(\omega g_j) d\omega$$

$$= \frac{1}{2\pi i} \int_{-i\infty}^{i\infty} \exp(-vA) \prod_{j=1}^{n} C(-ivg_j) dv. \qquad (13)$$

The last line is obtained from the first by substituting $\omega = -iv$. The second integral of this last equation is now evaluated approximately by using the method of steepest descents. It is known that the normalized steepest descents approximation has a relative error of $n^{-3/2}$ rather than the $n^{-1/2}$ of the central limit theorem and that successive corrections are proportional to powers of n^{-1} rather than $n^{-1/2}$ [26,27]. The increased accuracy of the steepest-descents approximations had been verified for $P\bar{1}$ in which $C(\xi) = J_0(\xi)$ as in eq. (6). In this case the equation for the steepest descents root, $\bar{v}$, is

$$\sum_{i=1}^{n} g_j \left[I_1(\bar{v}g_j) / I_0(\bar{v}g_j) \right] = A \qquad (14)$$

and the unnormalized approximation to $g(A)$ is

$$g_{\mathrm{SD}}(A) \approx \left(2\pi \sum_{j=1}^{n} g_j^2 \psi(\bar{v}g_j) \right)^{-1/2} \exp(-\bar{v}A) \prod_{j=1}^{n} I_0(\bar{v}g_j), \qquad (15)$$

where

$$\psi(x) = 1 - \frac{1}{x} \frac{I_1(x)}{I_0(x)} - \left(\frac{I_1(x)}{I_0(x)} \right)^2 \qquad (16)$$

and the $I_j(x)$'s are Bessel functions. The normalized steepest-descents approximation is

$$g_{\text{norm}}(A) \approx g_{\text{SD}}(A) / \int_{-F_m}^{F_m} g_{\text{SD}}(u) du, \quad A^2 < F_m^2, \qquad (17)$$
$$= 0, \qquad A^2 > F_m^2,$$

where

$$F_m = \sum_{j=1}^{n} g_j \qquad (18)$$

is the maximum value of the structure factor. An interesting feature of the approximation shown in eq. (17) is that it differs from zero only in the interval $A^2 < F_m^2$. Whether this feature is reproduced by the steepest-descents approximation for more complicated space groups is not known at the present time.

A bonus of the rotation of the axis of integration that leads from the first line of eq. (13) to the second is that it allows one to find an asymptotic expansion of the pdf of A in the neighborhood of its maximum value, $A^2 \approx F_m^2$. This may be seen by rewriting eq. (13) as

$$g(A) = \frac{1}{2\pi i} \int_{-i\infty}^{i\infty} \exp[v(F-A)] \left\{ \prod_{j=1}^{n} \exp(-vg_j) \, C(-ivg_j) \right\} dv$$
$$= \mathscr{L}^{-1} \left\{ \prod_{j=1}^{n} \exp(-vg_j) \, C(-ivg_j) \right\}, \qquad (19)$$

where $\mathscr{L}^{-1}$ denotes the inverse Laplace transform. Since the resulting function depends on $F_m - A$ only we can determine the behavior of $g(A)$ for $A \approx F_m$ by using Tauberian methods, and expanding the Laplace transform for large $|v|$. In the case of the space group $P\bar{1}$, $C(-ivg_j) = I_0(vg_j)$ so that to lowest order

$$\exp(-vg) \, I_0(vg) \approx (2\pi g v)^{-1/2}. \qquad (20)$$

Thus to lowest order $g(A)$ can be expanded as

$$g(A) \approx \left\{ \prod_{j=1}^{n} (2\pi g_j)^{-1/2} \right\} \frac{(F_m - A)^{n/2 - 1}}{\Gamma(n/2)} \qquad (21)$$

for $|F_m - A|$ small. Higher order terms can be calculated [16].

From the crystallographer's point of view perhaps the most important method available for approximate evaluation of pdf's is truncation of an exact expansion. The original formulation of this technique is due to

Barakat [27,28], who applied it to two- and three-dimensional Pearson random walks. It has been very effectively applied to some crystallographic problems by Shmueli et al. [17]. The basic idea is that since, for example, $g(A)$ is nonzero only in $A^2 < F_m^2$ one can expand $g(A)$ in a Fourier series within the interval, finding

$$g(A) = \frac{1}{2F_m}\left\{1 + 2\sum_{l=1}^{\infty} C\left(\frac{\pi l}{F_m}\right)\cos\left(\frac{\pi l A}{F_m}\right)\right\}, \qquad (22)$$

where $C(\omega)$ is the characteristic function. This Fourier expansion is appropriate for the analysis of centrosymmetric crystals. When the crystal is not symmetric so that we are dealing with a nondegenerate two-dimensional random walk, an expansion in a Fourier–Bessel function series is the more convenient one. Using the parameter F_m introduced in the last equation we find

$$f(|E|) = 2F_m^{-2}|E|\sum_{j=1}^{\infty} D_j J_0\left(\gamma_j F_m^{-1}|E|\right), \quad 0 < |E| < |E|_{\max}, \qquad (23)$$

where the γ are successive roots of $J_0(\gamma) = 0$ and

$$D_j = \frac{1}{J_1^2(\gamma_j)}\prod_{k=1}^{n} C\left(\frac{\gamma_j g_k}{F_m}\right) \qquad (24)$$

which is again expressible in terms of the characteristic function. An alternative approach to the calculation of $f(|E|)$ is to expand the joint pdf of A and B in a double Fourier series. Denoting this pdf by $p(A, B)$ we have

$$p(A, B) = \frac{1}{4F_m^2}\sum_{r=-\infty}^{\infty}\sum_{s=-\infty}^{\infty} C\left(\frac{\pi r}{F_m}, \frac{\pi s}{F_m}\right)\exp\left[\frac{\pi i}{F_m}(rA + sB)\right], \qquad (25)$$

where

$$C(\omega_1, \omega_2) = \int_{-F_m}^{F_m}\int p(A, B)\exp\left[-\frac{\pi i}{F_m}(\omega_1 A + \omega_2 B)\right]dA\,dB$$

$$= \int_{-\infty}^{\infty}\int p(A, B)\exp\left[-\frac{\pi i}{F_m}(\omega_1 A + \omega_2 B)\right]dA\,dB \qquad (26)$$

is the joint generating function. To find the pdf of $|E| = (A^2 + B^2)^{1/2}$ we introduce polar coordinates into eq. (25), $A = |E|\cos\theta$, $B = |E|\sin\theta$, and integrate over θ. This leads to the expansion

$$f(|E|) = \frac{\pi}{2F_m^2}\sum_r\sum_s C\left(\frac{\pi r}{F_m}, \frac{\pi s}{F_m}\right)J_0\left(\frac{\pi E}{F_m}(r^2 + s^2)^{1/2}\right). \qquad (27)$$

We note parenthetically that if one is interested in calculating moments of $|E|$ from our exact formulae, eq. (23) is not a particularly useful starting point. It is possible, however, to relate the joint density of A and B, which we have shown to be a function only of $|E|$, to the density of the projection A, $g(A)$, and calculate the moments from that relation. This may be done in several ways; one of these is to eliminate the term

$$\prod_j J_0(\omega g_j)$$

from eq. (5) by taking the inverse Fourier transform of eq. (6). This leads to the relation

$$f(|E|) = -\frac{1}{\pi} \int_0^\infty g'(|E|\cosh\theta)\, d\theta. \tag{28}$$

The moments of $|E|$ are now calculated from those of the projection, which are much easier to find. The relation between moments is found to be

$$\langle |E|^m \rangle = \sqrt{\pi}\, \frac{\Gamma[(m+2)/2]}{\Gamma[(m+1)/2]} \int_{-F_m}^{F_m} |A|^m g(A)\, dA. \tag{29}$$

When m, the order of the moment of $|E|$, is even then it is proportional to a corresponding moment of A,

$$\langle |E|^{2s} \rangle = \frac{4^s (s!)^2}{(2s)!} \langle A^{2s} \rangle. \tag{30}$$

When m is odd, $\langle |E|^m \rangle$ is proportional to an absolute moment of A and the Fourier representation of $g(A)$ must be used to derive the relevant expression. For this case

$$\langle |E|^{2s+1} \rangle = F_m^{2s+1} \left\{ \frac{1}{2s+2} + 2 \sum_{j=1}^\infty C_n\left(\frac{\pi j}{F_m}\right) a_{2s+1}(j) \right\}, \tag{31}$$

where

$$C_n(\omega) = \prod_{j=1}^n J_0\left(\frac{g_j}{F_m}\omega\right), \qquad a_{2s+1}(j) = \int_0^1 \rho^{2s+1} \cos(\pi j \rho)\, d\rho.$$

The various forms of Fourier expansions just presented are exact, in contrast to approximation schemes discussed earlier. Several questions of a practical nature require an answer to render these techniques useful to the crystallographer. The first is whether it is always feasible to calculate the requisite characteristic functions, and the second is whether the resulting

Fourier series converge quickly enough to make their use preferable to approximations. Volume 1 of The International Tables for Crystallography [29] tabulates, among other things, the structure factors for all of the space groups. To see how these are used to calculate characteristic functions consider the space group Pm for which the Tables give

$$A = 2\cos[2\pi(hx + ky)]\cos(2\pi lz),$$
$$B = 2\sin[2\pi(hx + ky)]\cos(2\pi lz). \qquad (32)$$

On the assumption that at least one of the variables (h, k) differs from zero and that $l \neq 0$ we can assume that the variables $hx_j + ky_j$ are uniformly distributed in $(0, 1)$ for all values of j, and lz_j is also uniformly distributed in this interval as well as being independent of $hx_j + ky_j$. For this space group we find, for the two-dimensional characteristic function:

$$C(\omega_1, \omega_2) = \langle e^{i(\omega_1 A + \omega_2 B)} \rangle$$
$$= \frac{1}{(2\pi)^2} \int_{-\pi}^{\pi}\!\!\int \exp[2i\cos\theta_1(\omega_1\cos\theta_2 + \omega_2\sin\theta_2)]\, d\theta_1\, d\theta_2$$
$$= \frac{1}{(2\pi)} \int_{-\pi}^{\pi} J_0(2\omega\cos\theta_1)\, d\theta_1 = J_0^2(\omega), \qquad (33)$$

where

$$\omega^2 = \omega_1^2 + \omega_2^2. \qquad (34)$$

Another space group leading to a relatively simple expression for the characteristic function is Pmm2 for which

$$A = 4\cos(2\pi hx)\cos(2\pi ky)\cos(2\pi lz),$$
$$B = 4\cos(2\pi hx)\cos(2\pi ky)\sin(2\pi lz). \qquad (35)$$

When $h, k, l \neq 0$ we have

$$C(\omega_1, \omega_2) = \frac{1}{(2\pi)^3} \int\!\!\int_{-\pi}^{\pi}\!\!\int \exp[4i\cos\theta_1\cos\theta_2$$
$$\times (\omega_1\cos\theta_3 + \omega_2\sin\theta_3)]\, d\theta_1\, d\theta_2\, d\theta_3$$
$$= \frac{1}{(2\pi)^2} \int_{-\pi}^{\pi}\!\!\int J_0(4\omega\cos\theta_1\cos\theta_2)\, d\theta_1\, d\theta_2$$
$$= \frac{2}{\pi} \int_0^\pi J_0^2(2\omega\cos\theta)\, d\theta = \sum_{j=0}^{\infty} \frac{(-1)^j}{4^j}\frac{[(2j)!]^2}{(j!)^6}\omega^{2j}, \qquad (36)$$

where again ω is found from eq. (34). More complicated formulae can arise. For example, for P222 for which

$$A = 4\cos(2\pi hx)\cos(2\pi ky)\cos(2\pi lz),$$
$$B = 4\sin(2\pi hx)\sin(2\pi ky)\sin(2\pi lz) \tag{37}$$

we have after integrating over one angle variable

$$C(\omega_1, \omega_2) = \frac{1}{(2\pi)^2} \int_0^{2\pi}\int J_0\Big(4\big(\omega_1^2\cos^2\theta_1\cos^2\theta_2 + \omega_2^2\sin^2\theta_1\sin^2\theta_2\big)^{1/2}\Big)\,d\theta_1\,d\theta_2$$

$$= \sum_{s=0}^{\infty}\sum_{t=0}^{\infty}(-1)^{s+t}\frac{[(2s)!(2t)!]\,\omega_1^{2s}\omega_2^{2t}}{4^{s+t}[s!t!(s+t)!]^3}. \tag{38}$$

Not all space groups lead to easily calculable characteristic functions because the trigonometric expressions for A and B can be extremely complicated. The convergence of the Fourier series in eq. (22) or (23) is sufficiently rapid that it is quite practical to use it routinely. Even with extreme heterogeneity we never needed to compute more than 40 terms of either eqs. (22) or (23) for intensity statistics. Direct methods which require the evaluation of multiple Fourier series may lead to more serious convergence problems, but then acceleration methods might be tried to speed up the convergence rate. Such methods have been developed for Fourier series in one variable [30], and it is likely that analogous methods could be developed for higher dimensional Fourier series.

Higher-order classes of space groups can also be treated by these probabilistic techniques. As an example, the bicentric $P\bar{1}$ space group is one in which in addition to the crystallographic center of symmetry there are two symmetrically located noncrystallographic centers of symmetry at $\pm d$ respectively. For this space group the normalized intensity E_h is a scalar:

$$E_h = \sum_{j=1}^{n/4} g_j\big\{\cos(2\pi\mathbf{h}\cdot\mathbf{r}_j) + \cos\big[2\pi\mathbf{h}\cdot(2\mathbf{d}-\mathbf{r}_j)\big]\big\}$$

$$= \sum_{j=1}^{n/4} g_j\big\{\cos(2\pi\mathbf{h}\cdot\mathbf{r}_j)[1+\cos(4\pi\mathbf{h}\cdot\mathbf{d})]$$

$$+ \sin(2\pi\mathbf{h}\cdot\mathbf{r}_j)\sin(4\pi\mathbf{h}\cdot\mathbf{d})\big\}. \tag{39}$$

The summation range is from 1 to $n/4$ because there are four symmetrically located subgroups of atoms and the different terms in eq. (39) arise from the analysis in fig. 1. To evaluate the characteristic function for this space group one makes the assumption that both the $\mathbf{h}\cdot\mathbf{r}_j$ and $\mathbf{h}\cdot\mathbf{d}$ are uniformly distributed over the integral (0, 1). On performing the averages over the $\theta_j = 2\pi \mathbf{h}\cdot\mathbf{r}_j$ we find that the characteristic function can be expressed as

$$C(\omega_1) = \frac{2}{\pi} \int_0^{\pi/2} \left\{ \prod_{j=1}^{n/4} J_0(4\omega_1 g_j \cos \phi) \right\} d\phi \qquad (40)$$

which is easily calculated by numerical integration. It is not too difficult to show that in the case of identical atoms, $g_1 = g_2 = \cdots = g_n = g$ the resulting Fourier series can be summed approximately to yield

$$p(|E|) \approx \frac{1}{\pi^{3/2}} \exp(-E^2/8) K_0(E^2/8), \qquad (41)$$

a result first found by Rogers and Wilson in 1953 [31,36]. However, eq. (40) together with the Fourier series of eq. (22) represents the first attempt to take heterogeneity into account. More complicated symmetries can also be treated by similar techniques.

The study of so-called direct methods of structure determination requires the evaluation of joint densities of the intensity evaluated at two or more scattering vectors, $\mathbf{h}_1, \mathbf{h}_2, \ldots,$ [32]. Consider for example the joint density of

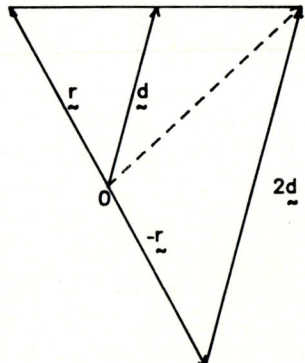

Fig. 1. The vector diagram needed to derive the expression for E_h in eq. (39). The origin, assumed to be a crystallographic center of symmetry, is labelled 0, and one of the auxiliary centers of symmetry is at $\mathbf{d}$. For every atom located at $\mathbf{r}$ the construction indicates that there is another atom at $2\mathbf{d} - \mathbf{r}$. Since there must be another symmetry center at $-\mathbf{d}$, the four groups of atoms are accounted for.

E_h and E_{2h} for the $P\bar{1}$ space group as first discussed in the framework of the central limit theorem by Cochran and Woolfson [33], and Klug [34]. For simplicity let $E = E_h$, $G = E_{2h}$ and denote the joint density of E and G by $h(E, G)$. Since both of these variables are bounded by F_m we will expand $h(E, G)$ in a double Fourier series as in eq. (25).

The explicit expressions for E and G are

$$E = \sum_j g_j \cos \theta_j, \qquad G = \sum_j g_j \cos 2\theta_j, \qquad (42)$$

where $\theta_j = 2\pi h \cdot r_j$. Thus, the coefficients of the Fourier series are

$$C(\omega_1, \omega_2) = \prod_j \langle e^{ig_j(\omega_1 \cos \theta_j + \omega_2 \cos 2\theta_j)} \rangle, \qquad (43)$$

which can be shown to be a product of the functions

$$C(\omega_1, \omega_2) = J_0(g\omega_1) J_0(g\omega_2) + 2 \sum_{j=1}^{\infty} (-1)^j J_{4j}(g\omega_1) J_{2j}(g\omega_2)$$

$$+ 2i \sum_{s=0}^{\infty} (-1)^{s+1} J_{4s+2}(g\omega_1) J_{2s+1}(g\omega_2) \qquad (44)$$

[35]. While it is not appropriate to discuss problems that arise from direct methods in crystallography in detail, we have shown how the exact solution of these random walks can replace at least some analyses in the crystallographic literature whose validity for heterogeneous crystals has never been tested.

4. Some numerical results

In our concluding section we present some specific examples of results achievable using approximations currently in use in crystallography. The most widely used approximation is the Edgeworth expansion. We also examine a second approximation in which the lowest-order term is furnished by the steepest-descents method and correction terms consist of a series in the polynomials orthogonal with respect to that function. Approximation of the pdf by an Edgeworth expansion works well when all of the g_j are roughly equal. When they are not, the performance of the approximation can be considerably degraded. Figure 2 shows the results of Edgeworth approximations applied to an extreme case of a $P\bar{1}$ structure consisting of a unit cell with 15 atoms, 14 of which have $g_j = 1$, and

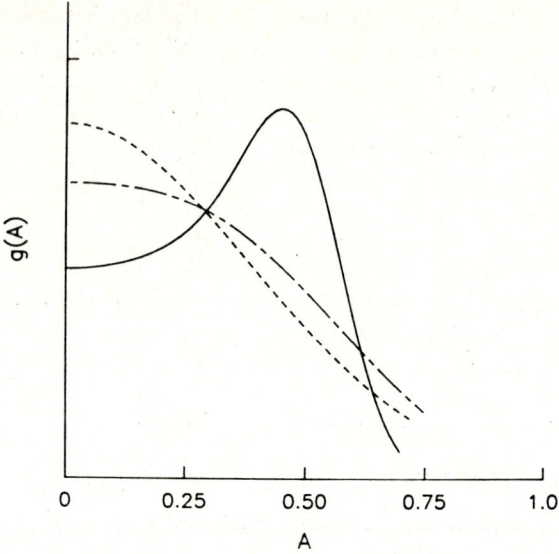

Fig. 2a. Approximations to the exact pdf $g(A)$ (———) for a unit cell in space group $P\bar{1}$, consisting of 14 atoms with scattering factor 1 and one atom with scattering factor $15\frac{1}{3}$, corresponding to 14 carbon atoms and a uranium atom. The unit of length has been chosen so that $A_{max} = 1$. The approximations are a Gaussian (— — —) and a Gaussian corrected with two moments (-— - —-).

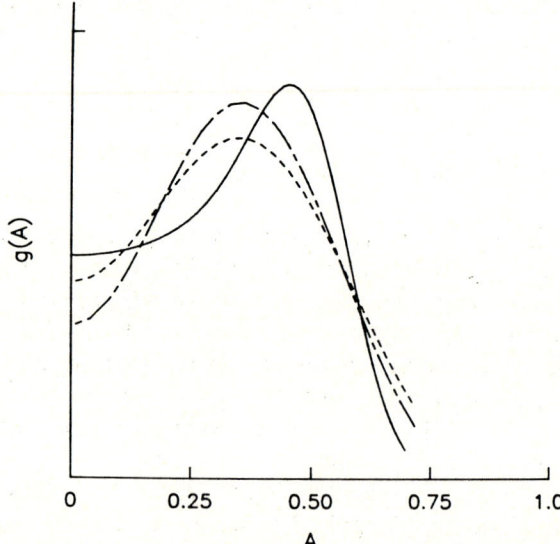

Fig. 2b. Approximations to the same pdf as in fig. 2a by a Gaussian with 4 moments (— — —) and one with 8 moments (-— - —-). Notice that the approximations in fig. 2a do not reproduce the peak at $A \approx 0.46$, while those in fig. 2b do reproduce it but at a lower value of A.

$g_{15} = 15\frac{1}{3}$. This roughly corresponds to 14 carbon atoms and a uranium atom. As one can see from fig. 2a, with zero or two moments the Edgeworth expansion is monotonic and is a very poor approximation. With four or eight moments the Edgeworth expansion gives qualitatively correct results but the peaks are consistently at a lower value of A than the exact position. On the other hand, the expansion in terms of polynomials orthogonal with respect to the steepest-descents approximation gives both qualitatively and quantitatively good agreement with the exact pdf when at least two moments are used. This can be seen from the curves in fig. 3. It should be noted though, that the assumed heterogeneity is considerable in this case and that a smaller amount of heterogeneity leads to a correspondingly better fit. The calculation using the steepest-descents method suggests that the zeroth-order approximation plays a crucial role in determining the accuracy of approximations that use higher moments. Figure 2 shows that while the zeroth-order steepest-descents approximation hardly resembles the exact pdf it still leads to far greater accuracy from the expansion in terms of orthogonal polynomials. This further suggests that one might try an arbitrary pdf which is easily calculated but has the essential qualitative feature

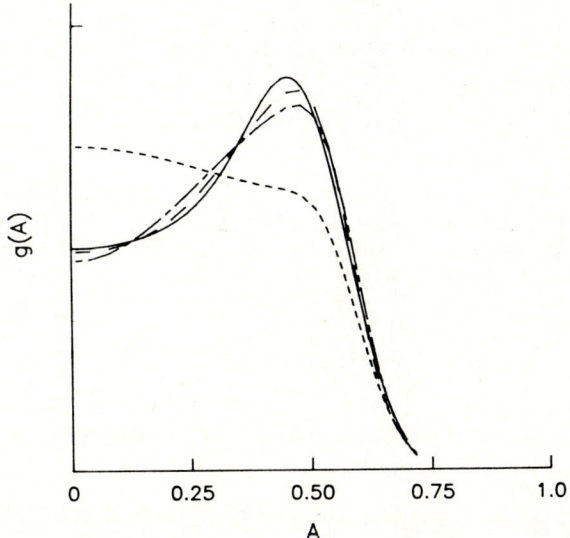

Fig. 3. Approximation of the same pdf as in fig. 2 by an expansion in terms of polynomials orthogonal with respect to the steepest-descents result: (— — —), zero-moment result; (- — - — -), that corrected by two moments: (— — —), that corrected by four moments.

contained in the exact pdf, as a weight function to generate orthogonal polynomials. We have not as yet tried to pursue this line of investigation.

As a measure of the agreement of different approximations we have calculated the quantities

$$I_N = \int_0^{A_{max}} [g(A) - g_N(A)]^2 dA \qquad (45)$$

where $g_N(A)$ is the approximation construed with N moments. Results of this calculation are given in table 1. It is evident that for equal scattering factors both the Edgeworth expansion and the steepest descents approximation perform very well. The Edgeworth expansion gives good results because the Gaussian pdf is close to the exact expression so that only small corrections beyond that are required. On the other hand, when the heterogeneity is great enough so that the Gaussian is not a useful approximation the steepest descents approximation leads to far more convincing agreement, both qualitative and quantitative, with the exact result. It is interesting to observe that the addition of moments beyond the fifth does not appear to change the accuracy of the approximation. While these results indicate the overall superiority of the steepest-descents method for approximating to the pdf, one needs to calculate the characteristic function as a first step. This can be difficult, if not practically impossible for many space groups (although the calculations are indeed possible for the more frequently occurring space groups). An Edgeworth expansion based on more readily computable moments then becomes a tempting alternative.

Table 1

The mean square measure of agreement, I_N, for the Edgeworth expansion and steepest-descents approximation for a unit cell containing 15 atoms. Case A consists of equivalent atoms, $g_1 = g_2 = \cdots = g_{15}$; case B consists of $g_1 = g_2 = \cdots = g_{14} = 1$, $g_{15} = 15\frac{1}{3}$.

N	A		B	
	Edgeworth	Steepest descents	Edgeworth	Steepest descents
0	1.2 (−5)	4.8 (−7)	3.5 (−1)	5.6 (−2)
1	1.2 (−5)	3.5 (−9)	3.5 (−1)	1.4 (−2)
2	9.5 (−9)	9.7 (−12)	3.2 (−1)	2.5 (−3)
3	6.4 (−8)	2.0 (−11)	1.8 (−1)	2.8 (−3)
4	1.3 (−9)	1.5 (−11)	1.1 (−1)	1.2 (−3)
5	8.5 (−10)	6.6 (−12)	7.9 (−2)	3.1 (−4)
6	1.5 (−10)	3.3 (−12)	6.1 (−2)	3.2 (−4)
7	8.8 (−12)	3.3 (−12)	6.0 (−2)	3.0 (−4)
8	1.4 (−11)	3.6 (−12)	6.3 (−2)	3.0 (−4)

Another possibility is to use the Fourier expansion of eq. (22), for example, with an approximation based on expanding the characteristic function in terms of moments,

$$C(\omega) = \int_{-F_m}^{F_m} g(A) \exp(i\omega A) \, dA = \sum_{n=0}^{\infty} \frac{(-1)^n}{(2n)!} \omega^{2n} \mu_{2n}, \qquad (46)$$

truncating the expansion, and fitting the result to an Edgeworth series. This form of Edgeworth approximation would have the possible advantage that the tails of the density function might be better reproduced. We have not carried out any detailed calculations with this idea as yet.

References

[1] E.W. Montroll and B.J. West, in: Fluctuation Phenomena, Studies in Statistical Mechanics VII, eds. E.W. Montroll and J.L. Lebowitz (North-Holland, Amsterdam, 1979).
[2] E.W. Montroll, J. Chem. Phys. 41 (1950) 2241.
[3] E.W. Montroll, J. Soc. Ind. Appl. Math. 4 (1956) 209.
[4] E.W. Montroll, in: Proc. 16th Symp on Applied Mathematics (American Mathematical Society, Providence, RI, 1964) p. 36.
[5] E.W. Montroll and G.H. Weiss, J. Math. Phys. 6 (1965) 167.
[6] H. Scher and E.W. Montroll, Phys. Rev. B12 (1975) 2455.
[7] U. Landman, E.W. Montroll and M.F. Shlesinger, Proc. Nat. Acad. Sci. 74 (1977) 430.
[8] H. Scher, S. Alexander and E.W. Montroll, Proc. Nat. Acad. Sci. 72 (1980) 3758.
[9] B.D. Hughes, E.W. Montroll and M.F. Shlesinger, J. Stat. Phys. 30 (1983) 273.
[10] R. Srinivisan and S. Parthasarathy, Some Statistical Applications in X-ray Crystallography (Pergamon, New York, 1976).
[11] M.G. Kendall and A. Stuart, The Advanced Theory of Statistics, Vol. 1 (Hafner, New York, 1963).
[12] E.W. Montroll, J. Chem. Phys. 10 (1942) 218.
[13] E.W. Montroll, J. Chem. Phys. 11 (1943) 481.
[14] E.W. Montroll and D.C. Peaslee, J. Chem. Phys. 12 (194) 98.
[15] C. Domb, A.A. Maradudin, E.W. Montroll and G.H. Weiss, Phys. Rev. 115 (1959) 18, 24.
[16] G.H. Weiss and J.E. Kiefer, J. Phys. A16 (1983) 489.
[17] U. Shmueli, G.H. Weiss, J.E. Kiefer and A.J.C. Wilson, Acta Crystallogr. A40 (1984) 551.
[18] A.J.C. Wilson, Nature 150 (1942) 152.
[19] A.J.C. Wilson, Acta Crystallogr. 2 (1949) 318.
[20] K. Pearson, Nature 72 (1905) 294.
[21] J.C. Kluyver, Kon. Akad. Wet. Amsterdam 8 (1906) 325.
[22] M.M. Woolfson, An Introduction to X-ray Crystallography (Cambridge University Press, Cambridge, 1970).
[23] U. Shmueli and U. Kaldor, Acta Crystallogr. A39 (1983) 601.
[24] J. Karle, Acta Crystallogr. B28 (1972) 3362.
[25] J. Karle and R.D. Gilardi, A29 (1973) 401.
[26] H.E. Daniels, Ann. Math. Stat. 25 (1954) 631.

[27] R. Barakat, J. Phys. A6 (1973) 796.
[28] R. Barakat, Opt. Acta 21 (1974) 903.
[29] International Tables for X-ray Crystallography, vol. 1, eds. N.F.M. Henry and K. Lonsdale (Kynoch, Birmingham, 1965).
[30] J.E. Kiefer and G.H. Weiss, Comput. and Math. 7 (1981) 327.
[31] D. Rogers and A.J.C. Wilson, Acta Crystallogr. 6 (1953) 439.
[32] C. Giacovazzo, Direct Methods in Crystallography (Academic, New York, 1980).
[33] W. Cochran and M.M. Woolfson, Acta Crystallogr. 8 (1955) 1.
[34] A. Klug, Acta Crystallogr. 11 (1958) 515.
[35] W.A. Hendrickson and E.E. Lattman, Acta Crystallogr. B26 (1970) 136.
[36] U. Shmueli, G.H. Weiss and J.E. Kiefer, Acta Crystallogr., to be published.

On the Quantum Langevin Equation: The Linear Oscillator

Bruce J. WEST

Center for Studies of Nonlinear Dynamics *
La Jolla Institute
P.O. Box 1434, La Jolla, CA 92038
U.S.A.

and

Katja LINDENBERG

Department of Chemistry
University of California at San Diego
La Jolla, CA 92093
U.S.A.

---

* Affiliated with the University of California, San Diego.

The Wonderful World of Stochastics
A Tribute to Elliott W. Montroll
Eds. M.F. Shlesinger and G.H. Weiss

© Elsevier Science Publishers B.V. 1985

Contents

1. Introduction — 191
2. RWA equations of motion — 193
   2.1. Traditional RWA — 193
   2.2. Modified RWA — 196
   2.3. Additional comments — 197
3. Fully coupled equations of motion — 199
4. Summary and conclusions — 201
References — 203

1. Introduction

Elliott W. Montroll was without peer in implementing simple mathematical concepts to model the salient features of complex physical systems. Two of his favorite tools were random walk models [1] to phenomenologically represent the evolution of complex systems, and the harmonic approximation [2] to make the analysis of such systems more or less tractable. He used these ideas with dispatch to study such phenomena as photosynthesis [3], population dynamics [4], the effect of defects on the relaxation of excitations in lattices [5], thermal conductivity in solids [6], quantum statistics [7], as well as to establish quite general nonequilibrium statistical mechanical properties of physical systems [8]. Here we study a model of which he was quite fond: that of a harmonic oscillator coupled to a heat bath [8,9]. We combine a number of Elliott's interests and techniques in this study, as well as those of a number of his friends who are present here today [10–13], and develop a quantum Langevin equation. It is surprising how elusive has been the proper description of the relaxation of a quantum oscillator interacting with a heat bath [10–23].

There has been a recent burst of activity in the study of the thermal relaxation of quantum systems [18–24]. In large part this activity is due to the recognition that the emerging data in such diverse areas as tunneling across Josephson junctions [18], the vibrational relaxation of polyatomic molecules [25–28], exciton transport at low temperatures [29], and laser noise [29,30] cannot be fully understood without a clear picture of the quantum-mechanical source of fluctuations and dissipation caused by the interaction of the system with the surrounding heat bath [8,22]. Many sophisticated techniques have been applied to the description of this problem [17,31,32]. We find that the Brownian motion of a quantum oscillator bilinearly coupled to a quantum-mechanical heat bath of linear oscillators is sufficient to expose and indeed to resolve many of the difficulties previously encountered. We are confident that Elliott would have been pleased that so many questions could be answered using such a simple model.

Instead of proceeding with a historical introduction to this problem, let us merely point out that in most of the preceding analyses one or more of the following approximations have been made: (i) the *Markovian limit* in which the dissipation is taken to be instantaneous and the fluctuations are delta-correlated in time; (ii) the *weak-coupling limit* in which the coupling between the Brownian oscillator and the heat-bath oscillators is assumed to be the weakest energy parameter in the system; (iii) the *rotating-wave approximation* (RWA) in which rapidly oscillating terms coupling the oscillator to the heat bath are neglected. In this paper we review our analysis of the meaning and interrelation of these several ubiquitous concepts [22,23]. The three most salient consequences of our analysis are: (a) the time scales for the correlation of fluctuations and for the dissipation can be quite distinct; (b) the traditional implementation of the RWA only gives valid results in the strict weak-coupling limit; (c) a reformulation of the RWA valid at arbitrary coupling strengths and hence valid at arbitrarily low temperatures is possible [22,23].

The dynamics of the quantum Brownian oscillator are characterized by four energy parameters:

(1) The natural frequency $E/\hbar$ of the isolated oscillator.

(2) The coupling strength λE between the oscillator and the heat bath.

(3) The memory time γ^{-1} of the dissipation of the oscillator energy by the heat bath.

(4) The temperature T of the heat bath.

The several approximations (or limiting forms) noted above involve choosing limiting values of these four parameters. Thus, for instance, instantaneous dissipation corresponds to taking $\gamma \to \infty$ while the weak-coupling limit implies $\lambda \to 0$. Prior to the recent work discussed here these various parameter variations have in the main been done independently, i.e. the results have been assumed to be independent of the order in which the limits are taken. We here display the interdependence of the various approximations and, in particular, of the special role played by the temperature in the quantum oscillator system. Towards this end we present a fully quantum-mechanical derivation of a generalized Langevin equation. We obtain a quantum fluctuation–dissipation relation which differs from the classical one and reduces to it in the high temperature limit [33,34]. The main consequence of this relation is that *spontaneous fluctuations in the quantum system can only be taken as Markovian under special circumstances*. Thus, even in the limit $\gamma \to \infty$ where the dissipation is delta-correlated in time, the fluctuations necessarily remain correlated over a *finite* time that depends on the temperature [22]. This in turn induces macroscopic effects whose origin is quantum-mechanical and which are absent in the classical limit.

2. RWA equations of motion

Consider a quantum oscillator with creation and annihilation operator $a^\dagger$ and a, respectively, with natural frequency $E/\hbar$ and Hamiltonian $H_S = E a^\dagger a$ for the isolated oscillator. The oscillator is assumed to be in contact with a heat bath of bosons labeled by index ν, which are created and annihilated by $b_\nu^\dagger$ and b_ν, respectively. The analysis is restricted to a bath of harmonic bosons of frequency ω_ν and Hamiltonian $H_B = \sum_\nu \hbar \omega_\nu b_\nu^\dagger b_\nu$. The coupling between the oscillator and the heat bath is assumed to be bilinear. The Hamiltonian for the composite system is

$$H = H_S + H_B + H_{SB} \tag{2.1}$$

and the interaction Hamiltonian H_{SB} distinguishes one linear coupled system from another. In this section we discuss the choice [14–17,25–30],

$$H_{SB} = \lambda^{1/2} \sum_\nu \Gamma_\nu \left(a^\dagger b_\nu + a b_\nu^\dagger \right), \tag{2.2}$$

which we refer to as the RWA-oscillator [cf. eq. (3.1)]. The Γ's are real coupling coefficients and λ is a parameter that measures the average strength of the interaction.

2.1. Traditional RWA [12,13]

The dynamical equations for the system operators are given by

$$\dot{a} = \frac{\mathrm{i}}{\hbar}[H, a] \tag{2.3}$$

and its Hermitian conjugate. To eliminate the bath operators from (2.3) we similarly write the dynamical equations for the set $\{b_\nu\}$ and solve these in terms of the initial state of the bath. Because the system is linear these latter equations can be *solved exactly* and substituted into the former to yield the traditional RWA equations of motion

$$\dot{a} = -\mathrm{i}\frac{E}{\hbar} a - \mathrm{i}\bar{f}_{\text{RWA}}(t) - \int_0^t \mathrm{d}\tau \bar{K}_{\text{RWA}}(t - \tau) a(\tau), \tag{2.4}$$

where

$$\tilde{f}_{RWA}(t) = \frac{\lambda^{1/2}}{\hbar} \sum_\nu \Gamma_\nu b_\nu(0) e^{-i\omega_\nu t} \qquad (2.5a)$$

and

$$\bar{K}_{RWA}(t) = \frac{\lambda}{\hbar^2} \sum_\nu \Gamma_\nu^2 e^{-i\omega_\nu t}. \qquad (2.5b)$$

The operator $\tilde{f}_{RWA}(t)$ is interpreted as fluctuating because of the uncertainty in the specification of the initial state of the bath. The distribution of *initial* bath states is taken to be canonical [8,10], i.e.

$$\bar{\rho}_0 = Z^{-1}(\beta) \exp(-\beta H_B), \qquad (2.6)$$

where $\beta = 1/kT$ and $Z(\beta)$ is the partition function $Z(\beta) = \text{Tr}[\exp(-\beta H_B)]$. The fluctuations are thus zero-centered and Gaussian, with correlation functions ($\langle \theta \rangle \equiv \text{Tr}[\bar{\rho}_0 \theta]$)

$$\langle \tilde{f}_{RWA}(t) \tilde{f}_{RWA}(t') \rangle = 0, \qquad (2.7a)$$

$$\langle \tilde{f}^\dagger_{RWA}(t) \tilde{f}_{RWA}(t') \rangle = \frac{\lambda}{\hbar^2} \sum_\nu \Gamma_\nu^2 n_\nu e^{i\omega_\nu(t-t')}, \qquad (2.7b)$$

$$\langle \tilde{f}_{RWA}(t) \tilde{f}^\dagger_{RWA}(t') \rangle = \frac{\lambda}{\hbar^2} \sum_\mu \Gamma_\nu^2 (n_\nu + 1) e^{-i\omega_\nu(t-t')}, \qquad (2.7c)$$

where n_ν is the Bose population $[e^{\beta\hbar\omega_\nu} - 1]^{-1}$.

The standard interpretation of (2.4) is to consider the integral term to be dissipative, and to construct a fluctuation–dissipation relation between $\tilde{f}_{RWA}(t)$ and the memory kernel $\bar{K}_{RWA}(t)$. It is argued that the correlation functions (2.7) always occur as integrands multiplied by $\exp\{\pm iE(t-t')/\hbar\}$ in integrals over time. This procedure, it is said, essentially filters the correlation function and selects only the $\hbar\omega_\nu = E$ component. The set (2.7) is consequently replaced by the "equivalent" set

$$\langle \tilde{f}^\dagger_{RWA}(t) \tilde{f}_{RWA}(t') \rangle = \frac{\lambda}{\hbar^2} n_E \sum_\nu \Gamma_\nu^2 e^{i\omega_\nu(t-t')}, \qquad (2.8a)$$

$$\langle \tilde{f}_{RWA}(t) \tilde{f}^\dagger_{RWA}(t') \rangle = \frac{\lambda}{\hbar^2} (n_E + 1) \sum_\nu \Gamma_\nu^2 e^{-i\omega_\nu(t-t')}. \qquad (2.8b)$$

The argument thus culminates in the "fluctuation–dissipation relation"

$$\langle \tilde{f}_{RWA}(t) \tilde{f}^\dagger_{RWA}(t') \rangle = (n_E + 1) \bar{K}_{RWA}(t-t') \qquad (2.9)$$

in direct analogy with the classical result in which the fluctuation correlation function is directly proportional to the dissipative memory kernel in the generalized Langevin equation. Finally, it is also customary to make the Markovian approximation $\bar{K}_{RWA}(t-\tau) = 2\lambda E \delta(t-\tau)$. With these assumptions it is easily shown that the equilibrium population ($t \to \infty$) of the oscillator is n_E and that the relaxation toward this value is exponential and of rate $2\lambda E/\hbar$.

We emphasize that no explicit restriction has been placed on the temperature in formulating the preceding arguments. Let us consider the situation in which the memory kernel has a very broad and flat frequency spectrum. In this case $\bar{K}_{RWA}(t-\tau)$ is sharply peaked and the n_ν population factor appearing in the correlation functions (2.7) causes a logarithmic divergence at low frequencies ($\Gamma_\nu^2 n_\nu \sim \Gamma_\nu^2/\omega_\nu$ for small ω_ν). Conversely, a sharply peaked correlation function corresponds to a $\bar{K}_{RWA}$ which is unphysical. The "fluctuation–dissipation relation" (2.9) so often assumed is therefore based on an approximation which replaces a divergent quantity with a constant. The fallacy in the argument is that the proper multiplication factor in the integrands containing correlation functions is not $\exp[\pm iE(t-t')/\hbar]$ but is instead $\exp[(\pm iE - \lambda E)(t-t')/\hbar]$. The damping factor induces a contribution to the integral from all frequency components and not just the $\hbar\omega_\nu = E$ component. Especially important are the low frequencies since these provide a divergent contribution when $\bar{K}_{RWA}(t)$ is assumed to be sharply peaked.

The traditional arguments thus present us with the following dilemma: either the memory kernel is sharply peaked and consequently the correlation function of the fluctuations diverges except in the *strict* $\lambda \to 0$ limit, or the correlation function is sharply peaked and therefore the dissipative memory takes on the unphysical form of the derivative of a delta function. The traditional choice of a delta-correlated memory kernel thus leads to incorrect results (except in the *strict* $\lambda \to 0$ limit). We therefore conclude that $\bar{K}_{RWA}(t-\tau)$ is in general *not* a dissipative memory kernel and thus is not necessarily related to $\bar{f}_{RWA}(t)$. The errors introduced by the traditional arguments are most serious at low temperatures ($kT \ll \lambda E$) since in this regime the limit $\lambda \to 0$ is obviously not met. Therefore the Bose distribution and the simple exponential relaxation obtained in the RWA may not be (and in fact are not) valid for these temperatures.

2.2. Modified RWA [22,23]

The ambiguities and inconsistencies in the traditional procedure are consequences of a misinterpretation of the terms in the original dynamic equation (2.4). The proper interpretation of (2.4) depends on the identification of the dissipation. Consider the rate of change of the momentum operator $p = iE^{1/2}(a^\dagger - a)$, which using (2.4) is

$$\dot{p} = -\left(\frac{E}{\hbar}\right)^2 q + 2E^{1/2}\,\text{Re}\,\tilde{f}_{\text{RWA}}(t) - \int_0^t d\tau\,\bar{K}_{\text{RWA}}(t-\tau)\,p(t). \quad (2.10)$$

The term that had been interpreted as the dissipation is proportional to the *momentum* of the oscillator. The *correct* dissipative force is proportional to the oscillator *velocity* $\dot{q}$. In the RWA, p and $\dot{q}$ are not proportional to one another and the distinction becomes crucial.

The equality $\dot{q} = (\hbar/E)^{1/2}(\dot{a}^\dagger + \dot{a})$ suggests that in (2.4) we integrate by parts to obtain:

$$\dot{a}(t) = -i\frac{\varepsilon}{\hbar}a - if_{\text{RWA}} - i\hbar\int_0^t d\tau\,K_{\text{RWA}}(t-\tau)\dot{a}(\tau), \quad (2.11)$$

where ε is the shifted energy, $\varepsilon = E - \hbar^2 K_{\text{RWA}}(0)$ and where

$$f_{\text{RWA}}(t) = \frac{\lambda^{1/2}}{\hbar}\sum_\nu \Gamma_\nu\left[b_\nu(0) + \lambda^{1/2}\frac{\Gamma_\nu}{\hbar\omega_\nu}a(0)\right]e^{-i\omega_\nu t}$$

$$= \sum_\nu f_\nu^{\text{RWA}}(t) \quad (2.12a)$$

and

$$K_{\text{RWA}}(t-\tau) = \frac{\lambda}{\hbar^2}\sum_\nu \frac{\Gamma_\nu^2}{\hbar\omega_\nu}e^{-i\omega_\nu(t-\tau)}. \quad (2.12b)$$

If we use these definitions to construct the force law $\dot{p}$ from (2.11), we obtain an equation analogous to (2.10) but with $p(\tau)$ in the integrand replaced with $\dot{q}(\tau)$. We now have some confidence that $K_{\text{RWA}}(t)$ is a "dissipative" kernel.

We can see that (2.12a) differs from its predecessor (2.5a), but we can again interpret it as fluctuating due to the uncertainty in the initial state of the bath. To specify the statistical properties of the fluctuations we consider an ensemble of initial states in which the system operators are fixed at the values $(a^\dagger(0), a(0))$ and the initial bath operators are drawn from an ensemble that is canonical relative to the system [35]. The initial distribution is then the conditional density matrix [8,10,22,36,37]

$$\rho_c = Z_c^{-1}(\beta)\exp\{-\beta H_B^C\}, \quad (2.13)$$

where

$$H_B^C = \sum_\nu \hbar\omega_\nu \left(b_\nu^\dagger + \frac{\lambda^{1/2}\Gamma_\nu a^\dagger}{\hbar\omega_\nu}\right)\left(b_\nu + \frac{\lambda^{1/2}\Gamma_\nu a}{\hbar\omega_\nu}\right). \quad (2.14)$$

With this choice of distribution function the statistics of $f_{\text{RWA}}(t)$ are identical to those of $\bar{f}_{\text{RWA}}(t)$ defined with respect to (2.6), i.e. it is a zero-centered Gaussian process with correlation functions given by (2.7).

Now we can relate the "dissipative" kernel to the fluctuation correlation function

$$\Phi_\nu^{\text{RWA}}(t-\tau) = \lambda \langle f_\nu^{\text{RWA}}(t) f_\nu^{\text{RWA}\dagger}(\tau) + f_\nu^{\text{RWA}\dagger}(\tau) f_\nu^{\text{RWA}}(t)\rangle \quad (2.15)$$

by means of the quantum-mechanical fluctuation–dissipation relation (FDR)

$$K_{\text{RWA}}(t-\tau) = \sum_\nu \Phi_\nu^{\text{RWA}}(t-\tau) \frac{\tanh(\hbar\beta\omega_\nu/2)}{\hbar\omega_\nu}. \quad (2.16)$$

In the classical limit ($\hbar \to 0$) this relation reduces to the familiar form

$$K_{\text{RWA}}(t-\tau) = \beta \Phi_{\text{RWA}}(t-\tau). \quad (2.17)$$

Note that (2.16) is valid *regardless of the value of the coupling coefficient* λ.

The quantum-mechanical FDR has been obtained by many authors in a variety of forms [11,12,21,34,35], but, with one notable exception [16], its consequences seem not to have been explored. In particular, an analysis of (2.16) reveals the remarkable property that the correlation time τ_c of the fluctuations and the decay time γ^{-1} of the dissipative kernel are in general quite different. It is only in the high-temperature limit ($kT \gg \hbar\gamma$) that these time scales coincide ($\tau_c \approx \gamma^{-1}$). At low temperatures $\tau_c \sim \hbar\beta$, i.e. the correlation function decays on a temperature-dependent time scale. We interpret this result to mean that the bath can *dissipate* quanta of energy in the range $(0, \hbar\gamma)$ while the *spontaneous fluctuations* occur only in the energy range $(0, kT)$ if $kT < \hbar\gamma$. The correlation time of the fluctuations is therefore the longer of $\hbar/kT$ and γ^{-1}. As the temperature is lowered, the correlations in the fluctuations become increasingly long lived even for an infinitely broad band bath spectrum.

2.3. Additional comments

A touchstone that has been used to test the validity of various heuristic equations of motion is the condition that the correct commutation relations must be satisfied on the average at all times. These relations are $\langle[a, a^\dagger]\rangle = 1$, $\langle[a, a]\rangle = 0$ or their equivalent $\langle[q, p]\rangle = i\hbar$. Now consider the traditional RWA oscillator in the "Markovian limit" [cf. (2.10)]

$$\dot{q} = p, \tag{2.18a}$$

$$\dot{p} = -\left(\frac{E}{\hbar}\right)^2 q - \lambda \frac{E}{\hbar} p + F(t), \tag{2.18b}$$

where $F(t) \equiv 2E^{1/2} \operatorname{Re} \tilde{f}_{\text{RWA}}(t)$. For the fluctuations Senitzky [15] chooses a correlation function which results in the commutation relation

$$\langle[F(t), F(\tau)]\rangle = \frac{2i\lambda E^2}{\hbar \pi} P\left[\frac{1}{t-\tau}\right], \tag{2.19}$$

where $P(\cdot)$ indicates the principal value. Lax [12] shows that with this commutation rule one obtains the incorrect commutator

$$\langle[q, p]\rangle = i\hbar\left[1 - \frac{\lambda}{\pi}\right] + O(\lambda^2). \tag{2.20}$$

On this basis he concludes that the correct commutator for the Langevin force should be

$$\langle[F(t), F(\tau)]\rangle = 2i\lambda E \delta(t - \tau) \tag{2.21}$$

since this leads *exactly* to the correct commutation relation $\langle[q, p]\rangle = i\hbar$. This latter choice is also made by Benguria and Kac [19] and by us [22,23]. We note that the forms (2.19) and (2.21) are consequences of different assumptions for the coupling coefficients in the Hamiltonian [22].

The difficulty does not lie in the choice (2.19). The problem instead is that the "dissipation" associated with the fluctuations in (2.18b) does not contain a similar principal value [38]. Said differently, the choice made by Senitzky for the fluctuations is inconsistent with the FDR associated with (2.18). In the RWA if one wishes to use (2.19) it is necessary to replace (2.18) with

$$\dot{p} = -\left(\frac{E}{\hbar}\right)^2 q - \lambda \frac{E}{\hbar} p + \frac{i\lambda E}{\pi \hbar} P\left(\frac{1}{t-\tau} p(\tau)\right) + F(t) \tag{2.22}$$

as the proper equation to associate with Senitzky's choice of fluctuation commutator. Thus both choices [i.e. (2.18b) with (2.2) *or* (2.22) with (2.19)] yield the *exact* commutation relations $\langle[q, p]\rangle = i\hbar$. Both choices of commutator are correct along with their respective FDR's, but they correspond to different couplings in the Hamiltonian [22].

As a final comment on the RWA oscillator we note that the equilibrium population of the oscillator to $O(\lambda)$ can be written as

$$n_{\text{RWA}}(T) = \lim_{t \to \infty} \langle a^\dagger(t) a(t) \rangle_{\text{RWA}} = \lambda \sum_\nu \frac{\Gamma_\nu^2 n_\nu}{\lambda^2 \varepsilon^2 + (\varepsilon - \hbar \omega_\nu)^2}. \qquad (2.23)$$

At high temperatures ($kT \gg \varepsilon$) we find that $n_{\text{RWA}}(T) \to (e^{\beta \varepsilon} - 1)^{-1}[1 + O(\lambda)]$, i.e. the Bose population, and one thus recovers the usual "weak-coupling" result. At low temperatures, however, the temperature dependence of the population is quite different and has the form

$$n_{\text{RWA}}(T) \propto \lambda \left(\frac{kT}{\varepsilon}\right)^2 [1 + O(kT/\varepsilon)]. \qquad (2.24)$$

This deviation from the "anticipated" Bose distribution should be readily observable.

3. Fully coupled (FC) equations of motion

We now consider the more general ("fully coupled") interaction Hamiltonian [12,22,28]

$$H_{\text{SB}} = \lambda^{1/2} \sum_\nu \Gamma_\nu (a^\dagger + a)(b_\nu^\dagger + b_\nu) \qquad (3.1)$$

which differs from the RWA Hamiltonian (2.2) in that terms of the form $a^\dagger b_\nu^\dagger$ and ab_ν are here included. These are the rapidly oscillating terms that are omitted in the RWA. Equation (3.1) is a bilinear interaction between the oscillator and the heat bath involving only the displacement, i.e. the interaction is proportional to qq_ν. Again we solve the equations for the heat bath and substitute these back into the dynamic equations for the oscillator to obtain

$$\dot{a} = -i\frac{\varepsilon}{\hbar} a + i\Delta a^\dagger - if_{\text{FC}} - i\hbar \int_0^t d\tau K_{\text{FC}}(t - \tau)[\dot{a}(\tau) + \dot{a}^\dagger(\tau)] \qquad (3.2)$$

and its Hermitian conjugate. Here $\Delta = 2\lambda \sum_\nu \Gamma_\nu^2 / \hbar^3 \omega_\nu$; ε is the shifted energy $\varepsilon = E - \hbar \Delta$ and

$$f_{\text{FC}}(t) = \frac{\lambda^{1/2}}{\hbar} \sum_\nu \Gamma_\nu [F_\nu^\dagger(t) + F_\nu(t)] = \lambda^{1/2} \sum_\nu f_\nu^{\text{FC}}(t), \qquad (3.3a)$$

where

$$F_\nu(t) \equiv \left\{ b_\nu^\dagger(0) + \frac{\lambda^{1/2} \Gamma_\nu}{\hbar \omega_\nu} [a^\dagger(0) + a(0)] \right\} e^{i\omega_\nu t}. \qquad (3.3b)$$

The kernel in (3.2) is given by

$$K_{FC}(t-\tau) = \frac{2\lambda}{\hbar^2} \sum_\nu \frac{\Gamma_\nu^2}{\hbar\omega_\nu} \cos\omega_\nu(t-\tau). \tag{3.3c}$$

Note that $f_{FC}(t)$ and $K_{FC}(t)$ are real for the FC oscillator and that the latter contains no principal value contributions.

Recall that $q = (\hbar/E^{1/2})(a^\dagger + a)$ so that the memory kernel is multiplied by the velocity operator $\dot{q}$. The last term is recognized as a dissipation. We interpret $f_{FC}(t)$ as a fluctuating operator, again due to the uncertainty in the specification of the initial state of the heat bath conditioned on $(a^\dagger(0), a(0))$ [35]. The initial distribution is then the conditional density matrix [8,10,36,37]

$$\rho_0 = Z_{FC}^{-1}(\beta) \exp\{-\beta H_{FC}\}, \tag{3.4}$$

where $Z_{FC}(\beta) = \text{Tr}[\exp(-\beta H_{FC})]$ is the partition function and [using (3.6b)]

$$H_{FC} = \sum_\nu \hbar\omega_\nu \left[b_\nu^\dagger + \frac{\lambda^{1/2}\Gamma_\nu}{\hbar\omega_\nu}(a^\dagger+a)\right]\left[b_\nu + \frac{\lambda^{1/2}\Gamma_\nu}{\hbar\omega_\nu}(a^\dagger+a)\right]. \tag{3.5}$$

In this ensemble the fluctuations $F_\nu(t)$ are zero-centered and Gaussian, i.e.

$$\langle F_\nu(t)\rangle = 0, \tag{3.6}$$

where $\langle\theta\rangle = \text{Tr}(\rho_0\theta)$. The correlation functions of the fluctuations are

$$\langle F_\nu(t) F_{\nu'}(t')\rangle = 0, \tag{3.7a}$$

$$\langle F_\nu^\dagger(t) F_{\nu'}(t')\rangle = n_\nu e^{i\omega_\nu(t-t')}\delta_{\nu\nu'}, \tag{3.7b}$$

$$\langle F_\nu(t) F_{\nu'}^\dagger(t')\rangle = (n_\nu+1)e^{-i\omega_\nu(t-t')}\delta_{\nu\nu'}, \tag{3.7c}$$

and the symmetrized correlation function is

$$\Phi_\nu^{FC}(t-\tau) = \lambda\langle f_\nu^{FC}(t) f_\nu^{FC}(\tau) + f_\nu^{FC}(\tau) f_\nu^{FC}(t)\rangle. \tag{3.8}$$

The quantum-mechanical FDR for the fully coupled oscillator is calculated to be

$$K_{FC}(t-\tau) = \sum_\nu \Phi_\nu^{FC}(t-\tau)\frac{\tanh(\hbar\beta\omega_\nu/2)}{\hbar\omega_\nu} \tag{3.9}$$

just as in the case of the modified RWA [cf. (2.16)]. Again we note that the FDR is independent of the coupling parameter λ. We could again go to the classical limit ($\hbar \to 0$) in which the dissipation kernel is proportional to the correlation function of the fluctuations. A straightforward analysis shows that in the case $kT \gg \gamma \gg E/\hbar$, $\Phi^{FC}(t) \equiv \Sigma_\nu \Phi_\nu^{FC}(t)$ decays on the time scale γ^{-1}, i.e. $\Phi^{FC}(t-\tau) \sim \lambda kT\gamma e^{-\gamma|t-\tau|}$. At lower temperatures ($kT \ll \hbar\gamma$) an analysis of (3.9) reveals that the correlation function decays on a time scale $\hbar/kT$ rather than γ^{-1}, i.e. essentially

$$\Phi^{FC}(t-\tau) \sim \lambda \frac{(kT)^2}{\hbar} e^{-kT|t-\tau|/\hbar}.$$

Here again the idea is that *fluctuations and dissipation can have quite distinct time scales*, as for the modified RWA oscillator.

The time- and temperature-dependent population of the FC oscillator,

$$n_{FC}(T, t) = \langle a^\dagger(t) a(t) \rangle_{FC}, \tag{3.10}$$

has been calculated [22] and exponentially relaxes at a rate $2\lambda E/\hbar$. Here we merely record the zero-temperature equilibrium population [22]:

$$n_{FC}(0, \infty) = \frac{\lambda}{\pi} \ln(\hbar\gamma/E) - \lambda/2 + O(\lambda E/\hbar\gamma) \tag{3.11}$$

for weak but finite coupling parameter λ. The small-T behavior is given by

$$n_{FC}(T, \infty) - n_{FC}(0, \infty) = \frac{4\lambda}{\pi} \left(\frac{kT}{E}\right)^2 \left[\zeta(2) + O\left((kT/E)^2\right)\right] \tag{3.12}$$

where $\zeta(\cdot)$ is the zeta function. Equation (3.12) is to be contrasted with the weak-coupling result

$$n_{FC}(T, \infty) - n_{FC}(0, \infty) = (e^{\beta E} - 1)^{-1} [1 + O(\lambda)] \tag{3.13}$$

inadvisedly used at arbitrarily low temperatures.

We emphasize that in the above we took λ to be small, a condition often called the "weak-coupling limit". However, from this analysis it is clear that weak coupling as it is traditionally used is achieved only at temperatures $kT \gg \lambda E$ in addition to λ being small. At sufficiently low temperatures ($kT \ll \lambda E$) the effect of the coupling is macroscopically observable as a deviation from the Bose population (3.13).

4. Summary and conclusions

A number of general implications about quantum-mechanical dynamical systems are contained in the foregoing analysis. Firstly the utility of a Langevin description for the dynamics of a system lies in the relation one can construct between the fluctuations and the dissipation. This relation insures that a thermodynamically closed system will equilibrate. Only with the proper identification of the dissipative terms in the dynamical equations is one able to construct a generalized FDR [cf. (3.9)]. At high temperatures, (2.16) and (2.9) reduce to the classical relation

$$K(t-\tau) = \beta \Phi(t-\tau)$$

so that the correlation of the fluctuations and the dissipative memory kernel decay on the same time scale. At arbitrary temperatures this is not the case, however. If γ^{-1} is the decay time of $K(t-\tau)$, then the correlations in the fluctuations relax on the *longer* of the two time scales $(\hbar\beta)^{-1}$ or γ^{-1}. Thus at low temperatures ($kT \ll \hbar\gamma$) the fluctuations have long persistence times even if the dissipation is instantaneous.

The second implication is a consequence of the fact that the notions of temperature and of statistical mechanical fluctuations enter in the analysis through the specification of bath initial conditions. We note the remarkable fact that in the *classical* limit with the choices (2.13) and (3.4) the oscillator relaxes to a canonical distribution proportional to $\exp[-\beta H_S]$ *regardless of the coupling strength* λ. In the quantum case the canonical distribution is attained only in the weak-coupling limit.

A third general conclusion has to do with the non-uniqueness of the quantum-mechanical fluctuations and dissipation. In the literature the veracity of different heuristic equations of motion has been assessed by determining whether the correct commutation relations are satisfied on the average at all times by the oscillator operators if they are satisfied initially (cf. section 2.3) [12,13]. We have shown that this test is in turn satisfied if the proper FDR is contained in the model. The apparent freedom in the choice of commutation relation for the fluctuations is illusory. It arises only in phenomenological or heuristic discussions of the equations of motion. When the equations are derived from a Hamiltonian and the full Hamiltonian dynamics are preserved at all stages of the analysis, the commutation relations are *predetermined*. Different forms of coupling in the Hamiltonian lead to different commutation relations for the fluctuations.

A final point of interest is that the quantum Langevin equation does not in general have a Markovian limit because the fluctuations cannot be delta-correlated in time (even if the dissipative kernel is). This is a direct consequence of the quantum FDR, since $K(t-\tau)$ and $\Phi(t-\tau)$ can only have the same time scale in the classical limit ($\hbar \to 0$).

Thus we conclude that:

(1) the time scales for the correlation of fluctuations and for the dissipation can be quite distinct;

(2) the traditional implementation of the RWA only gives valid results in the strict weak-coupling limit;

(3) a reformulation of the RWA valid at arbitrary coupling strengths, and hence at arbitrarily low temperatures, is possible.

References

[1] E.W. Montroll and B.J. West, in: Fluctuation Phenomena, Studies in Statistical Mechanics VII, eds. E.W. Montroll and J.L. Lebowitz (North-Holland, Amsterdam, 1979) pp. 62–175.
E.W. Montroll and M.F. Shlesinger, in: Nonequilibrium Phenomena II, From Stochastics to Hydrodynamics, Studies in Statistical Mechanics XI, eds. J.L. Lebowitz and E.W. Montroll (North-Holland, Amsterdam, 1984) p. 1.
E.W. Montroll and G.H. Weiss, J. Math. Phys. 6 (1965) 178.
[2] A.A. Maradudin, E.W. Montroll, G.H. Weiss and I.P. Ipatova, Theory of Lattice Dynamics in the Harmonic Approximation, 2nd Ed. (Academic, New York, 1971).
[3] E.W. Montroll, J. Math. Phys. 10 (1969) 753.
[4] N.S. Goel, S.C. Maitra and E.W. Montroll, Rev. Mod. Phys. 43 (1971) 231.
E.W. Montroll and B.J. West, in: Synergetics, Cooperative Phenomena in Multi-Component Systems, ed. H. Haken (Teubner, Stuttgart, 1973) pp. 143–156.
[5] E.W. Montroll and R.B. Potts, Phys. Rev. 100 (1955) 525; 102 (1955) 72.
[6] E.W. Montroll, Quart. Appl. Math. V (1947) 223.
[7] E.W. Montroll and J. Ward, Phys. Fluids 1 (1958) 55.
[8] E.W. Montroll, in: Lectures in Theoretical Physics, Vol. 3 (Wiley–Interscience, New York, 1961) pp. 221–325.
[9] E.W. Montroll and K.E. Shuler, J. Chem. Phys. 26 (1957) 454.
[10] R. Zwanzig, in: Lectures in Theoretical Physics, Vol. 3 (Wiley–Interscience, New York, 1961) pp. 106–141; J. Stat. Phys. 7 (1973) 215.
[11] G.W. Ford, M. Kac and P. Mazur, J. Math. Phys. 6 (1965) 504.
[12] M. Lax, Phys. Rev. 129 (1963) 2342; 145 (1966) 109.
[13] H. Haken, Laser Theory, in: Encyclopedia of Physics, Vol. XXV/2C (Springer, Berlin, 1970); Rev. Mod. Phys. 42 (1975) 67.
[14] J. Schwinger, J. Math. Phys. 2 (1961) 407.
[15] I.R. Senitzky, Phys. Rev. 119 (1960) 670; 124 (1961) 642.
[16] P. Ullersma, Physica 32 (1966) 27.
[17] G.S. Agarwal, Phys. Rev. 4A (1971) 739.
[18] A.O. Caldeira and A.J. Leggett, Ann. Phys. 149 (1983) 374.
[19] R. Benguria and M. Kac, Phys. Rev. Lett. 46 (1980) 1.
[20] H. Grabert, U. Weiss and P. Talkner, to appear in Z. Phys. B.
F. Haake and R. Reibold, to be published.
[21] K. Lindenberg and B.J. West, Phys. Rev. Lett 51 (1983) 1370; also in: Random Walks and their Applications in the Physical and Biological Sciences eds. M. Shlesinger and B.J. West, AIP Conf. Proc. 109 (1983) 111.
[22] K. Lindenberg and B.J. West, Phys. Rev. A30 (1984) 568.
[23] B.J. West and K. Lindenberg, Phys. Lett. 102A (1984) 189.
[24] P. Ruggiero and M. Zannetti, Phys. Rev. 28A (1977) 987.
[25] A. Nitzan and J. Jortner, Mol. Phys. 25 (1973) 713.
[26] A. Nitzan and R. Silbey, J. Chem. Phys. 60 (1974) 4070.

[27] S.H. Lin and H. Eyring, Ann. Rev. Phys. Chem. 25 (1974) 39.
[28] M.J. Ondrechen, A. Nitzan and M.A. Ratner, Chem. Phys. 16 (1976) 49.
[29] W.H. Louisell, Radiation and Noise in Quantum Electronics (McGraw-Hill, New York, 1964); Quantum Statistical Properties of Radiation (Wiley, New York, 1973).
[30] M. Sargent III, M.O. Scully and W.E. Lamb Jr., Laser Physics (Addison–Wesley, Reading, MA, 1974).
[31] H. Mori, Prog. Theor. Phys. (Kyoto) 34 (1965) 423.
[32] P. Mazur and I. Oppenheim, Physica 50 (1970) 241.
[33] W. Bernard and H.B. Callen, Rev. Mod. Phys. 31 (1959) 1017.
[34] R. Kubo, Rep. Prog. Phys. 29 (1966) 255.
[35] For a discussion of the significance of the choice of the ensemble of initial bath states see E. Cortes, K. Lindenberg and B.J. West, to be published.
[36] K. Lindenberg and V. Seshadri, Physica 109A (1981) 483.
[37] B.J. West, Phys. Rev. 25A (1982) 1683.
[38] Such a principal value contribution to the memory kernel $K(t-t')$ is not really dissipative; hence the quotation marks.

CHAPTER 10

Some Inequalities for Anisotropic Rotators *

J. BRICMONT

Institut de Physique Théorique
2 Chemin du Cyclotron, B-1348 Louvain-La-Neuve
Belgium

J.L. LEBOWITZ

Department of Mathematics and Physics
Rutgers University, New Brunswick, NJ 08903
USA

C.E. PFISTER

Département de Mathématique
Ecole Polytechnique Fédérale, CH-1015 Lausanne
Switzerland

* Supported in part by NSF Grant N^r DMR 81-14726-02

© Elsevier Science Publishers B.V. 1985

The Wonderful World of Stochastics
A Tribute to Elliott W. Montroll
Eds. **M.F. Shlesinger** and G.H. Weiss

Contents

1. Introduction ... 207
2. Results ... 208
   2.1. Decay of correlations ... 208
   2.2. Phase diagram ... 209
3. Remarks ... 211
References ... 213

1. Introduction

We consider a three-component rotator system on a d-dimensional lattice with ferromagnetic interactions,

$$H = -J \sum_{\langle ij \rangle} \left[s_x(i) s_x(j) + s_y(i) s_y(j) + \lambda s_z(i) s_z(j) \right],$$

$$s_x^2(i) + s_y^2(i) + s_z^2(i) = 1 \quad \forall i \in \mathbb{Z}^d, \quad J, \lambda \geq 0. \tag{1}$$

We shall be interested mainly in the case $\lambda < 1$ (easy-plane anisotropy) and $d = 2$. This case appears to be related to physical situations (e.g., $\lambda = 0.99$ seems to be a good model for the planar magnet K_2CuF_4 [1]) and was investigated numerically in ref. [2]. Evidence was found in favour of the following behaviour (see fig. 1):

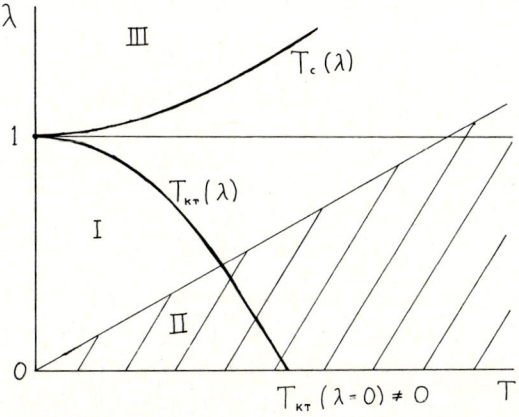

Fig. 1. (I) KT phase (conjectured). (II) Region where it is proven (without any assumption) that $\langle s_z(0) s_z(i) \rangle$ decays exponentially. (Inequality 3.) (III) Spontaneous magnetization along the z component (proven) with $\lambda_c(T) - 1 \leq e^{-c\beta}$ [16].

(a) There is a Kosterlitz–Thouless (KT) transition at a temperature

$$T_{KT}(\lambda) \approx T_{KT}(\lambda = 0)/\ln(1-\lambda)^{-1/2} \quad (2)$$

(for λ close to 1). For temperatures below $T_{KT}(\lambda)$, the correlation functions $\langle s_x(0) s_x(i)\rangle$ and $\langle s_y(0) s_y(i)\rangle$ have a (temperature-dependent) power-law decay.

(b) The third component, $\langle s_z(0) s_z(i)\rangle$ appears to have an exponential decay at all temperatures for $\lambda < 1$. The correlation length associated with that component appears to have two finite peaks, one of which is around $T_{KT}(\lambda)$ [2].

Let us first note why (b) may be a little bit surprising at first sight (discussions about this point with A. Bishop, one of the authors of [2], were in fact the motivation for this note): The components s_x, s_y, s_z are coupled via the constraint $s_x^2 + s_y^2 + s_z^2 = 1$ and therefore, if s_x and s_y possess massless excitations, s_z too could "feel" this massless mode. A good example where such a phenomenon occurs is the "Goldstone boson" for our model (1) when $\lambda = 1$ and $d \geqslant 3$. In this case there is a spontaneous magnetization at low temperatures and, since a continuous symmetry is broken, massless correlations appear. Suppose that we orient the magnetization so that $\langle s_z \rangle \neq 0$, $\langle s_x \rangle = \langle s_y \rangle = 0$. The Mermin–Wagner argument [5] implies that $\langle s_x(0) s_x(i)\rangle$, $\langle s_y(0) s_y(i)\rangle$, do not decay exponentially [6]. However, in this situation, it is also true that $\langle s_z(0) s_z(i)\rangle - \langle s_z(0)\rangle^2$ does not decay exponentially because of the following inequality [7,9]:

$$\langle s_x(0) s_x(i)\rangle^2 \leqslant \langle s_z(0) s_z(i)\rangle^2 - \langle s_z(0)\rangle^2 \langle s_z(i)\rangle^2.$$

Although we do not, strictly speaking, prove (a) or (b), we shall give below some simple rigorous arguments supporting both claims.

2. Results

2.1. Decay of correlations

We now consider the case $\lambda < 1$, $d = 2$: By correlation inequalities [8,9], $\langle s_z(0) s_z(i)\rangle(\lambda)$ is *increasing* in λ, for $\lambda \geqslant 0$ (using e.g. open or periodic boundary conditions). However, for $\lambda = 1$, i.e. for the isotropic case, one expects that the correlation length is finite at all temperatures. While we are not aware of any "rigorous" proof of this statement, it is supported by strong theoretical arguments [3] as well as by numerical evidence [4]. From this it immediately follows that $\langle s_z(0)\, s_z(i)\rangle(\lambda)$ decays exponentially for all λ, $0 \leq \lambda \leq 1$, and for all temperatures. Moreover, even without assuming anything about the isotropic case we can prove the same result if $\lambda\beta$ is not too large (see fig. 1): indeed, by other correlation inequalities [10,11],

$$\langle s_z(0)\, s_z(i)\rangle(\lambda) \leq \langle \sigma(0)\, \sigma(i)\rangle(\lambda\beta J), \tag{3}$$

where the rhs is the correlation function of an Ising model, $\sigma(j) = \pm 1$ with coupling λJ at inverse temperature β. The latter decays exponentially if $\lambda\beta J < \beta_c \cong 0.44$, the Onsager critical temperature.

We note further that for $\lambda\beta$ small enough it is possible to make an expansion of $\langle s_z(0)\, s_z(i)\rangle(\lambda)$ in powers of $\lambda\beta$ which yield an exponentially decaying upper bound.

In carrying out this expansion, use is made of the local symmetry $s_z(j) \leftrightarrow -s_z(j)$ (for *each* j) of the expectations in the $\lambda = 0$ case. This symmetry implies that

$$\left\langle \prod_j s_z(j)^{n_j} \right\rangle(\lambda = 0) = 0$$

if some n_j is odd. Therefore, if we expand the numerator and the denominator in $\langle s_z(0)\, s_z(i)\rangle$ in powers of $\lambda\beta$ we can repeat Fisher's argument [12] giving an upper bound in terms of a sum over self-avoiding random walks ω joining 0 to i, each walk having a weight $(\lambda\beta)^{|\omega|}$. Thus for $\lambda\beta$ small enough, $\langle s_z(0)\, s_z(i)\rangle$ decays exponentially.

We remark that the exponential decay of $\langle s_z(0)\, s_z(i)\rangle$ implies a similar decay of

$$\langle f(\{s_x(j)\})\, s_z(0)\, s_z(i)\, f(\{s_x(i+j)\})\rangle,$$

where $f(\cdot)$ is any polynomial with positive coefficients in the variables $s_x(j)$. Indeed, by correlation inequalities [8,9], the latter quantity is less than

$$\langle s_z(0)\, s_z(i)\rangle \langle f(\{s_x(j)\})\, f(\{s_x(i+j)\})\rangle.$$

2.2. Phase diagram

Now we turn to point (a), namely the Kosterlitz–Thouless transition. The occurrence of the latter has been proven by Fröhlich and Spencer [13] for the two-component rotator model. It is not entirely straightforward to extend their proof to the anisotropic three-component model. However, for $\lambda = 0$, we can use a correlation inequality due to Wells [14] in order to prove the presence of a KT transition: Let us integrate out the $s_z(i)$ component for each site i; then we obtain a two-component system with Hamiltonian

$$H = -J \sum_{\langle ij \rangle} r_i r_j \cos(\phi_i - \phi_j) \tag{4}$$

and with one site distribution:

$$(1 - r_i^2)^{1/2} \, \mathrm{d}r_i \, \mathrm{d}\phi_i, \quad 0 \leqslant r_i \leqslant 1, \quad 0 \leqslant \phi_i \leqslant 2\pi.$$

The result of Fröhlich and Spencer refers to the case where $r_i = 1$, $\forall i \in \mathbb{Z}^2$. We can however easily extend their result as follows: Let $\nu(r_i)$ be any measure on $\mathbb{R}_+$ which is not concentrated at 0. Then, by Wells' inequality [14], there exists an $a > 0$ such that

$$\langle f(r, \phi) \rangle^{(\nu)} \geqslant \langle f(r, \phi) \rangle^{(a)} \tag{5}$$

where $\langle \ \rangle^{(\nu)}$ is the expectation with respect to the measure

$$\mathrm{d}\mu \cong \exp(-\beta H) \prod_i \mathrm{d}\nu(r_i) \, \mathrm{d}\phi_i,$$

H given by (4); f is any polynomial with positive coefficients in the variables $\{r_i\}$ and $\{\cos(\sum_i m_i \phi_i), m_i \in \mathbb{Z}\}$. In $\langle \ \rangle^{(a)}$ we take the single-spin distribution $\nu(r) = \delta(r - a)$. Of course, by scaling, a model with $r = a$ at inverse temperature β is the same as one with $r = 1$ at inverse temperature βa^2.

Applying inequality (5) to the function

$$f(r, \phi) = s_x(i) s_x(j) + s_y(i) s_y(j) = r_i r_j \cos(\phi_i - \phi_j)$$

implies, using [13], the presence of a KT phase, for β large, in the model (1) with $\lambda = 0$.

Unfortunately this simple argument does not prove the KT transition for $\lambda \neq 0$. It seems however very reasonable to expect, as is generally done, that the qualitative structure of the phase diagram will be as in fig. 1. The precise behaviour of the KT transition line near $\lambda = 1$ is however a more delicate question. We can nevertheless argue as follows in support of the very slow approach of $T_{\mathrm{KT}}(\lambda)$ to 0, when λ increases to one, given by (2), i.e.

$$1 - \lambda_{\mathrm{KT}}(T) \cong \exp(-c\beta) \tag{6}$$

with $\beta = T^{-1}$ and $c = 2T_{\mathrm{KT}}(\lambda = 0)$.

To do this, let us consider first the case $\lambda > 1$. One can then prove [15,8,16] that at sufficiently low temperatures there is a spontaneous magnetization for the s_z component. Moreover, the line $\lambda_c(T) - 1$, defined by the point where the spontaneous magnetization vanishes, satisfies an upper bound of the form (6) with some $c > 0$ [16] (see fig. 1). In other words, a very small perturbation $\lambda - 1$ of the isotropic case is able to produce a spontaneously magnetized phase. Similarly one can expect that a small perturbation of the opposite sign ($\lambda < 1$) produces a KT phase. Both phenomena are related to the fact that the mass of the isotropic model is bounded from above by $\exp(-c\beta)$, for some $c > 0$, when $\beta \to \infty$ [17]. The smallness of this mass suggests that a perturbation of a similar size can destroy it either by producing an ordered phase ($\lambda > 1$) or a KT phase ($\lambda < 1$).

3. Remarks

(1) In lattice dimension $d \geq 3$, the phase diagram of our system (1) is as follows (fig. 2): For all $\lambda \geq 0$, there is long-range order in the x–y plane at low enough temperatures [18]. Moreover, inequality (3) proves that the s_z component decays exponentially for $\lambda\beta$ small enough. We expect (in analogy with the $d = 2$ case) that this remains true for all $\lambda < 1$. For $\lambda = 1$, of course, all three components are magnetized at low temperatures, while for $\lambda > 1$ only the z component remains magnetized, the x and y components decaying exponentially. We emphasize that this exponential decay holds also at the critical point: since $\langle s_z(0) \rangle(\lambda)$ is increasing in λ, we have

$$T_c(\lambda) \geq T_c(\lambda = 1) \quad \text{for} \quad \lambda \geq 1, \tag{7}$$

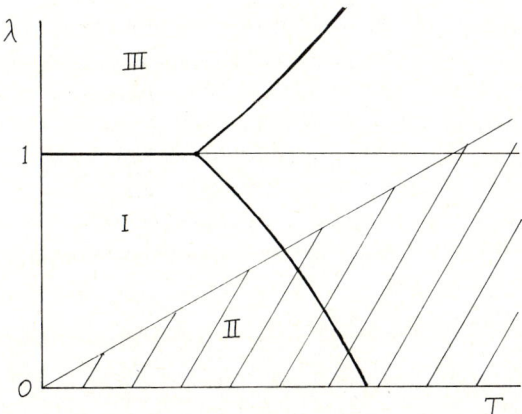

Fig. 2. (I) Spontaneous magnetization in the $x - y$ directions. (II, III) As in fig. 1.

both $\langle s_x(i)\,s_x(j)\rangle(\lambda)$, $\langle s_y(i)\,s_y(j)\rangle(\lambda)$ are decreasing in λ [8,9]. Therefore, for $\lambda > 1$, at $T_c(\lambda)$, the x and y correlation functions are smaller than their value at $\lambda = 1$, but there, one is above $T_c(1)$ [assuming that inequality (7) is strict for $\lambda > 1$], and the decay is exponential. For a similar reason $\langle s_z(i)\,s_z(j)\rangle$ decays exponentially also at $T_c(\lambda)$ for $\lambda < 1$.

(2) For $\lambda > 1$ a convergent low temperature expansion can be constructed [19]. It is formally based on the change of variables:

$$s_z = +\sqrt{1 - s_x^2 - s_y^2}\,.$$

Introducing new variables $s'_x = s_x/\sqrt{T}$, $s'_y = s_y/\sqrt{T}$, βH takes the form

$$\beta H = \tfrac{1}{2} J \Bigg(\sum_{\langle ij\rangle} \left(s'_x(i) - s'_x(j)\right)^2 + \left(s'_y(i) - s'_y(j)\right)^2$$

$$+ (\lambda - 1) 2d \sum_i \left(s'_x(i)^2 + s'_y(i)^2\right) + \mathrm{O}(T) \cdots \Bigg)$$

while

$$\int \delta\!\left(s_x^2 + s_y^2 + s_z^2 - 1\right) \mathrm{d}s_z = \exp\!\left[-\tfrac{1}{2}\ln\!\left(1 - T s'^2_x - T s'^2_y\right)\right].$$

Using this convergent expansion, one can prove the following [19]: There are two pure phases with spontaneous magnetization along the z component. In these phases, all correlation functions decay exponentially with a rate $m(\lambda, T)$ which, for small T and small $\lambda - 1$, is approximately equal to $\sqrt{\lambda - 1}$. Moreover, the free energy and the correlation functions are analytic in T away from $T = 0$. All correlation functions have an asymptotic expansion in powers of T. In addition to these translation-invariant states one can also, for $d \geqslant 3$, construct non translation-invariant states with $\langle s_z(i)\rangle > 0$ for $i = (i_1, i_2, i_3)$ and $i_1 \to +\infty$ but $\langle s_z(i)\rangle < 0$ for $i_1 \to -\infty$. For small T, one can define, in a natural way, an interface between the positively and negatively magnetized phases whose width is approximately $1/\sqrt{\lambda - 1}$. This width diverges as λ decreases to 1, as it should, since these non translation-invariant states do not occur when $\lambda = 1$ [20].

Acknowledgements

It is a pleasure to thank A.R. Bishop and E. Brezin for interesting discussions. J.L. would like to thank the Theoretical Physics group at Saclay for its kind hospitality.

References

[1] K. Hirakawa, in: Proc. Int. Conf. on Magnetism and Magnetic Materials, Atlanta, USA (1981).
[2] C. Kawabata and A.R. Bishop, Solid State Commun. 42 (1982) 595; J. Phys. Soc. Jpn. 52 Suppl. (1983) 27.
[3] A.M. Polyakov, Phys. Lett. 59B (1975) 79, 82.
A.A. Migdal, Sov. Phys. JETP 42 (1976) 743.
Wiegman, Nordita Preprint.
[4] S.H. Shenker and J. Tobochinik, Phys. Rev. B22 (1980) 4462.
[5] D. Mermin, J. Math. Phys. 6 (1967) 1061.
[6] J.L. Lebowitz and O. Penrose, Phys. Rev. Lett. 35 (1975) 549.
[7] F. Dunlop and C. Newman, Commun. Math. Phys. 44, (1975) 223.
[8] H. Kunz, C.E. Pfister and P.A. Vuillermot, J. Phys. A9 (1976) 1673.
[9] F. Dunlop, Commun. Math. Phys. 49 (1976) 247.
[10] C.J. Thompson, Phys. Lett. 43A (1973) 259.
[11] J. Bricmont, Phys. Lett. 57A (1976) 411.
[12] M.E. Fisher, Phys. Rev. 162 (1967) 480.
[13] J. Fröhlich and T. Spencer, Commun. Math. Phys. 81 (1981) 527.
[14] D. Wells, Thesis (Indiana University, 1977); preprint, unpublished. For a published proof, see
J. Bricmont, J.L. Lebowitz and C.E. Pfister, J. Stat. Phys. 24 (1981) 269.
[15] V.A. Malyshev, Commun. Math. Phys. 40 (1975) 75.
[16] J. Bricmont and J.R. Fontaine, Commun. Math. Phys. 87 (1982) 417.
[17] J. Fröhlich and T. Spencer. Phase transitions in statistical mechanics and quantum field theory, in: New developments in quantum field theory, eds. M. Levy and P. Mitter (Plenum, New York, 1977).
[18] J. Fröhlich, B. Simon and T. Spencer, Commun. Math. Phys. 50 (1976) 527.
[19] J. Bricmont, J.L. Lebowitz and C.E. Pfister, unpublished.
[20] J. Fröhlich and C.E. Pfister, Commun. Math. Phys. 89 (1983) 303.

IV. Elliott W. Montroll
Selected Reprints *

---

* The handwritten corrections in the first preprint are Elliott Montroll's own writing.

A typical view of Elliott W. Montroll.

A Preprint of an Abstract of a Doctor's Dissertation, UNIVERSITY OF PITTSBURGH BULLETIN, Volume 37, Number 3, January 15, 1941

SOME NOTES AND APPLICATIONS OF THE CHARACTERISTIC VALUE THEORY OF INTEGRAL EQUATIONS

BY ELLIOTT W. MONTROLL, B.S. IN CHEM., PH.D.

ACKNOWLEDGMENT

Sincere thanks are due Dr. J. E. Mayer for his suggestions, discussions, and hospitality during the author's stay at Columbia University. The encouragement and discussions of Drs. J. S. Taylor, A. E. Staniland, and M. M. Culver are greatly appreciated.

I. INTRODUCTION

The equation

$$\varphi(x) = \lambda \int_a^b K(x, y) \varphi(y) dy \tag{1}$$

is an integral equation if $K(x, y)$ is a given function, $\varphi(x)$ an unknown function, and λ a parameter independent of x and y. The kernel, $K(x, y)$, which shall be assumed to be continuous and not identically zero in the region $(a \leqslant x, y \leqslant b)$, is Hermitian if $K(x, y) = \overline{K(y, x)}$ and symmetric if $K(x, y) = K(y, x)$.

Fredholm's determinant, $D(\lambda)$, is defined by

$$D(\lambda) = 1 + \sum_{n=1}^{\infty} \frac{(-1)^n D_n \lambda^n}{n!}, \tag{2}$$

where

$$D_1 = \int_a^b K(x, x) dx, \quad D_2 = \int_a^b \int_a^b \begin{vmatrix} K(x, x) & K(x, y) \\ K(y, x) & K(y, y) \end{vmatrix} dx\, dy$$

and

$$D_n = \int_a^b \cdots \int_a^b \begin{vmatrix} K(x_1, x_1) & \cdots & K(x_1, x_n) \\ \cdot & \cdot & \cdot \\ K(x_n, x_1) & \cdots & K(x_n, x_n) \end{vmatrix} dx_1 dx_2 \cdots dx_n.$$

Equation (1) has nontrivial solutions only when λ is a root of [1]

$$D(\lambda) = 0. \tag{3}$$

It can be shown that $D(\lambda)$ converges for all values of λ, that it has at least one zero, and that under the given conditions the zeros of $D(\lambda)$ form a finite or denumerably infinite ordered set $\{\lambda_i\}$.[2] The elements

---

[1] Whittaker, E. T., and Watson, G. N., *A Course in Modern Analysis*, Chapter XI, p. 219, Cambridge (1920).
[2] Courant, R., and Hilbert, D., *Methoden der Mathematischen Physik*, Chapter III, pp. 104–128, Berlin, Springer (1924).

of the set $\{\lambda_i\}$ are called characteristic values of (1) and the set of solutions of (1), $\varphi_i(x)$ (which may be made into an orthonormal set), each $\varphi_i(x)$ of which corresponds to a λ_i, is called the set of characteristic functions of (1).

II. Properties of Fredholm's Determinant

It can be shown that $D(\lambda)$ is an integral function of order < 2; therefore

$$D(\lambda) = \exp\left\{-\lambda \int_a^b K(x,x)dx\right\} \prod_{i=1}^{\infty}\left(1 - \frac{\lambda}{\lambda_i}\right)e^{\lambda/\lambda_i} \quad (4)$$

and $\sum_{n=1}^{\infty} 1/\lambda_n^i$ converges absolutely when $i \geqslant 2$.

The function

$$K_n(x,y) = \int_a^b \cdots \int_a^b K(x,x_1)K(x_1,x_2) \cdots K(x_{n-1},y)dx_1 \cdots dx_{n-1}$$

is called the nth iterated kernel of (1) and can, according to the Hilbert-Schmidt theory,[3] be expressed as a uniformly and absolutely convergent series

$$K_n(x,y) = \sum_{i=1}^{\infty} \frac{\varphi_i(x)\overline{\varphi_i(y)}}{\lambda_i^n} \quad (n = 2, 3 \cdots). \quad (5)$$

By the orthonormality of the φ_i's

$$\int_a^b K_n(x,x)dx = \sum_{i=1}^{\infty} \frac{1}{\lambda_i^n} \quad (n = 2, 3 \cdots), \quad (6)$$

and if the order of $D(\lambda)$ is less than 1,

$$\int_a^b K(x,x)dx = \sum_{i=1}^{\infty} \frac{1}{\lambda_i}. \quad (7)$$

In this case

$$D(\lambda) = \prod_{i=1}^{\infty}\left(1 - \frac{\lambda}{\lambda_i}\right) = 1 - \lambda \sum_{i=1}^{\infty} \frac{1}{\lambda_i} + \lambda^2 \sideset{}{'}\sum_{i,j=1}^{\infty} \frac{1}{\lambda_i\lambda_j} - \cdots. \quad (8)$$

By equating like powers of λ in (2) and (8):

$$D_1 = \sum_{i=1}^{\infty} \frac{1}{\lambda_i} = \int_a^b K(x,x)dx,$$

$$D_2 = 2 \sideset{}{'}\sum_{i,j=1}^{\infty} \frac{1}{\lambda_i\lambda_j} = \int_a^b \int_a^b \begin{vmatrix} K(x,x) & K(x,y) \\ K(y,x) & K(y,y) \end{vmatrix} dx\, dy,$$

and

[3] *Ibid.*, pp. 120.

$$D_n = n! \sum_{\alpha_1} \sum_{\alpha_2} \cdots \sum_{\alpha_n}{}' \frac{1}{\lambda_{\alpha_1} \lambda_{\alpha_2} \cdots \lambda_{\alpha_n}}$$

$$= \int \cdots \int \begin{vmatrix} K(x_1, x_1) & \cdots & K(x_1, x_n) \\ \cdot & & \cdot \\ \cdot & & \cdot \\ K(x_n, x_1) & \cdots & K(x_n, x_n) \end{vmatrix} dx_1 \cdots dx_n$$

(the primes on the summation signs indicate that terms in which $\alpha_i = \alpha_j$ are not included). Using (7) and Bessel's inequality one can show that the sum of products involved in D_n converges absolutely. By expanding D_n,

$$\frac{(-1)^n D_n}{n!} = \sum_{r_t} \prod_t \frac{(-A_t)^{r_t}}{t^{r_t} r_t!},$$
$$(\Sigma t r_t = n)$$

where A_t is the integral of the tth iterated kernel

$$A_t = \int_a^b \cdots \int_a^b K(x_1, x_2) K(x_2, x_3) \cdots K(x_t, x_1) dx_1 \cdots dx_t.$$

This means that $\frac{(-1)^n D_n}{n!}$ can be written as a contour integral, for

$$\sum_{r_t} \prod_t \left(-\frac{A_t}{t}\right)^{r_t} \frac{1}{r_t!}$$ is the coefficient of z^n in the power series expansion
$(\Sigma t r_t = n)$

of $\exp.\left\{\sum_{t=1}^{\infty} -\frac{A_t}{t} z^t\right\}$. The theory of residues yields

$$\frac{(-1)^n D_n}{n!} = \frac{1}{2\pi i} \int_C \frac{\exp.\left\{\sum_{t=1}^{\infty} -\frac{A_t}{t} z^t\right\}}{z^{n+1}} dz$$

if the contour of integration is a closed path enclosing the origin, and

$$D(\lambda) = \sum \frac{D_n(-\lambda)^n}{n!} = \frac{1}{2\pi i} \int_C \frac{\exp.\left\{\sum -\frac{A_t}{t} z^t\right\}}{(z - \lambda)} dz$$

if the path of integration is a closed curve about the origin but enclosed in a circle of radius λ whose center is at the origin.

An asymptotic expression for Fredholm's determinant of a negative definite kernel follows from a theorem due to Hille and Tamarkin [4] on integral equations with negative definite kernels and one on integral functions with negative zeros due to Titchmarsh, Payley, and Wiener.[5] This expression is included in:

Theorem: If (a) the characteristic functions $\varphi_i(x)$ of

$$\varphi(x) = \lambda \int_a^b K(x, y) \varphi(y) dy$$

[4] Hille, E., and Tamarkin, J. D., "On the Characteristic Values of Linear Integral Equations," *Acta Mathematica*, **57**, 1 (1931).
[5] Payley, R. E. A. C., and Wiener, N., *Fourier Transforms in the Complex Domain*, p. 78, American Mathematical Society, New York (1934).

are uniformly bounded; (b) $K(x, y)$ is a negative definite Hermitian kernel but not minimal and $\int_a^b K(x, x)dx = \sum_{i=1}^{\infty} 1/\lambda_i$; (c) the number of characteristic values λ_i of $K(x, y)$ with absolute values less than r becomes asymptotically $N(r) \sim \mu r^\rho$ $(0 < \rho < 1)$, then

$$\log D(\lambda) \sim (\pi\mu \cdot \lambda^\rho \cdot \operatorname{cosec} \pi\rho)$$

and $\log |D(xe^{i\theta})| \sim (x^\rho \cdot \pi\mu \operatorname{cosec} \pi\rho \cos \theta\rho)$ $(|\theta| < \pi)$. And under conditions (a) and (b), if $D(\lambda)$ has the asymptotic representation $\log D(\lambda) \sim (\pi\rho\lambda^\rho \operatorname{cosec} \pi\rho)$ for large real values of λ, then the number of characteristic values with absolute values less than r is asymptotically $N(r) \sim \mu r^\rho$ $(0 < \rho < 1)$.

III. Integration of a Type of Multiple Integral Occurring in Statistical Mechanics

In order to find the physical properties of imperfect gases and condensing systems, according to the Mayer-Ursell[6,7] development, one must calculate integrals of the form

$$p_n(0) = \frac{1}{V}\int \cdots \int f(r_{12})f(r_{23}) \cdots f(r_{n+1,1})d\tau_1 d\tau_2 \cdots d\tau_{n+1}$$

where $d\tau_i = dx_i dy_i dz_i$, $f(r_{ij}) = \exp\left\{-\frac{V(r_{ij})}{KT}\right\} - 1$, and the integration extends over the volume of the vessel in which the particles of the system are confined. $V(r_{ij})$ is the mutual potential energy of a pair of particles whose distance of separation is r_{ij}; K is Boltzmann's constant and T is the absolute temperature of the system at equilibrium.

If the integration of $p_n(0)$ is taken over a cube with edges of length a, $p_n(0)$ is the integral of the $(n+1)$st iterated kernel of the integral equation

$$\varphi(x_1, y_1, z_1) = \lambda \int_{-a/2}^{a/2} \int_{-a/2}^{a/2} \int_{-a/2}^{a/2} f(r_{12})\varphi(x_2, y_2, z_2)dx_2 dy_2 dz_2.$$

From equation (7)

$$p_n(0) = \frac{1}{a^3} \sum_{i,j,k=-\infty}^{\infty} (a^{3/2} c_{i,j,k})^{n+1}$$

if

$$c_{l,m,n} = \frac{1}{a^{3/2}} \int_{-a/2}^{a/2} \int_{-a/2}^{a/2} \int_{-a/2}^{a/2} f(x, y, z)e^{-(2\pi i/a)(xl+ym+zn)}dx\, dy\, dz.$$

Thus, the $3(n+1)$-fold integral has been reduced to a triple sum.

[6] Mayer, J. E., "Statistical Mechanics of Condensing Systems," *Journal of Chemical Physics*, 6, p. 87 (1938).
[7] Ursell, H. D., "The Evaluation of Gibbs' Phase Integral for Imperfect Gases," *Proceedings of Cambridge Philosophical Society*, 23, p. 685 (1926).

VALUE THEORY OF INTEGRAL EQUATIONS

If we remember that $f(r)$ is practically zero for all values of r greater than one that is much smaller than a, we have

$$a^{3/2}c_{l,m,n} = \frac{4\pi}{t}\int_0^\infty r \sin tr f(r) dr = g(t)$$

when

$$t = \frac{2\pi}{a}(l^2 + m^2 + n^2)^{1/2}.$$

Therefore

$$p_n(0) = \frac{1}{2\pi^2}\int_0^\infty t^2 g^{n+1}(t) dt.$$

Using some properties of Fourier Transforms, one can obtain an asymptotic expression for $p_n(0)$ that is valid for large values of n:

$$p_n(0) \sim \frac{(6\pi)^{3/2}(4\pi)^{n+1}\mu_2^{n+5/2}}{8\pi^3[\mu_4(n+1)]^{3/2}}\left[1 + 0\left(\frac{1}{n}\right)\right],$$

$$\mu_m = \int_0^\infty r^m f(r) dr,$$

and the term of order $1/n$ is

$$\frac{135\mu_2^2}{\mu_4^2(n+1)}\left[\frac{\mu_6}{5!\mu_2} - \frac{\mu_4^2}{72\mu_2^2}\right].$$

The more complicated integrals

$$s_n(0) = \frac{1}{v}\int\cdots\int f_0(r_{12})f_1(r_{23})\cdots f_n(r_{n+1,1})d\tau_1\cdots d\tau_{n+1}$$

can be evaluated in a manner similar to that used for $p_n(0)$ and one obtains

$$s_n(0) = \frac{1}{2\pi^2}\int_0^\infty t^2 \prod_{i=0}^n g_i(t) dt,$$

where

$$g_i(t) = \frac{4\pi}{t}\int_0^\infty rf_i(r)\sin tr\, dr.$$

If we define

$$\mu_{i,m} = \int_0^\infty r^m f_i(r) dr,$$

for large values of n:

$$s_n(0) \sim \frac{(4\pi)^{n+1}(6\pi)^{3/2}\prod_{i=0}^n \mu_{i,2}}{\left(\sum_{i=0}^n \frac{\mu_{i,4}}{\mu_{i,2}}\right)^2}.$$

The methods discussed here for the evaluation of $p_n(0)$ and $s_n(0)$ make the calculation of virial coefficients of an imperfect gas a reasonable problem, and it is possible that they will be useful in developing a theory of liquids.

VITA

Elliott W. Montroll was born on May 4, 1916, in Pittsburgh, Pennsylvania, and received his elementary and high school education at the Dormont Public Schools. In 1933, he entered the University of Pittsburgh and, in 1937, he received the degree of B.S. in Chemistry. From 1937 until 1939, he was a graduate assistant in the mathematics department of the University of Pittsburgh; and, during the first semester of the school year 1939–40, he did research in the chemistry department of Columbia University.

Statistical Mechanics of Imperfect Gases*

ELLIOTT W. MONTROLL,† *Sterling Chemistry Laboratory, Yale University, New Haven, Connecticut*

AND

JOSEPH E. MAYER, *Chemistry Department, Columbia University, New York, New York*

(Received May 27, 1941)

The application of statistical mechanical equations to the calculation of thermodynamic properties of imperfect gases has been hindered by the occurrence of highly multiple integrals in these equations. By observing that some of these multiple integrals are related to the iterated kernels of an integral equation involving the potential energy function of a pair of molecules, a technique is developed which expresses these integrals in terms of the characteristic values of the integral equation. This technique is applied to the calculation of third virial coefficients and the molecular distribution function (which is essentially the probability of finding two specified molecules in two small volume elements a distance r from each other) at various temperatures for imperfect gases with Lennard-Jones potential energy functions.

I. INTRODUCTION

RECENTLY, using methods first investigated by Ursell,[1] formal expressions have been derived[2-7] for the physical properties of imperfect gases and condensing systems. In spite of the exactness of these equations, their use in making calculations has been limited by the fact that they are expressed in terms of so called "irreducible integrals"

$$\beta_n = \frac{1}{Vn!} \int\int \cdots \int \sum_{n+1} \prod_{i>j\geq 1} f_{ij} d\{\mathbf{n}+1\} \tag{1}$$

(sum over all products with all molecules more than singly connected)

and

$$H_{\nu m}\{\mathbf{v}\} = \frac{1}{m!} \int\int \cdots \int \sum_{\substack{m\geq i>j\geq 1 \\ m\geq k\geq 1 \\ \nu\geq \kappa\geq 1}} \prod_{(\nu>i)} f_{ij} f_{\kappa k} d\{\mathbf{m}\} \tag{2}$$

(sum over all connected products for which each molecule of the set $\{\mathbf{m}\}$ is connected to at least two of the set $\{\mathbf{v}\}$ by an independent path)

whose direct evaluation is a Herculean task. A fundamental statistical mechanical function, the Helmholtz free energy (from which all thermodynamic quantities can be calculated) is given by

$$A = RT\left[\log\left(\frac{h^2}{2\pi mkT}\right)\frac{1}{ev} - \sum_{n\geq 1}\frac{1}{n+1}\beta_n v^{-n}\right] \tag{3}$$

and the equation of state of the system is

$$\frac{Pv}{kT} = 1 - \sum_{n\geq 1}\frac{n}{n+1}\beta_n v^{-n}. \tag{4}$$

---

* Publication assisted by the Ernest Kempton Adams Fund for Physical Research of Columbia University.
† Sterling Research Fellow, Yale University.
[1] H. D. Ursell, Proc. Camb. Phil. Soc. **23**, 685 (1927).
[2] J. E. Mayer, J. Chem. Phys. **5**, 67 (1937).
[3] J. E. Mayer and P. G. Ackermann, J. Chem. Phys. **5**, 74 (1937).
[4] J. E. Mayer and S. F. Harrison, J. Chem. Phys. **6**, 87 (1938).
[5] S. F. Streeter and J. E. Mayer, J. Chem. Phys. **7**, 1025 (1939).
[6] B. Kahn and G. E. Uhlenbeck, Physica, **4**, 299 (1938).
[7] M. Born and K. Fuchs, Proc. Roy. Soc. **166**, 391 (1938).

As has become customary in empirical equations of state, the coefficient of v^{-n} (v is the volume per molecule) in (4) is called the $(n+1)$st virial coefficient. The distribution function $F_2\{2\}$ (defined in such a manner that $F_2\{2\}d\tau_1 d\tau_2/V^2$ is the probability that any specified molecule 1 is in the volume element $d\tau_1$ at the position x_1, y_1, z_1 at the same time 2 is in $d\tau_2$ at x_2, y_2, z_2) is given by [8,9]

$$F_2\{2\} = \exp(-V(r_{12})/kT)\left[1 + \sum_{m \geq 1} v^{-m} H_{2,m}\{2\}\right]. \tag{5}$$

In Eqs. (1–5) $\{n\}$ represents a set of n specified molecules and the set of $3n$ coordinates of these molecules:

$$\{n\} = x_1, y_1, z_1, \cdots, z_n,$$
$$d\{n\} = d\tau_1 d\tau_2 \cdots d\tau_n = dx_1 \cdots dz_n;$$

$f(r_{ij}) = f_{ij} = [\exp\{-V(r_{ij})/kT\}] - 1$; and the integration extends over the volume V of the space in which the molecules of the system are confined. The mutual potential energy $V(r_{ij})$ of a pair of molecules whose distance of separation is r_{ij} becomes infinite as $r_{ij} \to 0$, has a negative minimum when r_{ij} is the equilibrium distance between the molecules, and rapidly approaches zero as the molecules separate from their equilibrium positions. k is Boltzmann's constant and T is the absolute temperature of the system at equilibrium.

The simplest irreducible integrals are

$$\beta_1 = \int_0^\infty 4\pi r^2 f(r) dr, \quad \beta_2 = \frac{1}{2V} \iiint f_{12} f_{23} f_{31} d\{3\}, \quad \beta_3 = \frac{1}{3!V}(3\beta_{30} + 6\beta_{31} + \beta_{32}), \tag{6}$$

where

$$\beta_{30} = \frac{1}{V} \iiiint f_{12} f_{23} f_{34} f_{41} d\{4\}, \quad \beta_{31} = \frac{1}{V} \iiiint f_{12} f_{13} f_{14} f_{23} f_{34} d\{4\},$$

$$\beta_{32} = \frac{1}{V} \iiiint f_{12} f_{13} f_{14} f_{23} f_{24} f_{34} d\{4\}$$

and which correspond to "graphs"

The corresponding H integrals leave at least two volume elements unintegrated. Graphs with every particle connected to exactly two others (for example, β_2 and β_{30}) are said to represent "ring integrals," the most general of which is a numerical factor multiplied by

$$p_n(0) = \frac{1}{V} \iint \cdots \int f_{12} f_{23} \cdots f_{n,n+1} f_{n+1,1} d\tau_1 \cdots d\tau_{n+1}. \tag{7}$$

Integrals whose graphs are linear chains with the positions of the two end particles fixed are called "chain integrals" and are of the form

$$p_n(r_{1,n+2}) = \iint \cdots \int f_{12} f_{23} \cdots f_{n,n+1} f_{n+1,n+2} d\tau_2 \cdots d\tau_{n+1}. \tag{8}$$

Since the integration over the last volume element in Eq. (7) contributes V as a factor, as $r_{n+2} \to r_1$, $p_n(r_{1,n+2}) \to p_n(0)$.

[8] J. E. Mayer and E. Montroll, J. Chem. Phys. **9**, 2 (1941).
[9] J. De Boer and A. Michels, Physica **6**, 97 (1939).

Inasmuch as β_1 is an integral over a single variable, its integration is not difficult;[10] however, passage to higher β's and H's introduces severe complications. Even the evaluation of β_2 necessitates the integration of a ninefold integral whose integrand is a rather complicated function.

The purpose of this paper is to demonstrate the applicability of some properties of integral equations and Fourier transforms to the calculation of the β's and H's. We shall calculate the third virial coefficient and the distribution function $F_2\{2\}$ for gases with Lennard-Jones potentials.

II. Mathematical Techniques

1. Development of $p_n(r)$ in terms of characteristic functions and characteristic values of a linear homogeneous integral equation. First consider the integration of $p_n(r_{1, n+2})$ to be taken over a cube with edges of length a. Then

$$p_n(r_{1, n+2}) = \int_{-a/2}^{a/2}\int_{-a/2}^{a/2}\cdots\int_{-a/2}^{a/2} f_{12}f_{23}\cdots f_{n+1, n+2}d\tau_2\cdots d\tau_{n+1} \qquad (9)$$

$$r_{ij} = [(x_i-x_j)^2 + (y_i-y_j)^2 + (z_i-z_j)^2]^{\frac{1}{2}}.$$

Now suppose f_{ij} is expanded as a bilinear form

$$f_{ij} = \sum_{\alpha, \beta} a_{\alpha\beta}\psi_\alpha(x_i, y_i, z_i)\psi^*_\beta(x_j, y_j, z_j) \qquad (10)$$

of a complete set of functions $\{\psi_\alpha(x, y, z)\}$, orthonormal in the cube. The orthonormality condition means

$$\int_{-a/2}^{a/2}\int_{-a/2}^{a/2}\int_{-a/2}^{a/2} \psi_\alpha(i)\psi^*_\beta(i)d\tau_i = \delta_{\alpha\beta}$$

($\delta_{\alpha\beta} = 1$ if $\alpha = \beta$ and zero otherwise).

By multiplying both sides of (10) by $\psi^*_\alpha(i)\psi_\beta(j)$ and integrating over the entire cubes, the orthonormality condition yields

$$a_{\alpha\beta} = \int\int f(r_{ij})\psi^*_\alpha(i)\psi_\beta(j)d\tau_i d\tau_j.$$

One might ask whether there exists a set $\{\psi_\alpha\}$ such that $a_{\alpha\beta} = 0$ when $\alpha \neq \beta$. The answer is yes, for if $\{\psi_\alpha\}$ is the set of characteristic functions of the integral equation

$$\psi(x_i, y_i, z_i) = \lambda \int_{-a/2}^{a/2}\int_{-a/2}^{a/2}\int_{-a/2}^{a/2} f(r_{ij})\psi(x_j, y_j, z_j)d\tau_j, \qquad (11)$$

then

$$a_{\alpha\beta} = \frac{1}{\lambda_\beta}\int_{-a/2}^{a/2}\int_{-a/2}^{a/2}\int_{-a/2}^{a/2} \psi^*_\alpha(i)\psi_\beta(i)d\tau_i = \frac{1}{\lambda_\beta}\delta_{\alpha\beta}.$$

Since the kernel $f(r_{ij})$ in gas theory is generally a function of $(x_i-x_j)^2 + (y_i-y_j)^2 + (z_i-z_j)^2$, its characteristic functions are[11]

$$\psi_{k, l, m} = a^{-\frac{3}{2}}\exp\left\{\frac{2\pi i}{a}(kx+ly+mz)\right\}, \qquad (k, l, m = 0, \pm 1, \pm 2, \cdots). \qquad (12)$$

---

[10] See, for example, Lennard-Jones, Proc. Roy. Soc. **106**, 463 (1924), where it is calculated for the potential energy function $V(r) = Ar^{-s} - Br^{-t}$.
[11] Cf. Hille and Tamarkin, Acta Math. **57**, 1 (1931).

This implies that $f(r)$ can be expanded formally as

$$f(x, y, z) \sim \sum_{k, l, m=-\infty}^{\infty} c_{k, l, m} a^{-\frac{3}{2}} \exp\left\{\frac{2\pi i}{a}(kx+ly+mz)\right\}, \tag{13}$$

where the coefficients $c_{k, l, m}$ can be evaluated by the usual method of multiplying both sides of (13) by $\psi^*_{k, l, m}$ and integrating over the entire cube. Thus

$$c_{k, l, m} = a^{-\frac{3}{2}} \int_{-a/2}^{a/2} \int_{-a/2}^{a/2} \int_{-a/2}^{a/2} f(x, y, z) \exp\left\{-\frac{2\pi i}{a}(kx+my+nz)\right\} dx dy dz. \tag{14}$$

Substituting (12) into (10), the characteristic values of $f(r_{12})$ are

$$\lambda_{k, l, m} = (1/a)^{\frac{3}{2}} c_{k, l, m}.$$

If we observe that (9) is the nth iterated kernel of the integral equation (11), we can express it in the well-known form[12]

$$p_n(r_{1, n+2}) = \sum_{k, l, m=-\infty}^{\infty} \frac{\psi_{k, l, m}(1) \psi^*_{k, l, m}(n+2)}{\lambda^{n+1}_{k, l, m}}$$

$$= \frac{1}{a^3} \sum_{k, l, m=-\infty}^{\infty} (a^{\frac{3}{2}} c_{k, l, m})^{n+1} \exp \frac{2\pi i}{a} \{kx_{1, n+2} + ly_{1, n+2} + mz_{1, n+2}\} \tag{15}$$

($x_{i, j} = x_i - x_j$, etc.). The same result could have been achieved by merely substituting (12) into (9); however, by using the method given above, the theory of integral equations insures the convergence of the series as long as $n \geq 2$ and $\iint |f(r_{12})|^2 d\tau_1 d\tau_2 < \infty$. This integrability condition is satisfied in most statistical mechanical problems, the most important exceptions being those involving Coulombic forces (electrolyte problems) in which cases the developments of this paper are not valid.

In general, a triple sum is inconvenient to handle, so (15) will now be reduced to a definite integral over one variable.

2. *Reduction of $p_n(r)$ to a simple integral.* First an expression for $c_{k, l, m}$ which is simpler and more useful than (14) will be derived. If $f(r) \sim 0$ when r is greater than some small fraction of a (as is the case in the theory of imperfect gases and condensing systems) the limits of integration on (14) can, with negligible error, be made infinite. Transforming to spherical polar coordinates yields

$$c_{k, l, m} = a^{-\frac{3}{2}} \int_0^\infty r^2 f(r) dr \int_0^\pi \left[\exp\left\{-\frac{2\pi i r m \cos\theta}{a}\right\}\right] \sin\theta d\theta$$

$$\times \int_0^{2\pi} \left[\exp\left\{-\frac{2\pi i r}{a} \sin\theta(k\cos\varphi + l\sin\varphi)\right\}\right] d\varphi \tag{16}$$

but, as will be shown in the Appendix,

$$I = \int_0^\pi \int_0^{2\pi} \exp\left\{-\frac{2\pi i r}{a}(k\cos\varphi \sin\theta + l\sin\varphi \sin\theta + m\cos\theta)\right\} \sin\theta d\theta d\varphi$$

$$= \frac{2a}{r(k^2+l^2+m^2)^{\frac{1}{2}}} \sin\frac{2\pi r}{a}(k^2+l^2+m^2)^{\frac{1}{2}}. \tag{17}$$

---

[12] Cf. R. Courant and D. Hilbert, *Methoden der Mathematischen Physik I* (Berlin, 1931), second edition, p. 116.

Therefore,
$$c_{k,l,m} = \frac{2}{a^{\frac{3}{2}}(k^2+l^2+m^2)^{\frac{1}{2}}} \int_0^\infty rf(r) \sin\left[\frac{2\pi r}{a}(k^2+l^2+m^2)^{\frac{1}{2}}\right] dr. \quad (18)$$

This equation shows that $c_{k,l,m}$ is a function of $(k^2+l^2+m^2)^{\frac{1}{2}}$ with spherical symmetry about the origin of a Cartesian k, l, m coordinate system. As has been mentioned before, the main contribution to $c_{k,l,m}$ originates from very small values of r (say when $r < 10^{-6}a$). If k, l, m are small integers it is apparent that their corresponding terms in $p_n(r_{1,n+2})$ will be practically independent of k, l, m. Since the terms in the series which depend on k, l, m are those for which these integers are so large that the summand in (15) can be considered as a continuous function of k, l, m, the summation in (15) can be changed to an integration over k, l, m space. Writing

$$t = (2\pi/a)(k^2+l^2+m^2)^{\frac{1}{2}}, \quad (19a)$$

$$a^3 c_{k,l,m} = 4\pi \int_0^\infty r^2 f(r)(\sin tr/tr) dr = g(t), \quad (19b)$$

the integral representation of $p_n(r_{1,n+2})$ becomes (in spherical polar coordinates)

$$p_n(r_{1,n+2}) = \frac{1}{a^3} \int_0^\infty t^2 g^{n+1}(t) dt \int_0^\pi \left[\exp\left\{-\frac{2\pi i t}{a} z_{1,n+2} \cos\theta\right\}\right] \sin\theta d\theta$$

$$\times \int_0^{2\pi} \left[\exp\left\{-\frac{2\pi i t \sin\theta}{a}(x_{1,n+2}\cos\varphi + y_{1,n+2}\sin\varphi)\right\}\right] d\varphi.$$

Applying (17) to the integration over the angle variables

$$p_n(r) = \frac{1}{2\pi^2} \int_0^\infty t^2 g^{n+1}(t)(\sin tr/tr) dt \quad (20)$$

and as $r_{n+2} \to r_1$,

$$p_n(0) = (1/2\pi^2) \int_0^\infty t^2 g^{n+1}(t) dt. \quad (21)$$

Thus the highly multiple chain and ring integrals have been reduced to simple integrals which have the same general form for all values of n. For very large n's one can derive an asymptotic expression for these integrals by the methods discussed in the next section.

3. *Asymptotic formula for* $p_n(r_{1,n+2})$. If $g(t)$ has a greatest maximum at $t=0$ (as it does in most cases of interest in the theory of condensing systems; see Fig. 1) we can find an asymptotic expansion[13] for $p_n(r)$ that is valid for large n. Defining

$$\mu_m = \int_0^\infty r^m f(r) dr, \quad g(t) = 4\pi \sum_{m=0}^\infty \frac{(-1)^{m+1} t^{2m-2} \mu_{2m}}{(2m-1)!}. \quad (22)$$

Then, assuming that

$$g(t) = 4\pi \mu_2 \left(\sum_{m=0}^\infty a_m t^m\right) \exp\left(-\frac{t^2 \mu_4}{6\mu_2}\right)$$

and comparing the coefficients of t^m with those in (22),

$$a_1 = a_2 = a_3 = a_5 = \cdots = a_{2n+1} = 0$$

and

$$a_4 = (\mu_6/5!\mu_2) - (\mu_4^2/72\mu_2^2).$$

---

[13] This approach is somewhat similar to that used in some problems in the theory of probability; cf. J. A. Carroll, Quart. J. Math. (Oxford series) **9**, 176 (1938).

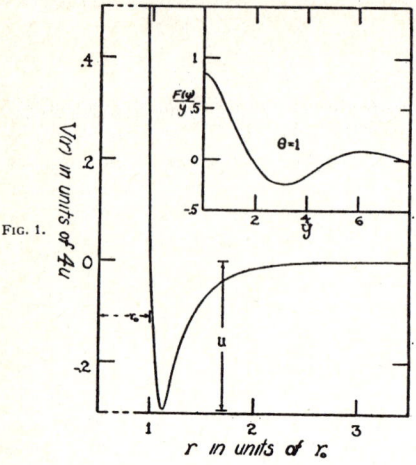

Fig. 1. Plot of the potential between pairs of molecules $V(r)$ Eq. (32), against r in units of r_0, and of $F(y)/y$, Eq. (35) against y.

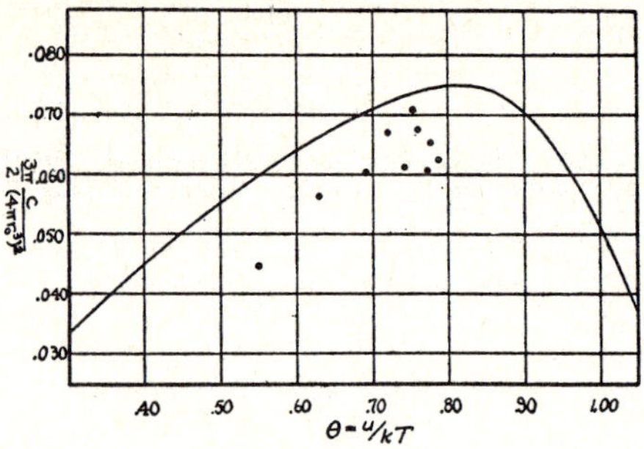

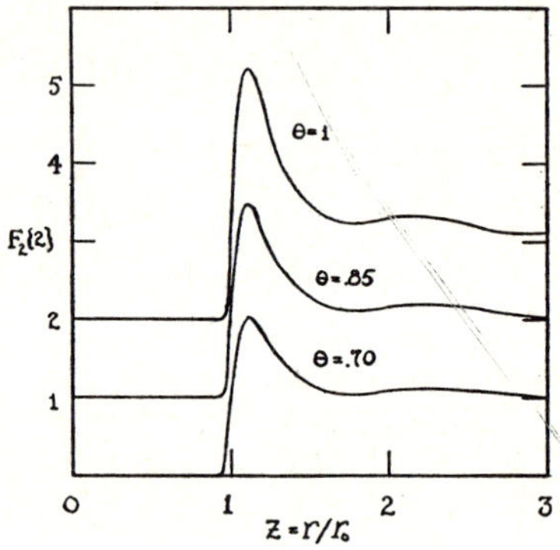

Fig. 2. Plot of the third virial coefficient C in units of $(3\pi/2)/(4\pi r_0^3)^2$ against $\theta = u/kT$.

Fig. 3. The radial distribution function $F_2(r)$ as a function of $Z = r/r_0$ when the volume per molecule, v, is $2\pi r_0^3/3$, which is about $(2/3) v$ (critical). $\theta = 1$ corresponds to about $0.8\,T$ (critical).

Thus

$$g(t) = 4\pi\mu_2(1 + a_4 t^4 + \cdots)\exp(-t^2\mu_4/6\mu_2)$$

and

$$\frac{tg^{n+1}(t)}{(2\pi)^{\frac{3}{2}}} = \frac{t(4\pi\mu_2)^{n+1}}{(2\pi)^{\frac{3}{2}}}[1 + (n+1)a_4 t^4 + \cdots]\exp\left\{-\frac{(n+1)t^2\mu_4}{6\mu_2}\right\}.$$

Substituting this in (20) and remembering that

$$\int_0^\infty t\exp(-Mt^2)\sin trdt = \frac{\pi^{\frac{1}{2}}}{4M^{\frac{3}{2}}}\exp(-r^2/4M),$$

$$\int_0^\infty t^5\exp(-Mt^2)\sin trdt = \frac{15\pi^{\frac{1}{2}} r\exp(-r^2/4M)}{16M^{7/2}}[1 + 0(1/M)],$$

we have

$$p_n(r_{1,\,n+1}) \sim \frac{(2/\pi)^{\frac{1}{2}}(4\pi)^n 3^{\frac{3}{2}}\mu_2^{n+5/2}}{[(n+1)\mu_4]^{\frac{3}{2}}}\exp\left[-\frac{3\mu_2 r_{1,\,n+2}}{2(n+1)\mu_4}\right]\left[1 + 0\left(\frac{1}{n}\right)\right] \quad (23a)$$

and

$$p_n(0) \sim \left(\frac{2}{\pi}\right)^{\frac{1}{2}}\frac{(4\pi)^n 3^{\frac{3}{2}}\mu_2^{n+5/2}}{[(n+1)\mu_4]^{\frac{3}{2}}}[1 + 0(1/n)]. \quad (23b)$$

The leading term of $0(1/n)$ in $p_n(0)$ is

$$\frac{135\mu_2^2}{(n+1)\mu_4^2}\left[\frac{\mu_6}{5!\mu_2} - \frac{\mu_4^2}{72\mu_2^2}\right].$$

4. More general integrals. Other components of irreducible integrals can be treated in a manner similar to that developed for ring and chain integrals. For example consider the part of β_4,

$$\frac{1}{V}\int\cdots\int f_{15}f_{12}f_{23}f_{34}f_{45}f_{13}d\{5\}, \tag{24a}$$

which corresponds to the graph of a square having one side in common with a triangle. This integral, which can be written as

$$\frac{1}{V}\int\int\int f(r_{12})f(r_{23})f(r_{31})p_2(r_{13})d\{3\}, \tag{24b}$$

is a special case of the more general class of integrals

$$s_n(0)=\frac{1}{V}\int\cdots\int f_0(r_{12})f_1(r_{23})\cdots f_n(r_{n+1,1})d\{\mathbf{n+1}\}. \tag{25a}$$

Here, for all i, $f_i(r_{kl})=f_i(r_{lk})$, and each $f_i(r)$ is a function of the same type as $f(r)$. If, as before, we consider all the particles of the system as being contained in a cube with sides of length a,

$$s_n(r_{1,n+2})=\int_{-a/2}^{a/2}\cdots\int_{-a/2}^{a/2} f_0(r_{12})f_1(r_{23})\cdots f_n(r_{n+1,n+2})d\tau_2\cdots d\tau_{n+1}. \tag{25b}$$

The characteristic functions of the integral equations

$$\varphi(x_1,y_1,z_1;i)=\int_{-a/2}^{a/2}\int_{-a/2}^{a/2}\int_{-a/2}^{a/2} f_i(r_{12})\varphi(x_2,y_2,z_2;i)dx_2dy_2dz_2$$
$$(i=0,1,2\cdots,n)$$

are

$$\varphi_{k,l,m}(x,y,z;i)=a^{-\frac{3}{2}}\exp\{(2\pi i/a)(kx+ly+mz)\}$$
$$(k,l,m=0,\pm 1,\pm 2,\cdots).$$

So, if the development of $f_i(r)$ in terms of its characteristic functions is

$$f_i(r)=\sum_{k,l,m=-\infty}^{\infty} c_{k,l,m;i}a^{-\frac{3}{2}}\exp\{(2\pi i/a)(kx+ly+mz)\},$$

then

$$s_n(r_{1,n+2})=\frac{1}{a^3}\sum_{k,l,m}\prod_{i=0}^{n}(a^{-\frac{3}{2}}c_{k,l,m;i})\exp\frac{2\pi i}{a}[kx_{1,n+2}+ly_{1,n+2},mz_{1,n+2}]. \tag{26}$$

Following the same procedure as in the case of all f's alike; if we let

$$a^{\frac{3}{2}}c_{k,l,m;i}=4\pi\int_0^\infty r^2 f_i(r)(\sin tr/tr)dr=g_i(t) \tag{27}$$

and

$$t=\frac{2\pi}{a}(k^2+l^2+m^2)^{\frac{1}{2}},$$

we have

$$s_n(r)=\frac{1}{2\pi^2}\int_0^\infty \frac{t^2\sin tr}{tr}\prod_{i=0}^{n}g_i(t)dt \tag{28a}$$

and

$$s_n(0)=\frac{1}{2\pi^2}\int_0^\infty t^2\prod_{i=0}^{n}g_i(t)dt. \tag{28b}$$

Writing
$$g_0(t) = 4\pi \int_0^\infty r^2 f(r)(\sin tr/tr)dr$$
and
$$g_1(t) = 4\pi \int_0^\infty r^2 f(r)p_2(r)(\sin tr/tr)dr,$$
the special case (24a) becomes
$$\frac{1}{2\pi^2}\int_0^\infty t^2 g_0{}^3(t)g_1(t)dt,$$

and the original integration over fifteen variables has been converted to an integral over one variable.

An additional simplification is possible when the number of particles to be integrated over is large. If all the $g_i(t)$'s have their greatest maximum at $t=0$, it is easy to show, using reasoning similar to that of Section 3, that

$$s_n(r) = \frac{(4\pi)^{n+1}}{8\pi^2}\left(\prod_{i=0}^n \mu_{i,2}\right)\frac{\exp(-r^2/4M)}{M^{\frac{3}{2}}}\pi^{\frac{1}{2}}\left(1 + \frac{15r^4}{4M^2}\sum_{i=0}^n a_{i,4} + \cdots\right). \tag{29}$$

Here

$$a_{i,4} = (\mu_{i,6}/5!\mu_{i,2}) - (\mu_{i,4}^2/72\mu_{i,2}^2) \quad \text{and} \quad M = \sum_{i=0}^n (\mu_{i,4}/6\mu_{i,2}).$$

Suppose there is a largest $a_{i,4}$, and that its absolute value is A; also suppose there is a smallest $\mu_{i,4}/\mu_{i,2}$ and that its absolute value is a. Then

$$\left|\frac{\sum a_{i,4}}{M^2}\right| = \left|\frac{\sum a_{i,4}}{\frac{1}{36}\left(\sum \frac{\mu_{i,4}}{\mu_{i,2}}\right)^2}\right| \leqslant \frac{36nA}{n^2 a^2} = O\left(\frac{1}{n}\right).$$

So, as $n \to \infty$, we have the asymptotic expressions

$$s_n(r) \sim \frac{(4\pi)^{n+1}}{8\pi^2}\left(\prod_{i=0}^n \mu_{i,2}\right)\frac{\exp(-r^2/4M)}{M^{\frac{3}{2}}}\pi^{\frac{1}{2}}\left[1 + O\left(\frac{1}{n}\right)\right] \tag{30a}$$

and

$$s_n(0) \sim 3(6/\pi)^{\frac{1}{2}}(4\pi)^n \prod_{i=0}^n \mu_{i,2} \bigg/ \left(\sum_{i=0}^n \mu_{i,4}/\mu_{i,2}\right)^2. \tag{30b}$$

II. Application to the Calculation of Thermodynamic Functions

In order to calculate the thermodynamic properties of an imperfect gas it is sufficient to calculate

$$s_0 = \sum_{k=1}^\infty \beta_k v^{-k} = 4\pi v^{-1}\int_0^\infty r^2 f(r)dr + \sum_{k=2}^\infty \beta_{k,0} v^{-k} + \sum_{k=3}^\infty \beta_{k,1} v^{-k} + \cdots,$$

where the $\beta_{k,0}$'s are integrals over rings, the $\beta_{k,1}$'s correspond to rings with one internal connection, etc. By the proper number of differentiations and integrations of s_0 with respect to v, one can evaluate the Helmholtz free energy, the equation of state, and other thermodynamic functions.

Now, if n_{k+1} is the number of ways a $(k+1)$-membered ring can be constructed from $k+1$ molecules,

$$\beta_{k,0} = \frac{n_{k+1}}{Vk!}\int\cdots\int f_{12}\cdots f_{k+1,1}d\{\mathbf{k}+\mathbf{1}\}.$$

Since $n_{k+1} = k!/2$ and $\beta_{k,0} = p_k(0)/2$, Eq. (21) implies that

$$\sum_{k=2}^{\infty} \beta_{k,0} v^{-k} = \frac{1}{4\pi^2} \int_0^{\infty} t^2 \sum_{k=2}^{\infty} g^{k+1}(t) v^{-k} dt = \frac{1}{(2v\pi)^2} \int_0^{\infty} \frac{t^2 g^3(t) dt}{1 - g(t)/v}.$$

To find the contribution to s_0 of rings with one internal connection, we define

$$\gamma_{m,k} = \frac{1}{V} \int \cdots \int f_{12} \cdots f_{m+1,m+2} f_{m+2,1} d\{m+1\} \int \cdots \int f_{m+2,m+3} \cdots f_{k+1,1} d\tau_{m+3} \cdots d\tau_{k+1}$$

and recall Eq. (19). Thus

$$\gamma_{m,k} = \frac{1}{V} \int \cdots \int f_{12} \cdots f_{m+1,m+2} f_{m+2,1} p_{k-m-1}(r_{1,m+2}) d\{m+2\}.$$

In s_0 each γ must be weighted with the number of ways it can be formed from $(k+1)$ molecules. This number is the product of the number of ways of forming a ring of $(k+1)$ molecules multiplied by the number of ways of inserting a single connecting bond of the desired type in the given ring. If $k+1 \neq 2m+2$, this number is $(k+1)!/2$; and if $k+1 = 2m+2$, $(k+1)!/4$. Thus

$$\sum_{k=3}^{\infty} \beta_{k,1} v^{-k} = \sum_{k=1}^{\infty} \frac{(k+1)}{2} \gamma_{k,2k+1} v^{-(2k+1)} + \sum_{m=1}^{\infty} \sum_{k=2m+3}^{\infty} \frac{(k+1)}{2} \gamma_{m,k} v^{-k}.$$

But if we write

$$g_k(t) = 4\pi \int_0^{\infty} r^2 f(r) p_k(r) (\sin tr/tr) dr$$

and apply (28a),

$$\sum_{k=3}^{\infty} \beta_{k,1} v^{-k} = \sum_{k=1}^{\infty} \frac{k+1}{4\pi^2} v^{-(2k+1)} \int_0^{\infty} t^2 g^{k+1}(t) g_k(t) dt + \sum_{m=1}^{\infty} \sum_{k=2m+1}^{\infty} \frac{(k+1)}{4\pi^2} v^{-k} \int_0^{\infty} t^2 g^{m+1}(t) g_{k-m-1}(t) dt + \cdots.$$

In a similar manner expressions involving only functions of single integrals can be obtained for rings with more internal connections; however, one complication prevents the complete expansion of s_0 as a sum of simplified integrals. If one tries to apply the Fourier series method to configurations which are "tightly connected" (i.e., a bond connects every molecule in the configuration with every other one; for example β_{32}) new expressions result, but they are sums and integrals of the same "tightly connected" form in the reciprocal k, l, m space as the original ones in the x, y, z space.

To calculate the molecular distribution function $F_2\{2\}$ for imperfect gases, we will employ the expressions for chain integrals, (20) and (28a). Let the components of $H_{2,m}$ in (15) be ordered in sets $H^0_{2,m}$, $H^1_{2,m}$, $\cdots$, such that the H^0's correspond to integrals over chains without internal connections, H^1's to those with one internal connection, etc. Since the integral over a chain with n doubly connected molecules and two singly connected end molecules is $p_{n+1}(r_{12})$ when the distance between the two singly connected end molecules is r_{12}, and since this integral remains invariant under any of the $n!$ interchanges of the internal molecules,

$$H^0_{2,n} = p_{n+1}(r_{12}) = \frac{1}{2\pi^2} \int_0^{\infty} \frac{t^2 g^{n+1}(t) \sin tr_{12}}{tr_{12}} dt.$$

Thus

$$F_2\{2\} = \exp(-V(r_{12})/kT) \left[1 + \frac{1}{2\pi^2} \int_0^{\infty} \frac{t^2 \sin tr_{12}}{vtr_{12}} \sum_{m=0}^{\infty} v^{-m} g^{m+2}(t) dt + \sum_{m=2}^{\infty} v^{-m} H^1_{2,m} + \cdots \right]$$

$$= \exp(-V(r_{12})/kT) \left[1 + \frac{1}{2\pi^2 v} \int_0^{\infty} \frac{t^2 g^2(t)}{1 - g(t)/v} \frac{\sin tr_{12}}{tr_{12}} dt + \sum_{m=2}^{\infty} v^{-m} H^1_{2,m} + \cdots \right]. \tag{31}$$

TABLE I. $F(y, \theta)/y$ as a function of θ and y.

y	$\theta=1$	$\theta=0.85$	$\theta=0.754$	$\theta=0.695$	$\theta=0.55$	$\theta=0.40$	y	$\theta=1$	$\theta=0.85$	$\theta=0.754$	$\theta=0.695$	$\theta=0.55$	$\theta=0.40$
0.00	0.8460	0.6342	0.5082	0.4409	0.2695	0.1042	5.00	0.0096	0.0111	0.0116	0.0117	0.0101	0.0100
0.50	.7191	.5320	.4262	.3544	.2042	.0613	5.50	.0618	.0540	.0492	.0464	.0399	.0332
1.00	.4569	.3194	.2381	.2004	.0810	−.0207	6.00	.0838	.0715	.0632	.0601	.0509	.0417
1.50	.1745	.0952	.0467	.0191	−.0443	−.1010	6.50	.0785	.0665	.0596	.0562	.0452	.0380
2.00	−.0516	−.0831	−.1015	−.1114	−.1335	−.1532	7.00	.0541	.0456	.0416	.0382	.0324	.0272
2.50	−.1930	−.1875	−.1846	−.1840	−.1764	−.1692	7.50	.0218	.0185	.0167	.0159	.0140	
3.00	−.2374	−.2171	−.2027	−.1946	−.1743	−.1547	8.00	−.0098	−.0070	−.0059	−.0052	−.0034	
3.50	−.2193	−.1899	−.1716	−.1623	−.1394	−.1181	8.50	−.0300	−.0243	−.0216	−.0193	−.0150	
4.00	−.1506	−.1268	−.1127	−.1049	−.0872	−.0701	9.00	−.0454	−.0327	−.0289	−.0266	−.0217	
4.50	−.0570	−.0534	−.0459	−.0416	−.0330	−.0259							

The sum over $H^1_{2,m}$ can be treated in the same manner as that over $\beta_{k,1}$ but inasmuch as it will not be used in the calculations of the next section, we do not write it out explicitly. In fact, it is possible to sum most of the series that do not involve "tightly connected" configurations, but the resulting expressions are very complicated.

III. Imperfect Gases with Lennard-Jones Potentials

The mutual potential energy function, $V(r)$, between two molecules whose distance of separation is r becomes infinite as $r \rightarrow 0$, has a negative minimum when r is the equilibrium distance of the molecules, and rapidly approaches zero as the molecules tend to infinite separation. London[14] has shown from quantum-mechanical considerations that the potential energy contribution of the attractive forces between two molecules with no permanent electric moment has a dominant term proportional to r^{-6}. Since, to find the total potential energy, a repulsive term that becomes infinite at $r=0$ must be added, Lennard-Jones[15] assumed this term to be proportional to r^{-m}. Then after calculation of the theoretical second virial coefficient from

$$B(T) = \tfrac{1}{2} \int_0^\infty 4\pi r^2 (1 - e^{-V(r)/kT}) dr$$

with such a general $V(r)$, he found the proportionality constants and m values that gave the best agreement with experiments. Here m will be chosen to be 12 (this number gives the best results for the most kinds of molecules). Thus if u and r_0 are constants for a given molecular species

$$V(r) = 4u[(r_0/r)^{12} - (r_0/r)^6], \quad (32)$$

in which the minimum value of $V(r)$ is $-u$ and occurs at $r = (\sqrt[6]{2})r_0$. The potential $V(r)$ is zero at $r = r_0$, positive for $r < r_0$ and negative for $r > r_0$. For argon[15] $r_0 = 3.408$ A.U. and $u = 165.0 \times 10^{-16}$ erg; for neon $r_0 = 2.743$ A.U. and $u = 48.81 \times 10^{-16}$ erg. With this potential function,

$$f(r) = \exp[-4\theta(z^{-12} - z^{-6})] - 1, \quad (33a)$$

and

$$g(t/r_0) = (4\pi r_0^3/t) \int_0^\infty z\{\exp[-4\theta(z^{-12} - z^{-6})] - 1\} \sin tz \, dz, \quad (33b)$$

where $\theta = u/kT$ and $z = r/r_0$. The complexity of the integrand in (33b) makes the analytical evaluation of $g(t/r_0)$ difficult, so numerical integration is resorted to.

One of the simplest quantities which can be calculated from $g(t/r_0)$ and which can be compared with experimental data is the third virial coefficient. By definition this quantity (which shall henceforth be designated by C) is given by the equation

$$Pv/kT = 1 + B/v + C/v^2 + \cdots,$$

---

[14] For an up-to-date review see H. Margenau, Rev. Mod. Phys. **11**, 1, 25 (1939).
[15] J. E. Lennard-Jones, Physica **4**, 941 (1937).

TABLE II. Values of $\int_0^\infty y^2 \left\{ \frac{F(y, \theta)}{y} \right\}^2 \frac{\sin yz}{yz} dy$.

z	$\theta=1.0$	$\theta=0.85$	$\theta=0.70$	z	$\theta=1.0$	$\theta=0.85$	$\theta=0.70$
1.0	0.150	0.075	0.041	2.2	0.147	0.084	0.042
1.2	.068	.010	−.014	2.4	.134	.080	.044
1.4	.023	−.023	−.046	2.6	.086	.055	.030
1.6	.020	−.013	−.049	2.8	.051	.034	.016
1.8	.059	−.006	−.023	3.0	.036	.012	
2.0	.115	.055	.015				

from which it follows (Eq. (4)) that $C = -2\beta_2/3$. But

$$\beta_2 = \tfrac{1}{2} p_2(0) = \int_0^\infty t^2 g^3(t) dt / 4\pi^2.$$

therefore

$$C = -\frac{2}{3} \frac{(4\pi r_0^3)^2}{\pi} \int_0^\infty y^2 \left\{ \frac{F(y, \theta)}{y} \right\}^3 dy, \tag{34}$$

where

$$F(y, \theta) = \int_0^\infty z \{\exp[-4\theta(z^{-12} - z^{-6})] - 1\} \sin yz \, dz, \tag{35}$$

and $y = tr_0$. The integration of $F(y, \theta)$ is described in the Appendix and $F(y, \theta)/y$ is tabulated for various values of θ in Table I. The function

$$3\pi C/2(4\pi r_0^3)^2$$

as calculated from (34) is plotted in Fig. 2 as a function of $\theta = u/kT$, and is compared with the corresponding quantity obtained from the experimental third virial coefficients of argon.[16] Considering that the experimental C values are the result of least squaring isotherm data to fit an equation with a small number of virial coefficients while the theoretical ones are terms in an infinite series, the agreement is as good as can be expected. Actually, the deviations are of the same order as those resulting from the comparison of the data of different experimentalists. Also, a small error in r_0 would cause a rather large error in converting an experimental C to $(3\pi/2)C/(4\pi r_0^3)^2$ because r_0 is raised to such a high power in the conversion factor.

Another fundamental quantity that can easily be calculated from the $F(y, \theta)/y$ tables is the distribution function $F_2\{2\}$. Since Eqs. (31) and (35) imply (here z is written for r_{12}) that

$$F_2\{2\} = e^{-V(z)/kT} \left[1 + \frac{4\pi r_0^3}{\pi v} \int_0^\infty y^2 \left\{ \frac{F(y, \theta)}{y} \right\}^2 \frac{\sin yz}{yz} dy + \cdots \right], \tag{36}$$

the values of the integral

$$\int_0^\infty y^2 \left\{ \frac{F(y, \theta)}{y} \right\}^2 \frac{\sin yz}{yz} dy$$

as tabulated in Table II can be used to calculate $F_2\{2\}$ up to fairly high densities. The graphs of $F_2\{2\}$ in Fig. 3 (note that the curve for $\theta = 0.85$ has been raised one unit and that for $\theta = 1$ by two units) behave qualitatively in the manner one would expect from physical considerations. The first peak results from the tendency of nearest neighbors of any molecule to be at a distance which corresponds to the minimum in the potential well of Fig. 1 and which is smoothed out a bit by thermal agitation. The second peak corresponds to the tendency of next nearest neighbors to occupy positions that correspond to the minimum in their mutual potential energy well with respect to the first nearest neighbors. As θ increases, the peaks sharpen as one would expect from the diminishing of thermal

[16] J. Otto, *Wien-Harms Handbuch der Experimentalphysik* (Leipzig, 1929), Vol. VIII-2, p. 151.

agitation. Experimentally, $F_2\{2\}$ is related to the Fourier transform of the intensity of x-rays scattered from the gas.[17] The plot is made at a density greater than the critical, for which, at the lower temperatures, the equations are not expected to be valid, since the series of Eq. (31) diverge.

The authors wish to thank Dr. F. Nachod for his aid with some of the calculations.

Appendix I

Integration of $I = \int_0^\pi \int_0^{2\pi} \exp(-2\pi i r/a)(k \cos\varphi \sin\theta + l \sin\varphi \sin\theta + m \cos\theta) \sin\theta d\theta d\varphi.$

If one defines a vector

$$\mathbf{h} = \frac{k}{a}\mathbf{i} + \frac{l}{a}\mathbf{j} + \frac{m}{a}\mathbf{k}, \qquad \exp[-(2\pi i r/a)(k \cos\varphi \sin\theta + - -)] = \exp(-2\pi i \mathbf{r}\cdot\mathbf{h}).$$

Choosing a new polar coordinate system with h as polar axis, $\mathbf{r}\cdot\mathbf{h} = rh \cos\gamma$,

$$h = (1/a)(k^2 + l^2 + m^2)^{\frac{1}{2}}$$

and

$$I = \int_0^\pi \int_0^{2\pi} \exp(-2\pi i r h \cos\gamma) \sin\gamma d\varphi' d\gamma = 2\pi \int_{-1}^1 \exp(-2\pi i r h y) dy = (2/rh) \sin 2\pi rh.$$

For this method of integration, which is simpler than that originally proposed, the authors are indebted to Professor Kirkwood.

Appendix II

Evaluation of $F(y, \theta)/y$

By definition

$$F(y, \theta) = \int_0^\infty z\{\exp[-4\theta(z^{-12} - z^{-6})] - 1\} \sin yz dz.$$

It is possible to obtain an asymptotic expression when y is small by writing

$$\sin yz = yz - y^3z^3/3! + (\sin yz - yz + y^3z^3/3!),$$

for in that case (say $y < 0.4$)

$$F(y, \theta)/y \sim \mu_2 - \mu_4 y^2/3!.$$

In a manner described by Lennard-Jones,[10] if $n \leq 4$

$$\mu_n = \int_0^\infty z^n \{\exp[-4\theta(z^{-12} - z^{-6})] - 1\} dz = \frac{(4\theta)^{(1+n)/12}}{12} \sum_{l=0}^\infty \frac{(4\theta)^{l/2}}{l!} \Gamma\left(\frac{l}{2} - \frac{n+1}{12}\right).$$

For larger values of y it seems most convenient to divide the range of integration into three parts

$$F(y, \theta) = \int_0^{0.85} + \int_{0.85}^a + \int_a^\infty = I_1 + I_2 + I_3.$$

When $z \leq .85$, $\{\exp[-4\theta(z^{-12} - z^{-6})] - 1\} \sim -1$ and

$$I_1 = -(1/y^2)(\sin 0.85y - 0.85y \cos 0.85y).$$

If α is fairly large one can write

$$I_3 = \int_a^\infty z(4\theta z^{-6} - 4\theta(1-2\theta)z^{-12} + \cdots) \sin yz dz.$$

The α can be chosen to make the integral involving z^{-12} as small as the error (say ϵ) desired in $F(y, \theta)$. This choice is made by noting that

$$\epsilon = \left|\int_a^\infty z^{-11} \sin yz dz\right| \cdot |4\theta(1-2\theta)| < \int_a^\infty z^{-11} dz \cdot |4\theta(1-2\theta)| = |4\theta(1-2\theta)|/10\alpha^{10}$$

and

$$\alpha < [2\theta|(1-2\theta)|/5\epsilon]^{1/10}.$$

With such a choice of α, integration by parts yields[18]

$$I_3 = 4\theta \int_a^\infty z^{-5} \sin yz dz = \frac{\theta}{\alpha}\left\{\frac{\sin \alpha y}{\alpha}\left[\frac{1}{\alpha^2} - \frac{y^2}{6}\right] + \frac{y \cos \alpha y}{3}\left[\frac{1}{\alpha^2} - \frac{y^2}{2}\right] + \frac{y^4 \alpha}{6} \int_a^\infty \frac{\sin yz}{z} dz\right\}.$$

When y is large, integration by parts in the "other direction" gives the asymptotic expression

$$I_3 \sim \frac{4\theta}{y\alpha^5}\left\{\cos \alpha y\left[1 - \frac{5 \cdot 6}{(\alpha y)^2} + \frac{5 \cdot 6 \cdot 7 \cdot 8}{(\alpha y)^4}\right] + 5\frac{\sin \alpha y}{\alpha y}\left[1 - \frac{6 \cdot 7}{(\alpha y)^2} + \frac{6 \cdot 7 \cdot 8 \cdot 9}{(\alpha y)^4}\right] + \varphi \frac{5 \cdot 6 \cdot 7 \cdot 8 \cdot 9}{(\alpha y)^5}\right\},$$

where $|\varphi| < 1$.

Unfortunately it has not been possible to find simple analytical expressions for I_2, so it must be integrated numerically.

---

[17] Zernicke and Prins, Zeits. f. Physik **41**, 184 (1927); Debye and Mencke, Physik. Zeits. **31**, 797 (1932).

[18] The integral $\int_a^\infty \sin yz dz/z$ is obtainable from $Si(ay)$ tables, cf. *Mathematical Tables* (British Association for the Advancement of Science, London, 1931). Vol. 1.

ON THE THEORY OF MARKOFF CHAINS

By Elliott W. Montroll

University of Pittsburgh

1. Summary. Although there exists voluminous literature on the theory of probability of independent events, and powerful techniques have been developed for the analysis of most of the interesting problems in this field, the theory of probability of dependent events has been rather neglected. The first detailed investigations in this subject were published by A. Markoff [1]. S. Bernstein [2] has extended the fundamental limit theorems to chains of dependent events. The most extensive exposition of this field has been made by M. Fréchet [3].

In the present paper we shall develop methods of averaging functions over chains of dependent variables and find the probability distribution of these functions. It will be shown that for certain types of chains these averages and distribution functions can be expressed in terms of the characteristic values and vectors of a certain operator equation. Many of the methods discussed here have been applied to problems in statistical mechanics [4, 5, 6, 7, 8]. The most important application has been made by L. Onsager [8] who proved rigorously (on the basis of a simplified model) that Boltzmann's energy distribution in a solid with cooperative elements leads to a phase transition. The first explicit application of linear operator theory (through matrices and integral equations) to probability chains has apparently been made by Hostinsky [9].

2. Introductory Remarks. Suppose there exists a chain of events each of which might lead to one of ν possible results, and which are correlated in such a manner that the probability of n successive events leading to a chain of results

$$\alpha_1, \alpha_2, \cdots, \alpha_n$$

is proportional to

$$P_n(\alpha_1, \alpha_2, \cdots, \alpha_n).$$

The probability of a given function $F(\alpha_1, \alpha_2, \cdots, \alpha_n)$ having a value corresponding to the sequence of α's would be proportional to

$$F(\alpha_1, \alpha_2, \cdots, \alpha_n) P_n(\alpha_1, \cdots, \alpha_n)$$

and its average value over all configurations of the chain would be

(1) $$\bar{F} = F_1/F_0 = \sum_{\{\alpha_j\}} F(\alpha_1, \alpha_2, \cdots, \alpha_n) P_n(\alpha_1, \alpha_2, \cdots, \alpha_n) / \sum_{\{\alpha_j\}} P_n(\alpha_1, \cdots, \alpha_n)$$

where

(1a) $$F_m = \sum_{\{\alpha_j\}} [F(\alpha_1, \alpha_2, \alpha_3, \cdots, \alpha_n)]^m P_n(\alpha_1, \cdots, \alpha_n)$$

and the summation extends over all values of
$$\{\alpha_j\} = (\alpha_1, \alpha_2, \cdots, \alpha_n).$$

The probability of a result α_1 of the first event leading to a result α_n of the nth event is

(2) $$P_n(\alpha_1, \alpha_n) = (1/F_0) \sum_{\alpha_2,\cdots,\alpha_{n-1}} P_n(\alpha_1, \alpha_2, \cdots, \alpha_n).$$

In order to find the probability of a given function $F(\alpha_1, \cdots, \alpha_n)$ having a value between ξ and $\xi + h$ it is useful to know the moments and Thiele semi-invariants of $F(\alpha_1, \cdots, \alpha_n)$. Both of these functions of F can be calculated from

(3) $$Z_n(x) = \sum_{\{\alpha_j\}} P_n(\alpha_1, \cdots, \alpha_n) \exp\{xF(\alpha_1, \alpha_2, \cdots, \alpha_n)\}.$$

Obviously

(4) $$F_m = \lim_{x \to 0} \partial^m Z_n(x)/\partial x^m.$$

It is known [10] that the mth Thiele semi-invariant is given by

(5) $$\Lambda_m = \lim_{x \to 0} \partial^m \log Z_n(x)/\partial x^m.$$

In the notation of Cramér $Z_n(i\omega)/Z_n(0) = f(\omega)$, the characteristic function of F.

If $G(z)$ is defined so that $G(\xi + h) - G(\xi)$ is the probability that the function $F(\alpha_1, \cdots, \alpha_n)$ has a value between $\xi \le F(\alpha_1, \cdots, \alpha_n) < \xi + h$, then it is well known that [5] if $G(z)$ is continuous at $x = \xi$ and $x = \xi + h$

(6) $$G(\xi + h) - G(\xi) = \frac{1}{2\pi i} \lim_{T \to \infty} \int_{-T}^{T} \frac{(1 - e^{-i\omega h})e^{-i\omega\xi}}{\omega} \exp[\log f(\omega)] \, d\omega$$

where

(6a) $$\log f(\omega) = \sum_{m=1}^{\infty} \frac{\Lambda_m(i\omega)^m}{m!} = \sum_{m=1}^{k} \Lambda_m(i\omega)^m/m! + o(\omega^k).$$

When the derivative of $G(\xi)$ with respect to ξ exists, the probability of

$$F(\alpha_1, \cdots, \alpha_n)$$

having a value between ξ and $\xi + d\xi$ is

(6b) $$\varphi(\xi) \, d\xi = (\partial G/\partial \xi) \, d\xi = \frac{d\xi}{2\pi} \lim_{T \to \infty} \int_{-T}^{T} \exp\left\{\sum_{m=1}^{\infty} \Lambda_m(i\omega)^m/m!\right\} e^{-i\omega\xi} \, d\omega.$$

From (4)

(7) $$\sum_{m=1}^{\infty} \Lambda_m(i\omega)^m/m! = -\log Z_n(0) + \lim_{x \to 0} e^{-i\omega\partial/\partial x} \log Z_n(x).$$

Since, for a constant c independent of x,
$$e^{c\partial/\partial x} f(x) = f(x+c)$$
we have

(8) $$\sum_{m=1}^{\infty} \Lambda_m(i\omega)^m / m! = \log\{Z_n(i\omega)/Z_n(0)\},$$

and from (6)

(9) $$G(\xi + h) - G(\xi) = \frac{1}{2\pi i} \lim_{T\to\infty} \int_{-T}^{T} \frac{e^{-i\omega\xi}(1 - e^{-i\omega h})Z_n(i\omega)\,d\omega}{\omega Z_n(0)}.$$

Equations (3), (4), (5) and (9) indicate that much information concerning a chain of correlated events can be obtained from a knowledge of $Z_n(x)$. We shall now introduce procedures for the determination of $Z_n(x)$ for several general forms of $P(\alpha_1, \cdots, \alpha_n)$.

When α is a continuous variable, the results of this section and those to follow are easily generalized by replacing the summations operations over all values of the α's by integrals, and by replacing the matrix equations of the next section by integral equations.

3. Simple Chains, $P_n(\alpha_1, \cdots, \alpha_n) = \prod_{j=1}^{n-1} p(\alpha_j, \alpha_{j+1}).$

a. General theory. By a simple chain we shall mean a sequence of events, each of which leads to one of ν possible results and which occur in such a manner that if the result of the kth event is α_k, the probability of the $(k+1)$st one yielding a result α_{k+1} to proportional to $p(\alpha_k, \alpha_{k+1})$. This implies that the probability of the occurrence of the sequence of results
$$\alpha_1, \alpha_2, \cdots, \alpha_n$$
is

(10) $$\prod_{i=1}^{n-1} p(\alpha_i, \alpha_{i+1}) \bigg/ \sum_{\{\alpha_j\}} \prod_{i=1}^{n-1} p(\alpha_i, \alpha_{i+1}),$$

and the probability of a first result α_1, leading to an nth result α_n is

(11) $$P_n(\alpha_1, \alpha_n) = \sum_{\alpha_2,\cdots,\alpha_{n-1}} \prod_{j=1}^{n} p(\alpha_j, \alpha_{j+1}) \bigg/ \sum_{\{\alpha_j\}} \prod_{j=1}^{n} p(\alpha_j, \alpha_{j+1}).$$

The summations are to be extended over all ν possible values of each α_i indicated on the summation indices. Chains of this type are sometimes called simple Markoff chains after the first author who studied them systematically.

From (1), the average value of a function $F(\alpha_1, \cdots, \alpha_n)$ is

(12) $$F_1/F_0 = \frac{\sum_{\alpha_1}\cdots\sum_{\alpha_n} F(\alpha_1, \cdots, \alpha_n) \prod_{j=0}^{n-1} p(\alpha_j, \alpha_{j+1})}{\sum_{\alpha_1}\cdots\sum_{\alpha_n} \prod_{j=1}^{n-1} p(\alpha_j, \alpha_{j+1})}.$$

Many chain functions $F(\alpha_1, \cdots, \alpha_n)$ of interest are either additive or multiplicative and of one of the forms

(13a) a) $F_1(\alpha_1, \cdots, \alpha_n) = h(\alpha_1, \alpha_2) + h(\alpha_2, \alpha_3) + \cdots + h(\alpha_{n-1}, \alpha_n)$

(13b) b) $F_2(\alpha_1, \cdots, \alpha_n) = g(\alpha_1, \alpha_2) g(\alpha_2, \alpha_3) \cdots g(\alpha_{n-1}, \alpha_n)$.

In case (b) it is convenient to define a new function $h(\alpha_i, \alpha_j)$ by

(14) $\qquad g(\alpha_i, \alpha_j) = \exp[xh(\alpha_i, \alpha_j)]$

and in both cases to consider a function of the form

(15) $\qquad Z_n(x) = \sum_{\{\alpha_i\}} \prod_{j=1}^{n-1} p(\alpha_j, \alpha_{j+1}) \exp[xh(\alpha_j, \alpha_{j+1})]$,

for then the values of F_1 and F_2 averaged over the entire chain are given by

(16a) $\qquad <F_1>_{av.} = \lim_{x \to 0} \partial \log Z_n(x)/\partial x$

and

(16b) $\qquad <F_2>_{av.} = Z_n(1)/Z_n(0)$.

When n is large, the direct evaluation of (15) may become quite difficult because of the large number of variables involved. As an alternative we shall now introduce a procedure that is based on the observation that $Z_n(x)$ is the sum of the elements of the nth power of the matrix

(17) $\qquad \boldsymbol{P}_x = \begin{pmatrix} p_x(1, 1) & p_x(1, 2) & \cdots & p_x(1, \nu) \\ p_x(2, 1) & p_x(2, 2) & \cdots & p_x(2, \nu) \\ \cdots & \cdots & \cdots & \cdots \\ p_x(\nu, 1) & p_x(\nu, 2) & \cdots & p_x(\nu, \nu) \end{pmatrix}$

where the elements $p_x(\alpha, \beta)$ are defined as

(18) $\qquad p_x(\alpha, \beta) = p(\alpha, \beta) \exp[xh(\alpha, \beta)]$.

α and β range over the same set of values as one of the "result" parameters α_j; and each of the ν possible results is represented by a unique integer of the set $1, 2, \cdots, \nu$. Thus $Z_n(x) = $ sum of elements of $\boldsymbol{P}_x^{n-1}$. To employ this observation to advantage, let us consider the characteristic values and vectors of the matrix $\boldsymbol{P}_x$. It is well known that if the characteristic values are simple the characteristic vectors form a biorthogonal set; that is, if

(19a) $\qquad \Phi_{i,x} = \{\varphi_{i,x}(1), \varphi_{i,x}(2), \cdots, \varphi_{i,x}(\nu)\}, \quad (i = 1, 2, \cdots, \nu)$,

and

(19b) $\qquad \Psi_{i,x} = \begin{bmatrix} \psi_{i,x}(1) \\ \psi_{i,x}(2) \\ \psi_{i,x}(\nu) \end{bmatrix}$

satisfy the operator equations

(20a) $$\Phi_{i,x} \cdot P_x = \lambda_{i,x} \Phi_{i,x}$$

(20b) $$P_x \cdot \Psi_{i,x} = \lambda_{i,x} \Psi_{i,x}$$

where $\lambda_{i,x}$ is the ith characteristic value of (17), then

$$\Phi_{i,x} \cdot \Psi_{j,x} = \sum_{\alpha=1}^{\nu} \varphi_{i,x}(\alpha) \psi_{j,x}(\alpha) = 0 \quad \text{when} \quad i \neq j.$$

We shall for convenience always assume that the φ's and ψ's are normalized:

$$\Phi_{i,x} \cdot \Psi_{i,x} = 1$$

so that in general:

(21) $$\Phi_{i,x} \cdot \Psi_{j,x} = \delta_{ij} = \begin{cases} 0 & \text{when } i \neq j \\ 1 & \text{when } i = j. \end{cases}$$

It is well known from matrix theory that one can expand a matrix element as

(22) $$p_x(\alpha, \beta) = \sum_{i=1}^{\nu} \lambda_{i,x} \varphi_{i,x}(\beta) \psi_{i,x}(\alpha)$$

and that

(23) $$\lambda_{i,x} = \Phi_{i,x} \cdot P_x \cdot \Psi_{i,x}.$$

By substituting (22) into the expression for $Z_n(x)$ in terms of P_x^{n-1}, one can show that

(24) $$Z_n(x) = \sum_{i=1}^{\nu} \{\lambda_{i,x}\}^{n-1} \left\{ \sum_{\beta=1}^{\nu} \varphi_{i,x}(\beta) \right\} \left\{ \sum_{\alpha=1}^{\nu} \psi_{i,x}(\alpha) \right\}$$
$$= \sum_{i=1}^{\nu} \lambda_{i,x}^{n-1} (\Phi_{i,x} \cdot 1)(1 \cdot \Psi_{i,x}).$$

Therefore $Z_n(x)$ can be determined from a knowledge of the characteristic vectors and values of the matrix P_x.

If there exists a largest characteristic root $\lambda_{i,x}$ such that

(25) $$\lambda_{L,x} > |\lambda_{i,x}| \qquad \text{if } i \neq L,$$

one can obtain some interesting results. Before deriving these, we shall give a sufficient condition (which is satisfied in many chains) for the existance of this inequality. Frobenius [11] has shown that if all the elements of a finite matrix are > 0, then the characteristic value of largest absolute value of the matrix is real, positive, and simple (nondegenerate). Thus, as long as ν is finite and $p_x(\alpha, \beta) > 0$ for all α and β, (25) is valid.

We shall now prove that

(25a) $$\lim_{n \to \infty} \left\{ \frac{Z_n(x)}{\lambda_{L,x}^{n-1}(\Phi_{L,x} \cdot 1)(1 \cdot \Psi_{L,x})} - 1 \right\} = 0$$

that is,

(25b) $$Z_n(x) \sim \lambda_{L,x}^{n-1}(\Phi_{L,x} \cdot \mathbf{1})(\mathbf{1} \cdot \Psi_{L,x}).$$

First let us consider the case in which P_x is a symmetrical matrix. Then $\varphi_{j,x}(\alpha) = \psi_{j,x}(\alpha)$, all the characteristic values are real, and

$$Z_n(x) = \lambda_{L,x}^{n-1}(\Phi_{L,x} \cdot \mathbf{1})^2 + \sum_{i \neq L} \lambda_{i,x}^{n-1}(\Phi_{i,x} \cdot \mathbf{1})^2.$$

From Cauchy's inequality and (21)

$$|\Phi_{i,x} \cdot \mathbf{1}|^2 = \left| \sum_{\alpha=1}^{\nu} \varphi_{i,x}(\alpha) \right|^2 \leq \left[\sum_{\alpha=1}^{\nu} \varphi_{i,x}^2(\alpha) \right]\left[\sum_{\alpha=1}^{\nu} 1 \right] = \nu.$$

Therefore,

$$\left| \sum_{i \neq L} \lambda_{i,x}^{n-1}(\varphi_{i,x} \cdot \mathbf{1})^2 \right| \leq \nu \left| \sum_{i \neq L} \lambda_{i,x}^{n-1} \right| \leq \nu(\nu - 1) |\lambda_{s,x}^{n-1}|$$

where $\lambda_{s,x}$ is the characteristic value of P_x second largest in absolute value. This inequality yields

(25c) $$\left| \frac{Z_n(x)}{\lambda_{L,x}^{n-1}(\Phi_{L,x} \cdot \mathbf{1})^2} - 1 \right| \leq \frac{\nu(\nu - 1)}{(\Phi_{L,x} \cdot \mathbf{1})^2} \left| \left(\frac{\lambda_{s,x}}{\lambda_{L,x}} \right)^{n-1} \right|$$

and (25a) (since $\lambda_{s,x}/\lambda_{L,x} < 1$) follows. When P_x is not symmetrical, one can easily derive the analogous expression

$$\left| \frac{Z_n(x)}{\lambda_{L,x}^{n-1}(\Phi_{L,x} \cdot \mathbf{1})(\mathbf{1} \cdot \Psi_{L,x})} - 1 \right| \leq \frac{A(\nu - 1) |\lambda_{s,\alpha}/\lambda_{L,x}|^{n-1}}{(\varphi_{L,x} \cdot \mathbf{1})(\mathbf{1} \cdot \psi_{L,x})}$$

where

$$A = [\max \{|(\Phi_{i,x} \cdot \mathbf{1})|\}][\max \{|(\mathbf{1} \cdot \Psi_{i,x})|\}]$$

For brevity, when $x = 0$, we write $\lambda_{i,x}$ as λ_i, $\Psi_{i,x}$ as Ψ_i and $\Phi_{i,x}$ as Φ_i. By summing (10) over all α's except α_1, α_k and α_n we obtain the probability of an intermediate event leading to a result α_k if the results of the first and last events are known to have been α_1 and α_n. With the aid of (21) and (22) it is easy to show that this probability is exactly:

(26) $$\frac{\sum_{i,j=1}^{\nu} \lambda_j^{n-k} \lambda_i^{k-1} \psi_i(\alpha_1) \varphi_i(\alpha_k) \psi_j(\alpha_k) \varphi_j(\alpha_n)}{\sum_{i=1}^{\nu} \lambda_i^{n-1} \sum_{\alpha_1, \alpha_n} \psi_i(\alpha_1) \varphi_i(\alpha_n)}.$$

When n is very large, and when we have simultaneously $n \gg k \gg 1$, we can rewrite this equation to include λ_L, and neglect all terms containing other i's and j's. This leads to the results

a) If the number of events, n, in a simple chain is very large, the probability $P_n(\alpha_k)$ of a kth event far removed from the first and the last, yielding a result α_k when α_1, and α_n are unspecified is

(27) $$P_n(\alpha_k) \sim \psi_L(\alpha_k) \, \varphi_L(\alpha_k) / (\Phi_L \cdot \mathbf{1})(\mathbf{1} \cdot \Psi_L).$$

b) When $k = n$, the probability of the result $\alpha_1 \cdot$ of the first event leading to the result α_n of the nth event is

$$(28a) \qquad P_n(\alpha_1, \alpha_n) = \frac{\sum_{i=1}^{\nu} \lambda_i^{n-1} \psi_i(\alpha_1) \varphi_i(\alpha_n)}{\sum_{i=1}^{\nu} \lambda_i^{n-1} \sum_{\alpha_1, \alpha_n} \psi_i(\alpha_1) \varphi_i(\alpha_n)}.$$

So, as $n \to \infty$

$$(28b) \qquad P_n(\alpha_1, \alpha_n) \sim \frac{\psi_L(\alpha_1) \phi_L(\alpha_n)}{(\Phi_L \cdot 1)(1 \cdot \Psi_L)}.$$

c) When there exists no knowledge concerning the result of the first event, the probability of the nth event yielding the result α_n is

$$(29) \qquad P_n(\alpha_n) = \sum_{\alpha_1} P_n(\alpha_1, \alpha_n) \sim \Phi_L(\alpha_n)/(\mathbf{1} \cdot \Phi_L).$$

In chains of sufficient length for (25) to be valid, the probability of

$$F(\alpha_1, \cdots, \alpha_n)$$

having a value between ξ and $\xi + h$ has an especially simple asymptotic form. From (6) this probability is (if for a given n we let $T = an^{\frac{1}{2}}$)

$$(30) \qquad G(\xi + h) - G(\xi) = \frac{1}{2\pi i} \lim_{a \to \infty} \int_{-an^{1/2}}^{an^{1/2}} \left(\frac{d\omega}{\omega}\right) e^{-i\omega(\xi - \Lambda_1)}$$
$$(1 - e^{-i\omega h}) \exp\left\{-\tfrac{1}{2}\omega^2 \Lambda_2 - \frac{i\Lambda_3 \omega^3}{3!} + \cdots\right\}$$

and from (25) and (5)

$$(31) \qquad \Lambda_m \sim n \lim_{x \to 0} \partial^m \log \lambda_{L,x}/\partial x^m = nL_m$$

if

$$(32) \qquad L_m \equiv \lim_{x \to 0} \partial^m \log \lambda_{L,x}/\partial x^m.$$

Letting $y = \omega n^{\frac{1}{2}}$, (30) becomes

$$(33) \qquad G(\xi + h) - G(\xi) \sim \frac{1}{2\pi i} \lim_{a \to \infty} \int_{-a}^{a} \frac{dy}{y}$$
$$(e^{-iy\mu_1} - e^{-iy\mu_2}) e^{-\tfrac{1}{2}y^2 L_2} \left\{1 - \frac{L_3 y^3 i}{6n^{\frac{1}{2}}} + \cdots\right\}$$

where

$$(34a) \qquad \mu_1 = (\xi - \Lambda_1)/n^{\frac{1}{2}}$$
$$\mu_2 = (\xi + h - \Lambda_1)/n^{\frac{1}{2}}$$

$$(34b) \qquad \Lambda_1 = \text{average value of } F(\alpha_1, \cdots, \alpha_n) = \bar{F}.$$

Integrating (33)

(35) $$G(\xi + h) - G(\xi) \sim \frac{1}{(2\pi L^2)^{\frac{1}{2}}} \int_{\mu_1}^{\mu_2} e^{-\mu^2/2L_2}[1 + O(1/n)] \, d\mu.$$

As $n \to \infty$ and $h \to 0$

(35a) $$G(\xi + h) - G(\xi) \sim \frac{h}{(2\pi n L_2)^{\frac{1}{2}}} \exp\left(-\tfrac{1}{2}\right)[\xi - \bar{F}]/nL_2),$$

and the probability that $\xi \leq F < \xi + h$ becomes Gaussian.

b. *Examples of a simple chain.* As an example of a simple Markoff chain let us consider an event which can lead to either of two possible results, say "-1" or "1". Further, let us suppose that the probability of a given result being followed by an identical one is p and by one of another type is $(1-p)$; that is,

$$p(-1, -1) = p(1, \quad 1) = p$$
$$p(-1, \quad 1) = p(1, -1) = 1 - p.$$

This chain would be encountered in an analysis of a sequence of tosses of a coin with a "memory" so that the probability of two successive tosses showing the same face of the coin would be p and that of showing opposite faces $(1-p)$.

A question one might ask concerning such a chain is—What is the probability of the occurrence of a given number of transitions from one kind of result to another? In the chain of results

$$-1, -1, -1, 1, 1, -1, 1, -1, -1, -1$$

there would be four transitions, one corresponding to each -1 followed by a 1 and to each 1 followed by a -1. The function giving the number of transitions in a sequence of n events is

(36) $$F(\alpha_1, \cdots, \alpha_n) = \sum_{i=1}^{n-1} h(\alpha_i, \alpha_{i+1})$$

where

$$h(-1, -1) = h(1, \quad 1) = 0$$
$$h(-1, \quad 1) = h(1, -1) = 1.$$

Even though the α's are dependent, in this special case, $h(\alpha_i, \alpha_{i+1})$ and $h(\alpha_{i+1}, \alpha_{i+2})$ are independent so that (40) could have been obtained on this basis.

To apply the methods described in the beginning of this section we must find the characteristic values and vectors of the matrix

(37) $$P_x = \begin{pmatrix} p & (1-p)e^x \\ (1-p)e^x & p \end{pmatrix}$$

(the configuration index α has the value either -1 or 1 in this case instead of

"1" and "2" as given in (17)). The characteristic values are the roots of the equation

$$\begin{vmatrix} p - \lambda & (1-p)e^x \\ (1-p)e^x & p - \lambda \end{vmatrix} = 0$$

that is,

(38)
$$\lambda_{1,x} = p + (1-p)e^x$$
$$|\lambda_{2,x}| = |p - (1-p)e^x| < \lambda_{1,x}$$

and the characteristic vectors are

$$\psi_{1,x} = 2^{-\frac{1}{2}} \begin{pmatrix} 1 \\ 1 \end{pmatrix} \quad \text{and} \quad \psi_{2,x} = 2^{-\frac{1}{2}} \begin{pmatrix} 1 \\ -1 \end{pmatrix}.$$

The ψ and φ vectors have the same components in this case because of the symmetry of the P_x matrix. Clearly

$$\lambda_L = \lambda_1 = \lambda_{1,0} = 1; \quad \lambda_2 = \lambda_{2,0} = 2p - 1$$
$$\psi_1(\alpha) = 2^{-\frac{1}{2}} \quad \text{and} \quad \psi_2(\alpha) = -\alpha \cdot 2^{-\frac{1}{2}}.$$

From (26) we see that if the result of the first event in the chain is α_1, and that of the nth event is α_n, the probability of the kth event yielding the result α_k is

$$\frac{[(2p-1)^{k-1}\alpha_1\alpha_k + 1][1 + (2p-1)^{n-k}\alpha_k\alpha_n]}{2[1 + (2p-1)^{n-1}\alpha_1\alpha_n]}.$$

As k, n_1 and $(n-k)$ simultaneously get very large, $P_n(\alpha_k) \sim \frac{1}{2}$, independently of α_k.

The probability of an initial result α_1 leading to a final result α_n is (from 28a)

$$P_n(\alpha_1, \alpha_n) = (\tfrac{1}{4})\{1 + (2p-1)^{n-1}\alpha_1\alpha_n\}$$

so that

$$P_n(1, 1) = P_n(-1, -1) = (\tfrac{1}{4})\{1 + (2p-1)^{n-1}\}$$
$$P_n(-1, 1) = P_n(1, -1) = (\tfrac{1}{4})\{1 - (2p-1)^{n-1}\}.$$

Now, to answer our original question regarding the probability distribution of the transition function (36)

(39)
$$F(\alpha_1, \cdots, \alpha_n) = \sum_{i=1}^{n-1} h(\alpha_i, \alpha_{i+1}),$$

we use the expression for $Z_n(x)$ determined from (24)

(39)
$$Z_n(x) = 2[p + (1-p)e^x]^{n-1}$$

From (9) the probability of there being between ξ and $\xi + h$ transitions in a sequence of $n + 1$ events is

(40)
$$G(\xi + h) - G(\xi) = \frac{1}{2\pi i} \int_{-\infty}^{\infty} e^{i\omega\xi}(1 - e^{i\omega\xi})\{p + (1 - p)e^{i\omega}\}^n \, d\omega/\omega$$

$$= \frac{1}{2\pi i} \int_{-\infty}^{\infty} (e^{-i\omega\xi} - e^{-i\omega(\xi-h)}) \sum_{k=0}^{n} \frac{n!(1-p)^k p^{n-k}}{(n-k)!k!}.$$

Letting $x = \omega h/2$ and rearranging

$$G(\xi + h) - G(\xi) = \frac{1}{\pi} \sum_{k=0}^{n} \frac{n!(1-p)^k p^{n-k}}{(n-k)!k!} D\left(1 + \frac{2}{h}(\xi + h)\right),$$

where $D(\lambda)$ is the Dirichlet integral

$$D(\lambda) = \frac{1}{\pi} \int_{-\infty}^{\infty} \frac{\sin x \cos \lambda x}{x} \, dx = \begin{cases} 0 & \text{if } |\lambda| > 1 \\ \tfrac{1}{2} & |\lambda| = 1 \\ 1 & |\lambda| < 1. \end{cases}$$

We therefore have, when $[\xi + h] \leq n$

(41)
$$G(\xi + h) - G(\xi) = \sum_{k=[\xi+1]}^{[\xi+h]} \frac{n!(1-p)^k p^{n-k}}{(n-k)!k!}.$$

Here $[x]$ denotes the greatest integer not exceeding x. The sum is zero if $[\xi + h] < [\xi + 1]$. When $[\xi + h] > n$

(42)
$$G(\xi + h) - G(\xi) = \sum_{k=[\xi+1]}^{n} \frac{n!(1-p)^k p^{n-k}}{k!(n-k)!}.$$

When n is large it is difficult to get a clear picture of the function $G(\xi)$ from (41) and (42), so we shall develop asymptotic results for large n by using (6) instead of (9).

By employing (5), we see that (this section will be developed on the basis of $n + 1$ trials instead of n)

$$\Lambda_1 = \bar{F} = n(1 - p)$$

$$\Lambda_2 = np(1 - p)$$

$$\Lambda_3 = np(1 - p)(2p - 1) \text{ etc.}$$

Therefore, from (6)

$$\Delta G = G(\xi + h) - G(\xi) = \frac{1}{2\pi i} \int_{-\infty}^{\infty} \frac{e^{-i\omega(\xi-\Lambda_1)}(1 - e^{-i\omega h})}{\omega}$$

$$\exp\left[-\tfrac{1}{2}np(1-p)\omega^2 - inp(1-p)(2p-1)\omega^3/6 - \cdots\right] d\omega.$$

Letting $u = \omega n^{\frac{1}{2}}$, we have

$$\Delta G = \frac{1}{2\pi i}\int_{-\infty}^{\infty}\frac{du}{u}[e^{-iu(\xi-\Lambda_1)/n^{\frac{1}{2}}} - e^{-iu(\xi+h-\Lambda_1)/n^{\frac{1}{2}}}]$$

$$\left[1 - \frac{ip(1-p)(2p-1)u^3}{6n^{\frac{1}{2}}} + O\left(\frac{u^4}{n}\right)\right]e^{-\frac{1}{2}u^2 p(1-p)}$$

$$= \frac{1}{2\pi}\int_{\mu_1}^{\mu_2}d\lambda\int_{-\infty}^{\infty}e^{-iu\lambda}\left[1 - \frac{ip(1-p)(2p-1)u^3}{6n^{\frac{1}{2}}} + O\left(\frac{u^4}{n}\right)\right]e^{-\frac{1}{2}u^2 p(1-p)}\,du$$

where

$$\mu_1 = (\xi + h - \Lambda_1)/n^{\frac{1}{2}}$$
$$\mu_2 = (\xi - \Lambda_1)/n^{\frac{1}{2}}.$$

Since

$$\int_{-\infty}^{\infty}e^{-au^2}e^{-i\lambda u}\,du = (\pi/a)^{\frac{1}{2}}\exp(-\lambda^2/4a)$$

$$i\int_{-\infty}^{\infty}u^3 e^{-au^2}e^{-i\lambda u}\,du = \frac{3\lambda\pi^{\frac{1}{2}}}{4a^{5/2}}\left(1 - \frac{\lambda^2}{6a}\right)e^{-\lambda^2/4a},$$

we have for large n

(43a)
$$\Delta G \sim \frac{1}{[2\pi p(1-p)]^{\frac{1}{2}}}\int_{-\mu_1}^{\mu_2}e^{-\lambda^2/2p(1-p)}$$
$$\left\{1 - \frac{(2p-1)\lambda}{2p(1-p)n^{\frac{1}{2}}}\left(1 - \frac{\lambda^2}{3p(1-p)}\right) + O\left(\frac{1}{n}\right)\right\}d\lambda.$$

As $n \to \infty$ and $h \to 0$, this becomes

(43b)
$$G(\xi + h) - G(\xi) \sim \frac{h\exp\{-[\xi - \bar{F}]^2/2p(1-p)n\}}{[2\pi np(1-p)]^{\frac{1}{2}}}$$
$$\left\{1 - \frac{(2p-1)(\xi - \bar{F})}{2p(1-p)n} + O\left(\frac{1}{n^2}\right)\right\}.$$

A similar problem which occurs in statistics of high polymers can be stated abstractly as follows. Suppose there exists a sequence of events each of which leads to a translation of length a of a point either to the right or to the left, and that the probability of a translation continuing in the same direction as its predecessor is p while that of changing its direction is $(1-p)$. After n translations what is the probability of a point being displaced a distance ξ from its origin.

If "-1" represents a translation to the left and "$+1$" a translation to the right,

$$p(-1, -1) = p(1, \ 1) = p$$
$$p(-1, \ 1) = p(1, -1) = (1-p)$$

The function giving the distance of the point from its origin after n displacements is (when $\alpha = \pm 1$)

$$F(\alpha_1, \cdots, \alpha_n) = a \sum_{j=1}^{n} \alpha_j = \tfrac{1}{2}a\alpha_1 + h(\alpha_1, \alpha_2) + \cdots + h(\alpha_{n-1}, \alpha_n) + \tfrac{1}{2}a\alpha_n$$

where

$$h(1, 1) = a, \quad h(-1, -1) = -a$$
$$h(1, -1) = h(-1, 1) = 0.$$

Neglecting the terms $a\alpha_1/2$ and $a\alpha_n/2$ in $F(\alpha_1, \cdots, \alpha_n)$, one can answer questions concerning this problem by evaluating $Z_n(x)$ as defined by (15). In this case P_x has the form

$$P_x = \begin{pmatrix} pe^{ax} & 1-p \\ 1-p & pe^{-ax} \end{pmatrix}.$$

Its characteristic roots are

$$\lambda_{1,x} = p \cosh ax + [p^2 \cosh^2 ax + (1-2p)]^{\frac{1}{2}} = \lambda_{L,x}$$
$$|\lambda_{2,x}| = |p \cosh ax - [p^2 \cosh^2 ax + (1-2p)]^{\frac{1}{2}}| < \lambda_{1,x}.$$

and its characteristic vectors:

$$\psi_{1,x} = [(p-1)^2 + (pe^{ax} - \lambda_1)^2]^{-\frac{1}{2}} \begin{pmatrix} p-1 \\ pe^{ax} - \lambda_1 \end{pmatrix}$$

$$\psi_{2,x} = [(p-1)^2 + (pe^{ax} - \lambda_2)^2]^{-\frac{1}{2}} \begin{pmatrix} p-1 \\ pe^{ax} - \lambda_2 \end{pmatrix}.$$

Since

$$\bar{F} = \Lambda_1 = \lim_{x \to 0} \partial \log Z_n(x)/\partial x,$$

one can show in the present problem that $\bar{F} = 0$. Therefore, the probability of the translated point being a distance between ξ and $\xi + h$ from the origin after $(n+1)$ translations, is, as $n \to \infty$ and $h \to 0$

$$F(\xi + h) - F(\xi) \sim h(2\pi n L_2)^{-\frac{1}{2}} e^{-\xi^2/2nL_2}$$

where L_2 is by (32):

$$L_2 = \lim_{x \to 0} \partial^2 \log \lambda_{L,x}/\partial x = a^2 p/(1-p).$$

Thus,

$$F(\xi + h) - F(\xi) \sim h[a^2 2\pi n p/(1-p)]^{-\frac{1}{2}} e^{-\xi^2(1-p)/2na^2 p}.$$

When $p = 2/3$ this problem is equivalent to the determination of the probability distribution of the components in an arbitrary direction of the distance between the ends of a linear polymer. In this case

$$F(\xi + h) - F(\xi) \sim h(4a^2\pi n)^{-\frac{1}{2}} \exp(-\xi^2/4na^2)$$

a result obtained by Tobolsky [12] after a lengthy and complicated combinatory calculation.

Another type of simple chain is encountered in the determination of the "life span" of a particle which is displaced a unit distance to the right or left per unit time along a straight line until it collides with an absorbing boundary either $-(q+1)$ or $(p+1)$ units from the starting point. This problem has been analyzed by M. Kac using the methods discussed in the present paper. We shall generalize his results to include the effect of an attraction of the particle toward one end of the line so that displacements toward that end are more probable than those in the other direction.

Following the notation of Kac [13] we let X_j represent the jth displacement, m_j its length, and $\delta(m)$ the probability of a given displacement having the length m. Then,

$$\delta(m) = \begin{cases} s & \text{if } m = 1 \\ 1 - s & \text{if } m = -1 \\ 0 & \text{otherwise.} \end{cases}$$

If N represents the life span of a particle, the probability of its exceeding n is

$$\text{Prob } \{N > n\} = \text{Prob } \{-q \leq X_1 \leq p, -q \leq X_1 + X_2 \leq p, \cdots,$$
$$-q \leq X_1 + X_2 + \cdots + X_n \leq p\} = \Sigma \, \delta(m_1)\delta(m_2) \cdots \delta(m_n)$$

where the summation extends over all integers $m_1, m_2, \cdots, m_n$ such that $-q \leq m_1 \leq p, -q \leq m_1 + m_2 \leq p, \cdots, -q \leq m_1 + m_2 + \cdots + m_n \leq p$.

Defining the new set of variables

$$\alpha_j = q + m_1 + m_2 + \cdots + m_j \qquad (j = 1, 2, \cdots n)$$

we see that

$$\text{Prob } \{N > n\} = \sum_{\alpha_1, \cdots, \alpha_n = 0}^{p+q} \delta(\alpha_1 - q)\delta(\alpha_2 - \alpha_1) \cdots \delta(\alpha_n - \alpha_{n-1}).$$

As before, if we introduce the P matrix (of $p + q + 1$ rows and columns)

$$P = (\delta(\alpha - \beta)) = \begin{pmatrix} 0 & 1-s & 0 & 0 & \cdots \\ s & 0 & 1-s & 0 & \cdots \\ 0 & s & 0 & 1-s & \cdots \\ \cdots & \cdots & \cdots & \cdots & \cdots \end{pmatrix}$$

we obtain after applying the equivalent of (22)

$$\text{Prob}\{N > n\} = \sum_{j=1}^{p+q+1} \lambda_j^n \varphi_j(q) \sum_{\alpha_n=0}^{p+q} \psi_j(\alpha_n).$$

Where λ_j is the jth characteristic value of P, and ψ_j and φ_j are its associated characteristic vectors as defined by (19) and (20) (here the range of α starts from 0 instead of 1 as in (17) and (19)).

It is easy to show that the characteristic values of P are

$$\lambda_j = 2[s(1-s)]^{\frac{1}{2}} \cos \zeta_j \quad (j = 1, 2, \cdots, p+q+1)$$

where

$$\zeta_j = \pi j/(p+q+2)$$

and that the components of the characteristic vectors are

$$\psi_j(\alpha) = [2/(p+q+2)]^{\frac{1}{2}}[s/(1-s)]^{\frac{1}{2}\alpha} \sin(\alpha+1)\zeta_j \quad (\alpha = 0, 1, \cdots, p+q)$$

and

$$\varphi_j(\alpha) = [2/(p+q+2)]^{\frac{1}{2}}[(1-s)/s]^{\frac{1}{2}\alpha} \sin(\alpha+1)\zeta_j.$$

Since

$$\sum_{\alpha_n=0}^{p+q} \psi_j(\alpha_n) = \frac{\sqrt{2}(1-s)}{\sqrt{p+q+2}} \frac{\{1 - 1(-1)^j[s/1-s]^{\frac{1}{2}(p+q+2)}\} \sin \zeta_j}{1 - 2[s(r-s)]^{\frac{1}{2}} \cos \zeta_j}$$

we finally have

$$\text{Prob}\{N > n\} = \frac{(1-s)^{\frac{1}{2}(n+q+2)} 2^{n+1} s^{\frac{1}{2}(n-q)}}{p+q+2}$$
$$\sum_{j=1}^{p+q+1} \frac{\{1-(-1)^j(s/1-s)^{\frac{1}{2}(p+q+2)}\} \cos^n \zeta_j \sin \zeta_j \sin(q+1)\zeta_j}{1 - 2\sqrt{s(1-s)} \cos \zeta_j}$$

When $s = \frac{1}{2}$ this reduces to the result of Kac: (* means summation is only over even j's

$$\text{Prob}\{N > n\} = \frac{2}{p+q+2} \sum_{j=1}^{p+q+1} * \cos^n \zeta_j \sin(q+1)\zeta_j \cot \tfrac{1}{2}\zeta_j.$$

4. Simple Chains with Restrictions.

Often when studying chains of dependent events, certain functions averaged over the entire chains are known to be restricted between definite limits. That is, there might exist k functions $g_j(\alpha_1, \alpha_2, \cdots, \alpha_n)$ such that

(44) $\qquad -\Delta G_j < G_j - g_j(\alpha_1, \cdots, \alpha_n) < \Delta G_j, \quad (j = 1, 2, \cdots k),$

where the G_j's and ΔG_j's are preassigned constants. To calculate averages of other functions (1) is no longer valid, for it is an unrestricted sum over all sets of α's, including those incompatible with (44). All unrestricted sums in this formula (and other similar ones) must be replaced by sums over only those sets of α's compatible with (44). Since it is sometimes more difficult to evaluate restricted sums than unrestricted ones, we shall apply an idea of Markoff [1] to the reduction of the former to the latter type.

Let us seek an explicit expression for a function $P_n^*(\alpha_1, \alpha_2, \cdots, \alpha_n)$ which has the property:

$$P_n^*(\alpha_1, \cdots, \alpha_n) = P_n(\alpha_1, \cdots, \alpha_n) \quad \text{when } \alpha\text{'s are chosen so that (44) is satisfied of all } j.$$
$$ 0 \quad \text{otherwise.}$$

Since the Dirichlet integrals

$$\delta_j = \frac{1}{\pi} \int_{-\infty}^{\infty} \frac{\sin (\rho_j \Delta G_j)}{\rho_j} \exp (i\rho_j \gamma_j) \, d\rho_j$$

have the property

$$\delta_j = 1 \text{ when } -\Delta G_j < \gamma_j < \Delta G_j$$
$$0 \text{ otherwise,}$$

$$P_n^*(\alpha_1, \cdots, \alpha_n) = \delta_1 \delta_2 \cdots \delta_k P_n(\alpha_1, \cdots, \alpha_n)$$

has the required character provided

$$\gamma_j = G_j - g_j(\alpha_1, \cdots, \alpha_n).$$

The average value of a function $F(\alpha_1, \cdots, \alpha_n)$ can be written in terms of the unrestricted sum

$$\bar{F} = \sum_{\{\alpha_e\}} F(\alpha_1, \cdots, \alpha_n) P_n^*(\alpha_1, \cdots, \alpha_n) \Big/ \sum_{\{\alpha_e\}} P_n^*(\alpha_1, \cdots, \alpha_n),$$

where the summation extends over the complete set of $\{\alpha_e\}$'s

$$\{\alpha_e\} = (\alpha_1, \alpha_2, \cdots, \alpha_n).$$

As in the case of chains without auxiliary restrictions, a useful function is

(45)
$$Z_n(x) = \sum_{\{\alpha_e\}} P_n^*(\alpha_1, \cdots, \alpha_n) \exp \{xF(\alpha_1, \cdots, \alpha_n)\}$$
$$= \frac{1}{\pi^k} \int_{-\infty}^{\infty} \cdots \int_{-\infty}^{\infty} S_n(x, \rho_1, \cdots, \rho_k) \prod_{m=1}^{k} \left\{ \frac{\sin (\rho_m \Delta G_m)}{\rho_m} e^{i\rho_m G_m} \, d\rho_m \right\}$$

where

$$S_n(x, \rho_1, \cdots, \rho_k) = \sum_{\{\alpha_e\}} P_n(\alpha_1, \cdots, \alpha_n)$$
$$\exp \left\{ xF(\alpha_1, \cdots, \alpha_n) - i \sum_{j=1}^{k} \rho_j g_j(\alpha_1, \cdots, \alpha_n) \right\}.$$

When $F(\alpha_1, \cdots, \alpha_n)$ and $\{g_j(\alpha_1, \cdots, \alpha_n)\}$ are all additive or multiplicative functions of the form (13a) and (13b), say

$$F(\alpha_1, \cdots, \alpha_n) = \sum_{k=1}^{n-1} h(\alpha_k, \alpha_{k+1})$$

$$g_j(\alpha_1, \cdots, \alpha_n) = \sum_{k=1}^{n-1} g_j(\alpha_k, \alpha_{k+1})$$

and the probability chain is a simple one, $Z_n(x)$ reduces to a simple form.

Suppose

$$P_n(\alpha_1, \cdots, \alpha_n) = \sum_{j=1}^{n-1} p(\alpha_j, \alpha_{j+1})$$

then following the derivation of (24), we have

(46) $$S_n(x, \rho_1, \cdots, \rho_k) = \sum_{l=1}^{\nu} \{\lambda_{l,x,\rho}\}^{n-1}(\Phi_{l,x,\rho}\cdot\mathbf{1})(\mathbf{1}\cdot\Psi_{l,x,\rho})$$

where $\lambda_{l,x,\rho}$, $\Phi_{l,x,\rho}$ and $\Psi_{l,x,\rho}$ are characteristic values and vectors of the matrix

$$P_{x,\rho} = \begin{pmatrix} p_{x,\rho}(1,1) & \cdots & p_{x,\rho}(1,\nu) \\ \cdots\cdots\cdots\cdots\cdots\cdots\cdots \\ p_{x,\rho}(\nu,1) & \cdots & p_{x,\rho}(\nu,\nu) \end{pmatrix}$$

and

$$p_{x,\rho}(\alpha, \beta) = p(\alpha, \beta) \exp\{xh(\alpha, \beta) - i \sum_j \rho_j g_j(\alpha, \beta)\}.$$

Substitution of (46) into (45) allows one to calculate $Z_n(x)$.

5. More Complicated Chains. In a chain of N events in which the result of each event depends on those of its n predecessors ($n << N$), the calculation of $Z_n(x)$ proceeds in essentially the same manner as in the case of a simple chain. Let the N events be divided into N/n sets of "grand events" of n simple events each (for simplicity we assume N is divisible by n, this can easily be avoided). Thus, if each simple event could lead to any one of ν possible results, a grand event could lead to any one of ν^n possible results and a complicated chain becomes a simple chain of grand events with the result of each grand event depending on the preceeding grand event. Quantitative calculations thus proceed formally in the same manner as in a simple chain.

6. Continuous Case. In this section we generalize, by studying an example, to the case in which each event in a simple chain may lead to any one of a continuum of results. The example is a problem arising in statistical mechanics of molecular chains.

Consider a linear chain of n identical molecules whose centers of mass remain at a set of fixed regularly spaced positions, but which may rotate about their centers of mass in a plane. Suppose, that the potential energy of interaction between neighboring pairs of molecules is a function of the angles a specified axis of the molecules makes with the line connection the centers of mass of the molecules; that is, the potential energy of interaction between pairs of adjacent molecules can be written as $V(\theta_j, \theta_{j+1})$. Assuming that forces are sufficiently short ranged for interaction between more distant neighbors can be neglected, Boltzmann's theorem states that the probability of the axis of the first molecule making an angle between θ_1 and $\theta_1 + a\theta_1$ with the line of centers of the chain, the second between θ_2 and $\theta_2 + \alpha\theta_2$ and the nth between θ_n and $\theta_n + d\theta_n$ is proportional to

$$\exp\left[-kT\left\{V(\theta_1, \theta_2) + V(\theta_2, \theta_3) + \cdots + V(\theta_{n-1}, \theta_n)\right\}\right] d\theta_1 \cdots d\theta_n$$

where k is Boltzmann's constant and T is the absolute temperature. The contribution of the interaction to the thermodynamic properties of the chain can be derived from the partition function

$$(47) \quad Z_n = \int_0^{2\pi} \int_0^{2\pi} \cdots \int_0^{2\pi} \exp\left\{-\frac{1}{kT}\left[V(\theta_1, \theta_2) + \cdots + V(\theta_{n-1}, \theta_n)\right]\right\} d\theta_1 \cdots d\theta_n.$$

For example, the internal energy is

$$\bar{E} = \partial \log Z_n / \partial(-1/kT)$$

and the specific heat is $c = \partial E / \partial T$.

It is to be noted that Z_n is exactly the integral of the iterated kernel of the integral equation

$$(48) \quad \lambda \psi(\theta_1) = \int_0^{2\pi} \psi(\theta_2) \exp\left\{-\frac{1}{kT} V(\theta_1, \theta_2)\right\} d\theta_2.$$

If $V(\theta_1, \theta_2)$ is symmetrical in θ_1 and θ_2, this linear homogeneous integral equation has a set of orthonormal characteristic functions $\{\psi_j(\theta)\}$ such that

$$(49) \quad \int_0^{2\pi} \psi_j(\theta) \psi_k(\theta) \, d\theta = \delta_{jk}.$$

To each of these characteristic functions there corresponds a characteristic value λ_j. Now it is well known that the kernel of (48) can be expanded as a series in its characteristic functions

$$\exp\left\{-\frac{1}{kT} V(\theta_1, \theta_2)\right\} = \sum_j \lambda_j \psi_j(\theta_1) \psi_j(\theta_2).$$

Introduction of this expression into (47) and applying the orthogonality conditions (49) one obtains

$$(47\text{a}) \quad Z_n = \sum_j \lambda_j^{n-1} \left\{\int_0^{2\pi} \psi_j(\theta) \, d\theta\right\}^2.$$

Probably the most interesting example of a molecular chain of the type described above is a chain of magnetic dipoles which are restricted to rotate only in a plane. In that case

$$V(\theta_j, \theta_{j+1}) = \frac{\mu^2}{r^3}[\cos(\theta_j - \theta_{j+1}) - 3\cos\theta_j \cos\theta_{j+1}].$$

Where μ is the magnetic moment of each dipole and r is the distance between a pair of adjacent centers of mass. This potential function leads to the integral equation

$$\lambda\psi(\theta_1) = \int_0^{2\pi} \psi(\theta_2) \exp\left\{-\frac{\mu^2}{r^3 kT}[\cos(\theta_1 - \theta_2) - 3\cos\theta_1 \cos\theta_2]\right\} d\theta_2.$$

Since this equation is rather complicated to solve, we shall devote the rest of the section to a potential function of less physical interest, but which leads to a less formidable integral equation.

In studying hindered rotation of molecules, one sometimes uses potential functions of the form:

$$V(\theta_j, \theta_{j+1}) = -\beta \cos(\theta_j - \theta_{j+1})$$

where β is a constant. With this potential function (48) becomes

(50) $$\lambda\psi(\theta_1) = \int_0^{2\pi} \psi(\theta_2) \exp\{J \cos(\theta_1 - \theta_2)\} d\theta_2$$

where

$$J = \beta/kT.$$

The characteristic functions and characteristic values of (50) are easily found with the aid of the Fourier Series for $\exp(J \cos \theta)$:

(51) $$\exp(J \cos \theta) = I_0(J) + 2\sum_{m=1}^{\infty} I_m(J) \cos m\theta$$

where $I_m(J)$ is the mth Bessel function of imaginary argument:

$$I_m(J) = \sum_{m=0}^{\infty} \frac{(\frac{1}{2}J)^{2k+m}}{(m+k)!k!}.$$

From (51)

$$\exp[J \cos(\theta_1 - \theta_2)] = I_0(J) + 2\sum_{m=1}^{\infty} I_m(J)(\cos m\theta_1 \cos m\theta_2 + \sin m\theta_1 \sin m\theta_2).$$

Substituting this expression into (50) we have

$$\lambda\psi(\theta_1) = \int_0^{2\pi} \psi(\theta_2) \left\{I_0(J) + 2\sum_{m=1}^{\infty} I_m(J)(\cos m\theta_1 \cos m\theta_2 + \sin m\theta_1 \sin m\theta_2)\right\} d\theta_2.$$

Because of the orthogonality of the trigonometric functions, one can verify by direct substitution that the characteristic functions are

$$\psi_0(\theta) = 1/(2\pi)^{\frac{1}{2}}$$

$$\psi_m^{(1)}(\theta) = \pi^{-\frac{1}{2}} \sin m\theta; \qquad \psi_m^{(2)} = \pi^{-\frac{1}{2}} \cos m\theta, \ (m = 1, 2, \cdots)$$

and the corresponding characteristic values are

$$\lambda_0 = 2\pi I_0(J)$$

$$\lambda_m^{(1)} = \lambda_m^{(2)} = 2\pi I_m(J) \qquad\qquad m > 0.$$

Introduction of these characteristic functions and values into (47a) we obtain the simple formula for the partition function:

$$Z_n = 2\pi \{2\pi I_0(J)\}^{n-1}.$$

The internal energy of the molecular chain is therefore

$$\bar{E} = \partial \log Z_n / \partial(-1/kT)$$
$$= -\beta(n-1) \, I_1(J)/I_0(J),$$

and the specific heat is:

$$C = \partial \bar{E}/\partial T = \tfrac{1}{2} k(n-1) J^2 \left\{ 1 + \frac{I_2(J)}{I_0(J)} - 2\left[\frac{I_1(J)}{I_0(J)}\right]^2 \right\}.$$

REFERENCES

[1] A. MARKOFF, *Wahrscheinlichkeitsrechnung*, Leipzig, 1912.
[2] S. BERNSTEIN, "Sur l'extension du théorème limite du calcul des probabilités aux sommes de quantités dependantes," *Math. Ann.*, Vol. 97 (1927), p. 1.
[3] M. FRECHÉT, *Recherces Theoretiques Moderns sur La Theorie des Probabilites*, Vol. 2, Paris (1937).
[4] H. KRAMERS AND G. WANNIER, "Statistics of the two-dimensional ferromagnet: Part I," *Phys. Rev.*, Vol. 60, (1941), p. 252.
[5] E. MONTROLL, "Statistical mechanics of nearest neighbor systems," *Jour. Chem. Phys.*, Vol. 9 (1941), p. 708; Vol. 10 (1942), p. 61.
[6] E. LASSETTRE AND J. HOWE, "Thermodynamic properties of binary solid solutions on the basis of the nearest neighbor approximation," *Jour. Chem. Phys.*, Vol. 9 (1941), p. 747.
[7] J. ASHKIN AND W. E. LAMB, "The propagation of order in crystal lattices," *Phys. Rev.*, Vol. 64 (1943), p. 159.
[8] L. ONSAGER, "Crystal statistics I. A two-dimensional model with an order-disorder transition," *Phys. Rev.*, Vol. 65, (1944), p. 117.
[9] M. HOSTINSKY, *Methodes generales du Calculu des Probabilites*, Paris, 1931.
[10] H. CRAMÉR, *Random Variables and Probability Distributions*, Cambridge Univ. Press, 1937, Chap. 4.
[11] G. FROBENIUS, "Über Matrizen aus positiven Elementen. II.," *Preuss. Acad. Wiss. Sitz.*, (1909), p. 514.
[12] A. TOBOLSKY, POWELL AND H. EYRING, an article in *Chemistry of Large Molecules*, Interscience Publishers, 1943, pp. 156, 182.
[13] M. KAC, "Random walk in the presence of absorbing barriers," *Annals of Math. Stat.* Vol. 14, (1945), p. 62.

Frequency Spectrum of Crystalline Solids

Elliott W. Montroll*
Sterling Chemistry Laboratory, Yale University,† New Haven, Connecticut
(Received January 2, 1942)

It is generally accepted that a normal crystalline solid can be pictured at absolute zero as an assembly of molecules arranged at periodically placed lattice points. Since at higher temperatures each molecule becomes a harmonic oscillator about its lattice point, in order to calculate thermodynamic properties of the crystal it is necessary to know the distribution of its internal normal modes of vibration. On the basis of the Born-von Kármán model these normal modes of vibration are roots of a secular determinant. In this paper it is shown that the $2n$th moment of the distribution function of normal modes is proportional to the trace of the nth power of the matrix of the Born-von Kármán determinant. By expressing the distribution function as a linear combination of Legendre polynomials it is shown that the coefficient of each polynomial is a linear combination of the moments. The frequency distribution function of a two-dimensional simple cubic lattice is calculated by the above method and turns out to have two maxima. Usually the equation for a thermodynamic function $F(T)$ involves the integral of the product of the frequency distribution function $g(\nu)$ and a known function $K(T, \nu)$. We show here that when $F(T)$ is a known function of T an integral equation results with $g(\nu)$ under the integral sign. This integral equation can be solved for $g(\nu)$ by use of Fourier transforms.

I. INTRODUCTION

ONE of the failures of classical mechanics and classical statistics was their inability to explain even qualitatively the decrease in specific heat of solids with decreasing temperatures; and one of the selling points of Planck's original quantum theory had its source in the success of Einstein's[1] application of Planck's "quantized oscillators" to the lattice vibrations of a solid. By Einstein's method the vibrational factor of the partition function of a normal crystal is

$$Q = \prod_{i=1}^{3N} \frac{\exp(-h\nu_i/2kT)}{1-\exp(-h\nu_i/kT)} \quad (1)$$

and the Einstein assumption of equal numerical value of all $3N$ frequencies of a crystal explains the observed decrease of the specific heat below the Dulong-Petit value of $3NkT$ at low temperatures. Here h = Planck's constant, k = Boltzmann's constant, T = absolute temperature, ν_i = frequency of the ith normal mode of vibration, N = number of lattice points.

Refinements in specific heat measurements in the first decade of this century demonstrated that the agreement between experiments and Einstein's theory was not quantitative at low temperatures; and Debye[2] and Born and von Kármán[3] pointed out that the discrepancies were probably due to the existence of a distribution of normal modes of vibration instead of a single "Einstein frequency." Thus, for example, the temperature dependent vibrational contribution per lattice point to the free energy should be written[4]

$$F_V(T) = -kT \int_0^\infty g(\nu) \log(1-e^{-h\nu/kT})d\nu, \quad (2)$$

where $g(\nu)d\nu$ is the fraction of normal modes with frequencies between ν and $\nu+d\nu$.

Debye's proposition to treat the solid as a continuum with the same elastic properties led to a frequency distribution proportional to ν^2 and to equations of physical quantities in terms of a parameter $\Theta = h\nu_L/k$ which theoretically should be a constant for a given crystal. The term ν_L is the largest frequency of the normal modes of vibration, and Θ is usually referred to as the Debye or characteristic temperature. Unfortunately this development which mathematically is quite simple is insufficient, for experiments

---

* Sterling Research Fellow in Chemistry.
† Present address: Baker Chemistry Laboratory, Cornell University, Ithaca, New York.
[1] A. Einstein, Ann. d. Physik **22**, 180 (1906); **34**, 170 (1911).
[2] P. Debye, Ann. d. Physik **39**, 789 (1912).
[3] M. Born and T. von Kármán, Physik. Zeits. **13**, 297 (1912); **14**, 15 (1913).
[4] For a general discussion of theory of normal solids cf. E. Schroedinger, *Handbuch der Physik* (1926), Vol. 10; M. Born and M. Göppert-Mayer, *Handbuch der Physik* (1933), Vol. 24; R. Fowler, *Statistical Mechanics* (1936), Chap. 4.

show that Θ is not constant over a large temperature range.

The extension of Born and von Kármán of Einstein's work afforded a more accurate physical picture than that suggested by Debye. However, it led to a mathematical problem which could be solved analytically for a one-dimensional solid but which required tedious numerical calculations in two- and three-dimensional crystals. In the Born-von Kármán theory one sets up the equations of motion for a set of coupled oscillators with the same coupling and force constants as those of the crystal, and shows that the normal modes of vibration are the characteristic roots of a secular determinant of order N^3. Actually the determinant can be factored into secular determinants of very small order, but to find $g(\nu)$ by present methods the roots of all the resulting equations (and there are $O(N^3)$ of these) must be calculated and the number of roots in every small frequency interval must be determined. Even though it seems reasonable to base the $g(\nu)$ computations on the solution of a large random sample of cubic equations, the lack of popularity of the Born-von Kármán approach is understandable.

The first part of this paper will be devoted to developing an analytical method of finding $g(\nu)$ in a Born-von Kármán crystal. This method will depend on the matrix theorem that the sum of the kth powers of the characteristic values of a matrix is equal to the trace of the kth power of the matrix. Noticing that the sum of the kth powers of the characteristic values of the Born-von Kármán matrix is proportional to the kth moment of $g(\nu)$, and that a continuous distribution function defined in a finite interval can be expanded as a linear combination of Legendre polynomials, it is apparent that the coefficients of the Legendre polynomials can be expressed as linear combinations of the moments of $g(\nu)$. This means that $g(\nu)$ can be expressed analytically as a function of the lattice constants of the crystal.

These observations will be applied in detail to a two-dimensional cubic lattice, postponing to some later date the application (which is not much more difficult) to real three-dimensional crystals.

The second part of this paper will concern the determination of $g(\nu)$ from experimental specific heat data by use of Fourier transforms.

II. $g(\nu)$ FOR THE BORN-VON KÁRMÁN MODEL

1. Mathematical Techniques

The possible frequencies of normal modes of vibration in a Born-von Kármán crystal will be shown in the next section to be roots of a secular determinant (see Eq. (17a)) and two approaches to the problem of deriving their distribution from the determinant are: (a) by calculating all or a large random sample of its roots and counting the number in each small frequency interval; (b) by evaluating the moments of the distribution function and from the moments determining the distribution.

Blackman's[5] researches have demonstrated that the first method involves the solution of a large collection of quadratic or cubic equations and thus an enormous amount of tedious numerical work—and further, they show that the computation must be repeated for every set of lattice parameters, so method (a) cannot easily show the influence of changes in lattice parameters on changes in physical properties. Here we shall devise a technique based on the second approach which avoids tedious numerical work.

Let $\mathbf{M}$ be the matrix of a symmetrical secular determinant whose roots are $\{\lambda_i\}$. Then

$$\mathbf{M}\psi_i = \lambda_i \psi_i, \qquad (1)$$

where ψ_i is the ith characteristic vector of the matrix, and λ_i is the corresponding characteristic value. Multiplication of both sides of (1) by $\mathbf{M}$ yields

$$\mathbf{M}^2 \psi_i = \mathbf{M}\lambda_i \psi_i = \lambda_i \mathbf{M} \psi_i$$
$$= \lambda_i^2 \psi_i$$

and in general

$$\mathbf{M}^k \psi_i = \lambda_i^k \psi_i. \qquad (2)$$

Thus λ_i^k is the characteristic value corresponding to the characteristic vector ψ_i of $\mathbf{M}^k$. Now it is well known[6] that the trace of a matrix is the sum of its characteristic values, so

$$\text{Trace } \mathbf{M}^k = \sum_i \lambda_i^k. \qquad (3)$$

Since the total number of characteristic values (where a characteristic value of degree of degeneracy d is counted d times) is equal to the

---

[5] M. Blackman, Proc. Roy. Soc. **148**, 365, 384 (1935); **159**, 416 (1937); see also P. Fine, Phys. Rev. **56**, 355 (1939).
[6] Cf. Rojansky, *Introductory Quantum Mechanics* (New York, 1938), p. 324.

order of the matrix, which we shall assume to be m, the average values of powers of the characteristic values of **M** are

$$\sum \lambda_i^k/m = (\text{Trace } \mathbf{M}^k)/m. \qquad (4)$$

But this is exactly the kth moment of the distribution function of the characteristic values and will be abbreviated by μ_k.

We must now find a way of expressing the distribution function $g(\lambda)$ in terms of the set of moments $\{\mu_k\}$. Let λ_S and λ_L be, respectively, the smallest and largest characteristic values of **M** (since the matrix is finite they actually exist; also, since the matrix was assumed to be symmetrical, the λ's are all real) and let us assume that the elements of the set $\{\lambda_i\}$ are densely distributed.

Suppose the distribution function is expanded as a linear combination of Legendre polynomials

$$g(\lambda) = \sum_{n=0}^{\infty} a_n P_n\left(\frac{2\lambda - (\lambda_L + \lambda_S)}{\lambda_L - \lambda_S}\right), \qquad (5)$$

where

$$P_n(x) = \frac{1}{2^n n!}\frac{d^n}{dx^n}(x^2-1)^n \qquad (6a)$$

and

$$\int_{-1}^{1} P_k(x)P_j(x)dx = 2\delta_{kj}/(2k+1). \qquad (6b)$$

To evaluate the coefficients $\{a_k\}$ multiply both sides of (5) by

$$P_k\left(\frac{2\lambda - \lambda_L - \lambda_S}{\lambda_L - \lambda_S}\right)d\lambda$$

and integrate from λ_S to λ_L

$$\int_{\lambda_S}^{\lambda_L} g(\lambda)P_k\left(\frac{2\lambda-\lambda_L-\lambda_S}{\lambda_L-\lambda_S}\right)d\lambda$$

$$= \sum_{n=0}^{\infty} a_n \int_{\lambda_S}^{\lambda_L} P_n\left(\frac{2\lambda-\lambda_L-\lambda_S}{\lambda_L-\lambda_S}\right)P_k\left(\frac{2\lambda-\lambda_L-\lambda_S}{\lambda_L-\lambda_S}\right)d\lambda.$$

Letting $x = (2\lambda - \lambda_L - \lambda_S)/(\lambda_L - \lambda_S)$

$$\frac{(\lambda_L - \lambda_S)}{2}\int_{-1}^{1} g(\lambda)P_k(x)dx$$

$$= \sum_{n=0}^{\infty} a_n \int_{-1}^{1}\frac{(\lambda_L-\lambda_S)}{2}P_n(x)P_k(x)dx$$

$$= a_k(\lambda_L - \lambda_S)/(2k+1).$$

Therefore

$$a_k = \frac{(2k+1)}{2}\int_{-1}^{1} g\left(\frac{(1+x)\lambda_L + (1-x)\lambda_S}{2}\right)P_k(x)dx. \qquad (7)$$

If we define

$$u_k = \int_{-1}^{1} x^k g(\lambda)dx$$

$$= \frac{2}{(\lambda_L - \lambda_S)^{k+1}}\int_{\lambda_S}^{\lambda_L}(2\lambda - \lambda_L - \lambda_S)^k g(\lambda)d\lambda$$

$$= \frac{2}{(\lambda_L - \lambda_S)^{k+1}}\int_{\lambda_S}^{\lambda_L}(\text{polynomial in }\lambda)g(\lambda)d\lambda, \qquad (8)$$

it is apparent that the a_k's can be written as linear combinations of the u_k's. Furthermore, the definition of $g(\lambda)$ implies that its moments are

$$\mu_i = \int_{\lambda_S}^{\lambda_L} g(\lambda)\lambda^i d\lambda$$
$$= (\text{Trace } \mathbf{M}^i)/m, \qquad (9)$$

so the u_k's are linear combinations of μ_i's and therefore the coefficients a_k are also linear combinations of the moments (or the traces of $\mathbf{M}^i$) μ_i of $g(\lambda)$.

In the theory of frequency spectrum of solids the frequencies of normal modes are proportional to the square root of the roots of the secular equation, so the smallest frequency is the negative of the largest, and the odd moments of the distribution function are zero. Considering this case in detail we have:

$$g(\lambda) = \sum_{n=0}^{\infty} a_n P_n(\lambda/\lambda_L) \qquad (10a)$$

with

$$a_n = \frac{(2n+1)}{2}\int_{-1}^{1} g(x\lambda_L)P_n(x)dx. \qquad (10b)$$

If

$$u_k = \int_{-1}^{1} g(x\lambda_L)x^k dx, \qquad (11)$$

then

$$\mu = \int_{-\lambda_L}^{\lambda_L} g(\lambda)\lambda^k d\lambda$$

$$= \lambda_L^{k+1}u_k \qquad (12)$$

Therefore

$$a_0 = \frac{1}{2}\int_{-1}^{1} g(x\lambda_L) \cdot 1\, dx$$

$$= \mu_0/2\lambda_L$$

$$a_1 = \frac{3}{2}\int_{-1}^{1} g(x\lambda_L) \cdot x\, dx$$

$$= \frac{3}{2}(\mu_1/\lambda_L^2) = 0.$$

Similarly, since the odd Legendre polynomials involve only odd powers of x, and since the odd moments of $g(\lambda)$ are zero:

$$a_1 = a_3 = \cdots = a_{2n+1} = 0,$$

$$a_2 = \frac{5}{2}\int_{-1}^{1} g(x\lambda_L)\left(\frac{3}{2}x^2 - \frac{1}{2}\right)dx,$$

$$= 5(3\mu_2 - \mu_0\lambda_L^2)/4\lambda_L^3,$$

$$a_4 = 9(35\mu_4 - 30\mu_2\lambda_L^2 + 3\mu_0\lambda_L^4)/16\lambda_L^5,$$

$$a_6 = 13(231\mu_6 - 315\mu_4\lambda_L^2 + 105\mu_2\lambda_L^4 - 5\lambda_L^6)/32\lambda_L^7,$$

and in general

$$a_{2k} = \frac{4k+1}{(2k)!2^{2k+1}}\int_{-1}^{1} g(x\lambda_L)\left[\frac{d^{2k}}{dx^{2k}}(x^2-1)^{2k}\right]dx$$

$$= \frac{4k+1}{(2k)!2^{2k+1}\lambda_L}\left\{\frac{d^{2k}}{dy^{2k}}(y^2-1)^{2k}\right\}_{y^k = \mu_k/\lambda_L^k}, \quad (13)$$

where after the differentiation is performed in the last expression, y^k is replaced by μ_k/λ_L^k.

2. Two-Dimensional Cubic Lattice

Here we shall apply the Moment-Trace method to a two-dimensional one-component cubic lattice of the Born-von Kármán type which includes interactions between nearest and next nearest neighbors. Let α be the binding force between two nearest neighbors whose distance of separation is a and let γ be that for two next nearest neighbors whose distance of separation is $a\sqrt{2}$. If there are $2N \times 2N$ particles of mass M in our square lattice, each atom can be specified by two letters l, m ($1 \leq l, m \leq 2N$) and its displacement components along rows and columns can be represented by $u_{l,m}$ and $v_{l,m}$, respectively. The motion of the particle (l, m) is described by the simultaneous differential equations[5]

$$M\ddot{u}_{l,m} + \alpha(u_{l,m} - u_{l+1,m} + u_{l,m} - u_{l-1,m}) + \gamma(u_{l,m} + v_{l,m} - u_{l+1,m+1} - v_{l+1,m+1} + u_{l,m} + v_{l,m}$$

$$- u_{l-1,m-1} - v_{l-1,m-1} + u_{l,m} - v_{l,m} - u_{l+1,m+1} + v_{l+1,m-1} + u_{l,m} - v_{l,m} - u_{l-1,m+1} + v_{l-1,m+1}) = 0 \quad (14a)$$

and

$$M\ddot{v}_{l,m} + \alpha(v_{l,m} - v_{l+1,m} + v_{l,m} - v_{l-1,m}) + \gamma(u_{l,m} + v_{l,m} - u_{l+1,m+1} - v_{l+1,m+1} + u_{l,m} + v_{l,m}$$

$$- u_{l-1,m-1} - v_{l-1,m-1} + u_{l,m} - v_{l,m} - u_{l+1,m-1} + v_{l+1,m-1} + u_{l,m} - v_{l,m} - u_{l-1,m+1} + v_{l-1,m+1}) = 0. \quad (14b)$$

By assuming the periodic solutions

$$u_{l,m} = u' \exp i(2\pi\nu t + l\varphi_1 + m\varphi_2), \quad (15a)$$

$$v_{l,m} = v' \exp i(2\pi\nu t + l\varphi_1 + m\varphi_2), \quad (15b)$$

$$\varphi_1 = \pi a_1/N; \quad \varphi_2 = \pi a_2/N, \quad (15c)$$

where a_1 and a_2 are integers satisfying

$$-N \leq a_1, a_2 \leq N,$$

and substituting (15) into (14) one can easily show that the frequencies of the normal modes of vibration of the crystal must be roots of the characteristic equations

$$\begin{vmatrix} A(\varphi_1, \varphi_2) - 4\pi^2 M\nu^2 & B(\varphi_2, \varphi_1) \\ B(\varphi_1, \varphi_2) & A(\varphi_2, \varphi_1) - 4\pi^2 M\nu^2 \end{vmatrix} = 0, \quad (16)$$

when

$$A(\varphi_1, \varphi_2) = 2\alpha(1 - \cos\varphi_1) + 4\gamma(1 - \cos\varphi_1 \cos\varphi_2),$$
$$B(\varphi_1, \varphi_2) = 4\gamma \sin\varphi_1 \sin\varphi_2.$$

Henceforth it will be assumed that ν is measured in units of $2\pi\sqrt{M}$ so that the factor $4\pi^2 M$ can be omitted. If the matrix of the determinant (16) is represented by $M^2(a_1, a_2)$, the matrix of the entire characteristic determinant is

$$\mathbf{M}^2 = \begin{pmatrix} M^2(-N, -N) & 0 & \cdots & 0 & 0 \\ 0 & M^2(-N, -N+1) & \cdots & 0 & 0 \\ \cdots & \cdots & \cdots & \cdots & \cdots \\ 0 & 0 & \cdots & M^2(N, N-1) & 0 \\ 0 & 0 & \cdots & 0 & M^2(N, N) \end{pmatrix} \quad (17a)$$

(the submatrices are represented as squares because the roots of their characteristic determinants are proportional to the squares of the frequencies of the normal modes of vibration). Of course:

$$\mathbf{M}^{2k} = \begin{pmatrix} M^{2k}(-N, -N) & 0 & \cdots & 0 & 0 \\ 0 & M^{2k}(-N, -N+1) & \cdots & 0 & 0 \\ \cdots & \cdots & \cdots & \cdots & \cdots \\ 0 & 0 & \cdots & M^{2k}(N, N-1) & 0 \\ 0 & 0 & \cdots & 0 & M^{2k}(N, N) \end{pmatrix} \quad (17b)$$

Since all the small $M^2(a_1, a_2)$ matrices are of the same form, namely

$$M^2(a_1, a_2) = \begin{pmatrix} A & B \\ B & A^* \end{pmatrix} \quad (18)$$

(where $A = A(\varphi_1, \varphi_2)$; $A^* = A(\varphi_2, \varphi_1)$; $B = B(\varphi_1, \varphi_2) = B(\varphi_2, \varphi_1)$),

$$\text{Trace } \mathbf{M}^{2k} = \sum_{a_1=-N}^{N} \sum_{a_2=-N}^{N} \text{Trace } M^{2k}(a_1, a_2) \quad (19)$$

and we first need only to find traces for one general pair (a_1, a_2). Now

and

$$\text{Trace } M^2(a_1, a_2) = A(\varphi_1, \varphi_2) + A(\varphi_2, \varphi_1)$$

$$\text{Trace } \mathbf{M}^2 = \sum_{a_1, a_2} [A(\pi a_1/N, \pi a_2/N) + A(\pi a_2/N, \pi a_1/N)]$$

$$= 2 \sum_{a_1, a_2} A(\pi a_1/N, \pi a_2/N). \quad (20a)$$

Since

$$M^4(a_1, a_2) = M^2(a_1, a_2) \cdot M^2(a_1, a_2)$$

$$= \begin{pmatrix} A & B \\ B & A^* \end{pmatrix} \begin{pmatrix} A & B \\ B & A^* \end{pmatrix}$$

$$= \begin{pmatrix} A^2 + B^2 & B(A + A^*) \\ B(A + A^*) & B^2 + (A^*)^2 \end{pmatrix}$$

$$\text{Trace } \mathbf{M}^4 = \sum_{a_1, a_2} [A^2 + B^2 + B^2 + (A^*)^2]$$

$$= 2 \sum_{a_1, a_2} (A^2 + B^2). \quad (20b)$$

FREQUENCY SPECTRUM OF CRYSTALS

Continuation of this process leads to

$$\text{Trace } \mathbf{M}^6 = 2 \sum_{a_1, a_2} (A^3 + 3AB^2), \tag{20c}$$

$$\text{Trace } \mathbf{M}^8 = 2 \sum_{a_1, a_2} (A^4 + 4A^2B^2 + 2B^2AA^* + B^4), \tag{20d}$$

$$\text{Trace } \mathbf{M}^{10} = 2 \sum_{a_1, a_2} (A^5 + 5A^3B^2 + 5A^2B^2A^* + 5AB^4). \tag{20e}$$

These Traces will be sufficient to yield a $g(\nu)$ correct to the 10th Legendre polynomial. In general the sum of the diagonal elements of the matrix

$$M^{2k}(a_1, a_2) = \underbrace{\begin{pmatrix} A & B \\ B & A^* \end{pmatrix}\begin{pmatrix} A & B \\ B & A^* \end{pmatrix} \cdots \begin{pmatrix} A & B \\ B & A^* \end{pmatrix}}_{k \text{ matrices}}$$

is Trace $M^{2k}(a_1, a_2)$.

Actually

$$\text{Trace } \mathbf{M}^2 = 2 \sum_{a_1, a_2} \{2\alpha(1 - \cos \varphi_1) + 4\gamma(1 - \cos \varphi_1 \cos \varphi_2)\}$$

$$= 2\{\sum_{a_1, a_2}(2\alpha + 4\gamma) - 4\alpha \sum_{a_1, a_2} \cos \varphi_1 - 8\gamma (\sum_{a_1} \cos \varphi_1)^2\}.$$

But as shown in the appendix

$$\sum_{a_1=-N}^{N} \cos \varphi_1 = -1.$$

Therefore

$$\frac{\text{Trace } \mathbf{M}^2}{2(2N+1)^2} = (2\alpha + 4\gamma) + \frac{2\alpha}{(2N+1)} - \frac{4\gamma}{(2N+1)^2}.$$

Since N is very large, and since this is exactly the second moment of $g(\nu)$:

$$\mu_2 = \frac{\text{Trace } \mathbf{M}^2}{2(2N+1)^2} = (2\alpha + 4\gamma). \tag{21a}$$

To find μ_4 we must first calculate

$$\text{Trace } \mathbf{M}^4 = 2(\sum A^2 + \sum B^2).$$

For this calculation we find

$$\sum A^2 = \sum_{a_1, a_2} \{4\alpha^2(1 - 2\cos \varphi_1 + \cos^2 \varphi_2) + 16\alpha\gamma(1 - \cos \varphi_1)(1 - \cos \varphi_1 \cos \varphi_2)$$

$$+ 16\gamma^2(1 - 2\cos \varphi_1 \cos \varphi_2 + \cos^2 \varphi_1 \cos^2 \varphi_2)\}.$$

But from the appendix:

$$\sum_{a_2} \cos^2 \varphi_2 = N+1, \quad \sum_{a_1} \cos \varphi_1 = -1,$$

so

$$\sum_{a_1, a_2} A^2 = 4\alpha^2(2N+1)^2 + 4\alpha^2(2N+1)(N+1) + 16\alpha\gamma(2N+1)^2 + 16\gamma^2(2N+1)^2 + 16\gamma^2(N+1)^2 + O(N)$$

and

$$\sum A^2/(2N+1)^2 = 4(\alpha + 2\gamma)^2 + 2(\alpha^2 + 2\gamma^2) + O(1/N).$$

Also,

$$\sum_{a_1, a_2} B^2 = 16\gamma^2 (\sum_{a_1} \sin^2 \varphi_1)^2,$$

but from the appendix $\sum_{a_1} \sin^2 \varphi_1 = N$, therefore
$$\sum B^2/(2N+1)^2 = 4\gamma^2.$$
Since
$$\mu_4 = (\text{Trace } \mathbf{M}^4)/2(2N+1)^2, \quad \mu_4 = 4(\alpha+2\gamma)^2 + 2(\alpha^2+4\gamma^2) + O(1/N). \quad (21b)$$
In a similar manner some of the higher moments can be shown to be (to within terms of order $1/N$)
$$\mu_6 = 8(\alpha+2\gamma)^3 + 12(\alpha+2\gamma)(\alpha^2+4\gamma^2), \quad (21c)$$
$$\mu_8 = 16(\alpha+2\gamma)^4 + 48(\alpha+2\gamma)^2(\alpha^2+4\gamma^2) + 6(\alpha^2+4\gamma^2)^2 + 40\alpha^2\gamma^2, \quad (21d)$$
$$\mu_{10} = 32(\alpha+2\gamma)^5 + 160(\alpha+2\gamma)^3(\alpha^2+4\gamma^2) + 120\gamma^2(\alpha+2\gamma)(7\alpha^2+8\gamma^2)$$
$$+ 40\alpha^2\gamma^2(\alpha+\gamma) + 60\alpha^4(\alpha+2\gamma). \quad (21e)$$

Using the results of the previous section, if we assume
$$g(\nu) = \sum_{n=0}^{\infty} a_{2n} P_{2n}(\nu/\nu_L) \quad (22)$$
(ν_L = largest frequency of a normal mode) the first few coefficients are
$$a_0 = 0.5/\nu_L,$$
$$a_2 = 5(6\alpha + 12\gamma - \nu_L^2)/4\nu_L^3,$$
$$a_4 = 9[140(\alpha+2\gamma)^2 + 70(\alpha^2+4\gamma^2) - 60(\alpha+2\gamma)\nu_L^2 + 3\nu_L^4]/16\nu_L^5.$$

The number of parameters involved in the a's can immediately be reduced by one, for as Blackman[5] has shown, the largest normal mode ν_L results from the 2×2 determinant in which $\varphi_1 = 0$ and $\varphi_2 = \pi$ (or vice versa, $\varphi_2 = \pi$ and $\varphi_2 = 0$); therefore by (16)
$$\nu_L^2 = 4(\alpha+2\gamma).$$
This means
$$\nu_L a_0 = \tfrac{1}{2},$$
$$\nu_L a_2 = \tfrac{5}{8},$$
$$\nu_L a_4 = 9(9\alpha^2 - 104\alpha\gamma + 36\gamma^2)/8(\alpha+2\gamma)^2, \quad \text{etc.}$$

Since the purpose of this paper is more to introduce new methods than to obtain particular results, let us limit ourselves to the special case $\gamma = 0.05\alpha$. This α, γ relationship is of special interest because it is the one used by Blackman in his detailed calculations—and may therefore give us an opportunity to compare the results of the trace-moment approach with those obtained by Blackman in his detailed root calculations. In this case $\nu_L^2 = 4.4\alpha$.

The first six even moments in the special case $\gamma = 0.05\alpha$ are[7] (of course the odd moments are all zero)

$$\mu_0 = 1, \qquad \mu_6 = 0.28150826 \nu_L^6,$$
$$\mu_2 = 0.5 \nu_L^2, \qquad \mu_8 = 0.23560495 \nu_L^8,$$
$$\mu_4 = 0.35433884 \nu_L^4, \quad \mu_{10} = 0.20316225 \nu_L^{10}$$

and the corresponding Legendre coefficients of (22) are

$$\nu_L a_0 = 0.5, \qquad \nu_L a_6 = 0.3704,$$
$$\nu_L a_2 = 0.625, \qquad \nu_L a_8 = -0.3182,$$
$$\nu_L a_4 = 0.2260, \quad \nu_L a_{10} = -0.0133.$$

---

[7] The large number of significant figures is necessary because one must take differences of multiples of the μ's in order to find the a's.

The resulting function $\nu_L g(\nu)$ is plotted in Fig. 1. It qualitatively resembles the equivalent curve of Blackman as obtained by direct numerical computation of a random sample of normal modes. Blackman's first peak is sharper and is displaced a little more to the left. The quantitative disagreement may be the result of Blackman's smoothing process, or it may have its source in the existence of a large value for one or more of the coefficients of the higher Legendre polynomials that have been neglected here. Since our $g(\nu)$ is not quite zero at $\nu=0$ as it should be, we must be omitting some contribution of the higher polynomials.

The calculation of physical quantities from a frequency distribution of the form (22) offers no special difficulty. For example the specific heat is in general

$$C = k \int_0^{\nu_L} \frac{h^2 \nu^2 g(\nu) d\nu}{4k^2 T^2 \sinh^2 h\nu/2kT},$$

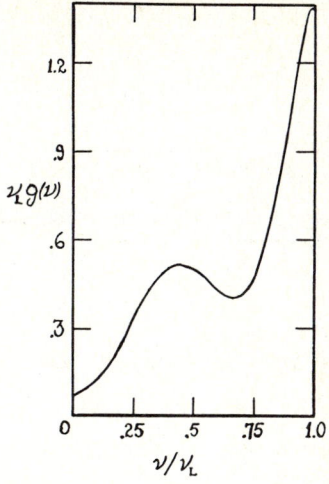

FIG. 1.

so in a low temperature region where one may expand $(\sinh^2 h\nu/2kT)^{-1}$ as an exponential series, elementary integrals of the form

$$\int_0^{\nu_L} e^{-nh\nu/kT} \text{(polynomial in } \nu) d\nu$$

are all that are involved, while at higher temperatures where it would be desirable to use a power series expansion for $(\sinh^2 h\nu/kT)^{-1}$ one encounters only sums of integrals of the form

$$\int_0^{\nu_L} \text{(polynomials in } \nu) d\nu.$$

Before concluding this section it might be worth while reminding the reader that the expressing of $g(\nu)$ as a linear combination of Legendre polynomials is somewhat arbitrary. Any other artifice which allows one to express $g(\nu)$ as a function of the moments would be equally valid. The Legendre development merely seemed to be the most straightforward and simplest approach which occurred to the author.

III. FREQUENCY DISTRIBUTION FROM EXPERIMENTAL DATA

If in a crystal of N structureless atoms or in a complex molecule of N atoms, the frequency distribution function $g(\nu)$ is defined so that $3Ng(\nu)d\nu$ is the number of normal modes of vibration in the frequency interval ν and $\nu+d\nu$, the logarithm of the partition function (per normal mode) is given by

$$\log Z(T) = -\int_0^\infty g(\nu) \log(1 - e^{-h\nu/kT}) d\nu \tag{23a}$$

under the restriction

$$\int_0^\infty g(\nu) d\nu = 1. \tag{23b}$$

Formally a knowledge of $Z(T)$ at all temperatures permits (23a) to be regarded as an integral equation in which $g(\nu)$ occurs as the unknown function. Also, the average energy per normal mode is

$$\bar{E}(T) = kT^2 \partial \log Z/\partial T$$

$$= \int_0^\infty \frac{h\nu g(\nu)d\nu}{e^{h\nu/kT}-1}. \tag{24}$$

(using for zero energy the state of lowest energy of the system including the residual energy $h\nu/2$ per normal mode) and by definition the specific heat is

$$C_v(T) = (\partial E/\partial T)_v$$

$$= k \int_0^\infty \left(\frac{h\nu}{2kT}\right)^2 \frac{g(\nu)d\nu}{\sinh^2(h\nu/2kT)}. \tag{25}$$

Making the substitution $\theta = h/kT$ and writing

$$\log Z(T) = \mathbf{Z}(\theta); \quad \mathbf{E}(\theta) = \theta E(T/h); \quad \mathbf{C}(\theta) = C_v(T)/k,$$

we have the three fundamental integral equations

$$\mathbf{Z}(\theta) = -\int_0^\infty g(\nu) \log(1-e^{-\theta\nu})d\nu, \tag{26}$$

$$\mathbf{E}(\theta) = \int_0^\infty g(\nu)\theta\nu d\nu/(e^{\theta\nu}-1), \tag{27}$$

$$\mathbf{C}(\theta) = \int_0^\infty \frac{\theta^2\nu^2 g(\nu)d\nu}{4\sinh^2(\theta\nu/2)},$$

all of which are of the form

$$f(\theta) = \int_0^\infty g(\nu)K(\theta\nu)d\nu. \tag{28}$$

It is quite clear that if the partition function, the internal energy or the specific heat of a given system could be found at all temperatures, and if the above integral equation could be inverted[8] so that $g(\nu)$ would be a function of the measurable quantities it would be possible to compute $g(\nu)$ and perhaps get a better understanding of the internal structure of a given crystal or complex molecule.

As has been demonstrated in Appendix II, the formal solution of (28) is

$$g(\nu) = \frac{1}{2\pi} \int_{-\infty}^{\infty} \frac{du}{I(u)} \int_0^\infty f(\theta)(\nu\theta)^{iu}d\theta, \tag{29}$$

where

$$I(u) = \int_0^\infty x^{iu} K(x)dx.$$

Inasmuch as[9]

$$\int_0^\infty x^{iu} \log(1-e^{-x})dx = -\zeta(2+iu)\Gamma(1+iu)$$

---

[8] Inversions of the partition function have been made before by S. Bauer, J. Chem. Phys. **6**, 403 (1938); **7**, 1097 (1939).
[9] These integrals are easily evaluated by expanding the transcendental functions in the integrands as sums of powers of e^{-x}.

($\zeta(z)$ is the Riemann zeta-function defined by $\zeta(z)=\sum_{n=1}^{\infty} n^{-z}$)

$$\int_0^\infty \frac{x^{iu+1}dx}{e^x-1}=\zeta(2+iu)\Gamma(2+iu), \quad \int_0^\infty \frac{x^{iu+2}dx}{4\sinh^2 x/2}=\zeta(2+iu)\Gamma(3+iu),$$

the formal equations for $g(\nu)$ are

$$g(\nu)=\frac{1}{2\pi}\int_{-\infty}^{\infty}\frac{du}{\zeta(2+iu)\Gamma(1+iu)}\int_0^\infty Z(\theta)(\nu\theta)^{iu}d\theta, \tag{30}$$

$$g(\nu)=\frac{1}{2\pi}\int_{-\infty}^{\infty}\frac{du}{\zeta(2+iu)\Gamma(2+iu)}\int_0^\infty E(\theta)(\nu\theta)^{iu}d\theta, \tag{31}$$

$$g(\nu)=\frac{1}{2\pi}\int_{-\infty}^{\infty}\frac{du}{\zeta(2+iu)\Gamma(3+iu)}\int_0^\infty C(\theta)(\nu\theta)^{iu}d\theta. \tag{32}$$

As is to be expected, substitution of the Debye specific heat formula

$$C(\theta)=C_v(T)/NK=3\left(\frac{12}{\theta^3\nu_L^3}\int_0^{\theta\nu_L}\frac{x^3dx}{e^x-1}-\frac{3\theta\nu_L}{e^{\theta\nu_L}-1}\right)$$

into (32) can be shown to yield the frequency distribution

$$g(\nu)=\begin{cases}9\nu^2/\nu_L^3 & \text{if } \nu<\nu_L \\ 0 & \text{if } \nu>\nu_L.\end{cases} \tag{33}$$

For actual calculation of $g(\nu)$ from (32) there seem to be several possible approaches. One might integrate (32) numerically after substitution of the experimental $C(\theta)$ values derived from specific heat measurements. However, since most specific heat curves are almost of the Debye form, one might divide $C(\theta)$ into a Debye and a residual portion. The Debye portion would contribute the amount (33) to $g(\nu)$ and the rest of $g(\nu)$ might be found either by numerical integration or by obtaining empirical equations for the residual $C(\theta)$. In the latter case one would attempt to fit a power series in θ to the high temperature part and a power series in $e^{-\theta}$ in the low temperature region. Then (32) could probably be integrated analytically. Another possibility would be to assume that $g(\nu)$ could be expanded in Legendre polynomials and finding the coefficients by applying the orthogonality property of the polynomials to the assumed equation (32).

In conclusion the author would like to thank Professor J. G. Kirkwood for his discussions concerning this problem.

APPENDIX I. SUMMATION FORMULA

In the main part of this paper the following trigonometric sums are needed

$$C_k=\sum_{a=-N}^{N}\cos^k\frac{\pi a}{N}; \quad S_k=\sum_{a=-N}^{N}\sin^k\frac{\pi a}{N}.$$

To sum C_k and S_k it is convenient to find first

$$E_k=\sum_{a=-N}^{N}\exp(ik\pi a/N).$$

If $x = \exp(ik\pi/N) \neq 1$ when $0 < k < N$,

$$E_k = \sum_{a=-N}^{N} x^a = x^{-N}(1-x^{2N+1})/(1-x)$$

Since

$$= e^{-i\pi k}(1 - e^{2\pi ik}e^{ik\pi/N})/(1-e^{ik\pi/N}).$$

$$e^{-i\pi k} = (-1)^k,$$

$$E_k = (-1)^k \quad \text{when} \quad 0 < k < N.$$

Obviously $E_0 = 2N+1$.

Now

$$C_k = \frac{1}{2^k} \sum_{a=-N}^{N} (e^{i\pi a/N} + e^{-i\pi a/N})^k$$

$$= \frac{1}{2^k} \sum_{m=0}^{k} \frac{k!}{m!(k-m)!} \sum_{a=-N}^{N} e^{i\pi(2m-k)/N}.$$

When k is odd, $2m - k$ is also odd and

$$\sum_{a=-N}^{N} \exp\{i\pi(2m-k)/N\} = -1.$$

So

$$C_{k(\text{odd})} = -\frac{1}{2^k} \sum_{m=0}^{k} \frac{k!}{m!(k-m)!} = -1.$$

When k is even $(2m-k)$ is also even and it is zero when $k = 2m$; so

$$C_{k(\text{even})} = \frac{1}{2^k} \left\{ \sum_{m=0}^{(k/2)-1} \frac{k!}{m!(k-m)!} + \sum_{(k/2)+1}^{k} \frac{k!}{m!(k-m)!} \right.$$

$$\left. + \frac{(2N+1)k!}{(k/2)!(k/2)!} \right\}$$

$$= \frac{1}{2^k} \left\{ \sum_{m=0}^{k} \frac{k!}{m!(k-m)!} + \frac{2Nk!}{(k/2)!(k/2)!} \right\}$$

$$C_{k(\text{even})} = 1 + 2^{1-k} N(k!)/(k/2)!(k/2)!.$$

To find S_k we proceed in essentially the same manner:

$$S_k = \frac{1}{(2i)^k} \sum_{a=-N}^{N} (e^{i\pi a/N} - e^{-i\pi a/N})^k$$

$$= \frac{1}{(2i)^k} \sum_{a=-N}^{N} \sum_{m=0}^{k} \frac{k!}{m!(k-m)!} (e^{i\pi a/N})^{k-m}(-e^{-i\pi a/N})^m$$

$$= \frac{1}{(2i)^k} \sum_{m=0}^{\infty} \frac{k!(-1)^m}{m!(k-m)!} \sum_{a=-N}^{N} e^{i\pi a(k-2m)/N}.$$

When k is odd $k - 2m$ is odd and

$$S_{k(\text{odd})} = \frac{-1}{(2i)^k} \sum_{m=0}^{k} \frac{(-1)^m k!}{m!(k-m)!}$$

$$= -\frac{1}{(2i)^k}(1-1)^k = 0.$$

If k is even $k - 2m$ is also even

$$S_{k(\text{even})} = \frac{1}{(2i)^k} \left\{ \sum_{m=0}^{k/2-1} \frac{k!(-1)^m}{m!(k-m)!} + \sum_{m=(k/2)+1}^{k} \frac{k!(-1)^m}{m!(k-m)!} \right.$$

$$\left. + \frac{(-1)^{k/2}(2N+1)k!}{(k/2)!(k/2)!} \right\}$$

$$= \frac{1}{(2i)^k} \left\{ \sum_{m=0}^{k} \frac{k!(-1)^m}{m!(k-m)!} + \frac{i^k(2N)k!}{(k/2)!(k/2)!} \right\}$$

$$S_{k(\text{even})} = \frac{N(k!)}{2^{k-1}(k/2)!(k/2)!}.$$

The first few C_k's and S_k's are

k	C_k	S_k
0	$2N+1$	$2N+1$
1	-1	0
2	$N+1$	N
3, 5, 7, 9	-1	0
4	$1 + (3N/4)$	$3N/4$
6	$1 + (5N/8)$	$5N/8$
8	$1 + (35N/64)$	$35N/64$

APPENDIX II

Formal Solution of $f(\theta) = \int_0^\infty g(\nu) K(\theta\nu) d\nu$

Here we shall find the formal solution $g(\nu)$ of the integral equation

$$f(\theta) = \int_0^\infty g(\nu) K(\theta\nu) d\nu, \tag{A}$$

when $f(\theta)$ and $K(\theta\nu)$ are assumed to be known functions. The Fourier transform method to be used is modeled on that of Payley and Wiener[10] (the formal solution is also a simple consequence of some operations with Mellin transforms of f, g, and K.[11] Let us make the substitutions

$$\theta = e^{-\eta}, \quad \nu = e^\alpha \tag{B}$$

in (A) yielding

$$e^{-\eta} f(e^{-\eta}) = \int_{-\infty}^{\infty} g(e^\alpha) e^{\alpha - \eta} K(e^{\alpha - \eta}) d\alpha. \tag{C}$$

Multiplying by

$$e^{-iu\eta} d\eta / \sqrt{2\pi}$$

and integrating:

$$\frac{1}{\sqrt{2\pi}} \int_{-\infty}^{\infty} e^{-\eta} f(e^{-\eta}) e^{-iu\eta} d\eta$$

$$= \frac{1}{\sqrt{2\pi}} \int_{-\infty}^{\infty} g(e^\alpha) e^{-i\alpha u} d\alpha \int_{-\infty}^{\infty} e^{\alpha - \eta} e^{-iu(\eta - \alpha)} K(e^{\alpha - \eta}) d\eta.$$

Now let $\eta - \alpha = \beta$

$$\frac{1}{\sqrt{2\pi}} \int_{-\infty}^{\infty} e^{-\eta} f(e^{-\eta}) e^{-iu\eta} d\eta$$

$$= \frac{1}{\sqrt{2\pi}} \int_{-\infty}^{\infty} g(e^\alpha) e^{-iu\alpha} d\alpha \int_{-\infty}^{\infty} e^{-\beta} e^{-iu\beta} K(e^{-\beta}) d\beta.$$

---

[10] R. Payley and N. Wiener, *Fourier Transforms* (New York, 1934), p. 38.
[11] Titchmarsh, *Theory of Fourier Transforms* (Oxford, 1937), p. 315.

Define

$$I(u) = \int_{-\infty}^{\infty} e^{-\beta(iu+1)} K(e^{-\beta}) d\beta$$

$$= \int_0^{\infty} x^{iu} K(x) dx.$$

Then

$$\frac{1}{\sqrt{2\pi}}\int_{-\infty}^{\infty} e^{-\eta} f(e^{-\eta}) e^{-iu\eta} d\eta = \frac{I(u)}{\sqrt{2\pi}} \int_{-\infty}^{\infty} g(e^{\alpha}) e^{-i\alpha u} d\alpha$$

and taking Fourier transforms

$$g(e^{\alpha}) = \frac{1}{2\pi} \int_{-\infty}^{\infty} \frac{e^{iu\alpha} du}{I(u)} \int_{-\infty}^{\infty} f(e^{-\eta}) e^{-\eta(1+iu)} d\eta.$$

By applying (B)

$$g(\nu) = \frac{1}{2\pi} \int_{-\infty}^{\infty} \frac{du}{I(u)} \int_0^{\infty} f(\theta)(\theta\nu)^{iu} d\theta.$$

Effect of Defects on Lattice Vibrations: Interaction of Defects and an Analogy with Meson Pair Theory*

ELLIOTT W. MONTROLL AND RENFREY B. POTTS†
University of Maryland, College Park, Maryland
(Received December 21, 1955)

An analysis is given of the determination of additive functions of the frequencies of the normal mode vibrations of a lattice. The method is applied to the problem of calculating the self-energies and interaction energies of defects in lattices of any dimension. In particular results are derived for the self-energies and interaction energies of isotopes, holes, and "source" defects in simple cubic monatomic and diatomic lattices. For example it is shown that two holes in a simple cubic lattice attract each other, the energy of interaction being inversely proportional to the cube of the distance of separation. The general method is also applied to the problem of the interaction of lattice defects with the boundaries of the lattice. Finally, if the lattice approaches the limit of a continuum, it is shown that the energy of interaction between two holes is just that obtained by Wentzel for the interaction between two fixed nucleons according to the scalar meson pair theory.

INTRODUCTION

THE influence of defects such as impurities and holes on the physical properties of crystals has been one of the most studied phases of solid state physics in recent years. This paper is the third report of a detailed mathematical investigation of the effect of localized irregularities on lattice vibrations[1] (the first report will be referred to as D-1 and the second as D-2). Although the authors are interested in the general defect problem, they have decided that various suitable mathematical techniques can more easily be applied to perturbations of lattice vibrations than to other degrees of freedom in a solid and have elected to examine that problem first. The work of Koster and Slater[2] on the theory of semiconductors via Wannier wave functions parallels our analysis to some extent (as do the brief remarks of Lax and Smith[3] on lattice vibrations). Work is now in progress on the influence of defects on the spin wave, Ising, and spherical models of magnetic materials.

D-1 is mostly concerned with those vibrational modes which are localized around lattice defects. It is shown that under certain conditions discrete normal modes exist which are displaced out of the continuum of modes of the unperturbed lattice. Only a few atoms in the immediate neighborhood of a defect participate in these modes. Generally motions of all atoms in a perfect monatomic crystal contribute equally to the energy in each normal mode. However, the atoms which participate in localized modes are responsible for more than their normal share of the internal energy of the crystal. Hence the region around a defect is equivalent to a "hot spot" in the lattice. A localized mode (either in the interior or on the surface of a crystal) might catalyze physical and chemical processes which would not normally occur at the existing temperature of the crystal. D-2 is concerned with localized modes in linear diatomic lattices.

It was also pointed out in D-1 that at low temperatures an attraction exists between "like" defects (for example a pair of isotopic impurities of the same mass) and a repulsion between "unlike" defects (for example, two isotopes of different mass, one heavier than a normal atom in the crystal and one lighter). This interaction is greatest at absolute zero. A consequence of the attraction between like defects would be a clustering tendency between atoms of like atomic weight in a mixed crystal of two isotopic species. Indeed one would expect a separation into two isotopic phases at $T=0$ (actually the equilibrium time for such a process might be very long). This effect has been discussed by Prigogine, Bingen, and Jeener.[4] It will be shown in Sec. 6 that light isotopes and holes are attracted to the free boundary of a crystal.

This paper is mainly concerned with the development of a formalism for the discussion of the effect of the interaction of defects on additive functions of the normal mode frequencies. The formalism will be applied to the calculation of the interaction energy between defects (as determined from the change in zero-point energy) and of that between defects and surfaces. Although thermodynamic quantities are additive functions of the frequencies, we shall postpone a discussion of their behavior until part 4 of this series. Detailed calculations are made here on simple cubic lattices with interactions (described through both central and noncentral forces) between nearest neighbors only. Both one- and two-component systems are analyzed.

Some remarks will be made concerning continuum field theory by letting our crystalline lattice spacings vanish. It was pointed out to the authors by Professor

---

* This research was supported by the Air Research and Development Command of the U. S. Air Force.
† On leave from the University of Adelaide, South Australia.
[1] E. W. Montroll and R. B. Potts, Phys. Rev. **100**, 525 (1955); Mazur, Montroll, and Potts, J. Wash. Acad. Sci. **46**, 2 (1956).
[2] G. F. Koster and J. C. Slater, Phys. Rev. **95**, 1167 (1954); G. F. Koster, Phys. Rev. **95**, 1436 (1954).
[3] M. Lax, Phys. Rev. **94**, 1391 (1954).

[4] Prigogine, Bingen, and Jeener, Physica **20**, 383 (1954); **20**, 516 (1954).

T. D. Lee that our methods are similar to those used by Wentzel[5] in his investigation of the meson pair theory of forces between nucleons. We show that pair theory is mathematically equivalent to the continuum limit of the theory of the interactions of holes in a crystal lattice.

The reader is referred to D-1 for a detailed discussion of the model[6] used here.

1. GENERAL FORMULAS FOR CALCULATION OF ADDITIVE FUNCTIONS OF NORMAL MODE FREQUENCIES

Let us suppose that the normal mode frequencies of a lattice are $\omega_1, \omega_2, \cdots$. Many quantities of interest can be expressed as sums of functions of the normal mode frequencies

$$S = \sum_j g(\omega_j). \quad (1.1)$$

For example, the zero-point energy of the lattice is given by S if $g(z) = \frac{1}{2}\hbar z$. The characteristic function, $E(\exp i\alpha\omega^2)$, whose Fourier transform is the frequency distribution function, corresponds to

$$g(z) = N^{-1} \exp(i\alpha z^2),$$

where N is the number of degrees of freedom of the lattice. Thermodynamic quantities are generally of the form of (1.1).

Let us also assume that the frequencies are roots of a characteristic equation

$$D(\omega) = 0.$$

It was pointed out in D-1 that if $g(z)$ is an analytic function inside of a closed counter-clockwise contour C and if $D(z)$ has all of its zeros but no poles inside the contour, then[7]

$$S = \sum_i g(\omega_j) = \frac{1}{2\pi i} \int_C g(z) \frac{d}{dz} \{\log D(z)\} dz. \quad (1.2)$$

We represent the function whose zeros are normal mode frequencies of a perfect lattice by $D_0(z)$, and represent the corresponding function associated with a lattice with defects enumerated by $\alpha, \beta, \cdots$ by $D(\alpha, \beta, \cdots; z)$.

The change in an additive function S which results from a single defect, α, is

$$\Delta S_\alpha = \frac{1}{2\pi i} \int_C g(z) \frac{d}{dz} \{\log D(\alpha; z) - \log D_0(z)\} dz$$

$$= \frac{1}{2\pi i} \int_C g(z) \frac{d}{dz} \log\{D(\alpha; z)/D_0(z)\} dz. \quad (1.3)$$

This quantity might be referred to as the self-S of the

[5] G. Wentzel, Helv. Phys. Acta **15**, 111 (1942).
[6] See also E. W. Montroll, Third Berkeley Symposium on Statistics and Probability, 1955.
[7] See E. C. Titchmarsh, *Theory of Functions* (Oxford University Press, Oxford, 1932), p. 116.

lattice. The "interaction-S" of a defect pair (α, β) is defined as the difference between the S of a system of two interacting defects and that of a pair of isolated defects and is given by

$$\Delta S_{\alpha,\beta} = \frac{1}{2\pi i} \int_C g(z) \frac{d}{dz}$$
$$\times \log\{D(\alpha, \beta; z)D_0(z)/D(\alpha; z)D(\beta; z)\} dz. \quad (1.4)$$

If $\epsilon_\alpha, \epsilon_\beta, \cdots$ are the parameters which characterize defects $\alpha, \beta, \cdots$, and $D(\alpha; z)$ is of the form

$$D(\alpha; z) = D_0(z)[1 + \epsilon_\alpha h_\alpha(z)] \quad (1.5a)$$

(as we shall show to be the case in a wide variety of situations), and

$$D(\alpha, \beta; z) = D_0(z)[1 + \epsilon_\alpha h_\alpha(z) + \epsilon_\beta h_\beta(z) + \epsilon_\alpha \epsilon_\beta h_{\alpha\beta}(z)], \quad (1.5b)$$

we find

$$\Delta S_\alpha = \frac{\epsilon_\alpha}{2\pi i} \int_C \frac{h_\alpha'(z)g(z)dz}{1 + \epsilon_\alpha h_\alpha(z)}, \quad (1.6a)$$

where the prime denotes the derivative with respect to the argument. Also

$$\Delta S_{\alpha\beta} = \frac{1}{2\pi i} \int_C g(z) \frac{d}{dz} \log\left\{1 + \frac{\epsilon_\alpha \epsilon_\beta (h_{\alpha\beta} - h_\alpha h_\beta)}{(1 + \epsilon_\alpha h_\alpha)(1 + \epsilon_\beta h_\beta)}\right\} dz, \quad (1.6b)$$

so that as ϵ_α and $\epsilon_\beta \to 0$

$$\Delta S_{\alpha\beta} \sim -\frac{\epsilon_\alpha \epsilon_\beta}{2\pi i} \int_C g(z)[h_\alpha(z)h_\beta(z) - h_{\alpha\beta}(z)]' dz$$

and the interaction S is of second order in $\epsilon_\alpha, \epsilon_\beta$ in the limit of "weak defects."

The total interaction S due to a large number of defects is $\sum \Delta S_{\alpha\beta}$ over all defect pairs and is a quadratic form in $\epsilon_\alpha, \epsilon_\beta, \cdots$ in the limit of weak defects, but cubic and higher order terms occur when "strong" defects (large ϵ's) exist.

2. ON THE GENERAL FORM OF THE CHARACTERISTIC EQUATIONS OF NORMAL MODE FREQUENCIES

The normal mode frequencies $\{\omega_j\}$ of a set of coupled harmonic oscillators are zeros of the characteristic determinant

$$D_0(\omega) = \begin{vmatrix} a_{11} + M_1\omega^2 & a_{12} & a_{13} & \cdots \\ a_{21} & a_{22} + M_2\omega^2 & a_{23} & \cdots \\ a_{31} & a_{32} & a_{33} + M_3\omega^2 & \cdots \\ \cdots & \cdots & \cdots & \end{vmatrix}. \quad (2.1)$$

Here the M_j's are particle masses and the a_{ij}'s are related to the force constants of the "springs." We shall assume that defects in a lattice alter the determinant through the introduction of a set of parameters $\delta_\alpha, \delta_\beta, \cdots$ at the αth, βth, $\cdots$ elements along the main diagonal so that these elements become $a_{\alpha\alpha}+M_\alpha\omega^2+\delta_\alpha$, etc. (Although it is sometimes necessary to perturb off-diagonal terms as well, the analysis of such cases is essentially the same as that given below.)

In the case of a single defect, the determinant can be expanded by the αth row in the usual manner to yield

$$D(\alpha;\omega)=D_0(\omega)+\delta_\alpha A_{\alpha\alpha}, \quad (2.2a)$$

where $A_{\alpha\alpha}$ is the cofactor of $a_{\alpha\alpha}+\omega^2 M_\alpha$ in the determinant $D_0(\omega)$. It is well known that if $\{a_{ij}{}^{(-1)}(\omega)\}$ is the set of elements of the inverse of the matrix of $D_0(\omega)$, then

$$A_{\alpha\alpha}=D_0(\omega)a_{\alpha\alpha}{}^{(-1)}(\omega), \quad (2.2b)$$

and

$$D(\alpha;\omega)=D_0(\omega)[1+\delta_\alpha a_{\alpha\alpha}{}^{(-1)}(\omega)], \quad (2.3)$$

which is of the form (1.5a) where h_α is to be identified with $a_{\alpha\alpha}{}^{(-1)}$.

Now let there be two defects, one at α and the other at β. Then, if we expand $D(\alpha,\beta;\omega)$ with respect to the βth row, we find

$$D(\alpha,\beta;\omega)=D(\alpha;\omega)+\delta_\beta A_{\beta\beta}(\alpha),$$

where $A_{\beta\beta}$ is the cofactor of the βth diagonal element of the determinant $D(\alpha;\omega)$. This is, however,

$$A_{\beta\beta}(\alpha)=A_{\beta\beta}+A\begin{Bmatrix}\alpha\alpha\\\beta\beta\end{Bmatrix},$$

where $A\begin{Bmatrix}\alpha\alpha\\\beta\beta\end{Bmatrix}$ is the second-order cofactor obtained by striking out the αth and βth rows and columns of $D_0(\omega)$. Since this cofactor is

$$A\begin{Bmatrix}\alpha\alpha\\\beta\beta\end{Bmatrix}=D_0(\omega)\begin{vmatrix}a_{\alpha\alpha}{}^{(-1)} & a_{\alpha\beta}{}^{(-1)}\\a_{\beta\alpha}{}^{(-1)} & a_{\beta\beta}{}^{(-1)}\end{vmatrix},$$

we have

$$D(\alpha,\beta;\omega)=D_0(\omega)\Bigg\{1+\delta_\alpha a_{\alpha\alpha}{}^{(-1)}(\omega)+\delta_\beta a_{\beta\beta}{}^{(-1)}(\omega)$$

$$+\delta_\alpha\delta_\beta\begin{vmatrix}a_{\alpha\alpha}{}^{(-1)} & a_{\alpha\beta}{}^{(-1)}\\a_{\beta\alpha}{}^{(-1)} & a_{\beta\beta}{}^{(-1)}\end{vmatrix}\Bigg\}$$

$$=D_0(\omega)\begin{vmatrix}1+\delta_\alpha a_{\alpha\alpha}{}^{(-1)} & (\delta_\alpha\delta_\beta)^{\frac{1}{2}}a_{\alpha\beta}{}^{(-1)}\\(\delta_\alpha\delta_\beta)^{\frac{1}{2}}a_{\beta\alpha}{}^{(-1)} & 1+\delta_\beta a_{\beta\beta}{}^{(-1)}\end{vmatrix}. \quad (2.4)$$

The reader can verify the three-defect formula,

$$D(\alpha,\beta,\gamma;\omega)=D_0(\omega)\begin{vmatrix}1+\delta_\alpha a_{\alpha\alpha}{}^{(-1)} & (\delta_\alpha\delta_\beta)^{\frac{1}{2}}a_{\alpha\beta}{}^{(-1)} & (\delta_\alpha\delta_\gamma)^{\frac{1}{2}}a_{\alpha\gamma}{}^{(-1)}\\(\delta_\beta\delta_\alpha)^{\frac{1}{2}}a_{\beta\alpha}{}^{(-1)} & 1+\delta_\beta a_{\beta\beta}{}^{(-1)} & (\delta_\beta\delta_\gamma)^{\frac{1}{2}}a_{\beta\gamma}{}^{(-1)}\\(\delta_\gamma\delta_\alpha)^{\frac{1}{2}}a_{\gamma\alpha}{}^{(-1)} & (\delta_\gamma\delta_\beta)^{\frac{1}{2}}a_{\gamma\beta}{}^{(-1)} & 1+\delta_\gamma a_{\gamma\gamma}{}^{(-1)}\end{vmatrix}, \quad (2.5)$$

and may generalize these results to any given number of defects.

Since the elements of the inverse matrices appear in all formulas independently of the specific nature of the defects, we derive formulas for these elements in the next section.

3. ELEMENTS OF INVERSE MATRIX FOR CERTAIN MODELS

We shall now find the elements of the inverse of the matrix D_0 which corresponds to one-component, n-dimensional simple cubic lattices with interactions between nearest neighbors only (and both central and noncentral forces). We assume the lattices to be cubes containing N^n lattice points. At the end of this section we discuss the case of a diatomic lattice.

The mechanics of these systems are discussed in detail in D-1 and in reference 6. A mathematically convenient (but physically somewhat unreal) feature of this model is the independence of the x, y, and z components of the motions of the lattice particles.

The equation of motion of the x component of the displacement of a particle at lattice point $(m_1,m_2,\cdots,m_n)$ is

$$M\ddot{x}(m_1,\cdots,m_n)=\sum_{j=1}^n\gamma_j[x(m_1,\cdots,m_j-1,\cdots,m_n)$$

$$-2x(m_1,\cdots,m_n)+x(m_1,\cdots,m_j+1,\cdots,m_n)], \quad (3.1a)$$

where γ_j is the force constant associated with displacements parallel to the jth coordinate axis. Similar equations exist for the other components of the displacement. We choose solutions of the form

$$x(m_1,\cdots,m_n)=e^{i\omega t}u(m_1,m_2,\cdots) \quad (3.1b)$$

and find D_0 to be the determinant of the coefficients of the u's in

$$\omega^2 Mu(m_1,m_2,\cdots)+\sum_{j=1}^n\gamma_j[u(\cdots,m_j-1,\cdots)$$

$$-2u(\cdots,m_j,\cdots)+u(\cdots,m_j+1,\cdots)]=0. \quad (3.2a)$$

We first assume the existence of periodic boundary conditions; later we discuss the cases of free and rigid boundaries. The characteristic vectors of the matrix of the coefficients of $u(m)$ are of the form

$$u_s(m)=N^{-\frac{1}{2}n}\exp(2\pi i\mathbf{s}\cdot\mathbf{m}/N), \quad (3.2b)$$

where

$$\mathbf{s}=(s_1,\cdots,s_n), \quad \mathbf{m}=(m_1,m_2,\cdots,m_n)$$

and the characteristic values

$$\lambda(s_1,s_2,\cdots)=M\omega^2-2\sum_{j=1}^n\gamma_j(1-\cos\varphi_j),$$

$$\varphi_j=2\pi s_j/N. \quad (3.)$$

The elements of our determinant can be expressed

$$a(\mathbf{m};\mathbf{m}')=N^{-n}\sum_{s_1,s_2,\cdots=1}^N\lambda(s_1,s_2,\cdots)$$

$$\times\exp\{2\pi i\mathbf{s}\cdot(\mathbf{m}-\mathbf{m}')/N\}, \quad (3.)$$

while those of the inverse are
$$a^{(-1)}(\mathbf{m};\mathbf{m}') = N^{-n} \sum_s \lambda^{-1}(\mathbf{s}) \times \exp\{2\pi i \mathbf{s} \cdot (\mathbf{m}-\mathbf{m}')/N\}. \quad (3.5)$$
In the limit as $N \to \infty$,
$$a^{(-1)}(\mathbf{m};\mathbf{m}') = \left(\frac{1}{2\pi}\right)^n \int_0^{2\pi} \cdots \int \times \frac{\exp[i(\mathbf{m}-\mathbf{m}') \cdot \boldsymbol{\varphi}] d^n\varphi}{[M\omega^2 - 2\sum_1^n \gamma_j(1-\cos\varphi_j)]}. \quad (3.6a)$$

These integrals are essentially the Green's functions discussed in D-1. In that notation,
$$a^{(-1)}(\mathbf{m};\mathbf{m}') = (\gamma_1 + \gamma_2 + \cdots + \gamma_n)^{-1} g(\mathbf{m}-\mathbf{m}'). \quad (3.6b)$$

In the one-dimensional case,
$$g(j) = \begin{cases} -\tfrac{1}{2}\operatorname{csch} y \exp(-|j|y) & \text{if } f^2 < 0 \\ \tfrac{1}{2}\operatorname{csch} y \exp\{-|j|(y+\pi i)\} & \text{if } f^2 > 1, \end{cases} \quad (3.7)$$
where $f = \omega/\omega_L$ and
$$\cosh y = |2f^2 - 1|. \quad (3.7a)$$

The case $0 < f^2 < 1$ corresponds to scattering problems and will not interest us here.

The elements of the inverse of the matrix for a two-dimensional square lattice has the form
$$a^{(-1)}(\mathbf{s}_1, \mathbf{s}_2) = \frac{\mu_{st}}{(2\pi)^2} \int_0^{2\pi}\!\!\int \frac{\exp(i\mathbf{s} \cdot \boldsymbol{\varphi}) d\varphi_1 d\varphi_2}{4b^2 + 2\gamma_1 \cos\varphi_1 + 2\gamma_2 \cos\varphi_2}, \quad (3.8)$$
where
$$\mu_{st} = \begin{cases} -1 & \text{if } f^2 = (\omega/\omega_L)^2 < 0 \\ (-1)^{s+t} & \text{if } f^2 > 1, \end{cases}$$
$$4b^2 = |M\omega^2 - 2(\gamma_1 + \gamma_2)|.$$

This integral can be expressed in terms of generalized hypergeometric functions of two variables.[6] This form is not particularly useful for our purpose. However, when $s_1 = s_2 = s$ a relatively simple expression exists for $a^{(-1)}(s,s)$:
$$a^{(-1)}(s,s) = \frac{\mu_{ss}}{(2\pi)^2} \int_0^\infty dx \int\!\!\int_0^{2\pi} \times \exp(-4b^2 x - 2x\gamma_1 \cos\varphi_1 - 2x\gamma_2 \cos\varphi_2) \times \exp(i\mathbf{s} \cdot \boldsymbol{\varphi}) d\varphi_1 d\varphi_2 \quad (3.9a)$$
$$= \frac{\mu_{ss}}{(2\pi)^2} \int_0^\infty \exp(-4b^2 x) I_s(2x\gamma_1) I_s(2x\gamma_2) dx$$
$$= \frac{\beta\mu_{ss}}{(\gamma_1 + \gamma_2)\pi} Q_{s-\frac{1}{2}}(1 + 2\beta^2 f^2 [f^2 - 1]),$$

where $Q_n(z)$ is the nth Legendre function of the second kind and
$$\beta = (\gamma_1 + \gamma_2)/(\gamma_1 \gamma_2)^{\frac{1}{2}}. \quad (3.9b)$$

An asymptotic expression for $a^{(-1)}(s_1, s_2)$ can be obtained when
$$s_1^2 \gamma_1^{-1} + s_2^2 \gamma_2^{-1}$$
is very large. However, since in the general case $a^{(-1)}(s_1, s_2, \cdots, s_n)$ can be considered as simply $a^{(-1)}(s_1, s_2)$, we proceed with the general case to find the asymptotic expression for $a^{(-1)}(s_1, s_2, \cdots, s_n)$ when
$$S = [s_1^2 \gamma_1^{-1} + \cdots + s_n^2 \gamma_n^{-1}]^{\frac{1}{2}} \quad (3.10)$$
is very large.

First let $\omega^2 < 0$. Then
$$a^{(-1)}(s_1, s_2, \cdots) = -\frac{1}{(2\pi)^n} \int_{-\pi}^{\pi} \cdots \int \times \frac{\exp(i\mathbf{s} \cdot \boldsymbol{\varphi}) d^n\varphi}{-M\omega^2 + 2(\gamma_1 + \cdots) - 2\gamma_1 \cos\varphi_1 - \cdots}. \quad (3.11)$$

When S is large, the integrand oscillates very rapidly except in the region of $|\varphi|$ close to the origin. Hence we can expand each of the cosines as a power series in φ and retain only the first few terms. Also, since the remote regions in φ space contribute practically nothing to the integral, after this expansion is made we can integrate over the entire φ space without significantly changing the integral (in the limit as $S \to \infty$). Hence
$$a^{(-1)}(s_1, s_2, \cdots) \simeq -\frac{1}{(2\pi)^n} \int_{-\infty}^{\infty} \cdots \int \times \frac{\exp i[(s_1 \gamma_1^{-\frac{1}{2}} \cdot \varphi_1 \gamma_1^{\frac{1}{2}}) + \cdots] d\varphi_1 \cdots d\varphi_n}{-M\omega^2 + \gamma_1 \varphi_1^2 + \cdots + \gamma_n \varphi_n^2}. \quad (3.12)$$

If we introduce new coordinates $x_j = \varphi_j \gamma_j^{\frac{1}{2}}$, the integral becomes an n-fold Fourier transform of a function of $r^2 = x_1^2 + \cdots + x_n^2$. Such integrals have been discussed by Bochner[8] and lead to the following result after a transformation to polar coordinates:
$$a^{(-1)}(s_1, s_2, \cdots) \simeq -\frac{(2\pi)^{\frac{1}{2}n}}{(2\pi)^n (\gamma_1 \gamma_2 \cdots \gamma_n)^{\frac{1}{2}} S^{\frac{1}{2}(n-2)}} \times \int_0^\infty r^{\frac{1}{2}n} [r^2 + (-M\omega^2)]^{-1} J_{\frac{1}{2}(n-2)}(Sr) dr. \quad (3.13)$$

Here the J function is a Bessel function of order $\frac{1}{2}(n-2)$, while the integral is a Hankel transform which

---

[8] S. Bochner, *Vorlesungen über Fouriersche Integral* (Chelsea Publishing Company, New York, 1948), p. 187.

is well known.[9] One finds

$$a^{(-1)}(s_1,s_2,\cdots) \simeq -\frac{(-M\omega^2)^{\frac{1}{2}(n-2)}}{(\gamma_1\cdots\gamma_n)^{\frac{1}{2}}(2\pi)^{\frac{1}{2}n}S^{\frac{1}{2}(n-2)}}$$
$$\times K_{\frac{1}{2}(n-2)}([-M\omega^2]^{\frac{1}{2}}S), \quad (3.14)$$

where $K_\nu(z)$ is that Bessel function which is commonly referred to as the K function. In particular,

$$K_{\frac{1}{2}}(z) = (\pi/2z)^{\frac{1}{2}}e^{-z},$$

while as $z \to \infty$

$$K_\nu(z) \sim (\pi/2z)^{\frac{1}{2}}e^{-z}$$

for all positive ν if $|\arg z| < \frac{3}{2}\pi$. Finally as $S \to \infty$ (with $\omega^2 < 0$), we have, when $n \geq 3$,

$$a^{(-1)}(s_1,s_2,\cdots) \sim -\frac{(-M\omega^2)^{\frac{1}{2}(n-3)}}{(2\pi)^{\frac{1}{2}n}(\gamma_1\cdots\gamma_n)^{\frac{1}{2}}}\frac{(\frac{1}{2}\pi)^{\frac{1}{2}}}{S^{\frac{1}{2}(n-1)}}$$
$$\times \exp\{-(-M\omega^2)^{\frac{1}{2}}S\}. \quad (3.15)$$

One can determine $a^{(-1)}(s_1,s_2,\cdots)$ in a similar manner when $(2f^2-1) > 1$. We replace the φ_j's by $\varphi_j + \pi$ to obtain

$$a^{(-1)}(s_1,s_2,\cdots) = \frac{(-1)^{s_1+\cdots+s_n}}{(2\pi)^n}\int_{-\pi}^{\pi}\cdots\int$$
$$\times \frac{\exp(i\boldsymbol{\varphi}\cdot\mathbf{s})d^n\varphi}{M\omega^2 - 2(\gamma_1 + \gamma_2 + \cdots) - 2\gamma_1\cos\varphi_1 + \cdots}$$

and (when $\omega^2 > \omega_L^2$)

$$a^{(-1)}(s_1,s_2,\cdots) \sim \frac{[M(\omega^2 - \omega_L^2)]^{\frac{1}{2}(n-2)}}{(\gamma_1\gamma_2\cdots\gamma_n)^{\frac{1}{2}}(2\pi)^{\frac{1}{2}n}S^{\frac{1}{2}(n-2)}}$$
$$\times K_{\frac{1}{2}(n-2)}([M(\omega^2-\omega_L^2)]^{\frac{1}{2}}S), \quad (3.16)$$

by using the arguments given in the foregoing.

It is to be noted that if we let

$$as = r, \quad k = \varphi/a,$$

a being the lattice spacing, (3.11) becomes

$$a^{(-1)}(\mathbf{r}_1,\mathbf{r}_2,\cdots) = -\frac{a^n}{(2\pi)^n}\int_{-\pi/a}^{\pi/a}\cdots\int$$
$$\times \frac{\exp(i\mathbf{r}\cdot\mathbf{k})d^n k}{-M\omega^2 + 2(\gamma_1 + \cdots) - \sum 2\gamma_j \cos a k_j},$$

so that, as $a \to 0$ in the continuum limit, (3.14) corresponds to an *exact* rather than asymptotic expression

---

[9] Erdelyi, Magnus, Oberhettinger, and Tricomi, *Tables of Integral Transforms* (McGraw-Hill Book Company, Inc., New York, 1954), Vol. 2, p. 23.

and

$$a^{(-1)}(\mathbf{r}_1,\cdots,\mathbf{r}_n) = -\frac{a^{\frac{1}{2}(n-2)}(-M\omega^2/\gamma_1)^{\frac{1}{2}(n-2)}\gamma_1^{n/2}}{(\gamma_1\cdots\gamma_n)^{\frac{1}{2}}R^{\frac{1}{2}(n-2)}(2\pi)^{\frac{1}{2}n}}$$
$$\times K_{\frac{1}{2}(n-2)}([-M\omega^2/\gamma_1]^{\frac{1}{2}}R/a), \quad (3.17a)$$

where now we define R (with units of length) by

$$R = [r_1^2 + r_2^2(\gamma_1/\gamma_2) + \cdots + r_n^2(\gamma_1/\gamma_n)]^{\frac{1}{2}}. \quad (3.17b)$$

Section 6 will be devoted to a discussion of the interaction of defects with crystal boundaries. For this purpose we shall record the inverses $a^{(-1)}(m,m')$ which correspond to rigid and free boundaries. We shall sketch the manner in which results were obtained by examining one-dimensional chains.

Let us consider a chain of $N+2$ masses with the end two held fixed at their equilibrium positions (the rigid boundary case). This corresponds to boundary conditions of (3.2a) (with $n=1$):

$$u(0) = u(N+1) = 0.$$

The components of the jth characteristic vector of the matrix whose elements are the coefficients of the u's in (3.2a) are

$$u_j(m) = [2/(N+1)]^{\frac{1}{2}}\sin[mj\pi/(N+1)],$$

the associated characteristic value being

$$\lambda_j = M\omega^2 - 2\gamma\{1 - \cos[j\pi/(N+1)]\}.$$

Hence the elements of the required inverse matrix are

$$a^{(-1)}(m;m')$$
$$= \frac{2}{N+1}\sum_{j=1}^{N}\frac{\sin[mj\pi/(N+1)]\sin[m'j\pi/(N+1)]}{M\omega^2 - 2\gamma\{1 - \cos[j\pi/(N+1)]\}}. \quad (3.18a)$$

The n-dimensional "rigid boundary" inverse is

$$a^{(-1)}(\mathbf{m};\mathbf{m}') = \left(\frac{2}{N+1}\right)^n \sum_{s_1=1}^{N}\cdots\sum_{s_n=1}^{N}$$
$$\times \frac{\prod_{k=1}^{n}\left\{\sin\frac{m_k s_k \pi}{N+1}\sin\frac{m'_k s_k \pi}{N+1}\right\}}{M\omega^2 - 2\sum_1^n \gamma_k\{1 - \cos[s_k\pi/(N+1)]\}}. \quad (3.18b)$$

The boundary conditions at a free boundary are

$$u(1) - u(0) = u(N) - u(N+1) = 0.$$

This is obtained by noting that the equation appropriate for an end-particle displacement $u(1)$ is

$$u(1)[M\omega^2 - \gamma] + \gamma u(2) = 0,$$

or

$$u(1)[M\omega^2 - 2\gamma] + \gamma u(2) + \gamma u(0) = 0,$$

the standard form if

$$u(1) - u(0) = 0.$$

The characteristic vectors which satisfy the boundary conditions have components

$$u_0(m) = (1/N)^{\frac{1}{2}},$$
$$u_j(m) = (2/N)^{\frac{1}{2}} \cos[(2m-1)\pi j/2N]$$
$$\text{if } j = 1, 2, \cdots, N-1.$$

with characteristic values

$$\lambda_j = M\omega^2 - 2\gamma[1 - \cos(\pi j/N)], \quad j = 0, 1, \cdots, N-1.$$

The elements of the required inverse are

$$a^{(-1)}(m; m') = \frac{2}{N} \sum_{j=1}^{N-1}$$
$$\times \frac{\cos[(2m-1)\pi j/2N] \cos[(2m'-1)\pi j/2N]}{M\omega^2 - 2\gamma[1 - \cos(\pi j/N)]} + \frac{1}{NM\omega^2}, \quad (3.19a)$$

with an n-dimensional generalization

$$a^{(-1)}(m; m') = \left(\frac{2}{N}\right)^n \sum_{s_1=1}^{N-1} \sum_{s_1=1}^{N-1} \frac{\prod_1^n \{\cos[(2m_k-1)\pi s_k/2N] \cos[(2m_k'-1)\pi s_k/2N]\}}{M\omega^2 - 2\sum_1^n \gamma_k[1 - \cos(\pi s_k/N)]} + O\left(\frac{1}{N}\right). \quad (3.19b)$$

The elements of the inverse matrix $a^{(-1)}(m; m')$ associated with a two-component system with nearest-neighbor interactions only can be discussed in a similar manner. The one-dimensional case will be developed in detail and the results merely stated for the general n-dimensional lattice. We postulate the even-numbered particles on our chain to be of mass M and the odd-numbered ones of mass m. Then the analog of (3.2a) is two sets of equation

$$\gamma u(2j+1) + (M\omega^2 - 2\gamma)u(2j) + \gamma u(2j-1) = 0, \quad (3.20a)$$
$$\gamma u(2j+2) + (m\omega^2 - 2\gamma)u(2j+1) + \gamma u(2j) = 0. \quad (3.20b)$$

If we let

$$v(2j) = (M\omega^2 - 2\gamma)^{\frac{1}{2}} u(2j), \quad (3.21)$$
$$v(2j+1) = (m\omega^2 - 2\gamma)^{\frac{1}{2}} u(2j+1), \quad (3.22)$$

we obtain the more compact single set of equations:

$$\gamma v(j-1) + (M^*\omega^2 - 2\gamma)v(j) + \gamma v(j+1) = 0, \quad (3.23)$$

where the mass M^* is defined by

$$\omega^2 M^* = 2\gamma + [(M\omega^2 - 2\gamma)(m\omega^2 - 2\gamma)]^{\frac{1}{2}}. \quad (3.24)$$

Clearly, if $m = M$ these equations reduce to the one-component ones and $M^* = M$.

It can be shown that in the n-dimensional case the new single set of equations is the same form as (3.2a) with the mass replaced by M^*, with

$$\omega^2 M^* = 2(\gamma_1 + \cdots + \gamma_n) + [(M\omega^2 - 2\gamma_1 - \cdots - 2\gamma_n) \times (m\omega^2 - 2\gamma_1 - \cdots - 2\gamma_n)]^{\frac{1}{2}}. \quad (3.25)$$

The normal mode frequencies of an n-dimensional monatomic lattice are

$$M\omega^2 = 2 \sum \gamma_j (1 - \cos\varphi_j), \quad \varphi_j = 2\pi s_j/N,$$

the s_j's being integers. In our diatomic lattice

$$M^*\omega^2 = 2 \sum \gamma_j(1 - \cos\varphi_j). \quad (3.26)$$

If we substitute (3.22) into this equation and solve for ω^2, we find two branches:

$$\omega^2 = (\gamma_1 + \cdots + \gamma_n)(M+m)/Mm$$
$$\pm (mM)^{-1}[(\gamma_1 + \cdots + \gamma_n)^2(M-m)^2$$
$$+ 4mM(\sum_j \gamma_j \cos\varphi_j)^2]^{\frac{1}{2}}. \quad (3.27)$$

If $M > m$, the largest frequency ω_L^2 is given by

$$\omega_L^2 = 2(\gamma_1 + \cdots + \gamma_n)(M+m)/Mm. \quad (3.28a)$$

The top edge of the lower band is at

$$\omega_1^2 = 2(\gamma_1 + \cdots + \gamma_n)/M, \quad (3.28b)$$

while the lower edge of the top band is at

$$\omega_2^2 = 2(\gamma_1 + \cdots + \gamma_n)/m. \quad (3.28c)$$

The Green's function (3.15) is valid for the diatomic lattice if $\omega^2 < 0$ and M is replaced by M^*.

4. CHARACTERIZATION OF DEFECTS

As in D-1, we shall be concerned mainly with changes in masses and force constants but not in equilibrium positions. If the mass of the particle at lattice point $\alpha = (\alpha_1, \alpha_2, \cdots, \alpha_n)$ is changed from M to M_α, Eq. (3.2a) with $m = \alpha$ can be put in the appropriate form by adding the term

$$\omega^2(M_\alpha - M)u(\alpha_1, \alpha_2, \cdots)$$

to the left-hand side. Then, we set (see 2.2a)

$$\delta_\alpha = -\omega^2 M \epsilon_\alpha, \quad (4.1)$$

where we define

$$\epsilon_\alpha = 1 - (M_\alpha/M). \quad (4.2)$$

The single mass defect function $D(\alpha; \omega)$ is then

$$D(\alpha; \omega) = D_0(\omega)\{1 - \omega^2 M \epsilon_\alpha a^{(-1)}(\alpha, \alpha)\}. \quad (4.3)$$

The characterization of a mass defect in a diatomic lattice is obtained from (3.20). Let the heavy mass M be replaced by M_α. Then Eq. (3.20a) has a correction term

$$\omega^2(M_\alpha - M)u(\alpha).$$

After the transformation (3.21) is made the new equation in $v(\alpha)$ has the term

$$-M\omega^2 \epsilon_\alpha (m\omega^2 - 2\gamma)^{\frac{1}{2}}/(M\omega^2 - 2\gamma)^{\frac{1}{2}}$$

added to the left-hand side of (3.23).

In general, the mass defect function is

$$D(\alpha; \omega) = D_0(\omega)\{1 - \omega^2 a^{(-1)}(\alpha; \alpha)$$
$$\times M\epsilon_\alpha[(m\omega^2 - 2\gamma)/(M\omega^2 - 2\gamma)]^{\frac{1}{2}}\} \quad (4.4)$$

if M is normal mass at α, and
$$D(\alpha;\omega)=D_0(\omega)\{1-\omega^2 a^{(-1)}(\alpha;\alpha) \\ \times m\epsilon_\alpha[(M\omega^2-2\gamma)/(m\omega^2-2\gamma)]^{\frac{1}{2}}\} \quad (4.5)$$
if m is normally at α. Generally,
$$\epsilon_\alpha=\left(1-\frac{\text{defect mass at }\alpha}{\text{normal mass at }\alpha}\right). \quad (4.6)$$

The defect in force constant as well as mass is first discussed in the one-dimensional monatomic case. Let the force constant associated with the interaction of α with $\alpha+1$ and $\alpha-1$ be changed from γ to γ'. Then (3.2a) becomes (in the cases $m=\alpha-1, \alpha, \alpha+1$):

$$[\omega^2 M - (\gamma'+\gamma)]u(\alpha-1)+\gamma'u(\alpha)+\gamma u(\alpha-2)=0,$$
$$[\omega^2 M_\alpha - 2\gamma']u(\alpha)+\gamma'u(\alpha-1)+\gamma'u(\alpha+1)=0,$$
$$[\omega^2 M - (\gamma'+\gamma)]u(\alpha+1)+\gamma'u(\alpha)+\gamma u(\alpha+2)=0.$$

If we replace $u(\alpha)$ by a new variable $(\gamma/\gamma')v(\alpha)$, the determinant of the coefficients of the u's and v is of the form (2.1) with additions to elements along the main diagonal:

$$\delta_{\alpha-1}=\delta_{\alpha+1}=-\gamma\tau_\alpha/(1-\tau_\alpha), \quad (4.7a)$$
$$\delta_\alpha=M\omega^2[(1-\epsilon_\alpha)(1-\tau_\alpha)^2-1]+2\gamma\tau_\alpha, \quad (4.7b)$$
where
$$\tau_\alpha=1-(\gamma/\gamma'). \quad (4.8)$$

If both the mass and the force constant are changed, three consecutive diagonal elements are changed in D_0.

If defects exist at $\alpha, \beta, \cdots$, the appropriate value of the new M's and γ's are substituted into (4.1–4.5) at the appropriate diagonal elements in (2.1).

Equation (4.7) is still valid in the n-dimensional case when $\gamma_1=\gamma_2=\gamma_3=\cdots$ and when the force constants between the αth particle and its nearest neighbors are all changed to γ'. However, if $\gamma_1\neq\gamma_2, \gamma_3, \cdots$, and only force constants in the m_1 direction are changed, the normal mode determinant (2.1) is changed in several off-diagonal elements as well as along the main diagonal. The functions (2.4) and (2.5) are somewhat more complicated but can be easily found.

In the n-dimensional case, with $\gamma_1=\gamma_2=\cdots$, a defect at $(\alpha,\beta,\cdots)$ yields
$$\delta_{\alpha,\beta,\cdots}=M\omega^2[(1-\epsilon_{\alpha,\beta,\cdots})(1-\tau_{\alpha,\beta,\cdots})^2-1] \\ +2n\gamma\tau_{\alpha,\beta,\cdots}, \quad (4.9)$$
$$\delta_{\alpha\pm 1,\beta,\cdots}=\delta_{\alpha,\beta\pm 1,\cdots}=\cdots=-\gamma\tau_{\alpha\beta\cdots}/(1-\tau_{\alpha\beta\cdots}), \quad (4.10)$$
where the defect mass at $(\alpha,\beta,\cdots)$ is $M_{\alpha,\beta,\cdots}$ and
$$\epsilon_{\alpha,\beta,\cdots}=(1-M^{-1}M_{\alpha,\beta,\cdots}), \quad (4.11)$$
while a change of the force constant to $\gamma_{\alpha,\beta,\cdots}$ between $(\alpha,\beta,\cdots)$ and its nearest neighbor yields
$$\tau_{\alpha,\beta,\cdots}=(1-\gamma\gamma_{\alpha,\beta,\cdots}^{-1}). \quad (4.12)$$

A defect which we shall discuss later, but which corresponds to neither a mass nor force constant change, is the defect "source" at α which we characterize by
$$\delta_\alpha=\kappa\gamma=\text{constant independent of }\omega.$$

5. SELF-ENERGY AND INTERACTION ENERGY OF DEFECTS

In this section, the formulas derived above will be applied to the calculation of the self-energy and interaction energies of various defects in monatomic and diatomic lattices.

The simplest type of defect is the source defect described by (4.13). Although it does not correspond to any attainable defect in a crystal lattice, we shall discuss it first to demonstrate the ideas involved in making more complicated calculations. We shall show at the end of this section that source defects are mathematically equivalent to holes in lattices if one is concerned with the interaction of holes separated by many lattice spacings.

The self-energy of a source defect of strength $\kappa\gamma$ is given by
$$\Delta E_S=\frac{\hbar}{4\pi i}\int_C \omega d\log[1+\kappa\gamma a^{(-1)}(\alpha,\alpha;\omega)], \quad (5.1)$$
where use has been made of (1.6a) and (4.13). From (3.6a),
$$a^{(-1)}(\alpha,\alpha;\omega) \\ =\frac{1}{(2\pi)^n}\int_0^{2\pi}\cdots\int\frac{d^n\varphi}{M\omega^2-2\sum_1^n\gamma_j(1-\cos\varphi_j)}.$$

The contour C has to contain the positive real axis of ω since the frequencies of interest are positive real numbers; it may be chosen to be the counter-clockwise contour about the right half-plane. Then the only non-vanishing contribution to (5.1) is the integration down the imaginary axis. Since the integrand is an even function, the logarithmic term being a function of ω^2, the integral reduces to
$$\Delta E_S=-\frac{\hbar}{2\pi}\int_0^\infty \omega d\log[1+\kappa\gamma a^{(-1)}(\alpha,\alpha;i\omega)].$$

In the case of the one-dimensional lattice, this integral can be evaluated in terms of elementary functions. The inverse $a^{(-1)}(\alpha,\alpha;i\omega)$ is given by $-1/\{4\gamma f(1-f^2)^{\frac{1}{2}}\}$ (where $f=\omega/\omega_L$), so that
$$\Delta E_S=-\frac{\hbar\omega_L}{2\pi}\int_0^\infty fd\log\{1-\tfrac{1}{4}\kappa[1/f(1-f^2)^{\frac{1}{2}}]\}.$$

If we let $f=\tan\vartheta$, we obtain after some manipulation
$$\Delta E_S=-\frac{\hbar\omega_L}{2\pi}\left\{\frac{\pi}{2}-\alpha\int_0^{\pi/2}\frac{d\vartheta}{\alpha-\sin\vartheta}-\int_0^{\pi/2}\frac{d\vartheta}{1+\alpha\sin\vartheta}\right\}, \quad (5.2a)$$
where α is related to κ by
$$\kappa=4\alpha/(1-\alpha^2), \quad -1<\alpha<1, \quad (5.2b)$$

so that by letting α range from -1 to $+1$, κ ranges from $-\infty$ to $+\infty$. Hence

$$\Delta E_S = -\frac{\hbar \omega_L}{2\pi}\left\{\frac{\pi}{2} + \frac{\alpha}{(1-\alpha^2)^{\frac{1}{2}}}\log\left|\frac{1+(1-\alpha^2)^{\frac{1}{2}}}{\alpha}\right| - (1-\alpha^2)^{-\frac{1}{2}}\cos^{-1}\alpha\right\}. \quad (5.3a)$$

As $|m-m'|\to\infty$, $a^{(-1)}(m,m';i\omega)\to 0$ so that at great distances we obtain (after integrating by parts), in the weak defect limit as κ and $\kappa'\to 0$,

$$\Delta E_I \simeq -\frac{\hbar\kappa\kappa'}{2\pi}\int_0^\infty \gamma^2 a^{(-1)}(m,m';i\omega)a^{(-1)}(m',m;i\omega)d\omega. \quad (5.3c)$$

Since

$$a^{(-1)}(m,m';i\omega) = a^{(-1)}(m',m;i\omega) \sim (M\omega_L^2)^{\frac{1}{2}(n-3)}$$
$$\times f^{\frac{1}{2}(n-3)}(\tfrac{1}{2}\pi)^{\frac{1}{2}}(2\pi\gamma)^{-\frac{1}{2}n}S^{\frac{1}{2}(n-1)}\exp\{-(M\omega_L^2)^{\frac{1}{2}}fS\},$$

where

$$S = \gamma^{-\frac{1}{2}}[s_1^2 + s_2^2 + \cdots + s_n^2]^{\frac{1}{2}},$$
$$s_j = m_j' - m_j, \quad \text{and} \quad M\omega_L^2 = 4n\gamma$$

we find

$$\Delta E_I \sim -\frac{\hbar\kappa\kappa'\gamma^2}{4(2\pi\gamma)^n}\frac{M^{\frac{1}{2}(n-3)}\omega_L^{n-2}}{s^{n-1}}\int_0^\infty f^{n-3}$$
$$\times \exp(-2M^{\frac{1}{2}}\omega_L Sf)df \quad (5.4)$$
$$= -\frac{\hbar\omega_L\kappa\kappa'(n-3)!}{2(4\pi)^n n^{\frac{1}{2}}|m'-m|^{2n-3}}.$$

In particular, if $n=3$,

$$\Delta E_I \simeq -\hbar\omega_L\kappa\kappa'/2(4\pi)^3\sqrt{3}|m'-m|^3; \quad (5.5)$$

or, if we let a be our lattice spacing and $R = a|m'-m|$, then

$$\Delta E_I \simeq -\hbar\omega_L\kappa\kappa'a^3/2(4\pi)^3\sqrt{3}R^3. \quad (5.6)$$

Two sources or "sinks" (we call the case $\kappa<0$ a sink)

$$\Delta E_I = -\frac{\hbar}{2\pi}\int_0^\infty \omega d\log\left[1 - \epsilon_\alpha\epsilon_\beta M^2\omega^4\left\{\frac{a^{(-1)}(\beta,\alpha;i\omega)a^{(-1)}(\alpha,\beta;i\omega)}{[1+\epsilon_\alpha M\omega^2 a^{(-1)}(\alpha,\alpha;i\omega)][1+\epsilon_\beta M\omega^2 a^{(-1)}(\beta,\beta;i\omega)]}\right\}\right]; \quad (5.10)$$

or, for ϵ_α, $\epsilon_\beta\to 0$ (in the general case the integration is more difficult but can be carried out either numerically or through various series expansions),

$$\Delta E_I \simeq -\frac{\hbar\epsilon_\alpha\epsilon_\beta}{2\pi}\int_0^\infty \omega d[M^2\omega^4 a^{(-1)}(\alpha,\beta;i\omega)a^{(-1)}(\beta,\alpha;i\omega)], \quad (5.11)$$

which after integrating by parts simplies to

$$\Delta E_I \approx -\frac{\hbar\epsilon_\alpha\epsilon_\beta}{2\pi}\int_0^\infty M^2\omega^4 a^{(-1)}(\alpha,\beta;i\omega)a^{(-1)}(\beta,\alpha;i\omega)d\omega. \quad (5.12)$$

The interaction energy between two source defects separated by a great distance can be obtained as follows for all numbers of dimensions ≥ 3. The case $n<3$ must be handled in a slightly different manner and will be omitted here. We consider the special case $\gamma_1+\gamma_2=\cdots =\gamma$. If one defect of strength κ is at $m=(m_1,m_2,\cdots)$ and the other κ' at $m'=(m_1',m_2',\cdots)$ we find

$$\Delta E_I = -\frac{\hbar}{2\pi}\int_0^\infty \omega d\log\left\{1 - \frac{\kappa\kappa'\gamma^2 a^{(-1)}(m,m';i\omega)a^{(-1)}(m',m;i\omega)}{[1+\gamma\kappa a^{(-1)}(m,m;i\omega)][1+\gamma\kappa' a^{(-1)}(m',m';i\omega)]}\right\}. \quad (5.3b)$$

attract each other, while a source and sink repel each other, with an energy of interaction inversely proportional to R^3.

A more realistic example is that of the isotopic defects in a lattice. We examine the behavior of such defects in both monatomic and diatomic lattices.

(a) Monatomic Lattice

The self-energy of an isotope of mass $(1-\epsilon_\alpha)M$ is given by

$$\Delta E_S = \frac{\hbar}{4\pi i}\int_C \omega d\log[1 - \epsilon_\alpha M\omega^2 a^{(-1)}(\alpha,\alpha;\omega)]. \quad (5.7)$$

As in the previous case, our integration can be carried from 0 to ∞:

$$\Delta E_S = -\frac{\hbar}{2\pi}\int_0^\infty \omega d\log[1 + \epsilon_\alpha M\omega^2 a^{(-1)}(\alpha,\alpha;i\omega)]. \quad (5.8)$$

In the case of the one-dimensional lattice, this integral can be evaluated explicitly; in fact using (3.7), (5.8) becomes

$$\frac{\Delta E_S}{\frac{1}{2}\hbar\omega_L} = -\frac{1}{\pi}\int_0^\infty fd\log[1 - \epsilon f(1+f^2)^{-\frac{1}{2}}] \quad (5.9a)$$
$$= \tfrac{1}{2}[(1-\epsilon^2)^{-\frac{1}{2}}-1] + \pi^{-1}(1-\epsilon^2)^{-\frac{1}{2}}\sin^{-1}\epsilon, \quad (5.9b)$$

in agreement with D-1, Eq. (4.15).

A similar analysis gives for the more interesting case of the interaction of two isotopes the formula

For the one-dimensional lattice, (5.12) gives

$$\frac{\Delta E_I}{\frac{1}{2}\hbar\omega_L} \approx \frac{\epsilon_\alpha\epsilon_\beta}{\pi}\int_0^\infty \frac{[(1+f^2)^{\frac{1}{2}}-f]^{2|\alpha-\beta|}}{f^2(1+f^2)} \quad (5.13)$$
$$= -\frac{\epsilon_\alpha\epsilon_\beta}{2\pi}\left\{\frac{4|\alpha-\beta|}{16(\alpha-\beta)^2-\frac{1}{4}} - \psi(\tfrac{3}{4}+2|\alpha-\beta|) + \psi(\tfrac{1}{4}+2|\alpha-\beta 1)\right\} \quad (5.14)$$

in agreement with D-1 (5.31).

Of more interest is the expression for the interaction energy for large distances. Then (3.15) may be inserted in (5.12) to yield

$$\Delta E_I \sim \frac{\hbar \epsilon_\alpha \epsilon_\beta}{2\pi} \int_0^\infty M^2 \omega^4 \frac{(M\omega^2)^{\frac{1}{2}(n-3)}}{(2\pi)^n (\gamma_1 \gamma_2 \cdots \gamma_n)^{\frac{1}{2}}} \frac{\pi}{2} S^{1-n}$$
$$\times \exp(-2\omega M^{\frac{1}{2}} S) d\omega \quad (5.15)$$

or

$$\frac{\Delta E_I}{\frac{1}{2}\hbar\omega_L} \sim \frac{-\epsilon_\alpha \epsilon_\beta}{4(\gamma_1 + \cdots + \gamma_n)^{\frac{1}{2}}(2\pi)^n (\gamma_1 \gamma_2 \cdots \gamma_n)^{\frac{1}{2}} S^{n-1}}$$
$$\times \int_0^\infty k^{n+1} e^{-2kS} dk$$
$$= \frac{-\epsilon_\alpha \epsilon_\beta (n+1)! S^{-(2n+1)}}{16(\gamma_1 + \cdots + \gamma_n)^{\frac{1}{2}}(\gamma_1 \gamma_2 \cdots \gamma_n)^{\frac{1}{2}}(4\pi)^n}. \quad (5.16)$$

For the case $\gamma_1 = \gamma_2 = \cdots = \gamma_n = \gamma$,

$$\gamma^{\frac{1}{2}} S = (s_1^2 + s_2^2 + \cdots + s_n^2)^{\frac{1}{2}} = R/a$$

and the energy of interaction is given by

$$\frac{\Delta E_I}{\frac{1}{2}\hbar\omega_L} \sim -\frac{\epsilon_\alpha \epsilon_\beta (n+1)! a^{2n+1}}{16 n^{\frac{1}{2}}(4\pi)^n R^{2n+1}}. \quad (5.17)$$

For one dimension, $\Delta E_I \propto R^{-3}$ as derived in D-1. For two dimensions $\Delta E_I \propto R^{-5}$ and for three dimensions $\Delta E_I \propto R^{-7}$.

An attraction exists when both M_α and M_β are larger or smaller than M (like defects), while a repulsion appears when $M_\alpha > M > M_\beta$ or $M_\alpha < M < M_\beta$ (unlike defects).

(b) Diatomic Lattice

The self-energy of an isotopic defect in a diatomic lattice of alternating masses m and M is given by (by a derivation similar to that for the monatomic lattice) by

$$\Delta E_S = -\frac{\hbar}{2\pi} \int_0^\infty \omega d \log[1 - D(\alpha; i\omega) a^{(-1)}(\alpha, \alpha; i\omega)], \quad (5.18)$$

where D is given by (4.4) or (4.5) and for the one-dimensional lattice

$$a^{(-1)}(\alpha, \alpha; i\omega) = -[(m\omega^2 + 2\gamma)(M\omega^2 + 2\gamma) - 4\gamma^2]^{-\frac{1}{2}}. \quad (5.19)$$

For an isotope of mass $(1-\epsilon_\alpha)m$, this integral becomes

$$\frac{\Delta E_S}{\frac{1}{2}\hbar\omega_L} = -\frac{1}{\pi\omega_L} \int_0^\infty \omega d$$
$$\times \log\left[1 - \frac{\epsilon m \omega^2 \{(m\omega^2 + 2\gamma)(M\omega^2 + 2\gamma)\}^{\frac{1}{2}}}{(m\omega^2 + 2\gamma)\{(m\omega^2 + 2\gamma)(M\omega^2 + 2\gamma) - 4\gamma^2\}^{\frac{1}{2}}}\right]. \quad (5.20)$$

This integral has been evaluated in D-2 for two special cases, namely, when $M = m(1+\eta)$, η small and $M = \zeta^2 m$, ζ small. In the first case,

$$\frac{\Delta E_S}{\frac{1}{2}\hbar(4\gamma/m)^{\frac{1}{2}}} = -\frac{1}{2} + \frac{1}{\pi(1-\epsilon^2)}[\frac{1}{2}\pi + \sin^{-1}\epsilon]$$
$$+ \frac{\epsilon\eta}{4\pi}\left[\frac{\sqrt{2}}{1+\epsilon^2}\log(1+\sqrt{2}) - \frac{\epsilon\pi\sqrt{2}}{2(1+\epsilon^2)} + \frac{\pi}{2\epsilon}\right.$$
$$\left. - \frac{(1-\epsilon^2)}{\epsilon(1+\epsilon^2)}\{\frac{1}{2}\pi + \sin^{-1}\epsilon\}\right] + O(\eta^2) \quad (5.21)$$

and in the second case,

$$\frac{\Delta E_S}{\frac{1}{2}\hbar(2\gamma/m)^{\frac{1}{2}}} = -\frac{1}{2} + \frac{1}{\pi(1-\epsilon^2)^{\frac{1}{2}}}\{\frac{1}{2}\pi + \sin^{-1}\epsilon\} + \frac{\zeta\epsilon}{4(1-\epsilon)}$$
$$- \zeta^2\left[\frac{\epsilon}{2\pi(1-\epsilon^2)^{\frac{1}{2}}} + \frac{\epsilon^2}{2\pi(1-\epsilon^2)^{\frac{3}{2}}}\{\frac{1}{2}\pi + \sin^{-1}\epsilon\}\right] + O(\zeta^3). \quad (5.22)$$

For the interaction energy between two isotopes,

$$\Delta E_I \approx -\frac{\hbar}{2\pi} \int_0^\infty \delta_\alpha(i\omega)\delta_\beta(i\omega)[a^{(-1)}(\alpha,\beta; i\omega)]^2 d\omega, \quad (5.23)$$

where δ_α and δ_β are given by (4.4) or (4.5).
If

$$u^2 = [(m\omega^2 + 2\gamma)(M\omega^2 + 2\gamma)]^{\frac{1}{2}} - 2\gamma, \quad (5.24)$$

then the asymptotic expression for the element of the inverse matrix can be inserted in (5.23), giving

$$\Delta E_I \sim -\frac{\hbar}{2\pi} \int_0^\infty \delta_\alpha(i\omega)\delta_\beta(i\omega) \frac{\pi}{2(2\pi)^n}$$
$$\times \frac{u^{n-3} \exp(-2uS)}{(\gamma_1\gamma_2\cdots\gamma_n)S^{n-1}} \frac{d\omega}{du} du. \quad (5.25)$$

But, from (5.24),

$$\omega^2 = \frac{2}{m+M} u^2 + O(u^4) \quad (5.26)$$

and

$$\delta_\alpha(i\omega)\delta_\beta(i\omega) = \frac{4\sigma_\alpha\sigma_\beta}{(m+M)^2} u^4 + O(u^6), \quad (5.27)$$

where $\sigma = \epsilon m$ or ϵM is the change in mass of the isotope from the normal mass. As above, the integration can be carried out with the final result:

$$\frac{\Delta E_I}{\frac{1}{2}\hbar\omega_L} \sim -\frac{\sigma_\alpha\sigma_\beta(mM)^{\frac{1}{2}}}{2(m+M)^3}$$
$$\times \frac{(n+1)! S^{-(2n+1)}}{(4\pi)^n(\gamma_1\gamma_2\cdots\gamma_n)(\gamma_1+\gamma_2+\cdots+\gamma_n)^{\frac{1}{2}}}. \quad (5.28)$$

If $m=M$, then

$$\sigma_\alpha \sigma_\beta (mM)^{\frac{1}{2}}/2(m+M)^3 = \epsilon_1\epsilon_2/16 \qquad (5.29)$$

and (5.28) is obtained.

For $\gamma_1=\gamma_2=\cdots=\gamma_n$, $\gamma^{\frac{1}{2}}S=R/a$, (5.24) becomes

$$\frac{\Delta E_I}{\frac{1}{2}\hbar\omega_L} \simeq -\frac{\sigma_\alpha\sigma_\beta(mM)^{\frac{1}{2}}}{2(m+M)^3}\frac{(n+1)!}{(4\pi)^n n^{\frac{1}{2}}}\frac{a^{2n+1}}{R^{2n+1}}. \qquad (5.30)$$

In particular, when $n=3$ our interaction energy varies inversely as the 7th power of the distance.

An interesting consequence of (5.30) is obtained if we consider a pair of defects whose masses lie between M and m. If these defects move on the same sublattice (the lattice of M's or m's), an attraction between the defects results; whereas, if in their motion one defect remains on one sublattice and one on the other sublattice, a continuing repulsion exists.

The interaction of two impurities (or holes) which do not distort the lattice equilibrium positions can be discussed in the same manner as that of the isotopic defects. We examine for simplicity the special case $\gamma_1=\gamma_2=\cdots=\gamma$ which corresponds to equal central and noncentral force constants.

Consider two similar defects, one at lattice point $(0,0,0)$ and the other at (l,m,n), characterized by a mass M' and springs with force constants γ' connecting these lattice points to their nearest neighbors. The determinant $D(\omega)$ is changed by the addition of

$$\delta_1 = \gamma - \gamma' \qquad (5.31)$$

to the diagonal elements $(-1,0,0)$, $(1,0,0)$, $(0,-1,0)$, $(0,1,0)$, $(0,0,-1)$, $(0,0,1)$, $(l-1,m,n)$, $(l+1,m,n)$, $(l,m-1,n)$, $(l,m+1,n)$, $(l,m,n-1)$, $(l,m,n+1)$ and by the addition of

$$\delta_2 = (\omega/\gamma')^2(M'\gamma^2-\gamma'^2)+(6\gamma/\gamma')(\gamma'-\gamma) \qquad (5.32)$$

to diagonal elements $(0,0,0)$ and (l,m,n). The energy of interaction between two defects is

$$\Delta E_I = -\frac{\hbar}{2\pi}\int_0^\infty \omega d\log\left\{\begin{vmatrix} A & B \\ B & A \end{vmatrix}|A|^{-2}\right\}, \qquad (5.33)$$

where A and B are 7×7 matrices given in Appendix I. The product of the determinants can be simply expressed as

$$|I-(A^{-1}B)^2|$$

and this can be easily evaluated by taking account of the symmetry of A and B. However, the resulting expression is rather complicated and will not be exhibited here. It simplifies considerably if only the first term in the expansion in inverse powers of the distance r between the defects is required. In this approximation, A can be taken as the unit matrix and (5.33) becomes

$$\Delta E_I = -\frac{\hbar}{2\pi}\int_0^\infty \text{Trace} B^2 d\omega. \qquad (5.34)$$

The first term in $\text{Tr}(B^2)$ is

$$[(6/\gamma')(\gamma'-\gamma^2)]^2[a^{(-1)}(0,l;i\omega)]^2, \qquad (5.35)$$

so that E_I is precisely the expression (5.36) obtained for the interaction energy between two source defects with

$$\kappa = (6/\gamma\gamma')(\gamma'-\gamma)^2. \qquad (5.36)$$

We have used the invariance relation $a^{(-1)}(m,m';i\omega) = a^{(-1)}(0,l;i\omega)$ if $l=m-m'$. It is to be noted that to this first order of approximation the interaction energy is independent of M'. When we set $M'=0$, our defects become holes in the lattice. We then find that the interaction energy of two holes is attractive and varies as the inverse third power of the separation distance (5.5). Furthermore, the discussion of the interaction of a pair of holes is equivalent to that between a pair of source defects when the source strength κ is defined by (5.36).

6. INTERACTION OF DEFECTS WITH BOUNDARIES

As a first example of the interaction of a defect with a boundary, we consider an isotopic impurity m lattice spacings from the end of a chain. In the case of a *rigid boundary* (end of chain held fixed) the characteristic determinant for normal modes is obtained by combining (4.3), (3.18a), and (2.3):

$$D(m;\omega) = D_0(\omega)\left[1-2\omega^2 M \epsilon N^{-1}\sum_{j=1}^{N}\left\{\sin^2\left(\frac{m\pi j}{N+1}\right)\right\}\bigg/\left\{M\omega^2-2\gamma\left[1-\cos\left(\frac{\pi j}{N+1}\right)\right]\right\}\right],$$

which, in the limit as $N\to\infty$, becomes for the case $f^2<0$:

$$D(m;\omega) = D_0(\omega)\{1-\epsilon(\tanh\tfrac{1}{2}z)(1-e^{-2mz})\},$$
$$\text{rigid boundary}, \qquad (6.1)$$

with $\cosh z = 1-2f^2$. As usual, $\epsilon = 1-(M'/M)$ with M' being the impurity mass. The corresponding determinant in the case of a free boundary is obtained by combining (3.19a) with (2.3) and (4.3)

$$D(m;\omega) = D_0(\omega)\{1-\epsilon(\tanh\tfrac{1}{2}z)(1+e^{-(2m-1)z})\},$$
$$\text{free boundary}. \qquad (6.2)$$

It is to be noted that as the defect recedes from the boundary the m dependent exponentials vanish and (6.1) and (6.2) reduce to the ordinary single-defect result:

$$D(\infty;\omega) = D_0(\omega)[1-\epsilon\tanh\tfrac{1}{2}z]. \qquad (6.3)$$

Furthermore, the sign of the interaction terms are such that if a defect is attracted to a free boundary it is repelled from a rigid boundary and vice versa. The interaction is of $O(\epsilon)$ rather than ϵ^2; hence interactions of isotopes with boundaries are not of the "image" type which exist in electrostatics and hydrodynamics. The ratio $D(m;\omega)/D(\infty;\omega)$, which is to be substituted into (1.3) to find the interaction, is then

$$\frac{D(m;\omega)}{D(\infty;\omega)} = \begin{cases} 1 + \dfrac{\epsilon e^{-2mz}\tanh\tfrac{1}{2}z}{1-\epsilon\tanh\tfrac{1}{2}z}, & \text{rigid boundary} \\[6pt] 1 - \dfrac{\epsilon e^{-(2m+1)z}\tanh\tfrac{1}{2}z}{1-\epsilon\tanh\tfrac{1}{2}z}, & \text{free boundary.} \end{cases} \quad (6.4)$$

Then the interaction with a free boundary is

$$\Delta E_{FB} = -\frac{\hbar\omega_L}{2\pi}\int_0^\infty f d\log\left\{1 - \frac{\epsilon e^{-(2m+1)z}\tanh\tfrac{1}{2}z}{1-\epsilon\tanh\tfrac{1}{2}z}\right\}, \quad (6.5a)$$

while that with a rigid boundary is

$$\Delta E_{RB} = -\frac{\hbar\omega_L}{2\pi}\int_0^\infty f d\log\left\{1 + \frac{\epsilon e^{-2mz}\tanh\tfrac{1}{2}z}{1-\epsilon\tanh\tfrac{1}{2}z}\right\}. \quad (6.5b)$$

If we integrate by parts in the rigid boundary case, we find

$$\Delta E_{RB} = \frac{\hbar\omega_L}{4\pi}\int_0^\infty (\cosh\tfrac{1}{2}z)\log\left[1 + \frac{\epsilon e^{-2mz}\tanh\tfrac{1}{2}z}{1-\epsilon\tanh\tfrac{1}{2}z}\right]dz$$

$$\simeq \epsilon\frac{\hbar\omega_L}{4\pi}\int_0^\infty e^{-2mz}\sinh\tfrac{1}{2}zdz \quad \text{as} \quad m\to\infty \quad (6.6a)$$

$$= \epsilon\hbar\omega_L/32\pi m^2,$$

while as $m\to\infty$ in the free boundary case

$$\Delta E_{FB} \sim -\epsilon\hbar\omega_L/32\pi m^2. \quad (6.6b)$$

Hence, in both the free boundary and rigid boundary cases, the interaction energy is inversely proportional to the square of the distance of the defect from the boundary. The interaction is one of attraction if

$$M < M' < \infty \quad \text{for rigid boundary,}$$
$$0 < M' < M \quad \text{for free boundary;}$$

while it is repulsive if

$$0 < M' < M \quad \text{for rigid boundary,}$$
$$M < M' < \infty \quad \text{for free boundary.}$$

This is qualitatively expected. An interaction of an isotope with a rigid boundary is equivalent to that between the isotope and a particle of infinite mass. If both are heavier than a normal lattice atom, an attraction exists; when the isotopic defect is lighter, it interacts with the "infinite mass" boundary and a repulsion ensues. On the other hand, a free boundary is equivalent to a very light impurity at the end of the chain. Our above results are consistent with Eq. (5.16).

These qualitative results are valid in three dimensions as well. Without any extra difficulty we can discuss the isotopic defect in an n-dimensional lattice. We postulate

it to be m lattice spacings in the m_1 direction (x direction when $n=3$) from a rigid surface. It is easy to show from (3.18b) by choosing $m_1=m$, $m_2=m_3=\cdots=\tfrac{1}{2}(N+1)$ and letting $N\to\infty$, that the generalization of the rigid boundary equation (6.4) becomes (after replacing ω by $i\omega$)

$$\frac{D(m,i\omega)}{D(\infty,i\omega)} = 1 - \frac{\omega^2 M\epsilon I(m,i\omega)}{1 + M\epsilon\omega^2 I(\infty,i\omega)}, \quad (6.7)$$

where

$$I(m,i\omega) = -\left(\frac{1}{2\pi}\right)^n \int_{-\pi}^{\pi}\cdots\int \frac{\exp is\cdot\varphi\, d\varphi_1\cdots d\varphi_n}{M\omega^2 + 2\sum \gamma_k(1-\cos\varphi_k)},$$

$$S = \gamma_1^{-\frac{1}{2}}(2m,0,0\cdots,0).$$

When m is fairly large, (3.15) is applicable with ω replaced by $i\omega$. Then the analog of (6.5b) which is appropriate here yields

$$\Delta E_{RB} \simeq -\frac{\hbar}{2\pi}\int_0^\infty \omega d$$

$$\times\log\left\{1 + \frac{\epsilon(M\omega^2)^{\frac{1}{2}(n+1)}(\tfrac{1}{2}\pi)^{\frac{1}{2}}\exp[-S(M\omega^2)^{\frac{1}{2}}]}{(2\pi)^{\frac{1}{2}n}(\gamma_1\cdots\gamma_n)^{\frac{1}{2}}S^{\frac{1}{2}(n-1)}}\right\}$$

$$\sim \frac{\hbar\omega_L\epsilon(\tfrac{1}{2}\pi)^{\frac{1}{2}}[\tfrac{1}{2}(n+1)]!}{(\gamma_1\cdots\gamma_n)^{\frac{1}{2}}S^{n+1}(2\pi)^{\frac{1}{2}(n+2)}(M\omega_L^2)^{\frac{1}{2}}}, \quad (6.8)$$

with $S=2m/\gamma_1^{\frac{1}{2}}$. It is easily verified that the case $n=1$ is exactly the same as (6.6a).

The isotopic defect in a three-dimensional lattice is repelled from a rigid boundary when $\epsilon > 0$ (light isotopic defect) and

$$\Delta E_{RB} \sim \frac{\hbar\omega_L\epsilon(\gamma_1/m\omega_L^2)^{\frac{1}{2}}}{(\gamma_2\gamma_3/\gamma_1^2)^{\frac{1}{2}}(2\pi)^2(2m)^4}. \quad (6.9)$$

A free boundary attracts a light isotopic defect with a force of the same magnitude but opposite in sign.

We therefore find that not only should an ordering process exist at absolute zero[1,4] temperature but that a coating or "frosting" of light isotope should develop in a solid isotopic mixture, leaving the heavier atomic species inside. It would be interesting to leave a hydrogen-deuterium mixture in a liquid helium bath at low temperatures for a long period to observe whether the separation process would require days or years. Note that the energy of the boundary attraction diminishes as the inverse fourth power of the distance while the interatomic attraction energy [Eq. (5.17)] varies as the inverse seventh power.

If the state of perfect order is to exist at low temperatures, we should expect holes in a lattice to be attracted to a free boundary and hence expelled from a crystal. A repulsion from a rigid boundary should also exist. We consider these effects by using the source defect model. This defect was shown in the last section to be equivalent to a hole when one deals with the interaction of a pair of holes separated by a distance large compared to a lattice spacing. We can also expect the model to apply to the interaction of a hole with a distant boundary. We restrict our discussion to the case $\gamma_1=\gamma_2=\cdots=\gamma$ and first analyze the effect of a rigid boundary.

We recall from (4.13) that a source defect is characterized by $\delta=\kappa\gamma$ and an isotopic defect by $\delta=-\omega^2 M\epsilon$. If $\omega\to i\omega$, this becomes $\omega^2 M\epsilon$ so that the expression equivalent to (6.7) for a source defect is

$$\frac{D(m,i\omega)}{D(\infty,i\omega)}=1-\frac{\kappa\gamma I(m,i\omega)}{1+\kappa\gamma I(\infty,i\omega)}.$$

The analog of (6.8) is

$$\Delta E_{RB}\simeq -\frac{\hbar}{2\pi}\int_0^\infty \omega d$$

$$\times\log\left\{1+\frac{\kappa\gamma(M\omega^2)^{\frac{1}{2}(n-3)}(\frac{1}{2}\pi)^{\frac{1}{2}}\exp[-S(M\omega^2)^{\frac{1}{2}}]}{(2\pi\gamma)^{\frac{1}{2}n}S^{\frac{1}{2}(n-1)}}\right\}$$

$$\sim\frac{\hbar}{2\pi}\frac{\kappa\gamma(\frac{1}{2}\pi)^{\frac{1}{2}}}{(2\pi\gamma)^{\frac{1}{2}n}S^{\frac{1}{2}(n-1)}}\int_0^\infty (M\omega^2)^{\frac{1}{2}(n-3)}\exp[-(M\omega^2)^{\frac{1}{2}}S]d\omega$$

$$=\frac{\hbar\kappa\gamma(\pi/2M)^{\frac{1}{2}}[\frac{1}{2}(n-3)]!}{2\pi(2\pi\gamma)^{\frac{1}{2}n}S^{n-1}}, \quad (6.10)$$

with $S=2m\gamma^{-\frac{1}{2}}$ and $M\omega_L^2=4n\gamma$. The sign is changed when the interaction is with a free boundary. In the three-dimensional case,

$$\Delta E=\begin{cases}+3^{-\frac{1}{2}}\hbar\omega_L\kappa/(8\pi m)^2, & \text{rigid boundary}\\ -3^{-\frac{1}{2}}\hbar\omega_L\kappa/(8\pi m)^2, & \text{free boundary.}\end{cases} \quad (6.11)$$

The hole corresponds to $\kappa\gamma=6(\gamma'-\gamma)^2/\gamma'$ and is attracted to a free boundary as was expected.

7. REMARKS ON THE CONTINUUM LIMIT AND ANALOGIES WITH QUANTUM FIELD THEORY

We shall now observe some consequences of letting our lattice spacings vanish and show the similarity of the continuum limit of a lattice with holes to Wentzel's[5] pair theory of the interaction of neutrons and protons.

One generally starts an analysis of a quantum field with the introduction of the proper Hamiltonian. Let us consider the Hamiltonian of a continuous medium of density ρ with propagation velocity c and a set of fixed point source defects of strength $\lambda_1, \lambda_2, \cdots$ at $r_1, r_2, \cdots$.

$$H=\frac{1}{2}\rho\int\dot\varphi^2 d\tau+\frac{1}{2}\rho c^2\int(\nabla\varphi)^2 d\tau+\frac{1}{2}\frac{m^2c^4}{\hbar^2}\rho\int\varphi^2 d\tau$$

$$+\frac{1}{2}c^2\sum_{\alpha=1}^n\lambda_\alpha\int\delta(r-r_\alpha)\varphi^2(r)d\tau. \quad (7.1)$$

Here a mass m is associated with quanta propagated from one source to another. The λ_α's (which have units of length) represent the coupling strengths of the defects with the medium in which they are immersed. The δ function is defined by the property

$$\int \delta(r)f(r)d\tau=\rho f(0).$$

If we divide our continuum into a simple cubic lattice with unit cell cube edges a, introduce a mass M with a single degree of freedom into the center of each cell, couple it to its nearest neighbors by a spring of spring constant γ, associate each coupling constant γ_α to a dimensionless constant κ_α, and finally relate $\gamma, M,$ and κ_α to the constants of the medium and sources $\rho, c,$ and λ_α by

$$\rho=M/a^3, \quad c^2=\gamma a^2/M, \quad \text{and} \quad \lambda_\alpha=\kappa_\alpha a, \quad (7.2)$$

we obtain

$$H=\tfrac{1}{2}M\sum\dot\varphi^2(l,m,n)$$
$$+\tfrac{1}{2}\gamma a^2\sum\{[\varphi(l+1,m,n)-\varphi(l,m,n)]^2/a^2+\cdots\}$$
$$+\frac{1}{2}\frac{m^2c^4}{\hbar^2}M\sum\varphi^2(l,m,n)+\frac{\gamma}{2}\sum_\alpha \kappa_\alpha u^2(l_\alpha,m_\alpha,n_\alpha), \quad (7.3)$$

where all summations except the last extend over all lattice points. Once the conjugate momentum

$$p(l,m,n)=M\dot\varphi(l,m,n)$$

is associated with $\varphi(l,m,n)$, the application of Hamilton's equations of motion yield

$$M\ddot\varphi(l,m,n)+\frac{m^2c^4}{\hbar^2}M\varphi(l,m,n)$$

$$=-\gamma\sum_{\substack{\epsilon_1,\epsilon_2,\epsilon_3=0,1\\ \epsilon_1+\epsilon_2+\epsilon_3=1}}[\varphi(l+\epsilon_1,m+\epsilon_2,n+\epsilon_3)$$

$$-2\varphi(l,m,n)+\varphi(l-\epsilon_1,m-\epsilon_2,n-\epsilon_3)] \quad (7.4)$$

except at $(l,m,n)=(l_\alpha,m_\alpha,n_\alpha)$ in which case the terms $\gamma\kappa_\alpha\varphi(l_\alpha,m_\alpha,n_\alpha)$ are added to right-hand side.

If we let

$$\varphi(l,m,n)=u(l,m,n)e^{i\omega t} \quad (7.5)$$

and define ω^* by

$$(\omega^*)^2=\omega^2-m^2c^4\hbar^{-2}, \quad (7.6)$$

the resulting equations are the same as (3.2a) when ω in (3.2) is replaced by ω^*, that is, if the lattice phonons are given a mass m.

It was shown at the end of Sec. 5 that the source defect is equivalent to a hole in the lattice if one considers the interaction of defects separated by a large number of lattice spacings. As the lattice spacing diminishes, this number increases for holes a fixed distance apart. Hence the Hamiltonian (7.1) can be interpreted either as that of the continuum limit of a set of holes in a simple cubic lattice (with nearest-neighbor interactions due to central and noncentral forces of equal magnitude) or of a set of particles interacting through a meson pair field.

When the mass m is included, our inverse

$$a^{(-1)}(r_1, r_2, r_3; i\omega)$$

given by (3.17a) must have ω replaced by ω^*. Then $a^{(-1)}(r_1, r_2, r_3; i\omega) = a^{(-1)}(R; i\omega)$

$$= -\frac{a}{\gamma R(2\pi)^{\frac{3}{2}}} \left(\frac{\pi}{2}\right)^{\frac{1}{2}} \exp\{-a^{-1}R[(\omega^2 + \mu^2 c^2)M/\gamma]^{\frac{1}{2}}\}, \quad (7.7)$$

where

$$R^2 = r_1^2 + r_2^2 + r_3^2 \quad \text{and} \quad \mu = mc/\hbar. \quad (7.8)$$

The interaction energy between two defects κ and κ' at two points separated by a distance $R = a|m - m'|$ is given by substituting (7.7) into (5.3c). This reduces the calculation of the interaction energy to quadratures *for all degrees of coupling*. In the weak coupling limit we can rederive Wentzel's result as follows:

$$\Delta E_I \sim -\frac{\hbar \kappa \kappa'}{(2\pi)^4} \frac{a^2}{R^2} \left(\frac{\pi}{2}\right)^{\frac{1}{2}} \int_0^\infty$$

$$\times \exp\{-2a^{-1}R[(\omega^2 + \mu^2 c^2)M/\gamma]^{\frac{1}{2}}\} d\omega. \quad (7.9)$$

This integral can be expressed in terms of the modified Bessel function of the third kind, K_1. If we let

$$\omega = \mu c \sinh x$$

and use the formula

$$K_1(z) = \int_0^\infty \exp(-z \cosh x) \cosh x \, dx,$$

we find

$$\Delta E_I \sim -\frac{\hbar \kappa \kappa' a^2 \mu c}{4(2\pi)^3 R^2} K_1(2Ra^{-1}\mu c[M/\gamma]^{\frac{1}{2}}),$$

or after employing (7.8) and (7.2) we find

$$\Delta E_I \sim -\frac{\lambda \lambda' mc^2}{32\pi^3 R^2} K_1(2R\mu). \quad (7.10)$$

This is exactly Wentzel's[5] result (with the exception of a factor of $\frac{1}{2}$ which was left out of his paper since he set $E = \sum \hbar \omega$ rather than $\sum \frac{1}{2} \hbar \omega$). The two limiting results follow:

$$\Delta E_I \sim -\frac{\lambda \lambda' \hbar c}{64\pi^3 R^3} \quad \text{if} \quad R \ll \mu^{-1}, \quad (7.11)$$

$$\Delta E_I \sim -\frac{\lambda \lambda' \hbar c}{64\pi^{5/2}} \left(\frac{\mu}{R}\right)^{\frac{1}{2}} \frac{e^{-2\mu R}}{R^2} \quad \text{if} \quad R \gg \mu^{-1}. \quad (7.12)$$

These finite convergent results are unique to a three-dimensional space. If we employ our n-dimensional inverse [Eq. (5.4)], we can show that in the n-dimensional case the weak-coupling approximation yields

$$\Delta E_I = -\frac{\hbar \lambda \lambda' a^{2n-6}}{(2\pi)^{n+1} R^{n-2}} \int_0^\infty [c^{-2}(\omega^2 + \mu^2 c^2)]^{\frac{1}{2}(n-2)}$$

$$\times K_{\frac{1}{2}(n-2)}^2 [Rc^{-1}(\omega^2 + \mu^2 c^2)^{\frac{1}{2}}] d\omega,$$

so that the lattice spacing a occurs only as a coefficient a^{2n-6} while all other parameters in the equation represent macroscopic properties of the medium or defect. When $n < 3$, the continuum limit $a \to 0$ gives an infinite interaction energy at all separation distances between sources. On the other hand, when $n > 3$ the interaction vanishes identically in the limit. Hence, if one were to take pair theory of nucleon forces seriously, one would have to conclude that only a three-dimensional universe could contain condensations of nucleons as we know them in atomic nuclei.

The continuum theory of the interaction of point defects in solids has been discussed by Eshelby.[10]

APPENDIX I. THE MATRICES A AND B OF EQ. (5.33)

$$A = \begin{bmatrix}
1+\delta_1(0,0,0) & \delta_1(2,0,0) & \delta_1(1,1,0) & \delta_1(1,1,0) & \delta_1(1,0,1) & \delta_1(1,0,1) & (\delta_1\delta_2)^{\frac{1}{2}}(1,0,0) \\
\delta_1(2,0,0) & 1+\delta_1(0,0,0) & \delta_1(1,1,0) & \delta_1(1,1,0) & \delta_1(1,0,1) & \delta_1(1,0,1) & (\delta_1\delta_2)^{\frac{1}{2}}(1,0,0) \\
\delta_1(1,1,0) & \delta_1(1,1,0) & 1+\delta_1(0,0,0) & \delta_1(0,2,0) & \delta_1(0,1,1) & \delta_1(1,0,1) & (\delta_1\delta_2)^{\frac{1}{2}}(0,1,0) \\
\delta_1(1,1,0) & \delta_1(1,1,0) & \delta_1(0,2,0) & 1+\delta_1(0,0,0) & \delta_1(0,1,1) & \delta_1(0,1,1) & (\delta_1\delta_2)^{\frac{1}{2}}(0,1,0) \\
\delta_1(1,0,1) & \delta_1(1,0,1) & \delta_1(0,1,1) & \delta_1(0,1,1) & 1+\delta_1(0,0,2) & \delta_1(0,0,2) & (\delta_1\delta_2)^{\frac{1}{2}}(0,0,1) \\
\delta_1(1,0,1) & \delta_1(1,0,1) & \delta_1(0,1,1) & \delta_1(0,1,1) & \delta_1(0,0,2) & 1+\delta_1(0,0,0) & (\delta_1\delta_2)^{\frac{1}{2}}(0,0,1) \\
(\delta_1\delta_2)^{\frac{1}{2}}(1,0,0) & (\delta_1\delta_2)^{\frac{1}{2}}(1,0,0) & (\delta_1\delta_2)^{\frac{1}{2}}(0,1,0) & (\delta_1\delta_2)^{\frac{1}{2}}(0,1,0) & (\delta_1\delta_2)^{\frac{1}{2}}(0,0,1) & (\delta_1\delta_2)^{\frac{1}{2}}(0,0,1) & 1+\delta_2(0,0,0)
\end{bmatrix}.$$

$$B = \begin{bmatrix}
\delta_1(l-2,m,n) & \delta_1(l,m,n) & \delta_1(l-1,m-1,n) & \delta_1(l-1,m+1,n) & \delta_1(l-1,m,n-1) & \delta_1(l-1,m,n+1) & (\delta_1\delta_2)^{\frac{1}{2}}(l-1,m,n) \\
\delta_1(l,m,n) & \delta_1(l+2,m,n) & \delta_1(l+1,m-1,n) & \delta_1(l+1,m+1,n) & \delta_1(l+1,m,n-1) & \delta_1(l+1,m,n+1) & (\delta_1\delta_2)^{\frac{1}{2}}(l+1,m,n) \\
\delta_1(l-1,m-1,n) & \delta_1(l+1,m-1,n) & \delta_1(l,m-2,n) & \delta_1(l,m,n) & \delta_1(l,m-1,n-1) & \delta_1(l,m-1,n+1) & (\delta_1\delta_2)^{\frac{1}{2}}(l,m-1,n) \\
\delta_1(l-1,m+1,n) & \delta_1(l+1,m+1,n) & \delta_1(l,m,n) & \delta_1(l,m+2,n) & \delta_1(l,m+1,n-1) & \delta_1(l,m+1,n+1) & (\delta_1\delta_2)^{\frac{1}{2}}(l,m+1,n) \\
\delta_1(l-1,m,n-1) & \delta_1(l+1,m,n-1) & \delta_1(l,m-1,n-1) & \delta_1(l,m+1,n-1) & \delta_1(l,m,n-2) & \delta_1(l,m,n) & (\delta_1\delta_2)^{\frac{1}{2}}(l,m,n-1) \\
\delta_1(l-1,m,n+1) & \delta_1(l+1,m,n+1) & \delta_1(l,m-1,n+1) & \delta_1(l,m+1,n+1) & \delta_1(l,m,n) & \delta_1(l,m,n+2) & (\delta_1\delta_2)^{\frac{1}{2}}(l,m,n+1) \\
(\delta_1\delta_2)^{\frac{1}{2}}(l-1,m,n) & (\delta_1\delta_2)^{\frac{1}{2}}(l+1,m,n) & (\delta_1\delta_2)^{\frac{1}{2}}(l,m-1,n) & (\delta_1\delta_2)^{\frac{1}{2}}(l,m+1,n) & (\delta_1\delta_2)^{\frac{1}{2}}(l,m,n-1) & (\delta_1\delta_2)^{\frac{1}{2}}(l,m,n+1) & \delta_2(l,m,n)
\end{bmatrix}.$$

$$(l,m,n) \equiv a^{(-1)}(0,0,0; l,m,n; i\omega).$$

---

[10] J. D. Eshelby, Acta Metallurgica **3**, 487 (1955).

Poincaré Cycles, Ergodicity, and Irreversibility in Assemblies of Coupled Harmonic Oscillators*

PETER MAZUR
Institute Lorentz, Leiden, The Netherlands

AND

ELLIOTT MONTROLL
Institute for Fluid Dynamics and Applied Mathematics, University of Maryland, College Park, Maryland
(Received January 10, 1960)

The transport coefficients (diffusion constant, electrical conductivity, etc.) associated with irreversible processes in an assembly of particles can be expressed as integrals over certain time relaxed correlation functions between small numbers of variables of the assembly. The scattering of slow neutrons is also a measure of time relaxed correlation functions.

Irreversibility is a consequence of the vanishing of the correlation coefficients as the relaxation time becomes infinite. On the other hand these coefficients have Poincaré cycles so that any value which they take on is repeated an infinite number of times. It is shown that, in the case of fluctuations of $0(N^{-\frac{1}{2}})$ from zero (N being the number of degrees of freedom), the period of Poincaré cycles is of the order of the mean period of normal mode vibrations while for fluctuations of a magnitude independent of N the period is of the order of C^N where C is a constant which is greater than 1.

The time relaxed correlation coefficients of a pair of particles separated by r lattice spacings decays as $t^{-m/2}$, m being the number of dimensions of the assembly. The statistics of the decay of the momentum of a particle from a preassigned initial value to its equipartition value are discussed.

1. INTRODUCTION

THIS year marks the 100th anniversary of the publication of the great Maxwell paper[1] entitled "Illustrations of the Dynamical Theory of Gases." The Maxwell velocity distribution was there first exhibited and the elementary Maxwell theory of transport processes of dilute gases there first presented. Maxwell's work was inspired by the now 101 year old paper of Clausius[2] on transport coefficients.

In celebration of this important anniversary the authors wish to make a small contribution to the development of the theory of irreversibility in statistical mechanics. A detailed analysis will be made of the manner in which irreversibility appears in the time development of a small number of variables embedded in an assembly of a large number of coupled harmonic oscillators. Some of the results and ideas presented are similar to those which appear in the works of van Hove[3] and Prigogine[4] and his collaborators. While this manuscript was in preparation, the authors received a copy of a doctoral dissertation by Hemmer[5] which also contains similar material. However, Hemmer's analysis is one dimensional and depends strongly on the analytical characteristics of linear chains of coupled oscillators. We have attempted to use rather general dynamical and statistical arguments which are independent of the dimensionality of the assembly under consideration.

At the turn of the century, certain sharp criticisms of the work of Boltzmann and Gibbs were made by Zermelo who claimed that a state of equilibrium of a mechanical assembly could not exist and therefore that statistical mechanics was nonsense. His argument was based on the existence of Poincaré cycles in closed dynamical assemblies. We shall now give a brief qualitative review of Zermelo's remarks, a more detailed proof being in Chandrasekhar's[6] review of the theory of stochastic processes.

Consider the total phase space available to an isolated closed mechanical assembly of total energy E, and fix attention on a subset of finite measure, or volume, $A(0)$ of the phase space. $A(0)$ represents a set of phase points associated with an ensemble of all assemblies which satisfy some special conditions: for example, all assemblies of N particles in the box (with perfectly reflecting walls) given in Fig. 1(b), which can be constructed so that all particles are in the cube marked A at time $t=0$ and the total energy is E. As time goes on, the phase points which were originally in $A(0)$ move through phase space. The shape of the envelope changes, but Liouville's equation (the equation of continuity in phase space) insures that the volume occupied by the

---

* This research was partially supported by the U. S. Air Force ARDC, through its European Office, and the Office of Scientific Research.
[1] J. C. Maxwell, Phil. Mag. (January, 1860).
[2] R. Clausius, Pogg. Ann. **105**, 239 (1859).
[3] L. van Hove, Phys. Rev. **95**, 249 (1954).
[4] G. Klein and I. Prigogine, Physica **19**, 74, 89, 1053 (1953); I. Prigogine and R. Bingen, Physica **21**, 299 (1955).
[5] P. C. Hemmer, "Dynamic and stochastic types of motion in the linear chain," Thesis, Trondheim, Norway, (1959).

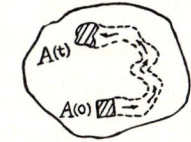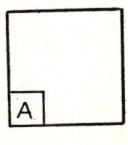

FIG. 1.

[6] S. Chandrasekhar, Revs. Modern Phys. **15**, 1 (1943).

points originally in $A(0)$ remains invariant. The trajectory of a phase point never crosses itself or any other trajectory in phase space (for otherwise the positions and momenta of an assembly at a given time would not yield a unique solution to the dynamical equations). Hence, eventually the set $A(t)$ must return to be identical with $A(0)$ and repeat its former course in a periodic way. At worst, the entire space would have to be filled with points generated by $A(t)$ before the repetition occurs. Since the entire phase space is finite and since the invariant volume of $A(t)$ is nonvanishing, the space eventually would be filled or a repetition of $A(0)$ would first occur. The length of time for a repetition is the Poincaré recurrence time of the set $A(0)$.

The existence of Poincaré cycles implied to Zermelo that no stationary equilibrium state was possible. On the other hand, through the investigation of certain probabilistic (nonmechanical) models, Ehrenfest and Smoluchowski showed that the existence of an equilibrium state with fluctuations need not be inconsistent with Poincaré cycles and indeed that small fluctuations from equilibrium should have small Poincaré cycles while large fluctuations should have very long Poincaré cycles.

Although the general discussion of Poincaré cycles of an entire assembly is philosophically interesting, we shall now exhibit several formulas which show that the Poincaré cycles which are most relevant in the analysis of typical experimental situations involve directly only a small number of variables.

Time dependent correlations exist between individual atoms even in an assembly at equilibrium. These can be observed through the scattering of slow neutrons or through certain magnetic resonance experiments. Van Hove[3] has shown that the scattering cross section depends only on a time relaxed *pair* distribution function which reflects the correlation of the position of two particles at different times. Two particle spin correlations can also be investigated by scattering experiments.

If time dependent external forces are applied to an assembly (electric or magnetic fields for example) or if flows, temperature gradients, concentration gradients, etc., are set up, the response of the assembly can be described through the solution of hydrodynamic equations. These equations are derived from conservation laws. The connection between the hydrodynamic equations and molecular dynamics is made through recently derived formulas, which relate the transport coefficients (diffusion constant, viscosity coefficients, electrical conductivity, etc.) of the hydrodynamic equations to averages over molecular motions. For example, the self-diffusion constant is[7-9]

$$D = \frac{1}{\beta n Z} \text{trace} \int_0^\infty d\tau \int_0^\beta e^{-\beta H} J_z(-i\lambda \hbar) J_z(\tau) d\lambda, \quad (1)\dagger$$

where J_z is the z component of the current operator,

$$J = \sum_j p_j e_j / m_j, \quad (2)$$

where p_j is the momentum of the jth particle, e_j its charge, and m_j its mass. Also Z is the partition function, n the number density and $\beta = 1/kT$. The operator $J(\tau)$ is related to $J(0)$ by the well-known formula,

$$J(\tau) = e^{iH\tau/\hbar} J(0) e^{-iH\tau/\hbar}, \quad (3)$$

where H is the Hamiltonian of the assembly. The current at an imaginary time $(-i\lambda \hbar)$ is defined by an application of (3)†

$$J(-i\lambda \hbar) = e^{\lambda H} J(0) e^{-\lambda H}.$$

If (2) is substituted into (1), it can be shown (since all particles are equivalent in a one component assembly) that

$$D = \frac{kTv}{Zm^2} \text{trace} \int_0^\infty d\tau \int_0^\beta \sum_{j,k} e^{-i\tau H/\hbar}$$
$$\times p_{j\mu} e^{i\tau H/\hbar} e^{-\lambda H} p_{k\mu} e^{-(\beta-\lambda)H} d\lambda$$
$$= \frac{kTv}{Zm^2} N \text{trace} \int_0^\infty d\tau \int_0^\beta e^{-i\tau H/\hbar}$$
$$\times p_{1\mu} e^{i\tau H/\hbar} e^{-\lambda H} p_{1\mu} e^{-(\beta-\lambda)H} d\lambda$$
$$+ \frac{N(N-1)kTv}{m^2 Z} \text{trace} \int_0^\infty d\tau \int_0^\beta e^{-i\tau H/\hbar}$$
$$\times p_{1\mu} e^{i\tau H/\hbar} e^{-\lambda H} p_{2\mu} e^{-(\beta-\lambda)H} d\lambda.$$

The first integral is merely an auto-correlation function and the second a pair correlation function. Hence the diffusion constant is a reflection of the direct correlation of at most *two* particles. This is also true for the electrical conductivity.[10] The viscosity coefficient is expressible in terms of *four* particle time relaxed correlations and the thermal conductivity six particle correlations. Although the more exotic transport coefficients such as thermal diffusion coefficient involve higher correlations, all the measureable coefficients involve direct correlations between very few variables. M. Green[11] has also derived formulas for the transport coefficients in terms of correlations.

In summary, if we wish to examine the effect of Poincaré cycles on observable quantities we need only be concerned with the cycles associated with small numbers of variables.

Fortunately, the complete dynamics of a set of coupled harmonic oscillators can be exhibited in a simple mathematical form. This paper is concerned with the calculation of time dependent harmonic oscillator correlation functions and the demonstration of how in the limit of a large number of degrees of freedom Poin-

---

[7] S. Nakajima, Progr. Theoret. Phys. **21**, 948 (1958).
[8] E. W. Montroll, Il Nuovo cimento (to be published).
[9] H. Mori, Phys. Rev. **112**, 1829 (1958).
† Equations are numbered beginning with (1) in each section.

If an equation in another section is referred to, it will contain the section number, for example, (1.1).
[10] R. Kubo, J. Phys. Soc. (Japan) **12**, 570 (1957).
[11] M. Green, J. Chem. Phys. **22**, 398 (1954).

caré cycles behave in the manner observed by Ehrenfest for special nonmechanical models.

It is well known from the theory of stochastic processes that the character of a stochastic process can be deduced from auto or joint correlation functions of variables whose time development is generated by the process. We shall show that the momentum of a single oscillator is generated by a Gaussian random process when the oscillator is coupled properly to an infinite set of other oscillators.

Before proceeding with the required analysis a few relevant theorems on ergodic and stochastic functions will be listed.

2. THEOREM ON ERGODIC FUNCTIONS

Consider a stationary stochastic process $x(t)$. Without restricting the generality of the argument, we shall assume that its average value is equal to zero,

$$E\{x(t)\}=0, \qquad (1)$$

where E is the symbol for the mathematical expectation or average value of $x(t)$. Averages are determined from an ensemble of observations. The process shall be normalized in such a way that the dispersion is equal to unity,

$$E\{|x(t)|^2\}=1. \qquad (2)$$

The stationarity of the process implies that the correlation function (x^* is complex conjugate of x)

$$\rho(\tau)=E\{x(t+\tau)x^*(t)\} \qquad (3)$$

depends only on τ and is an even function of τ. The process $x(t)$ is called continuous if $\rho(\tau)$ is continuous at $\tau=0$; i.e., if

$$\rho(0+)=1. \qquad (4)$$

Then $\rho(\tau)$ is an everywhere continuous function of τ as follows from the inequality of Schwarz. Khinchine[12] has shown that the correlation function of a continuous stationary process $x(\tau)$ may be represented as a Fourier-Stieltjes integral,

$$\rho(\tau)=\int_{-\infty}^{\infty} e^{i\tau\omega}dF(\omega), \qquad (5)$$

where $F(\omega)$, the spectrum of the process $x(\tau)$, is a never decreasing function of bounded variation. It follows from (5) that

$$\rho(0)=F(\infty)-F(-\infty)=1. \qquad (6)$$

On the other hand Cramer[13] has shown that the stationary process $x(t)$ itself has a spectral representation

$$x(t)=\int_{-\infty}^{\infty} e^{i\omega t}dy(\omega), \qquad (7)$$

[12] A. I. Khinchine, Math. Ann. **190**, 604 (1934).
[13] H. Cramer, Ack. Mat. Astr. Fys. **28B**, No. 12 (1942).

where the process $y(\omega)$ has orthogonal increments, with (see Appendix I)

$$E\{|y(b)-y(a)|^2\}=F(b)-F(a). \qquad (8)$$

From (5) and (7) the following results can be established (see Appendix II)

$$\lim_{T\to\infty}\frac{1}{T}\int_0^T x(t)dt = y(0+)-y(0-) \qquad (9)$$

$$\lim_{T\to\infty}\frac{1}{T}\int_0^T \rho(\tau)d\tau = F(0+)-F(0-)$$
$$= E\{|y(0+)-y(0-)|^2\}. \qquad (10)$$

The result (9) expresses the so-called "law of large numbers" for the stationary process $x(t)$. In deriving (10) we have employed (8).

The following theorem follows immediately from (9) and (10):

Theorem: If $\rho(\tau)\to 0$ as $\tau\to\infty$ the function $x(t)$ is ergodic. Indeed if $\rho(\tau)\to 0$ as $\tau\to\infty$ then from (10)

$$E\{|y(0+)-y(0-)|^2\}=0, \qquad (11)$$

and consequently from (9), for almost all initial conditions of the representative ensemble,

$$\lim_{T\to\infty}\frac{1}{T}\int_0^T x(t)dt = 0. \qquad (12)$$

This proves the ergodic nature of the function $x(t)$ since now the time average (12) of $x(t)$ is equal, for almost all initial conditions, to its ensemble average.

In statistical mechanics one is interested in the ergodic nature of some function $x\{p(t),q(t)\}$ depending on the canonical momenta $p(t)$ and coordinates $q(t)$ of a conservative mechanical system. If in the preceding discussion averages denoted by E represent averages over a surface of constant energy in phase space (over a micro-canonical ensemble), the foregoing theorem states that the dynamical function $x(p,q)$ is ergodic when its phase correlation function $\rho(\tau)$ (on the surface of constant energy) tends to zero as $t\to\infty$. In this last form, the theorem is contained in Khinchine's[14] monograph on the mathematical foundation of statistical mechanics.

3. DYNAMICS OF HARMONIC OSCILLATOR ASSEMBLIES

Most of the analysis in this paper will be concerned with harmonic oscillators coupled in periodic arrays. For simplicity, we associate one degree of freedom with each lattice point of an array. Generally, each atom in a crystal has three degrees of freedom. However, it is well known that the motions in the x, y, and z direction

[14] A. I. Khinchine, *Statistical Mechanics* (Dover Publications, New York, 1949).

of the atoms in simple cubic lattices are independent of each other.[15,16] Hence the one degree of freedom model would be applicable to a discussion of relaxation processes in such lattices. Motions in each direction would be treated separately. Actually, most of the results given in later sections can be generalized to other Bravais lattices in which motions in all directions are coupled.

We shall always discuss linear chains, square lattices, and simple cubic lattices by applying the Born-Karman periodic boundary conditions. The normal mode frequencies of an $N = M^n$, n-dimensional lattice are [15,16]

$$m\omega^2 = 2\sum_{j=1}^{n} \gamma_j\left(1 - \cos\frac{2\pi k_j}{M}\right) \quad k_j = 1, 2, \cdots, M$$

$$= 4\sum_j \gamma_j \sin^2 \pi k_j/M, \qquad (1)$$

where γ is the force constant in various directions and m is the mass of each particle. In an isotropic $3D$ lattice, γ_1 can be chosen to be the central force constant and $\gamma_2 = \gamma_3$ the noncentral force constants (with $\gamma_1 \gg \gamma_2 = \gamma_3$).

The frequency spectra[16] or normal mode distribution functions (as $N \to \infty$) associated with (1) are sketched in Fig. 2(a). Here $g(\omega)d\omega$ is the fraction of frequencies between ω and $(\omega + d\omega)$, while ω_L is the largest frequency. When $n = 1$ (linear lattice)

$$g(\omega) = 2/\pi(\omega_L^2 - \omega^2)^{\frac{1}{2}} \quad m\omega_L^2 = 4\gamma_1. \qquad (2)$$

When $n = 2$ (square lattice) with $\gamma_1 > \gamma_2$,

$$g(\omega) = \begin{cases} \dfrac{4\omega}{\pi^2[\omega^2(\omega_1^2+\omega_2^2-\omega^2)]^{\frac{1}{2}}} \\ \quad \times K\left(\dfrac{\omega_1\omega_2}{[\omega^2(\omega_1^2+\omega_2^2-\omega^2)]^{\frac{1}{2}}}\right) \quad \omega_2^2 < \omega^2 < \omega_1^2 \\ \dfrac{4\omega}{\pi^2\omega_1\omega_2}K\left(\dfrac{[\omega^2(\omega_1^2+\omega_2^2-\omega^2)]^{\frac{1}{2}}}{\omega_1\omega_2}\right) \\ \qquad\qquad \omega^2 < \omega_2^2; \quad \omega_1^2 < \omega^2 < \omega_L^2, \end{cases} \quad (3)$$

where

$$m\omega_1^2 = 4\gamma_1, \quad m\omega_2^2 = 4\gamma_2, \quad \text{and} \quad m\omega^2 = 4(\gamma_1 + \gamma_2) \quad (4)$$

and $K(k)$ is the complete elliptic integral of the second kind.

The $3D$ spectrum is sketched in Fig. 2(c). No simple formula exists in terms of standard functions. However,

[15] H. B. Rosenstock and G. F. Newell, J. Chem. Phys. **21**, 1607 (1953).
[16] E. W. Montroll, Proc. Third Berkeley Symposium on Math. Stat. and Prob. **3**, 209 (1957).

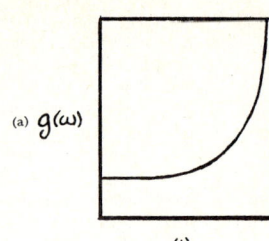
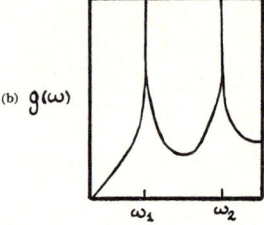
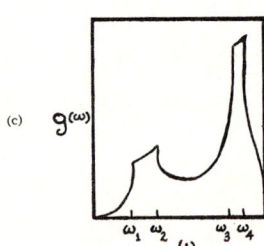

Fig. 2. Schematic frequency spectra of simple cubic lattice.

singularities exist at points $\omega_1, \cdots, \omega_4$, and ω_L, where

$$\begin{aligned} m\omega_1^2 &= 4\gamma_2 & m\omega_3^2 &= 4\gamma_1 \\ m\omega_2^2 &= 4(\gamma_3+\gamma_2) & m\omega_4^2 &= 4(\gamma_1+\gamma_3) \\ m\omega_L^2 &= 4(\gamma_1+\gamma_2+\gamma_3) & \gamma_1 &> \gamma_2 > \gamma_3. \end{aligned} \qquad (5)$$

The asymptotic form near $\omega = 0$, ω_L, and ω_j ($j = 1, \cdots, 4$) is summarized in

$$g(\omega) = \frac{m^{\frac{3}{2}}}{\pi^2(\gamma_1\gamma_2\gamma_3)^{\frac{1}{2}}}\{u(\omega) - (\omega_1^2-\omega^2)^{\frac{1}{2}}\}\omega^2/\omega_1$$
$$0 < \omega < \omega_1 \quad (6\text{a})$$

$$g(\omega) = \frac{\omega m^{\frac{3}{2}}u(\omega)}{\pi^2(\gamma_1\gamma_2\gamma_3)^{\frac{1}{2}}} \quad \omega_1 < \omega < \omega_2 \quad \text{and} \quad \omega_3 < \omega < \omega_4 \quad (6\text{b})$$

$$g(\omega) = \frac{\omega m^{\frac{3}{2}}}{\pi^2(\gamma_1\gamma_2\gamma_3)^{\frac{1}{2}}}\left\{u(\omega) - \frac{1}{2}\frac{(\omega^2-\omega_2^2)^{\frac{1}{2}}(\omega_3^2-\omega^2)^{\frac{1}{2}}}{(\omega_3^2-\omega_2^2)^{\frac{1}{2}}}\right\}$$
$$\omega_2 < \omega < \omega_3 \quad (6\text{c})$$

$$g(\omega) = \frac{\omega m^{\frac{3}{2}}}{\pi^2(\gamma_1\gamma_2\gamma_3)^{\frac{1}{2}}}\{u(\omega) - (\omega^2-\omega_4^2)^{\frac{1}{2}}\}\frac{(\omega_L^2-\omega^2)^{\frac{1}{2}}}{\omega_1^2},$$
$$\omega_4 < \omega < \omega_L. \quad (6\text{d})$$

Here $u(\omega)$ is a continuous function of ω with continuous derivatives.

The Hamiltonian of our "one degree of freedom per lattice point" model is

$$H = \frac{1}{2m}\sum_{j=1}^{N}|p_j|^2 + \sum \tfrac{1}{2}q_j A_{jk} q_k^*, \qquad (7)$$

where p_j is the momentum of the jth particle and q_j its displacement from its equilibrium position. The A_{jk} matrix depends on force constants and dimensionality of the lattice. Although the p's and q's are real, it is convenient in the case of periodic boundary conditions to use complex normal coordinates.

By transforming to normal coordinates $\{P_k\}$ and $\{Q_k\}$ through

$$p_j = \sum_k C_{jk} P_k \qquad (8a)$$

$$q_j = \sum_k C_{jk} Q_k, \qquad (8b)$$

one obtains a new Hamiltonian,

$$H = \tfrac{1}{2}\sum_{j=1}^{N}\left(\frac{1}{m}|P_j|^2 + m\omega_j^2 |Q_j|^2\right), \qquad (9)$$

where ω_j is the jth normal mode frequency and the matrix $C = (C_{jk})$ satisfying

$$\sum_j C_{jk} C_{ji}^* = \delta_{jl}. \qquad (10)$$

The variation of the normal mode Q_k's and P_k's with time is obtained by solving the equations of motion,

$$m\ddot{Q}_k + \omega_k^2 Q = 0. \qquad (11)$$

One finds

$$Q_k(t) = (P_k(0)/m\omega_k)\sin t\omega_k + Q_k(0)\cos t\omega_k \qquad (12)$$

$$P_k(t) = P_k(0)\cos t\omega_k - m\omega_k Q_k(0)\sin t\omega_k. \qquad (13)$$

The momentum of the jth particle at time t is

$$p_j(t) = \sum_k a_{jk} p_k(0) + \sum_k b_{jk} Q_k(0), \qquad (14)$$

where

$$a_{jk} = \sum_s C_{js} C_{ks}^* \cos t\omega_s, \qquad (15)$$

$$b_{jk} = -C_{jk} m\omega_k \sin t\omega_k. \qquad (16)$$

The orthogonality condition (10) implies that

$$\sum_j a_{jj} = \sum_k \cos t\omega_k. \qquad (17)$$

It also implies that

$$\sum_k \{|a_{jk}|^2 + (m\omega_k)^{-2}|b_{jk}|^2\} = 1. \qquad (18)$$

In an n-dimensional simple cubic lattice the C_{jk}'s have the form

$$C_{jk} = \frac{1}{M^{n/2}} \exp 2\pi i (\sum_{\alpha=1}^{n} j_\alpha k_\alpha)/M, \qquad (19)$$

where j and k are to be interpreted as vectors with components $\{j_\alpha\}$ and $\{k_\alpha\}$. Each k and j runs through the integers

$$j_\alpha = 1, 2, 3, \cdots, M,$$
$$k_\alpha = 1, 2, 3, \cdots, M.$$

It is sometimes convenient to transform to normal coordinates through an orthogonal transformation

$$(p_j, q_j) = \sum_k U_{jk}(P_k, Q_k) \qquad (20)$$

such that

$$\sum_j U_{jk} U_{jl} = \delta_{kl}. \qquad (21)$$

The diagonalized Hamiltonian is then

$$H = \tfrac{1}{2}\sum_{j=1}^{N}\left(\frac{1}{m}P_j^2 + m\omega_j^2 Q_j^2\right). \qquad (22)$$

Equation (14) is still valid, but the coefficients a_{jk} and b_{jk} are now defined by

$$a_{jk} = \sum_s U_{js} U_{ks} \cos t\omega_s \qquad (23)$$

$$b_{jk} = -U_{jk} m\omega_k \sin t\omega_k. \qquad (24)$$

The analog of (18) is

$$\sum_k \{a_{jk}^2 + (m\omega_k)^{-2} b_{jk}^2\} = 1. \qquad (25)$$

4. CLASSICAL STATISTICS OF HARMONIC OSCILLATOR ASSEMBLIES

The time relaxed correlation function between two particles separated by a lattice vector r in an assembly of $N = M^n$ oscillators is

$$\rho_N(t,r) = F_N(t,r)/F_N(0,0), \qquad (1)$$

where [using periodic boundary conditions and Eqs. (3.10) and (3.13)]

$$F_N(t,r) = \tfrac{1}{2}E\{p_{s+r}(t)p_s^*(0) + p_s(t)p_{s+r}^*(0)\} \qquad (2)$$

$$= \tfrac{1}{2}E\{\sum_{k,l}(C_{s+r,k} C_{s,l}^* + C_{s,k} C_{s+r,l}^*) \times [P_k(0)\cos t\omega_k - m\omega_k Q_k(0)\sin t\omega_k] P_l^*(0)\}. \qquad (3)$$

If, as is the case in an initially canonical or microcanonical ensemble, a particle has the same probability of possessing a momentum p as $-p$, and if initially equipartition exists,

$$E\{P_k(0) P_l^*(0)\} = mkT\delta_{kl}, \qquad (4)$$

$$E\{Q_k(0) P_l^*(0)\} = 0 \qquad (5)$$

so that, by use of (3.19),

$$F_N(t,r) = \frac{mkT}{M^n}\sum_k \cos\frac{2\pi r \cdot k}{M} \cos t\omega_k \qquad (6)$$

and

$$\rho_N(t,r) = \frac{1}{M^n}\sum_k \cos\frac{2\pi r \cdot k}{M} \cos t\omega_k. \qquad (7)$$

In particular, the autocorrelation function of any oscillator is found by setting $r = 0$,

$$\rho_N(t) = M^{-n}\sum_k \cos t\omega_k. \qquad (8)$$

The remainder of this section is devoted to general remarks about the autocorrelation function. The properties of the joint correlation function in the limit as $N \to \infty$ are discussed in Sec. 5. The summation extends over all normal mode frequencies $\{\omega_k\}$. The only statistical hypothesis made in the derivation of (8) concerns the nature of the ensemble at time $t=0$. Dynamics relates the properties at time t to those at $t=0$. Note that (8) is valid in both the initial microcanonical and canonical ensembles.

Various remarks can be made about $\rho_N(t)$ without employing detailed information about the ω's but only by using statistical properties of the set $\{\omega_j\}$. First, if the ω_k's are distinct,

$$f_N(t) = \sum_{k=1}^{N} \cos t\omega_k \qquad (9)$$

is almost periodic so that any value which is achieved once will be achieved an infinite number of times. The average frequency with which any value of such a sum of cosines will be achieved has been calculated by Kac[17] (a shortened proof for physicists is given in Appendix III).

Let $N_{\Delta T}(q)$ be the number of zeros of $(f_N(t)-q)$ in the interval ΔT. Then the mean frequency for the achievement of q by $f_N(t)$ is

$$L(q) = \lim_{\Delta T \to \infty} \frac{1}{\Delta T} N_{\Delta T}(q). \qquad (10)$$

For large N Kac has shown that

$$L(bN^{\frac{1}{2}}) \sim \frac{2}{\pi}\omega_0 \exp(-\tfrac{1}{2}b^2), \qquad (11)$$

where

$$\omega_0^2 = \frac{1}{N}\sum_{k=1}^{N}\omega_k^2. \qquad (12)$$

The formula being valid if as $N \to \infty$

$$\lim_{N\to\infty} \frac{1}{N^2}\sum_{k=1}^{N}\omega_k^4 = 0. \qquad (13)$$

On the other hand, N. B. Slater[18] has shown that when $(N-q)$ is small and N is large

$$L(q) \sim \frac{N^{\frac{1}{2}}\omega_0}{2\pi\Gamma(\tfrac{1}{2}N+\tfrac{1}{2})}\left(\frac{N-q}{2\pi}\right)^{\frac{1}{2}(N-1)}. \qquad (14a)$$

Application of Stirling's theorem and the introduction of the parameter α (with $0<\alpha<1$)

$$\alpha = (q-1)/(N-1) \sim (q/N) \qquad (14b)$$

---
[17] M. Kac, Am. J. Math. **65**, 609 (1943).
[18] N. B. Slater, Proc. Cambridge Phil. Soc. **35**, 56 (1939).

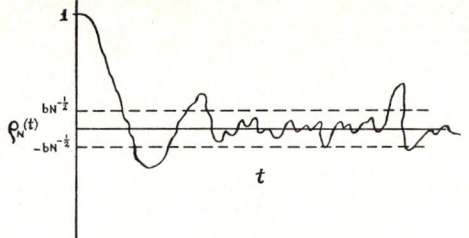

FIG. 3. Schematic autocorrelation function of momentum of particle in a simple cubic lattice.

yields

$$L(q) \sim \frac{\omega_0}{2\pi^{\frac{3}{2}}}\left(\frac{1-\alpha}{\pi e}\right)^{\frac{1}{2}(N-1)} \sim L(\alpha N). \qquad (14c)$$

If equilibrium is to be established as $t \to \infty$, $p(t)$ and $p(0)$ must become independent so that $\rho_N(t) \to 0$. On the other hand if N is finite but large, we know that Poincaré cycles exist, and therefore whatever value of $\rho_N(t)$ is achieved at some time is achieved over and over again. The passage of $\rho_N(t) \to 0$ and at the same time the existence of Poincaré cycles can be reconciled in the following manner:

First let N be very large and consider a value of $\rho_N(t) = (1/N)f_N(t)$ in the range $(-bN^{-\frac{1}{2}}, bN^{-\frac{1}{2}})$ where b is of $0(1)$ and independent of N. From Eq. (11) the frequency of a Poincaré cycle is

$$(2\omega_0/\pi)\exp(-\tfrac{1}{2}b^2). \qquad (15)$$

On the other hand, consider a value of $0<\alpha<1$ which is independent of N. Then the mean time between repetitions of a value (Fig. 3) of $f_N(t)=\alpha N$ or of $\rho_N(t)=\alpha$ is (from 14c)

$$\frac{2\pi^{\frac{3}{2}}}{\omega_0}\left(\frac{e\pi}{1-\alpha}\right)^{\frac{1}{2}(N-1)}. \qquad (16)$$

Hence as $N \to \infty$ the period of Poincaré cycles becomes enormously large for values of $\rho_N(t)$ outside the "fluctuation range" $\pm bN^{\frac{1}{2}}$ but are of the order of the reciprocal of the root mean square natural frequency ω_0 of an assembly for values of $\rho_N(t)$ inside the fluctuation range. As $N \to \infty$ the fluctuation range approaches zero so that a true equilibrium situation develops with $\rho_N(t)$ approaching zero as $t \to \infty$.

The function $\rho_N(t)$ is that associated with either a canonical or microcanonical ensemble and the mean recurrence times are also associated with the ensembles. As in the case of equilibrium statistical mechanics, one can show that fluctuations from the average ensemble behavior become small as the number of degrees of freedom N becomes large. This will be discussed elsewhere.

Let us now consider the statistics of the momentum of a particle which is known to have the momentum $p_j(0)$ at time $t=0$ while all other $(2N-1)$ variables required to describe the assembly are distributed initially according to a microcanonical distribution such that the total energy of the assembly is $\mathcal{E}$. Then (3.22) yields

$$\sum_{k \neq j} \frac{1}{2m} p_k{}^2(0) + \sum_{k=1}^{N} \tfrac{1}{2} m \omega_k{}^2 Q_k{}^2(0) = R^2 = \mathcal{E} - \frac{1}{2m} p_j{}^2(0). \quad (17)$$

Now

$$p_j(t) - a_{jj} p_j(0) = \sum_{k \neq j} a_{jk} p_k(0) + \sum_{k} b_{jk} Q_k(0), \quad (18)$$

where in an assembly with periodic boundary conditions we apply (3.17) (3.19), and (3.18) to find

$$a_{jj} = \sum_s C_{js} C_{js}{}^* \cos \omega_s = \frac{1}{N} \sum_s \cos \omega_s$$

$$= \rho_N(t). \quad (19)$$

The mean value (18) when averaged over (17) vanishes because $p_k(0)$ and $Q_k(0)$ have the same probability of achieving positive and negative values of the same absolute value.

The distribution function of

$$Y(t) = p_j(t) - p_j(0) \rho_N(t) \quad (20)$$

is the same as that of the right hand side of (18). We now find this distribution when every value of the set of p_k's and $Q_k(0)$'s consistent with (17) is given equal weight.

To this end we find the distribution function of

$$Y = \sum \alpha_j x_j \quad (21)$$

when the x_j's are distributed uniformly on the ellipsoid

$$\sum_{k=1}^{n} \beta_k x_k{}^2 = R^2. \quad (22)$$

The various α's and β's are to be identified with the parameters of (17) and (18). Our required distribution function is the fourier transform of the characteristic function $f(\alpha)/f(0)$, where

$$f(\alpha) = \int \cdots \int_{\sum \beta_k x_k{}^2 = R^2} \exp(i\alpha \sum \alpha_j x_j) dx_1 \cdots dx_n. \quad (23)$$

Let $x_j{}^2 = y_j{}^2/\beta_j$ and $\gamma_j = \alpha_j/\beta_j{}^{\frac{1}{2}}$ and convert the resulting spherically restricted integral (over $\sum y_k{}^2 = R^2$) $f(\alpha)$ to an unrestricted integral through the introduction of the Fourier integral representation of the δ function,

$$\delta(R^2 - \sum y_k{}^2) = \frac{1}{2\pi} \int_{-\infty}^{\infty} \exp i\beta(R^2 - \sum y_k{}^2) d\beta$$

$$= \frac{\exp(bR^2)}{2\pi} \int_{-\infty}^{\infty} \exp(i\beta R^2)$$

$$\times \exp\{-(b+i\beta) \sum y_k{}^2\} d\beta, \quad (24)$$

where b is any positive number.

Then, if we let $\sigma^2 = \sum \gamma_k{}^2$,

$$f(\alpha) = \frac{e^{bR^2}}{2\pi(\beta_1 \cdots \beta_n)^{\frac{1}{2}}} \int_{-\infty}^{\infty} e^{i\beta R^2} d\beta \prod_{j=1}^{n} \int_{-\infty}^{\infty} e^{i\alpha \gamma_j v_j}$$

$$\times \exp\{-y_j{}^2(b+i\beta)\} dy_j$$

$$= \frac{\pi^{n/2} \exp bR^2}{2\pi(\beta_1 \cdots \beta_n)^{\frac{1}{2}}} \int_{-\infty}^{\infty} d\beta \exp\{i\beta R^2 - \alpha^2 \sigma^2/4(b+i\beta)\}$$

$$\times (b+i\beta)^{-n/2}$$

$$= \frac{\pi^{n/2}}{(\beta_1 \cdots \beta_n)^{\frac{1}{2}}} \left(\frac{2R}{\alpha \sigma}\right)^{(n/2)-1} J_{(n/2)-1}(\alpha \sigma R). \quad (25)$$

One finds after employing the power series expansion for the Bessel function $J_n(x)$

$$f(0) = \frac{\pi^{n/2}}{(\beta_1 \cdots \beta_n)^{\frac{1}{2}}} \frac{R^{2(\frac{1}{2}n-1)}}{(\frac{1}{2}n-1)!}, \quad (26)$$

so that

$$f(\alpha)/f(0) = (\tfrac{1}{2}n-1)!(2/R\sigma)^{\frac{1}{2}n-1} J_{(n/2)-1}(\alpha \sigma R)$$

$$= 1 - (\alpha^2 \sigma^2 R^2/2n) + \cdots,$$

where the kth term in the expansion is, as $n \to \infty$,

$$\frac{(-1)^k (\alpha R)^{2k}}{2^{2k} k! (n/2)[(n/2)-1] \cdots [(n/2)+(k-1)]}$$

$$\sim \frac{(-1)^k \alpha^{2k} R^{2k}}{2^k n^k k!}.$$

Hence,

$$\lim_{n \to \infty} f(\alpha)/f(0) = \exp(-\tfrac{1}{2} \alpha^2 \sigma^2 R^2 n^{-1}). \quad (27)$$

The distribution function of Y is, as $n \to \infty$,

$$F(Y) = \frac{1}{2\pi} \int_{-\infty}^{\infty} e^{-i\alpha Y} \exp(-\tfrac{1}{2} \alpha^2 \sigma^2 R^2 n^{-1}) d\alpha$$

$$= \frac{1}{\sigma R (2\pi/n)^{\frac{1}{2}}} \exp - \frac{ny^2}{2\sigma^2 R^2}. \quad (28)$$

The values of α_k of (21) which correspond to the coefficients of $p_k(0)$ of (18) are a_{jk}, while those which are associated with the $Q_k(0)$'s are b_{jk}. The β_k's of (22) are similarly $(2m)^{-1}$ and $\tfrac{1}{2} m \omega_k{}^2$, the coefficients respectively of $p_k{}^2(0)$ and $Q_k{}^2(0)$ of 17. Then the $\gamma_k = \alpha_k \beta_k{}^{-1}$'s are

$$(2m)^{\frac{1}{2}} a_{jk} \quad \text{and} \quad b_{jk}/(\tfrac{1}{2} m \omega_k{}^2)^{\frac{1}{2}},$$

so that

$$\sigma^2 = \sum \gamma_k{}^2$$
$$= 2m \sum_k a_{jk}{}^2 + \sum_k b_{jk}{}^2/(\tfrac{1}{2} m \omega_k{}^2) - 2m a_{jj}{}^2. \quad (29)$$

The last term is subtracted because the term with $k = j$ is not included in (17) and (18). In view of (3.25) and (19)

$$\sigma^2 = 2m(1 - a_{jj}{}^2) = 2m\{1 - [\rho_N(t)]^2\}. \quad (30)$$

Also, from (17),

$$R^2 = \mathcal{E} - \frac{1}{2m} p_j{}^2(0). \quad (31)$$

Since the total energy $\mathcal{E}$ is $\tfrac{1}{2}(2N-1)kT$, as $N \to \infty$,

$$R^2 = NkT, \quad (32)$$

and the n of (28) is $(2N-1)$.

If the Y of (28) is associated with (18), and therefore with (20), the distribution function of $Y(t) = p_j(t) - p_j(0)\rho_N(t)$ is, when N is large,

$$P_N[p_j(t) | p_j(0)] = \{2\pi m k T [1 - \rho_N(t)]\}^{-\frac{1}{2}}$$

$$\times \exp - \left\{ \frac{[p_j(t) - p_j(0)\rho_N(t)]^2}{2mkT\{1 - [\rho_N(t)]^2\}} \right\}. \quad (33)$$

This function is just the probability for a transition $p_j(0) \to p_j(t)$ in the time t. It is not surprising that the distribution is Gaussian because as $N \to \infty$ the variables x_j in (21) become independent so that the central limit theorem is applicable.

We shall show below that as $N \to \infty$ and $t \to \infty$, $\rho_N(t) \to 0$. However, before proceeding with the proof of this fact let us investigate some of its consequences.

As $t \to \infty$ $P_n[p_j(t) | p_j(0)]$ approaches an equilibrium Maxwellian distribution independently of $p_j(0)$. The manner in which the kinetic energy achieves its equipartition value is observed by considering

$$E\left\{\frac{1}{2m} p_j{}^2(t)\right\} = E\left\{\frac{1}{2m}[p_j(t) - p_j(0)\rho_N(t)]^2\right\}$$

$$+ 2 p_j(0) \rho_N(t) E\left\{\frac{1}{2m}[p_j(t) - p_j(0)\rho_N(t)]\right\}$$

$$+ \frac{1}{2m} p_j{}^2(0) \rho_N(t) E\{1\}, \quad (34)$$

where the average is to be taken over the distribution (33). The first expectation value on the right (34) is just the dispersion of the distribution function (33)

$$\tfrac{1}{2} kT\{1 - [\rho_N(t)]^2\},$$

the second vanishes, and $E\{1\} = 1$. Hence the expecta-

tion value of the kinetic energy is

$$E\left\{\frac{1}{2m} p_j{}^2(t)\right\} = \tfrac{1}{2} kT\{1 - [\rho_N(t)]^2\} + \frac{1}{2m} p_j{}^2(0) \rho_N(t)$$

as $t \to \infty$ this approaches the equipartition value kT, while as $t \to 0$ it approaches $p_j{}^2(0)/2m$, since $\rho_N(t) \to 1$ as $t \to 0$.

As a further consequence of the fact that $\rho(\tau) \to 0$ as $\tau \to \infty$, we may apply the theorem quoted in Sec. 2 to conclude that $p_j(t)$ is ergodic; i.e., that for almost all initial conditions

$$\lim_{T \to \infty} \frac{1}{T} \int_0^T p_j(t) dt = E\{p_j(0)\} = 0,$$

the expectation value being taken over an initial canonical or microcanonical ensemble.

It also can be shown after some calculation that

$$R_n(t) = \frac{E\{(p_j{}^2(t) - \mathcal{E}_0)(p_j{}^2(0) - \mathcal{E}_0)\}}{E\{[p_j{}^2(0) - \mathcal{E}_0]\}^2} = [\rho_N(t)]^2,$$

where

$$\mathcal{E}_0 = E\{p_j{}^2(0)\}.$$

Hence, as $N \to \infty$, $\mathcal{E}_0 = mkT$ and

$$\lim_{t \to \infty} R_\infty(t) = 0.$$

Thus $p_j{}^2(t)$ is also ergodic so that

$$\lim_{T \to \infty} \frac{1}{T} \int_0^T p_j{}^2(t) dt = E\{p_j{}^2(0)\} = mkT.$$

One can finally show that the Gaussian character of $P_\infty[p_j(t) | p_j(0)]$ implies that any function which depends only on p_j and whose phase average exists is ergodic. Ergodicity does not appear as a general dynamical property of an assembly but as a property of a special class of functions of a small number of variables in an assembly composed of a large number of coupled degrees of freedom.

The average time that our oscillator spends in a specified momentum range can be obtained from a discussion of the function

$$D(p(t); a, b) = \begin{cases} 1 & \text{if } a < p(t) < b \\ 0 & \text{otherwise}. \end{cases}$$

The fraction of time f_{ab} spent by the jth oscillator in the momentum range

$$a < p_j(t) < b$$

is

$$f_{ab} = \lim_{T \to \infty} \frac{1}{T} \int_0^T D(p_j(t); a, b) dt.$$

Since the function D is ergodic in the limit as $N \to \infty$, f_{ab} can, for almost all initial conditions, be calculated from an initial ensemble average as well as from the time average. An initial canonical ensemble yields

$$f_{ab} = \frac{1}{(2\pi mkT)^{\frac{1}{2}}} \int_{-\infty}^{\infty} dp D(p; a,b) \exp(-p^2\beta/2m)$$

$$= \frac{1}{(2\pi mkT)^{\frac{1}{2}}} \int_a^b e^{-p^2\beta/2m} dp.$$

We thus see that for almost all initial conditions a single system will in a long time interval, spend most of the time within the "fluctuation region" of the momentum p_j, and spend only a small fraction of time in "improbable states" for which $p_j^2(t) \gg mkT$. This also implies that if a system is in such an improbable state it will on the average rapidly decay towards the "fluctuation region" $[p_j^2(t) = 0(mkT)]$ and remain there most of the time. Alternatively, we may say that the mean recurrence time for states within the fluctuation region will be small, whereas they will be very large for improbable states. This illustrates the irreversible behavior of the momentum $p_j(t)$ in a single system.

We note that all these conclusions hold for a single system in the limit $N \to \infty$.

For finite but very large N, the behavior of $p_j(t)$ will be practically the same.

The irreversible behavior of $p_j(t)$ may be further illustrated as follows: According to the Gaussian form of $P_\infty[p_j(t) | p_j(0)]$, the probability that $p_j(t)$ at time t will differ from its conditional average $\rho(t)p_j(0)$ by an amount much larger than $(mkT)^{\frac{1}{2}}$ is very small. Thus there is a high probability for a single system to decay close to the average path, in agreement with the macroscopic concept of irreversible behavior.

5. DETAILED CALCULATION OF MOMENTUM CORRELATION FUNCTIONS

The one dimensional correlation function is found by combining (4.7) with (3.1) when $n=1$. Then

$$\rho_N(t,r) = \frac{1}{N} \sum_{k=1}^{N} \cos\frac{2\pi r k}{N} \cos\left(\tau \sin\frac{\pi k}{N}\right) \quad (1)$$

with

$$\tau = t\omega_L \quad \text{and} \quad \omega_L = 2(\gamma/m). \quad (2)$$

In the limit as $N \to \infty$ let $\vartheta = \pi k/N$ and $d\vartheta = \pi/N$. Then

$$\rho_N(t,r) \sim \frac{1}{\pi} \int_0^\pi \cos 2r\vartheta \cos(\tau \sin\vartheta) d\vartheta$$

$$= \frac{2}{\pi} \int_0^{\pi/2} \cos 2r\vartheta \cos(\tau \sin\vartheta) d\vartheta = J_{2r}(\tau), \quad (3)$$

the Bessel function of order $2r$. The exact calculation of $\rho_N(t,r)$ is given in Appendix IV for arbitrary N.

It is merely a fortunate accident that $\rho_N(t,r)$ has such a simple form in the one dimensional lattice. In preparation for the 2D and 3D calculation of $\rho_N(t,0) \equiv \rho_N(t)$ we rewrite

$$\rho_N(t) = \frac{1}{N} \sum_i \cos t\omega_j \sim \int_0^{\omega_L} \cos t\omega\, g(\omega) d\omega, \quad (4)$$

where $g(\omega)d\omega$ is the fraction of circular frequencies between ω and $\omega + d\omega$. In the one-dimensional lattice,

$$g(\omega) = (2/\pi)(\omega_L^2 - \omega^2)^{-\frac{1}{2}}. \quad (5a)$$

Then as $N \to \infty$

$$\rho_N(t) \sim \frac{2}{\pi} \int_0^{\omega_L} \frac{\cos t\omega\, d\omega}{(\omega_L^2 - \omega^2)^{\frac{1}{2}}} = J_0(\tau). \quad (5b)$$

As $\tau \to \infty$ [we henceforth write $\rho(\tau) \equiv \rho_\infty(\tau)$],

$$\rho(\tau) \sim (2/\pi\tau)^{\frac{1}{2}} \sin\tau. \quad (6)$$

The value of the ω_0 which appears in the Poincaré cycle formulas (4.15) and (4.16) is, in the n-dimensional case,

$$\omega_0^2 = \frac{4}{mM^n} \sum_{j, k_j} \gamma_j \sin^2(\pi k_j/M).$$

$$= \frac{4}{m\pi} \sum_j \gamma_j \int_0^\pi \sin^2\vartheta d\vartheta = \frac{2}{m} \sum_j \gamma_j$$

$$= \frac{1}{2}\omega_L^2. \quad (7)$$

The short time behavior of $\rho(t)$ is found from the expansion

$$\rho_N(t) = \frac{1}{N} \sum_{j=1}^N \cos t\omega_j \sim 1 - t^2\mu_2/2! + t^4\mu_4/4! - \cdots, \quad (8a)$$

where

$$\mu_{2k} = \frac{1}{N} \sum_{j=1}^N \omega_j^{2k}. \quad (8b)$$

For a 2D lattice of N^2 lattice points,

$$\mu_{2k} = \frac{2^{2k}}{m^k N^2} \sum_{j_1, j_2=1}^M \left(\gamma_1 \sin^2\frac{\pi j_1}{M} + \gamma_2 \sin^2\frac{\pi j_2}{M}\right)^k$$

$$= \frac{2^{2k}}{\pi^2 m^k} \int_0^\pi \int (\gamma_1 \sin^2\varphi_1 + \gamma_2 \sin^2\varphi_2)^k d\varphi_1 d\varphi_2$$

$$= \frac{2^{2k}}{m^k} \sum_{l=0}^k \frac{k! \gamma_1^l \gamma_2^{k-l}}{l!(k-l)!} \left\{\frac{1}{\pi} \int_0^\pi \sin^{2l}\varphi_1 d\varphi_1\right\}$$

$$\times \left\{\frac{1}{\pi} \int_0^\pi \sin^{2(k-l)}\varphi_2 d\varphi_2\right\}$$

$$= \frac{1}{m^k} \sum_{l=0}^k \frac{\gamma_1^l \gamma_2^{k-l} k!(2l)!(2[k-l])!}{l!(k-l)! l! l!(k-l)!(k-l)!}. \quad (9)$$

The first few μ's are

$$\mu_2/2! = (\gamma_1+\gamma_2)/m; \quad \mu_4/4! = [\tfrac{1}{4}(\gamma_1^2+\gamma_2^2)+\tfrac{1}{3}\gamma_1\gamma_2]/m^2$$

$$\mu_6/6! = \left[\frac{1}{36}(\gamma_1^3+\gamma_2^3)+\frac{1}{20}\gamma_1\gamma_2(\gamma_1+\gamma_2)\right]\!\!\bigg/m^3. \quad (10)$$

As is well known in the theory of the asymptotic behavior of Fourier integrals, the asymptotic form of (4) for large t depends on the singularities of $g(\omega)$. As is shown in Appendix IV, the discontinuity of $g(\omega)$ at the end of a two-dimensional frequency spectrum yields, as $t \to \infty$,

$$\rho(t) \sim \frac{2\omega_L \sin t\omega_L}{\pi t \omega_1 \omega_2}, \quad (11a)$$

where

$$m\omega_1^2 = 4\gamma_1 \quad \text{and} \quad m\omega_2^2 = 4\gamma_2. \quad (11b)$$

In a 3D simple cubic lattice of N^3 lattice points,

$$\mu_{2k} = \frac{2^k}{m^k N^3} \sum_{j_1 j_2 j_3=1}^{M} \left(\gamma_1 \sin^2\frac{\pi j_1}{M} + \gamma_2 \sin^2\frac{\pi j_2}{M} + \gamma_3 \sin^2\frac{\pi j_3}{M}\right)^k. \quad (12)$$

The first few μ's are

$$\mu_2/2! = (\gamma_1+\gamma_2+\gamma_3)/m;$$

$$\mu_4/4! = [\tfrac{1}{4}(\gamma_1^2+\gamma_2^2+\gamma_3^2)+\tfrac{1}{3}(\gamma_1\gamma_2+\gamma_2\gamma_3+\gamma_3\gamma_1)]/m^2$$

$$\mu_6/6! = \left[\frac{1}{36}(\gamma_1^3+\gamma_2^3+\gamma_3^3)\right.$$
$$+\frac{1}{20}(\gamma_1\gamma_2^2+\gamma_2\gamma_3^2+\gamma_2\gamma_1^2+\gamma_3\gamma_2^2)$$
$$\left.+\frac{1}{15}\gamma_1\gamma_2\gamma_3\right]\!\!\bigg/m^3.$$

The long time behavior of $\rho(t)$ is a consequence of the four internal and one end singularity of the $g(\omega)$ function which is plotted in Fig. 2(c).

The three-dimensional correlation function (4.7) is in the limit $N \to \infty$

$$\rho(t,r) = \frac{1}{(2\pi)^3}\int\!\!\int\!\!\int_0^{2\pi} \cos(r\cdot\varphi)\cos[t\omega(\varphi)]d^3\varphi$$

$$= \tfrac{1}{4}\{[I(t,r)+I(t,-r)]+\text{cc}\}, \quad (13)$$

where

$$I(t,r) = \frac{1}{(2\pi)^3}\int\!\!\int\!\!\int_0^{2\pi} \text{exp}i[\varphi\cdot r+t\omega(\varphi)]d^3\varphi. \quad (14)$$

When t is very large this integral can be calculated by the method of stationary phase. The main contribution comes from the neighborhood of the stationary or critical points of $\omega(\varphi)$; those points for which $\text{grad}\omega(\varphi)=0$. These are of course the maxima, minima, and saddle points of the surfaces $\omega(\varphi)=\text{const}$. The critical points associated with the surfaces of constant frequency were first discussed by van Hove.[19]

We suppose that in the neighborhood of a critical point $\varphi^{(0)}$,

$$\omega = \omega_0 + (2m\omega_0)^{-1}\sum_{j=1}^{3}\epsilon_j\gamma_j'[\varphi_j-\varphi_j^{(0)}]^2$$
$$+0(\varphi_j-\varphi_j^{(0)})^3, \quad (15)$$

where γ_j' is a force constant and ϵ_j' is ± 1 (all are -1 in the neighborhood of a maximum and two are of one sign and one of the other in the neighborhood of a saddle point). Then, after substituting (15) into (14), we obtain the following contribution to I from the region of φ space near the critical point of interest (letting $\xi_j = \varphi_j - \varphi_j^{(0)}$),

$$\delta I(t,r) = \frac{\text{exp}i[r\cdot\varphi^{(0)}+t\omega_0]}{(2\pi)^3}$$

$$\times \int\cdot\int \text{exp}i[r\cdot\xi+(t/2m\omega_0)\sum\epsilon_j\gamma_j\xi_j^2]d^3\xi.$$

Now let

$$x_j = \xi_j(t\gamma_j/2m\omega_0)^{\frac{1}{2}} \quad \text{and} \quad s_j = r_j(2m\omega_0/t\gamma_j)^{\frac{1}{2}}.$$

Then, as $t \to \infty$,

$$\delta I(t,r) \sim \left(\frac{2m\omega_0}{t}\right)^{\frac{3}{2}}\frac{\text{exp}i[r\cdot\varphi^{(0)}+t\omega_0]}{(\gamma_1\gamma_2\gamma_3)^{\frac{1}{2}}}$$

$$\times \prod_{j=1}^{3}\int_{-\infty}^{\infty}\exp(i\epsilon_j x_j^2)dx_j.$$

If $\epsilon = \pm 1$,

$$\int_{-\infty}^{\infty}\exp(i\epsilon x^2)dx = \pi^{\frac{1}{2}}\exp(i\epsilon\pi/4).$$

Hence,

$$\delta I(t,r) \sim \left(\frac{2m\omega_0}{t}\right)^{\frac{3}{2}}\frac{\text{exp}i[r\cdot\varphi^{(0)}+t\omega_0+\tfrac{1}{4}\pi\sum\epsilon_j]}{(\gamma_1\gamma_2\gamma_3)^{\frac{1}{2}}}. \quad (16)$$

The contribution to $\rho(t,r)$ from the lth stationary point is, as $t \to \infty$,

$$\rho^{(l)}(t,r) \sim \left(\frac{2m\omega_l}{t}\right)^{\frac{3}{2}}(\gamma_1\gamma_2\gamma_3)^{\frac{1}{2}}$$

$$\times \cos(r\cdot\varphi^{(l)})\cos(t\omega_0+\tfrac{1}{4}\pi\sum\epsilon_j). \quad (17)$$

Hence the envelope of the correlation functions vanishes as $t^{-\frac{3}{2}}$ as $t \to \infty$.

The values of $\varphi^{(l)}$, $m\omega_l^2$ and $\sum\epsilon_j$ for the various critical points, with the exception of the minimum at $(0,0,0)$ which contributes a term of $0(t^{-2})$ to $\rho(t,r)$, are listed in Table I. The point (π,π,π) is a maximum. It

---

[19] L. van Hove, Phys. Rev. **89**, 1189 (1953).

TABLE I

$\varphi^{(l)}$	$m\omega_L^2$	$\Sigma \, \epsilon_j$
(π,π,π)	$4(\gamma_1+\gamma_2+\gamma_3)$	-3
$(0,0,\pi)$	$4\gamma_2$	1
$(0,\pi,0)$	$4\gamma_3$	1
$(\pi,0,0)$	$4\gamma_1$	1
$(\pi,0,\pi)$	$4(\gamma_1+\gamma_2)$	-1
$(\pi,\pi,0)$	$4(\gamma_1+\gamma_3)$	-1
$(0,\pi,\pi)$	$4(\gamma_2+\gamma_3)$	-1

yields the largest frequency $m\omega_L^2 = 4(\gamma_1+\gamma_2+\gamma_3)$. We generally assume that $\gamma_1 > \gamma_2 = \gamma_3$. These critical points were obtained from (3.1) when $n=3$. The total asymptotic expression for $\rho(t,r)$ is

$$\rho(t,r) = \sum_{l=1}^{7} \rho^{(l)}(t,r). \qquad (18)$$

Van Hove has made similar calculations in developing the theory of the time relaxed pair distribution function of crystals (see also Prigogine and Bingen[4]).

It is interesting to calculate $\rho(t)$ for the Debye spectrum,

$$g(\omega) = \begin{cases} 3\omega^2/\omega_L^3 & \omega < \omega_L \\ 0 & \omega > \omega_L \end{cases} \qquad (19)$$

Here

$$\rho(t) = \frac{3}{\omega_L^3} \int_0^{\omega_L} \omega^2 \cos\omega t \, d\omega$$

$$= \frac{3}{\tau^3}\{2\tau \cos\tau + (\tau^2-2)\sin\tau\}$$

with $\tau = t\omega_L$; As $t \to \infty$,

$$\rho(t) \sim (3/\omega_L t) \sin\omega_L t,$$

which is somewhat different from (18), being proportional to t^{-1} rather than $t^{-\frac{3}{2}}$. This has also been observed by Prigogine and Bingen.[4]

We close this section with an analysis of $\rho_N(t)$ for large but finite N in a one dimensional assemble. We use Eulers summation formula[20]

$$\sum_{n>a}^{n\leq b} f(n) = \int_a^b f(x)dx + u(b)f(b)$$

$$-f(a)u(a) - \int_a^b u(x)f'(x)dx,$$

which is valid when $f(x)$ is continuous and has a continuous first derivative in the interval (a,b). Here

$$u(x) = [x] - x + \tfrac{1}{2},$$

where $[x]$ is the integral part of the number x (the largest integer contained in x). In our problem $b=N$ and $a=0$.

[20] c.f., T. V. Uspensky, *Introduction to Mathematical Probability* (McGraw-Hill Book Company, Inc., New York, 1937), p. 347.

We have

$$\frac{1}{N}\sum_{k=1}^{N} \cos\tau\left(\sin\frac{\pi k}{N}\right) = \frac{1}{N}\int_0^N \cos\tau\left(\sin\frac{\pi x}{N}\right)dx$$

$$+ \frac{\tau\pi}{N^2}\int_0^N u(x) \cos\frac{\pi x}{N} \sin\left(\tau\sin\frac{\pi x}{N}\right)dx.$$

If we let $\vartheta = \pi x/N$ and apply Eqs. (1) and (3), we find

$$\rho_N(t) = J_0(\tau) + \frac{\tau}{N}\int_0^\pi u(N\vartheta/\pi)\cos\vartheta \sin(\tau\sin\vartheta)d\vartheta. \qquad (20)$$

Now $|u(x)| \leq \tfrac{1}{2}$, and the absolute values of $\sin\varphi$ and $\cos\varphi$ are ≤ 1. Hence, the integral component of $\rho_N(t)$ is $\leq \tau\pi/2N$. As long as $\tau \ll N$, $\rho_N(t)$ is well represented by $J_0(\tau)$. Actually it is still a good representation for much larger τ because when $\tau = 0(N)$, $\sin(\tau\sin\vartheta)$ and $u(N\vartheta/\pi)$ oscillate and change sign very rapidly so that the integral itself approaches zero as τ or as $N \to \infty$ unless as occasionally happens sign changes of $u(N\vartheta/\pi)$ and $\sin(\tau\sin\vartheta)$ cancel each others. It is just this occasional proper phasing of these two functions which leads to Poincaré cycles in $\rho_N(t)$ for values of $\rho_N(t)$ of $0(1)$. A complete expansion of (20) is given in Appendix V.

6. CALCULATION OF QUANTUM MECHANICAL AUTOCORRELATION FUNCTION

The ensemble average of a dynamical operator B is

$$E\{B\} = \text{trace} B\rho, \qquad (1)$$

where ρ is the density matrix associated with the ensemble. If we wish to find the quantum mechanical analog of the autocorrelation function (4.1) we must then find

$$E\{p_j(t)p_j(0)\} = \text{trace}\{p_j(t)p_j(0)\rho\}, \qquad (2)$$

where ρ is the density matrix associated with the appropriate ensemble. Since, H is the Hamiltonian of our assembly,

$$p_j(t) = e^{iHt/\hbar}p_j(0)e^{-iHt/\hbar}, \qquad (3)$$

the only operators which require an ensemble averaging are $p_j(0)$. If the assembly of interest belongs to a canonical ensemble at temperature T (with $\beta = 1/kT$), the density matrix is $\rho = Z^{-1}\exp(-\beta H)$, Z being the partition function. Then, if we abbreviate $p_j \equiv p_j(0)$ and employ periodic boundary conditions so that all particles are equivalent, we have

$$F_N(t) = \frac{1}{N}\sum_j E\{p_j(t)p_j(0)\}$$

$$= \frac{1}{N}\sum_j E\{e^{itH/\hbar}p_j(0)e^{-itH/\hbar}p_j(0)\}$$

$$= \frac{1}{NZ}\sum_j \text{trace}\{e^{iHt/\hbar}p_j e^{-iHt/\hbar}p_j e^{-\beta H}\}. \qquad (4)$$

As in the classical case, we transform to normal coordinates to find

$F_N(t)$

$$= \frac{1}{NZ} \sum_i \text{trace}\{e^{iHt/\hbar} \sum_k U_{jk}P_k e^{-iHt/\hbar} \sum_m U_{jm}P_m e^{-\beta H}\}$$

$$= \frac{1}{NZ} \sum_{k,m} \text{trace}\{e^{iHt/\hbar} P_k e^{-iHt/\hbar} P_m e^{-\beta H} \sum_j U_{jk}U_{jm}\}$$

$$= \frac{1}{NZ} \sum_k \text{trace}\{e^{iHt/\hbar} P_k e^{-iHt/\hbar} P_k e^{-\beta H}\} \quad (5)$$

(using 3.10). The following three identities:

$$\exp(-sH)\psi_j = \exp(-sE_j)\psi_j,$$

$$P_k \psi_j(P) = \sum_l \psi_l(P) \int \psi_l^*(P') P_k' \psi_j(P') d^N P',$$

and

$$\sum_m \psi_m^*(P')\psi_m(P'') = \delta(P'-P''),$$

yield

$$F_N(t) = \frac{1}{NZ} \int P_k'' P_k' \{\sum_l e^{-iE_l t/\hbar} \psi_l(P'') \psi_l^*(P')$$

$$\times \{\sum_j e^{-(\beta-it/\hbar)E_j} \psi_j^*(P'')\psi_j(P')\} d^N P'' d^N P'. \quad (6)$$

In view of the fact that the complete partition function Z factors into N single normal mode partition functions and that the wave functions factor into N independent wave function, we find, remembering that the energy levels of the kth oscillator are $\hbar\omega_k(j+\tfrac{1}{2})$ with $j=0, 1, 2\cdots$,

$$F_N(t) = \frac{1}{N} \sum_k Z_k^{-1} \sum_{l,j} \exp\{-i\omega_k t(l-j) - \beta\hbar\omega_k(j+\tfrac{1}{2})\}$$

$$\times (j|P_k|l)(l|P_k|j), \quad (7)$$

where (in a position representation)

$$(j|P|l) = \frac{\hbar}{i} \int_{-\infty}^{\infty} \varphi_j^*(x) \frac{\partial}{\partial x} \varphi_l(x) dx \quad (8)$$

is the matrix element of the momentum operator $P \equiv \hbar i^{-1} \partial/\partial x$. Here $\varphi_l(x)$ is the lth harmonic oscillator wave function. It is well known that

$$(\hbar/m\omega)^{\tfrac{1}{2}} d\varphi_n/dx = -[\tfrac{1}{2}(n+1)]^{\tfrac{1}{2}} \varphi_{n+1}(x) + (\tfrac{1}{2}n)^{\tfrac{1}{2}} \varphi_{n-1}(x). \quad (9)$$

Substitution of this expression into (8) and application of the orthonormality condition $\int \varphi_j^*(x)\varphi_l(x)dx = \delta_{jl}$ yields

$$(j|p|l) = (\hbar/i)(m\omega/\hbar)^{\tfrac{1}{2}}$$
$$\times \{-[\tfrac{1}{2}(l+1)]^{\tfrac{1}{2}} \delta_{j, l+1} + [\tfrac{1}{2}l]^{\tfrac{1}{2}} \delta_{j, l-1}\}. \quad (10)$$

The matrix element $(l|p|j)$ is obtained by interchanging j and l in this formula. Substitution of these matrix elements into (7) yields

$$F_N(t) = \frac{m\hbar}{2N} \sum_k (\omega_k/Z_k) e^{-\tfrac{1}{2}\beta\hbar\omega_k}$$

$$\times [e^{it\omega_k} + e^{\beta\hbar\omega_k} e^{-it\omega_k}] \sum_{j=0}^{\infty} j e^{-j\beta\hbar\omega_k}. \quad (11)$$

Since

$$\sum j e^{-j\theta} = -(\partial/\partial\theta) \sum_j e^{-j\theta}$$
$$= -(\partial/\partial\theta)(1-e^{-\theta})^{-1} = e^{-\theta}/(1-e^{-\theta})^2$$

and since the harmonic oscillator partition function is $Z = \exp(-\tfrac{1}{2}\hbar\omega\beta)/[1-\exp(-\hbar\beta\omega)]$, we finally obtain

$$F_N(t) = \frac{m\hbar}{2N} \{\sum_k \omega_k \coth(\tfrac{1}{2}\beta\hbar\omega_k) \cos t\omega_k$$
$$- i \sum \omega_k \sin t\omega_k\}, \quad (12)$$

which is complex, as is generally the case for quantum mechanical correlation functions.

Now the required autocorrelation function is

$$\rho_N(t) = F_N(t)/F_N(0)$$

$$= \frac{\{\sum_k \omega_k \coth(\tfrac{1}{2}\beta\hbar\omega_k) \cos t\omega_k - i \sum \omega_k \sin t\omega_k\}}{\sum_k \omega_k \coth(\tfrac{1}{2}\beta\hbar\omega_k)}. \quad (13)$$

It reduces to the classical form (4.8) as $\hbar \to 0$.

We hope to discuss quantum mechanical Poincaré cycles at a later date.

APPENDIX I

Let $y(\omega)$ and $x(t)$ be related in the manner described in Eq. (2.7). Then

$$y(b) - y(a) = \frac{1}{2\pi} \int_{-\infty}^{\infty} x(t) dt \int_a^b e^{-i\omega t} d\omega.$$

Hence,

$$|y(b) - y(a)|^2 = \frac{1}{4\pi^2} \int\int x(t)x^*(t') dt dt'$$

$$\times \int\int_a^b e^{-i(\omega t - \omega' t')} d\omega d\omega',$$

so that [in view of (2.1) and (2.5)]

$$E\{|y(b) - y(a)|^2\}$$

$$= \frac{1}{4\pi^2} \int\int_{-\infty}^{\infty} \rho(t-t') dt dt' \int\int_a^b e^{-i(\omega t - \omega' t')} d\omega d\omega'$$

$$= \frac{1}{4\pi^2} \int_{-\infty}^{\infty} \int\int_a^b dF(\omega'') d\omega d\omega'$$

$$\times \int\int e^{-i(t-t')(\omega-\omega'')} e^{it'(\omega'-\omega)} dt dt'.$$

Let $\tau=t-t'$ and integrate first with respect to τ. Then,

$$E\{|y(b)-y(a)|^2\}$$
$$=\int_{-\infty}^{\infty} dF(\omega'') \int_a^b \int_a^b d\omega d\omega' \delta(\omega-\omega'')\delta(\omega-\omega').$$

Since both ω and ω' are restricted to the same interval (a,b),

$$E\{|y(b)-y(a)|^2\}=\int_{-\infty}^{\infty} dF(\omega'') \int_a^b \delta(\omega-\omega'')d\omega.$$

Now the integral over ω is 1 if $a<\omega''<b$, and 0 otherwise. Hence

$$E\{|y(b)-y(a)|^2\}=F(b)-F(a).$$

In particular, as $\mathcal{E}\rightarrow 0$,

$$E\{|y(\omega+\mathcal{E})-y(\omega-\mathcal{E})|^2\}=F(\omega+\mathcal{E})-F(\omega-\mathcal{E})=dF(\omega).$$

APPENDIX II

Let
$$H(t)=\begin{cases} 1 & \text{if } |t|<T \\ 0 & \text{if } |t|>T. \end{cases}$$

Then $H(t)$ has the fourier integral representation

$$H(t)=\frac{1}{\pi}\int_{-\infty}^{\infty} \frac{\sin\omega T}{\omega} e^{-i\omega t} d\omega.$$

Hence,

$$\lim_{T\to\infty}\frac{1}{2T}\int_{-T}^{T} x(t)dt = \lim_{T\to\infty}\frac{1}{2\pi}\int_{-\infty}^{\infty} x(t)dt \int_{-\infty}^{\infty} d\omega \frac{\sin\omega T}{\omega T} e^{-i\omega T}.$$

Now we employ the fourier integral representation (2.7) of $x(t)$ and interchange orders of integration. Then

$$\lim_{T\to\infty}\frac{1}{2\pi}\int_{-\infty}^{\infty} d\omega \frac{\sin\omega T}{\omega T}\int_{-\infty}^{\infty} dy(u)\int_{-\infty}^{\infty} e^{-it(u-\omega)}dt$$
$$=\lim_{T\to\infty}\int_{-\infty}^{\infty} d\omega \frac{\sin\omega T}{\omega T} \int_{-\infty}^{\infty} \delta(u-\omega)dy(u)$$
$$=\lim_{T\to\infty}\int_{-\infty}^{\infty} \frac{\sin uT}{uT} dy(u) = \lim_{T\to\infty} I_1+I_2+I_3,$$

where, if we let $h=T^{-(1+\mathcal{E})}$,

$$(I_1,I_2,I_3)=\left(\int_{-h}^{h},\int_h^{\infty},\int_{-\infty}^{-h}\right)\frac{\sin uT}{uT}dy(u).$$

Since $|uT|<|T^{-\mathcal{E}}|$ in the interval $(-T^{-(1+\mathcal{E})}, T^{-(1+\mathcal{E})})$, one can by choosing T sufficiently large make $(\sin ut)/uT$ as close to unity as is desired throughout the entire interval. Hence,

$$\lim_{T\to\infty} I_1 = \lim_{h\to 0}\int_{-h}^{h} dy(u) = y(0+)-y(0-).$$

If $y(u)$ is of bounded variation it is easy to show by standard methods that $\lim(I_2+I_3)=0$. Hence

$$\lim\frac{1}{2T}\int_{-T}^{T} x(t)dt = y(0+)-y(0-).$$

APPENDIX III. STATISTICAL PROPERTIES OF FUNCTIONS WHICH CAN BE EXPRESSED AS FINITE FOURIER SERIES

We now derive the theorem of Kac concerning the average number of times a function

$$x(t)=\sum_{k=1}^{n} a_k \cos(t\omega_k+\alpha_k)$$

achieves a value q in a time interval of a given length. The a_k's and ω_k's are postulated to be real and the ω_k's to be linearly independent.

The required mean frequency can be written as

$$L(q)=\lim_{T\to\infty}\frac{1}{2T}N_T(q), \quad (2)$$

where $N_T(q)$ is the number of zeros of

$$F(t)=x(t)-q \quad (3)$$

in the interval $-T<t<T$.

In the calculation of $L(q)$ we use the Kronecker-Weyl theorem[21] which allows one to replace time averages by phase averages. Let a multiply periodic function $f(\vartheta_1,\cdots,\vartheta_n)\equiv f(\vartheta)$ be defined on an n-dimensional torus on which each ϑ_j ranges from 0 to 2π. $f(\vartheta')\equiv f(\vartheta'')$ if $\vartheta_j'\equiv\vartheta_j''(\text{mod}2\pi)$ for all j. Finally, if

$$\vartheta_j=t\omega_j+\alpha_j \quad (4)$$

and the ω_j's are linearly independent, i.e., if there exists no set of integral m_k's (positive or negative) such that

$$\sum \omega_k m_k = 0, \quad (5)$$

then one can prove the result of Weyl that

$$\frac{1}{2T}\int_{-T}^{T} f[\vartheta(t)]dt \rightarrow$$
$$\frac{1}{(2\pi)^n}\int_0^{2\pi}\cdots\int f(\vartheta_1\cdots\vartheta_n)d\vartheta_1\cdots d\vartheta_n \quad (6)$$

as $T\rightarrow\infty$.

We now show that in the case of linearly independent ω_k's the Kronecker-Weyl theorem and some delta function arguments lead immediately to a formula

---
[21] H. Weyl, Am. J. Math. **60**, 889 (1938).

derived more rigorously by Kac,

$$L(q) = \frac{1}{2\pi^2} \int\int_{-\infty}^{\infty} \eta^{-2} \cos q\alpha \{\prod_{k=1}^{n} J_0(|a_k|\alpha)$$

$$- \prod_{k=1}^{n} J_0(|a_k|[\alpha^2 + \eta^2\omega_k^2]^{\frac{1}{2}})\} d\alpha d\eta. \quad (7)$$

Let $F(t)$ be a function of t which is real for real values of t. Let its zeros be at $t_1, t_2 \cdots$. Then the number of zeros in the interval $(-T, T)$ is

$$\int_{-T}^{T} \sum_i \delta(t - t_i) dt. \quad (8)$$

Since complex roots occur in pairs, say at points $a+ib$ and $a-ib$ the delta function sum which go with the pair is

$$\delta([t-a]-ib) + \delta([t-a]+ib) = \text{const}\,\delta([t-a]^2 + b^2)$$

and

$$\int_{-\infty}^{\infty} \delta[(t-a)^2 + b^2] dt = 0,$$

(since the $[(t-a)^2 + b^2]$ cannot vanish for any real value of t). Hence, as we would expect, complex roots never contribute to (8).

The well-known delta function formula

$$F(t) = \sum_i \frac{\delta(t-t_i)}{|F'(t_i)|}$$

[the t_j's being the zeros of $F(t)$] implies that (8) is equivalent to

$$\int_{-T}^{T} |F'(t)| \sum_i \frac{\delta(t-t_i)}{|F'(t_i)|} dt = \int_{-T}^{T} |F'(t)| \delta[F(t)] dt. \quad (9)$$

Then, if we employ the Fourier integral representation of $\delta|F(t)|$, and interchange orders of integration, (2) becomes

$$L(q) = \lim_{T \to \infty} \frac{1}{4\pi T} \int_{-\infty}^{\infty} \int_{-T}^{T} |F'(t)| \exp i\alpha F(t) dt d\alpha. \quad (10)$$

Finally, the formula

$$|u| = \frac{1}{\pi} \int_{-\infty}^{\infty} \frac{1 - \cos \eta u}{\eta^2} d\eta$$

yields

$$L = \lim_{T \to \infty} \frac{1}{4\pi^2 T} \int\int \int_{-T}^{T} \eta^{-2} d\eta d\alpha \{1 - \cos[\eta F'(t)]\}$$

$$\times \exp[i\alpha F(t)] dt d\alpha$$

$$= \frac{1}{2\pi^2} \int\int_{-\infty}^{\infty} \eta^{-2} \{U(\alpha,0) - \tfrac{1}{2} U(\alpha,\eta) - \tfrac{1}{2} U(\alpha,-\eta)\} d\eta d\alpha,$$

where

$$U(\alpha, \beta) = E\{\exp i[\alpha F(t) + \beta F'(t)]\}. \quad (12)$$

This characteristic function is readily obtained for our special case $F(t) = x(t) - q$ since the Kronecker-Weyl theorem allows us to replace time averages by averages over our ϑ torus. Then

$$U(\alpha,\beta) = (\exp i\alpha q) \prod_{j=1}^{n} \frac{1}{2\pi} \int_0^{2\pi}$$

$$\times \exp i[\alpha a_j \cos \vartheta_j - \beta a_j \omega_j \sin \vartheta_j] d\vartheta_j$$

$$= (\exp i\alpha q) \prod_{j=1}^{n} J_0[|a_j|(\alpha^2 + \beta^2 \omega_j^2)^{\frac{1}{2}}]. \quad (13)$$

Substitution of (13) and (12) into (11) gives us the Kac formula (7).

Consider the special case $a_1 = a_2 = \cdots = a = 1$. Since

$$J_0(z) = \{1 - (\tfrac{1}{2}z)^2 + \tfrac{1}{2} \cdot \tfrac{1}{2}(\tfrac{1}{2}z)4 - \cdots\}$$
$$= \{1 - \tfrac{1}{2} \cdot \tfrac{1}{2}(\tfrac{1}{2}z)^4 + O(z^6)\} \exp{-(\tfrac{1}{2}z)^2} \quad (14)$$

is peaked at $z = 0$, the product (13) of Bessel functions (14) becomes even more peaked in the limit as $n \to \infty$. Then (13) becomes

$$U(\alpha, \beta) = \exp[i\alpha q - (n/4)(\alpha^2 + \beta^2 \omega_0^2)]$$

$$\times \left\{1 - \frac{1}{64} \sum_j (\alpha^2 + \beta^2 \omega_j^2)^2 + \cdots \right\},$$

where

$$\omega_0^2 = \frac{1}{n} \sum \omega_j^2.$$

Now let $q = bn^{\frac{1}{2}}$, $\alpha n^{\frac{1}{2}} = x$, $\eta n^{\frac{1}{2}} = y$, and let $n \to \infty$. Then, if

$$\lim \frac{1}{n^2} \sum \omega_j^4 = 0.$$

We have, as $n \to \infty$,

$$L(bn^{\frac{1}{2}}) \sim \frac{1}{2\pi^2} \int\int_{-\infty}^{\infty} \exp\{ibx - \tfrac{1}{4}x^2\}$$

$$\times (1 - \exp[-\tfrac{1}{4} y^2 \omega_0^2]) y^{-2} dx dy = (\omega_0/\pi) \exp -b^2,$$

which has also been derived by Kac.

APPENDIX IV. AUTOCORRELATION FUNCTION OF SQUARE LATTICE

The behavior of the autocorrelation function $\rho(t)$ as $t \to \infty$ can be obtained from (5.4) by employing the form (3.3) for the frequency spectrum (with $\gamma_1 > \gamma_2$). Logrithmic singularities exist at $\omega = \omega_1$, and $\omega = \omega_2$. If we substitute (3.3) into (5.4), break up the integral into three integrals of ranges $(0, \omega_1)$, (ω_1, ω_2), and (ω_2, ω_L),

and integrate by parts we find

$$\rho(t) = \lim_{\mathscr{E}\to 0} \Big\{ \frac{1}{t} \sin t\omega_2 [g([\omega_2{}^2 - \mathscr{E}]^{\frac{1}{2}}) - g([\omega_2{}^2 + \mathscr{E}]^{\frac{1}{2}})]$$

$$+ \frac{1}{t} \sin t\omega_1 [g([\omega_1{}^2 - \mathscr{E}]^{\frac{1}{2}}) - g([\omega_1{}^2 + \mathscr{E}]^{\frac{1}{2}})]$$

$$+ \frac{1}{t} \sin t\omega_L g([\omega_L{}^2 - \mathscr{E}]^{\frac{1}{2}}) - \int_0^{\omega_L} g'(\omega) \sin t\omega d t \Big\}.$$

However, it is easy to show that, as $\mathscr{E} \to 0$,

$$g([\omega_2{}^2 - \mathscr{E}]^{\frac{1}{2}}) \sim \frac{4}{\pi^2 \omega_1} K\{[(1 - \mathscr{E}\omega_2{}^{-2})(1 + \mathscr{E}\omega_1{}^{-2})]^{\frac{1}{2}}\}$$

$$= \frac{4}{\pi^2 \omega_1} K(1 + \tfrac{1}{2}\mathscr{E}[\omega_1{}^{-2} - \omega_2{}^{-2}])$$

$$g([\omega_2{}^2 + \mathscr{E}]^{\frac{1}{2}}) \sim \frac{4}{\pi^2 \omega_1} K([1 + \mathscr{E}\omega_2{}^{-2})(1 - \mathscr{E}\omega_2{}^{-2})]^{-\frac{1}{2}})$$

$$= \frac{4}{\pi^2 \omega_1} K(1 + \tfrac{1}{2}\mathscr{E}[\omega_1{}^{-2} - \omega_2{}^{-2}])$$

so that as $\mathscr{E} \to 0$

$$g([\omega_2{}^2 - \mathscr{E}]^{\frac{1}{2}}) - g([\omega_2{}^2 - \mathscr{E}]^{\frac{1}{2}}) \to 0.$$

Also

$$g([\omega_1{}^2 - \mathscr{E}]^{\frac{1}{2}}) \sim \frac{4}{\pi^2 \omega_2} K([(1 - \mathscr{E}\omega_1{}^{-2})(1 + \mathscr{E}\omega_2{}^{-2})]^{-\frac{1}{2}})$$

$$= \frac{4}{\pi^2 \omega_2} K(1 + \tfrac{1}{2}\mathscr{E}[\omega_1{}^{-2} - \omega_2{}^{-2}])$$

$$g([\omega_1{}^2 + \mathscr{E}]^{\frac{1}{2}}) \sim \frac{4}{\pi^2 \omega_2} K([(1 + \mathscr{E}\omega_1{}^{-2})(1 - \mathscr{E}\omega_2{}^{-2})]^{\frac{1}{2}})$$

$$= \frac{4}{\pi^2 \omega_2} K[1 + \tfrac{1}{2}\mathscr{E}(\omega_1{}^{-2} - \omega_2{}^{-2})],$$

so that, as $\mathscr{E} \to 0$,

$$g([\omega_1{}^2 - \mathscr{E}]^{\frac{1}{2}}) - g([\omega_1{}^2 - \mathscr{E}]^{\frac{1}{2}}) \to 0.$$

Then, since, as $\mathscr{E} \to 0$,

$$g([\omega_L{}^2 - \mathscr{E}]^{\frac{1}{2}}) \sim \frac{4\omega_L}{\pi^2 \omega_1 \omega_2} K(0) = \frac{2\omega_L}{\pi \omega_1 \omega_2},$$

we have

$$\rho(t) = -\frac{2\omega_L \sin t\omega_L}{\pi t \omega_1 \omega_2} - \frac{1}{t} \int_0^{\omega_L} g'(\omega) \sin \omega t d t$$

$$\sim (2\omega_L \sin t\omega_L)/\pi t \omega_1 \omega_2.$$

APPENDIX V

The correlation function,

$$\rho_N(t,r) = \frac{1}{M^n} \sum_k \cos \frac{2\pi r \cdot k}{M} \cos t\omega_k \quad (1)$$

has the property

$$\rho_N(t,0) = \delta_{r,0}. \quad (2)$$

It easily can be verified to satisfy the equation

$$m \frac{d^2 \rho_N}{dt^2} = \sum_i \gamma [_j\rho_N(t, \cdots, r_{j+1}, \cdots)$$
$$- 2\rho_N(t, \cdots r_j \cdots) + \rho_N(t, \cdots r_{j-1} \cdots)] \quad (3)$$

with boundary conditions,

$$\rho_N(t, r+sM) = \rho_N(t,r), \quad (4)$$

where s is any vector with integral components.
One can easily verify that

$$\rho_N(t,r) = \sum_{k=-\infty}^{\infty} J_{2(r+kN)}(t\omega_L) \quad (5)$$

is the solution of (2), (3), and (4) when $n=1$ (linear chain). Here $J_\alpha(t)$ is the Bessel function of order α.

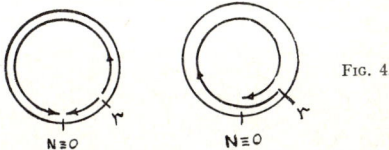

Fig. 4.

This result has a simple physical interpretation. The point 0 correlates with r on an infinite lattice through $J_{2r}(t\omega_L)$ (see Eq. 5.3). Hence, with our periodic boundary conditions the correlation of point 0 (which we identify with N) with r in a clockwise direction is also $J_{2r}(t\omega_L)$. In the counterclockwise direction, it is

$$J_{2(N-r)}(t\omega_L) \equiv J_{2(r-N)}(t\omega_L).$$

It is also correlated through multicirculation paths (see Fig. 4). The total correlation length in Fig. 4(b) is $(N+r)$ so that the term associated with this diagram is

$$J_{2(N+r)}(t\omega_L).$$

Finally, if one adds the contribution of all paths (clockwise and counterclockwise) which encircle our one dimensional lattice any number of times one merely obtains (5).

Reprinted from THE JOURNAL OF CHEMICAL PHYSICS, Vol. 26, No. 3, 454–464, March, 1957
Printed in U. S. A.

Studies in Nonequilibrium Rate Processes.* I. The Relaxation of a System of Harmonic Oscillators

ELLIOTT W. MONTROLL, *Institute for Fluid Dynamics and Applied Mathematics, University of Maryland, College Park, Maryland*

AND

KURT E. SHULER, *National Bureau of Standards, Washington, D. C.*

(Received April 13, 1956)

As a part of an investigation of nonequilibrium phenomena in chemical kinetics a theoretical study has been made of the collisional and radiative relaxation of a system of harmonic oscillators contained in a constant temperature heat bath and prepared initially in a vibrational nonequilibrium distribution. An exact solution has been obtained for the general relaxation equation applicable to this system and expressions have been derived for the relaxation of initial Boltzmann distributions, Poisson distributions, and δ-function distributions as well as for the relaxation of the moments of the distributions. Using the latter result, explicit expressions are given for the relaxation of the internal energy of the system of oscillators and for the time dependence of the dispersion of the distributions.

1. INTRODUCTION

IT has been recognized for many years that by its very nature a chemical reaction must produce a perturbation in the initial Maxwell-Boltzmann distribution of the reactant species.[1] The extent of the departure from equilibrium will depend upon the relative magnitudes of the rates of the elementary chemical reactions (i.e., the rate of transformation of reactants

---

* This research was partially supported by the U. S. Air Force through the Office of Scientific Research of the Air Research and Development Command and by the U. S. Atomic Energy Commission.

[1] See, e.g., R. H. Fowler and E. A. Guggenheim, *Statistical Thermodynamics* (Cambridge University Press, New York, 1949), Chap. XII, or Eyring, Walter, and Kimball, *Quantum Chemistry* (John Wiley and Sons, Inc., New York, 1944), Chap. XVI.

to products) and the rates of energy exchange between the various atomic and molecular species in the reaction system. If the rate of the chemical transformation is small compared with the rate of energy exchange, the perturbation of the initial equilibrium distribution will be small and the reaction system can be discussed in terms of equilibrium statistical mechanics. If, however, the rate of the chemical transformation exceeds the rate of intra- and intermolecular energy exchange, there may develop a considerable perturbation of the equilibrium Maxwell-Boltzmann distribution of energy during the course of the chemical reaction. Under these conditions the equilibrium hypothesis underlying the present collision and absolute rate theories of chemical kinetics may no longer be tenable. Recent experimental work on various rapid high-temperature chemical reactions has shown quite clearly that there is indeed in many cases a considerable perturbation of the initial equilibrium distribution of energy during the course of the chemical reaction.[2] It therefore becomes important to study the distribution of energy in a reaction system during the course of a chemical reaction so that a foundation can be laid for the development of a nonequilibrium theory of chemical kinetics.

The specific problem which we wish to consider in the above context concerns the relaxation of the distribution of a system of harmonic oscillators prepared initially in nonequilibrium vibrational distributions. The oscillators are excited to these distributions either by external perturbations, such as irradiation with short duration, high intensity light or by the passage of a shock wave, or internally by some specific chemical reaction.[3] After the external perturbation has been removed (i.e., after the light has been turned off or after the passage of the shock wave) or after the cessation of the reaction, the system of oscillators will relax to its final equilibrium distribution by inelastic collisions and by radiative transitions. We wish to study in detail the dynamic behavior of the distribution and of the moments of the distribution of the oscillators among their energy levels for various initial nonequilibrium distributions.

Our study of the relaxation of a system of harmonic oscillators is based on the following model:

(a) The oscillators are contained in a large excess of (chemically) inert gas which acts as a *constant temperature* heat bath throughout the relaxation process. This implies that the concentration of the excited oscillators is sufficiently small and the energy absorbed by them during their excitation is sufficiently small that the heat bath remains at its initial equilibrium temperature T throughout the relaxation process.

(b) The total concentration of excited oscillators is sufficiently small so that the relaxation process is first order with respect to the concentration of oscillators. The energy exchange which controls the relaxation thus takes place primarily between the oscillators and the heat bath.

(c) The excited oscillators can transfer their vibrational energy both by collision and by radiation. In the collisional transfer of energy, the vibrational energy of the excited oscillators can be exchanged with both the translational and the vibrational degrees of freedom of the heat bath molecules.

(d) The collisional transition probabilities for transitions between the vibrational levels i and j of the harmonic oscillators are to be calculated according to the prescription of Landau and Teller.[4] According to this prescription, the perturbations which induce the transitions are linear in the normal coordinate (i.e., the internuclear separation in the case of a harmonic oscillator) and sufficiently small for a first order perturbation calculation. With these assumptions, the matrix elements for collisional transitions are identical, except for a constant factor, with those for the radiative transitions of a harmonic oscillator. The same "selection rules" will thus hold for collisional transitions as for radiative ones in that the collision induced transitions of the oscillators will take place only between adjacent vibrational levels. The collisional transition probabilities per collision, $P_{i,j}$ are thus given by

$$P_{i,j}=P_{j,i}, \quad P_{i,j}=0 \quad \text{for} \quad j\neq \begin{cases} i+1 \\ i-1 \end{cases}$$

$$P_{i,j+1}=(i+1)P_{10},$$

(1.1)

where P_{10} is the collisional transition probability per collision for transitions between vibrational levels $i=1$ and $i=0$.

We now wish to derive the differential equations which govern the relaxation of the ensemble of harmonic oscillators in our model. It has been pointed out by Herzfeld[5] that an exact energy balance in a relaxation process of the type discussed here can be obtained when either (a) the excited system or the heat bath have a nearly continuous array of levels or (b) the excited system and the heat bath have equidistant energy levels and exchange only vibrational energy. Under either of these conditions, a transition $-\Delta E$ between two states in the excited system can be matched by a transition of a corresponding energy ΔE between two states of the heat bath. This latter case can readily be realized if one chooses for the relaxation system an

---

[2] For a more detailed discussion of this point see K. E. Shuler, J. Phys. Chem. **57**, 396 (1953); *5th Symposium (International) on Combustion* (Reinhold Publishing Corporation, New York, 1955), pp. 56–74.

[3] An example of the latter process is the formation of OH in the vibrational state $v=9$ in the reaction $H+O_3\rightarrow OH+O_2$ studied by A. B. Meinel, J. Astrophys. **111**, 207, 433, 555 (1950) and by McKinley, Garvin, and Boudart, J. Chem. Phys. **23**, 784 (1955).

[4] L. Landau and E. Teller, Physik. Z. Sowjetunion **10**, 34 (1936).

[5] K. F. Herzfeld in *Temperature, Its Measurement and Control in Science and Industry* (Reinhold Publishing Corporation, New York, 1955), p. 233.

ensemble of harmonic oscillators of which a small fraction are excited to an initial vibrational nonequilibrium distribution while the large excess of unexcited oscillators serves as the heat bath. If we let

$x_n(t)$ = fraction of excited oscillators in level n
y_i = concentration of heat bath oscillators in level i
$P_{n,n+1;i,i-1}$ = probability per collision for the energy transfer $n \to n+1$ as $i \to i-1$

the relaxation equation can be written as

$$\frac{dx_n(t)}{dt} = -Z[x_n(\sum_{i=0}^{\infty} y_i P_{n,n+1;i,i-1} + \sum_{i=0}^{\infty} y_i P_{n,n-1;i,i+1})$$

$$-x_{n+1}\sum_{i=0}^{\infty} y_i P_{n+1,n;i,i+1}$$

$$-x_{n-1}\sum_{i=0}^{\infty} y_i P_{n-1,n;i,i-1}] \quad (1.2)$$

where Z is the collision number, i.e., the number of collisions per second suffered by the oscillator in the level system $n = 0, 1, \cdots$ when the gas density is one molecule per unit volume. Using Eq. (1.1), the probabilities $P_{n,n+1;i,i-1}$ for concurrent collisional transitions can be written as

$$P_{n,n+1;i,i-1} = (n+1)i P_{10}$$
$$P_{n,n-1;i,i+1} = n(i+1) P_{10} \quad (1.3)$$

so that Eq. (1.2) becomes

$$\frac{dx_n(t)}{dt} = -ZP_{10}[(n+1)x_n\sum_{i=0}^{\infty} iy_i + nx_n\sum_{i=0}^{\infty}(i+1)y_i$$

$$-(n+1)x_{n+1}\sum_{i=0}^{\infty}(i+1)y_i - nx_{n-1}\sum_{i=0}^{\infty} iy_i] \quad (1.4)$$

$$n = 0, 1, 2, \cdots.$$

Since we assume that the heat bath remains in its initial Boltzmann distribution at temperature T throughout the relaxation process we can write, for all times t,

$$y_i = N(1-e^{-\theta})e^{-i\theta} \quad (1.5)$$

where N is the total concentration of oscillators in the heat bath and where $\theta = h\nu/kT$ and ν is the fundamental frequency of the oscillators. Substitution of (1.5) into (1.4) finally leads to

$$dx_n(t)/dt = k_{10}(1-e^{-\theta})^{-1}\{ne^{-\theta}x_{n-1}$$
$$-[n+(n+1)e^{-\theta}]x_n + (n+1)x_{n+1}\}$$
$$n = 0, 1, 2, \cdots \quad (1.6)$$

where $k_{10} = ZP_{10}N$ is the collisional transition probability per second for transitions between levels 1 and 0 of the oscillators. The set of differential difference equations (1.6) governs the relaxation of a system of excited harmonic oscillators contained in a harmonic oscillator heat bath (with $\nu_n = \nu_i$) when there is only vibrational energy exchange between the excited oscillators and the heat bath.

It is not possible to follow the method used above to obtain Eq. (1.6) when the relaxation proceeds by the interchange of the vibrational energy of the excited oscillators with the translational energy of the heat bath. In this case it is not possible to establish internal equilibrium by considering only the energy transfer between the excited oscillators and the heat bath as was done previously since the oscillators will give up their excitation energy only in quanta of $h\nu$ while the heat bath has a nearly continuous array of translational energy states. Furthermore, it is not possible to write down simple explicit expressions for the joint transition probabilities $P_{n;i}$ as was done in (1.3), where i now refers to the translational energy levels of the heat bath, within the framework of the Landau-Teller approximation used in our model. It has been shown, however, by Rubin and Shuler[6] that the set of differential difference equations governing the relaxation process now under discussion can be obtained by the method used by Fowler in discussing the equilibrium relationship between collisions of the first and second kind.[7] Using properties (a) to (d) of our model and applying the principle of detailed balancing at equilibrium, Rubin and Shuler showed that the relaxation equation for the case when the relaxation proceeds by the interchange of the vibrational energy of the excited oscillators with the translational energy of the heat bath has the form of (1.6) except for the absence of the factor $(1-e^{-\theta})^{-1}$ in front of the braces (see Appendix II).

A third relaxation mechanism involves the interchange of radiation between the excited oscillators and the heat bath. The relaxation equations for this case have been derived by Rubin and Shuler[8] by considering the interaction of the oscillators with a radiation heat bath in equilibrium with the heat bath at the temperature T. Using the Einstein coefficients A and B for spontaneous and induced emission and for absorption and Planck's radiation law for the density of the radiation, it could readily be shown that the relaxation equation for radiative transitions is again of the form of (1.6) but with k_{10} replaced by A_{10}, the Einstein coefficient for spontaneous emission between vibrational levels 1 and 0 of the oscillators.[9]

[6] R. J. Rubin and K. E. Shuler, J. Chem. Phys. 25, 59 (1956).
[7] R. H. Fowler, Phil. Mag. 47, 257 (1924).
[8] R. J. Rubin and K. E. Shuler, J. Chem. Phys. 26, 137 (1957).
[9] It should be noted that the case of radiative relaxation could also be discussed in terms of the transfer of photons with energy $h\nu$ between the excited oscillators and the heat bath oscillators by the method used above for the transfer of vibrational energy and without recourse to the radiation field. The exact correspondence between these two relaxation process explains the exact correspondence between the equations describing the two processes when the appropriate transition probabilities, i.e., k_{10} or A_{10}, are used in Eq. (1.1). It should also be noted that an internal energy balance can be maintained for this relaxation process.

The general relaxation equation applicable to the relaxation of a system of harmonic oscillators in a constant temperature heat bath can finally be written as

$$\frac{dx_n(t)}{dt} = \kappa\{ne^{-\theta}x_{n-1} - [n+(n+1)e^{-\theta}]x_n + (n+1)x_{n+1}\}$$
$$n = 0, 1, 2, \cdots \quad (1.7)$$

where

$$k = \begin{cases} k_{10}(1-e^{-\theta})^{-1} & \text{for collisional vibration-vibration exchange} \\ k_{10}' & \text{for collisional vibration-translation energy exchange} \\ A_{10}(1-e^{-\theta})^{-1} & \text{for radiative energy exchange} \end{cases} \quad (1.7a)$$

It is the object of this paper to obtain an exact solution of (1.7) subject to the condition $\sum x_n(t) = 1$ (closed system) for various initial distributions $x_n(0)$.

Rubin and Shuler[6] obtained a solution of (1.7) for the special case $\theta \ll 1$, which, in essence, corresponds to replacing the discrete set of energy levels by a quasi-continuum, by approximating the set of differential difference equations (1.7) by the related partial differential equation which then admitted of a solution in terms of a Fourier development in Laguerre polynomials. The choice of $\theta \ll 1$, made by Rubin and Shuler for mathematical convenience, is realized physically only for very few molecules and then only at rather high temperatures. Thus, for instance, one finds $\theta = h\nu/kT = 60/T$ for Cs_2 so that the inequality $\theta \ll 1$ can be fulfilled for $T > 10^3$ °K. For most diatomic species, however, $\theta > 1$ at ordinary temperatures (300–1000°K). Some examples are NO($\theta = 2.73 \cdot 10^3/T$), CO($\theta = 2.13 \cdot 10^3/T$), and OH($\theta = 5.37 \cdot 10^3/T$), where the frequencies ν correspond to the electronic ground states. For a heat bath at 300°K, one thus finds $\theta \sim 10$ to 20. We will show in the appendix that the general solution of (1.7), valid for all θ, reduces to the solution of Rubin and Shuler when $\theta \to 0$. The qualitative characteristics of relaxation from various initial distributions as determined by the small θ theory are in general agreement with the exact results derived below.

We shall show in Sec. 2 that the exact solution of (1.7) can be written in terms of the generating function

$$G(z,t) = \sum_{n=0}^{\infty} z^n x_n(t) \quad (1.8)$$

and the dimensionless time $\tau = \kappa t(1 - e^{-\theta})$ as

$$G(z,t) = \frac{(e^\theta - 1)}{(z-1)e^{-\tau} - (z - e^\theta)}$$
$$\times G_0\left[\frac{(z-1)e^{-\tau}e^\theta - (z - e^\theta)}{(z-1)e^{-\tau} - (z - e^\theta)}\right] \quad (1.9)$$

where $G_0(y) \equiv G(y,0)$ is determined by the initial condition (distribution) $x_n(0)$ and where $x_n(t)$, the fraction of the molecules in level n is the coefficient of z^n in (1.9). We consider in this study the relaxation of three initial nonequilibrium distributions which could readily be obtained in a physical system and the relaxation of their moments:

(1) An initial Boltzmann distribution with temperature $T_0 \neq T$ for which $x_n(0)$ is given by

$$x_n(0) = [1 - \exp(-\theta_0)]\exp(-n\theta_0) \quad (1.10)$$

where $\theta_0 = h\nu/kT_0$. Substitution of (1.10) into (1.9) yields the Boltzmann distribution (for details see Sec. 4)

$$x_n(t) = [1 - \exp(-\Theta)]\exp(-n\Theta) \quad (1.11)$$

with

$$\Theta = \log\left[\frac{e^{-\tau}(1-e^{-\theta_0}) - e^\theta(1-e^{-\theta_0})}{e^{-\tau}(1-e^{-\theta_0}) - (1-e^{-\theta_0})}\right]. \quad (1.12)$$

At early times

$$\Theta \sim \theta + \tau \frac{(1-e^{-\theta_0})}{(1-e^\theta)}(e^{\theta_0} - 1) + O(\tau^2) \quad (1.12a)$$

and as $\tau \to \infty$

$$\Theta \sim \theta + e^{-\tau}\left[\frac{(1-e^{-\theta})}{(1-e^{-\theta_0})}(1-e^{-\theta_0})\right] + O(e^{-2\tau}).$$

The initial Boltzmann distribution (1.10) thus relaxes to a final equilibrium Boltzmann distribution via the continuous sequence of Boltzmann distributions (1.11). Since the transient distribution of the relaxing oscillators is always canonical in this case, it is possible to characterize it by a "temperature" $T(t) = h\nu/k\Theta(t)$. To give an indication of the relaxation of this "temperature" we have plotted Θ^{-1} as a function of time for various initial and final temperatures T_0 and T in Fig. 1.[10]

An interesting feature of the curves in Fig. 1 is that the relaxation time associated with the temperature rise from $T_1 \to T_2(\theta_2 < \theta_1)$ is less than that for the corresponding temperature drop $T_2 \to T_1$ (see curves A and B). Qualitatively this is not surprising because more levels are available for occupation at the higher temperature equilibrium than at the lower. The system becomes "disordered" faster than it can be ordered.

(2) An initial Poisson distribution with

$$x_n(0) = e^{-a}a^n/n! \quad (1.13)$$

where a is the mean value $\bar{n}$ of the level number n. This represents a "peaked" initial distribution, $x_n(0)$, in which most of the excited oscillators are found initially in levels near $n = a$. The level population $x_n(t)$ resulting

---

[10] The persistence of the form of the Boltzmann distribution is a consequence of the Landau-Teller transition probabilities. It has been shown by Rubin and Shuler, J. Chem. Phys. **24**, 68 (1956), that other choices of transition probabilities will lead to a different relaxation behavior.

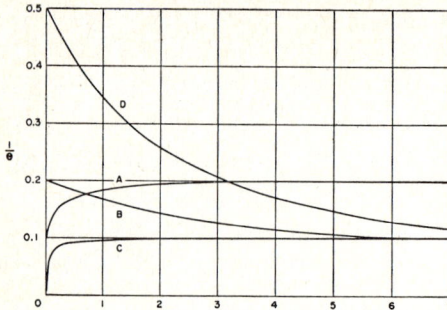

FIG. 1. The "temperature" $Tk/h\nu = \Theta^{-1}$ as a function of time τ for the relaxation of initial Boltzmann distributions [Eq. (1.12)]

Curve A: $\theta_0 = 10$, $\theta = 5$, $T_0 = \frac{1}{2}T$
Curve B: $\theta_0 = 5$, $\theta = 10$, $T_0 = 2T$
Curve C: $\theta_0 = 20$, $\theta = 10$, $T_0 = \frac{1}{2}T$
Curve D: $\theta_0 = 2$, $\theta = 10$, $T_0 = 5T$.

from an initial Poisson distribution is found to depend on the nth Laguerre polynomial [Eq. (5.6)]. We have plotted x_n as a function of τ for $\theta = 3$ and $a = 15$ in Fig. 2. Notice that the distribution narrows with time and shifts toward the equilibrium distribution as $t \to \infty$. We have also plotted $\log x_n$ vs t for the Poisson distribution (see Fig. 3) in order to gain some further information about the approach to equilibrium.

(3) An initial δ function distribution with all excited oscillators in state m:

$$x_n(0) = 1 \quad \text{when} \quad n = m$$
$$x_n(0) = 0 \quad \text{when} \quad n \neq m. \quad (1.14)$$

The level population $x_n(t)$ is given in terms of hypergeometric functions [see Eq. (6.5)]. The initially sharp distribution broadens and shifts to lower energy states

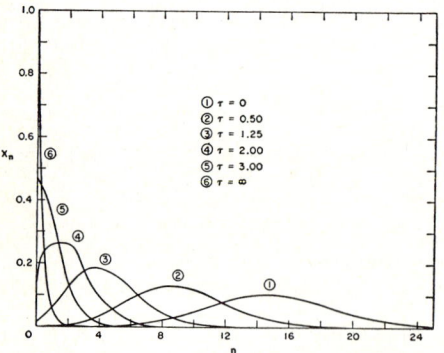

FIG. 2. The relaxation of an initial Poisson distribution $x_n(0) = e^{-a}a^n/n!$ with $a = \langle n \rangle = 15$ to a final Boltzmann distribution with $\theta = 3$. The ordinate x_n gives the fraction of oscillators in energy level n.

if $m > \bar{n}$ (in a manner similar to that plotted in reference 6 for the case $\theta \ll 1$).

(4) We have also obtained (see Sec. 3) a solution for the relaxation of the moments of the distribution. The transient behavior of the factorial moments of $x_n(t)$ [Eq. (3.1a)] defined by

$$f_m(t) = \sum_{n=0}^{\infty} n(n-1)\cdots(n-m+1)x_n(t)$$
$$m = 1, 2, \cdots \quad (1.15)$$

is described by (3.7):

$$\frac{df_m}{\kappa dt} + m(1 - e^{-\theta})f_m = m^2 e^{-\theta} f_{m-1}. \quad (1.16)$$

The internal energy $E(t)$ of our system of excited oscillators is related to the first moment f_1 by

$$E(t) = h\nu \sum_{n=0}^{\infty} nx_n(t) = h\nu f_1. \quad (1.17)$$

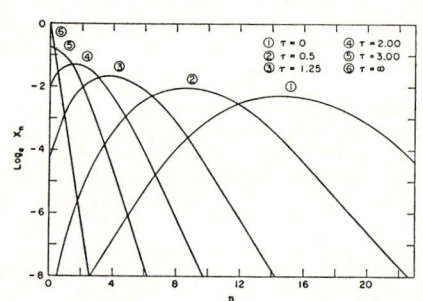

FIG. 3. A plot of $\log x_n$ vs n for the relaxation of the initial Poisson distribution shown in Fig. 2. The straight line portions of these curves for high n give a good indication of the adjustment of the initial Poisson distribution (at $\tau = 0$) to the equilibrium Boltzmann distribution at $\tau = \infty$.

The combination of Eqs. (1.16) and (1.17) readily leads to the remarkably simple expression of Bethe and Teller[11]

$$\frac{E(t) - E(\infty)}{E(0) - E(\infty)} = e^{-\tau} \quad (1.18)$$

for the relaxation of the internal energy where $E(\infty)$ is the internal energy which corresponds to the final Boltzmann distribution. It should be noted that according to (1.18), the magnitude of the internal energy at any time t depends only on $E(0)$ and not on the form of the initial distribution $x_n(0)$. The relaxation of the mean energy of a system of harmonic oscillators is therefore determined solely by the amount of energy added to the system and not by its distribution. Hence

[11] H. A. Bethe and E. Teller, "Deviations from thermal equilibrium in shock waves," Ballistic Research Laboratory, Report X-117, 1941. See also Rubin and Shuler, reference 6.

any description of nonequilibrium distributions (at least for the case of harmonic oscillators) which is based solely on the magnitude of the internal energy $E(t)$ can never give any information (other than $\bar{n}$) about the distribution $x_n(t)$ associated with this energy. From Eq. (1.18) and the definition of τ in (1.8) one obtains

$$t_{\text{relax}} = [\kappa(1-e^{-\theta})]^{-1} \quad (1.19)$$

for the relaxation time of the internal energy.

It is also possible to obtain the time dependence of the dispersion $\sigma^2(t)$, from Eq. (1.16) where

$$\sigma^2(t) = \langle (n-f_1)^2 \rangle_{\text{AV}} = f_2 + f_1(1-f_1). \quad (1.20)$$

A knowledge of the dependence of the dispersion of the distribution with time is of particular interest in connection with "peaked" initial distributions such as the δ function, Poisson, or Gaussian distributions since it gives some information about the "spreading" or "contraction" of the distribution as it tends toward the equilibrium Boltzmann distribution.

The "easy" problems of nonequilibrium statistical mechanics are those associated with physical systems which can be divided into two parts, (a) the large heat bath with many degrees of freedom which remains at equilibrium (and has the fluctuations expected in a system at equilibrium) and (b) the small nonequilibrium part with relatively few degrees of freedom which relaxes through interactions with the heat bath without disturbing the heat bath equilibrium.

The Einstein theory of Brownian motion is the classical example of this type of situation. The large Brownian particle with an initial δ-function distribution interacts with the surrounding fluid which remains at equilibrium. The theory developed in the present paper follows in the same spirit.[12] Mathematically the relevant equations associated with these processes are linear and can be discussed in considerable detail.

Those processes which do not permit the postulation of an equilibrium heat bath usually lead to nonlinear equations (for example the Boltzmann equation for the transport theory of gases) and have not been discussed in any really satisfactory manner. If the x_n's and y_i's in our Eq. (1.2) were both in a nonequilibrium state (for example by setting $x_i = y_i$) our differential equations would become nonlinear and little could be done with them.

2. GENERAL SOLUTION OF FUNDAMENTAL EQUATION

We solve (1.7) through the introduction of the generating function

$$G(z,t) = \sum_{n=0}^{\infty} z^n x_n(t) \quad (2.1a)$$

which is defined so that the coefficient of z^n is the fraction of molecules in the nth state at time t. We note that

$$\frac{\partial G}{\partial z} = z^{-1} \sum_{n=0}^{\infty} n z^n x_n(t). \quad (2.1b)$$

If we multiply (1.7) by z^n and sum from $n=0$ to $n=\infty$ we find that G satisfies the first-order partial differential equation

$$\frac{1}{\kappa}\frac{\partial G}{\partial t} = (z-1)e^{-\theta}\left\{\frac{\partial G}{\partial z}[(z-1)+(1-e^\theta)]+G\right\}. \quad (2.2)$$

This equation can be transformed into a somewhat simpler one by letting

$$y = z-1, \quad \lambda = \kappa t e^{-\theta}, \quad c = 1-e^\theta, \quad (2.3a)$$

and

$$H(y,\lambda) = (y+c)G. \quad (2.3b)$$

Then H satisfies

$$g\left(\frac{\partial H}{\partial \lambda}, \frac{\partial H}{\partial y}\right) \equiv \frac{\partial H}{\partial \lambda} - y(y+c)\frac{\partial H}{\partial y} = 0 \quad (2.4)$$

or, if we let $p_1 = \partial H/\partial \lambda$ and $p_2 = \partial H/\partial y$,

$$g(p_1, p_2) = p_1 - y(y+c)p_2 = 0. \quad (2.5)$$

Following the method of characteristics we consider[13]

$$\frac{dy}{\partial g/\partial p_2} = \frac{d\lambda}{\partial g/\partial p_1} \quad (2.6a)$$

or

$$d\lambda = -dy/y(y+c) \quad (2.6b)$$

whose solution

$$y e^{\lambda c}[y+c]^{-1} = \text{constant} \quad (2.6c)$$

implies that the general solution of (2.4) is

$$H(y,\lambda) = f(y[y+c]^{-1}e^{\lambda c}) \quad (2.7)$$

where f is an arbitrary function which is to be determined from the initial distribution $\{x_n(0)\}$.

The general solution of (2.2) is then

$$G(z,t) = (z-e^\theta)^{-1} \\ \times f\{[z-1][z-e^\theta]^{-1}\exp[-t\kappa(1-e^{-\theta})]\}. \quad (2.8)$$

It is to be noted that the definition of (2.1a) implies $G(1,t) = \sum x_n(t) = 1$. Hence

$$f(0) = (1-e^\theta). \quad (2.9)$$

The initial distribution $x_n(0)$ characterizes $G(z,0)$. The function f is related to $G_0(z) \equiv G(z,0)$ by setting

---

[12] The general theory of the "easy" problems has been discussed recently by many authors: M. Wang and G. E. Uhlenbeck, Revs. Modern Phys. **17**, 323 (1945); H. B. Callen and T. A. Welton, Phys. Rev. **83**, 34 (1951); P. Bergmann and J. Lebowitz, Phys. Rev. **99**, 578 (1955); and others. A brief review and bibliography of this subject has been prepared by E. Montroll and M. S. Green, Ann. Rev. Phys. Chem. **5**, 449 (1954).

[13] F. S. Woods, *Advanced Calculus* (Ginn and Company, New York, 1934), p. 292.

$t=0$ in (2.8) so that

$$f([z-1][z-e^\theta]^{-1}) = (z-e^\theta)G_0(z) \quad (2.10)$$

or

$$f(\eta) = \left(\frac{e^\theta-1}{\eta-1}\right)G_0\left(\frac{\eta e^\theta-1}{\eta-1}\right). \quad (2.11)$$

Clearly, if we let

$$\tau = \kappa t(1-e^{-\theta}) \quad (2.12)$$

we can express $G(z,t)$ as

$$G(z,t) = \frac{1-e^\theta}{(z-e^\theta)-(z-1)e^{-\tau}}$$

$$\times G_0\left(\frac{(z-1)e^{-\tau}e^\theta-(z-e^\theta)}{(z-1)e^{-\tau}-(z-e^\theta)}\right). \quad (2.13)$$

In principle our problem is solved because G_0 is determined by the initial conditions and our $x_n(t)$'s are coefficients of z^n in (2.13). As $t\to\infty$ (and hence as $\tau\to\infty$) we have, since $G(1,t)=1=G_0(1)$ for all t,

$$G(z,\infty) = (1-e^{-\theta})/(1-ze^{-\theta}). \quad (2.14a)$$

Hence by (2.1a) the equilibrium distribution is

$$x_n(\infty) = (1-e^{-\theta})e^{-n\theta} \quad (2.14b)$$

which is the Boltzmann distribution of a set of oscillators with $\theta = h\nu/kT$.

The more familiar type of expansion of the solution of (1.7) as a linear combination of orthogonal polynomials is presented for completeness in Appendix I.

3. MOMENTS OF THE LEVEL OCCUPATION DISTRIBUTION

We define the factorial moments of $x_n(t)$ by

$$f_m(t) = \sum_{n=0}^\infty n(n-1)\cdots(n-m+1)x_n(t)$$

$$m = 1, 2, \cdots \quad (3.1a)$$

$$f_0(t) = \sum_{n=0}^\infty x_n(t) = 1. \quad (3.1b)$$

Clearly the internal energy of our system is

$$E(t) = h\nu \sum n x_n(t) = h\nu f_1(t). \quad (3.2)$$

In general

$$f_m(t) = \frac{\partial^m}{\partial z^m}G(z,t)\bigg]_{z=1}. \quad (3.3)$$

A differential equation for f_m is readily derivable from (2.2), which we write as

$$\frac{1}{\kappa}\frac{\partial G}{\partial t} = (z-1)e^{-\theta}F \quad (3.4)$$

with

$$F = \frac{\partial G}{\partial z}[(z-1)+(1-e^\theta)]+G. \quad (3.5)$$

Then by differentiating (3.4) m times with respect to z and setting $z=1$ we find

$$\frac{1}{\kappa}\frac{\partial f_m}{\partial t} = me^{-\theta}\frac{\partial^{m-1}F}{\partial z^{m-1}}\bigg]_{z=1}.$$

But (3.5) yields

$$\frac{\partial^{m-1}F}{\partial z^{m-1}}\bigg]_{z=1} = m\frac{\partial^{m-1}G}{\partial z^{m-1}}\bigg]_{z=1} + (1-e^\theta)\frac{\partial^m G}{\partial z^m}\bigg]_{z=1}. \quad (3.6)$$

Hence

$$\frac{1}{\kappa}\frac{df_m}{dt} + m(1-e^{-\theta})f_m = m^2 e^{-\theta}f_{m-1}. \quad (3.7)$$

By using the steady state value of f_1

$$f_1(\infty) = \sum_{n=0}^\infty ne^{-n\theta}(1-e^{-\theta})$$

$$= e^{-\theta}/(1-e^{-\theta}) \quad (3.8)$$

(3.7) is equivalent to

$$\frac{df_m}{d\tau} + mf_m = m^2 f_{m-1}f_1(\infty) \quad (3.9)$$

and

$$f_m(\tau) = f_m(0)e^{-m\tau} + m^2 f_1(\infty)e^{-m\tau}\int_0^\tau f_{m-1}(x)e^{mx}dx. \quad (3.10)$$

Since by (3.1b) $f_0(t)=1$ we find

$$f_1(t) = f_1(0)e^{-\tau} + f_1(\infty)(1-e^{-\tau}). \quad (3.11)$$

The variation of the internal energy with time is obtained by combining (3.2) with (3.11)

$$\frac{E(t)-E(\infty)}{E(0)-E(\infty)} = e^{-\tau} \quad (3.12)$$

as has been previously obtained by Bethe and Teller.[11] Substitution of (3.11) into (3.10) when $m=2$ yields

$$f_2(t) = 2f_1^2(\infty) + 4f_1(\infty)[f_1(0)-f_1(\infty)]e^{-\tau} + [f_2(0) - 4f_1(0)f_1(\infty) + 2f_1^2(\infty)]e^{-2\tau}. \quad (3.13)$$

The dependence of the dispersion of our distribution on time is given by

$$\sigma^2(t) = \langle(n-f_1)^2\rangle_{Av} = f_2 + f_1(1-f_1)$$
$$= \sigma^2(\infty) + [\sigma^2(0) - \sigma^2(\infty)]e^{-2\tau}$$
$$+ [f_1(0)-f_1(\infty)][1+2f_1(\infty)]e^{-\tau}(1-e^{-\tau}) \quad (3.14)$$

where

$$\sigma^2(\infty) = f_1(\infty)[1+f_1(\infty)]. \quad (3.14a)$$

4. THE RELAXATION OF ONE BOLTZMANN DISTRIBUTION TO ANOTHER

It will now be shown that if our oscillators are initially distributed in their energy levels according to some Boltzmann distribution, then this distribution

will persist during the relaxation but its effective temperature will vary monotonically until the temperature of the heat bath is achieved.
Let
$$x_n(0) = (1-e^{-\theta_0})e^{-n\theta_0} \qquad (4.1)$$

$$G(z,t) = \frac{(1-e^{\theta})(1-e^{-\theta_0})}{[(e^{-\tau}-e^{\theta})+e^{(\theta-\theta_0)}(1-e^{-\tau})]+z[(1-e^{-\tau})-e^{-\theta_0}(1-e^{\theta-\tau})]} \qquad (4.3)$$

so that
$$x_n(t) = [1-\exp(-\Theta)]\exp(-n\Theta) \qquad (4.4a)$$
where
$$\Theta = \log\left\{\frac{e^{-\tau}(1-e^{\theta-\theta_0})-e^{\theta}(1-e^{-\theta_0})}{e^{-\tau}(1-e^{\theta-\theta_0})-(1-e^{-\theta_0})}\right\} \qquad (4.4b)$$

and τ is given by (2.12), $\tau = \kappa t[1-\exp(-\theta)]$.

Our distribution (4.4) is that of Boltzmann at all times and the "effective temperature" T varies with time as $T = h\nu/k\Theta(\tau)$. We have plotted $1/\Theta$ as a function of time for various initial and final temperatures in Fig. 1.

5. INITIAL POISSON DISTRIBUTION

We now examine the relaxation of the Poisson distribution
$$x_n(0) = e^{-a}a^n/n! \qquad (5.1)$$
where a is the mean value of n. We have
$$G_0(z) = \sum_{n=0}^{\infty} z^n x_n(0) = e^{-a(1-z)} \qquad (5.2)$$
so that by (3.3)
$$f_1(0) = \bar{n} = \partial G_0/\partial z = a \qquad (5.3a)$$
and
$$\sigma^2 = \langle(n-\bar{n})^2\rangle_{\text{Av}} = a. \qquad (5.3b)$$

Application of (2.13) yields
$$G(z,t) = \frac{1-e^{\theta}}{(z-e^{\theta})-(z-1)e^{-\tau}}$$
$$\times\exp\left\{-\frac{ae^{-\tau}(1-z)(1-e^{\theta})}{(z-e^{\theta})-(z-1)e^{-\tau}}\right\}. \qquad (5.4)$$

Since the generating function of Laguerre polynomials defined by[14]
$$n!L_n(y) = e^y(d/dy)^n(e^{-y}y^n)$$
$$= n!\sum_{\nu=0}^{n}\binom{n}{n-\nu}\frac{(-y)^\nu}{\nu!} \qquad (5.5a)$$

$$(1-\alpha)^{-1}\exp\{y\alpha/(\alpha-1)\} = \sum_{n=0}^{\infty}\alpha^n L_n(y) \qquad (5.5b)$$

---
[14] G. Szego, *Orthogonal Polynomials* (American Mathematical Society, 1939), p. 96.

where $\theta_0 = h\nu/kT_0$, T_0 being the temperature corresponding to the initial distribution. Then
$$G_0(z) = \sum z^n x_n(0) = (1-e^{-\theta_0})/(1-ze^{-\theta_0}). \qquad (4.2)$$

Hence, (2.13) yields

we obtain
$$x_n(t) = \left[\frac{1-e^{\theta}}{e^{-\tau}-e^{\theta}}\right]\left[\exp\left\{\frac{ae^{-\tau}(e^{\theta}-1)}{e^{-\tau}-e^{\theta}}\right\}\right]\left[\frac{1-e^{-\tau}}{e^{\theta}-e^{-\tau}}\right]^n$$
$$\times L_n\left\{\frac{ae^{-\tau}(e^{\theta}-1)^2}{(1-e^{-\tau})(e^{-\tau}-e^{\theta})}\right\}. \qquad (5.6)$$

Since $L_n(0) = 1$ it is easy to show that as $\tau \to \infty$, $x_n(t)$ tends to the final equilibrium Boltzmann distribution (2.14b). We have plotted $x_n(t)$ in Fig. 2 for $\theta = 3$ and $a = 15$.

6. ALL MOLECULES INITIALLY IN mTH STATE

Let
$$x_n(0) = \begin{cases} 1 & \text{if } n=m \\ 0 & \text{otherwise} \end{cases} \qquad (6.1)$$

This sharp distribution broadens and its peak is displaced toward $n=0$ with increasing time. Then
$$G_0(z) = z^m \qquad (6.2)$$
and
$$G(z,t) = \frac{(1-e^{\theta})[z(1-e^{-\tau+\theta})-e^{\theta}(1-e^{-\tau})]^m}{[z(1-e^{-\tau})+(e^{-\tau}-e^{\theta})]^{m+1}}$$
$$= \frac{(1-e^{\theta})e^{m\theta}(e^{-\tau}-1)^m[1-\alpha z]^m}{(e^{-\tau}-e^{\theta})^{m+1}[1-\beta z]^{m+1}} \qquad (6.3)$$
where
$$\alpha = \left(\frac{e^{-\tau}-e^{\theta}}{e^{-\tau}-1}\right) \text{ and } \beta = \left(\frac{e^{-\tau}-1}{e^{-\tau}-e^{\theta}}\right). \qquad (6.4)$$

Now, it is well known[15] that if $|y| < 1$ and $|y(1-s)| < 1$
$$(1-y)^{a-1}(1-y+sy)^{-a} = \sum_{n=0}^{\infty} y^n F(-n, a, 1; s) \qquad (6.5)$$

F being the hypergeometric function. Hence if we let $a = -m$
$$x_n(t) = \frac{(1-e^{\theta})e^{m\theta}}{(e^{-\tau}-e^{\theta})}\left(\frac{e^{-\tau}-1}{e^{-\tau}-e^{\theta}}\right)^{m+n} F(-n, -m, 1; u^2) \qquad (6.5a)$$
$$u = \frac{\sinh\frac{1}{2}\theta}{\sinh\frac{1}{2}\tau}.$$

---
[15] Erdelyi, Magnus, Oberhettinger, and Tricomi, *Higher Transcendental Functions* (McGraw-Hill Book Company, Inc., New York, 1953), Vol. 1, p. 82.

A standard transformation formula yields

$$x_n(t) = \frac{(1-e^\theta)e^{m\theta}}{(e^{-\tau}-e^\theta)}\left(\frac{e^{-\tau}-1}{e^{-\tau}-e^\theta}\right)^{m+n}$$
$$\times [1-u^2]^{1+m+n} F(1+n, 1+m, 1; u^2). \quad (6.5b)$$

Note that as $\tau \to \infty$ (t large), $u \to 0$ and $F \to 1$ and the Boltzmann distribution develops as is required. When $\tau > \theta$, (6.5b) converges rapidly and is suitable for making calculations.

At early times we expect the distribution to be close to a Gaussian,

$$x_n(t) \simeq \frac{1}{(2\pi\sigma^2)^{\frac{1}{2}}} \exp(n-\bar{n})^2/2\sigma^2. \quad (6.6)$$

Initially $\bar{n}(0) = m$ and $\sigma^2(0) = 0$. Hence from (3.11) and (3.14)

$$\bar{n}(\tau) = f_1(t) = me^{-\tau} + f_1(\infty)(1-e^{-\tau}) \quad (6.7a)$$

$$\sigma^2(\tau) = \sigma^2(\infty)[1-e^{-2\tau}]$$
$$+[m-f_1(\infty)][1+2f_1(\infty)]e^{-\tau}(1-e^{-\tau}) \quad (6.7b)$$

where

$$\sigma^2(\infty) = f_1(\infty)[1+f_1(\infty)]$$

and

$$f_1(\infty) = e^{-\theta}(1-e^{-\theta})^{-1}.$$

It is to be noted that the relaxation from any initial distribution can be expressed as a linear combination of the $x_n(t)$'s which result from initially sharp distributions.

APPENDIX I. FOURIER SERIES SOLUTION OF EQ. (1.7)

For completeness we shall discuss the solution of (1.7) as a linear combination of a certain set of eigenfunctions. This form of solution could be used as the basis of a perturbation theory or for the analysis of complicated initial distributions for which $G_0(z) = G(z,0)$ [see Eq. (2.1)] cannot be summed easily.

We seek solutions $x_n(t)$ which are a superposition of terms of the following type

$$a_\mu l_n(\mu) \exp\{-\mu t \kappa (1-e^{-\theta})\} = a_\mu l_n(\mu) e^{-\mu \tau}, \quad (I.1)$$

where μ is a positive number and a_μ a constant which is to be determined from the initial level population distribution $\{x_n(0)\}$. Direct substitution of (I.1) into (1.7) yields a difference equation in the numbers $\{l_n(\mu)\}$

$$(e^{-\theta}-1)\mu l_n = ne^{-\theta}l_{n-1}$$
$$-\{n+(n+1)e^{-\theta}\}l_n+(n+1)l_{n+1}$$
$$n = 0, 1, 2, \cdots. \quad (I.2)$$

This set of equations can be solved through the introduction of the generating function

$$F(w,\mu) = \sum_{n=0}^{\infty} l_n(\mu) w^n. \quad (I.3a)$$

Multiplication of (I.2) by w^n and summation with respect to n yields, after application of (I.3a) and

$$w\frac{\partial F}{\partial w} = \sum_{n=0}^{\infty} n l_n(\mu) w^n, \quad (I.3b)$$

$$F'(w)/F(w) = \mu(w-1)^{-1} - (\mu+1)(w-e^\theta)^{-1} \quad (I.4)$$

whose solution is

$$F(w,\mu) = (1-w)^\mu(1-we^{-\theta})^{-\mu-1} \quad (I.5)$$

if l_0 is chosen to be 1. Our required functions $l_n(\mu)$ are the coefficients of w^n in $F(w,\mu)$. These functions have been studied by Gottlieb and can be shown to be[16]

$$l_n(\mu) = e^{\theta\mu}\Delta^n\left\{\binom{\mu}{n}e^{-\theta\mu}\right\} \quad (I.6a)$$

or

$$l_n(\mu) = e^{-n\theta}\sum_{\nu=0}^{\infty}(1-e^\theta)^\nu\binom{n}{\nu}\binom{\mu}{\nu} \quad (I.6b)$$

where the standard binomial coefficient notation is used with

$$\binom{n}{\nu} = \begin{cases} n!/\nu!(n-\nu)! & \text{if } n \geq \nu \\ 0 & \text{if } n < \nu \text{ or } \nu < 0 \\ 1 & \text{if } \nu = 0. \end{cases} \quad (I.7)$$

The first few of these functions are

$$l_n(0) = e^{-n\theta} \quad (I.8a)$$

$$l_n(1) = e^{-n\theta}\{1+(1-e^\theta)n\}. \quad (I.8b)$$

These functions can be written in various ways in terms of hypergeometric functions; for example[17]

$$l_n(\mu) = F(-n, \mu+1, 1; 1-e^{-\theta}). \quad (I.9)$$

If μ is chosen to be an integer, it is clear by symmetry from (I.6b) that

$$e^{n\theta}l_n(\mu) = e^{\mu\theta}l_\mu(n). \quad (I.10)$$

A pair of useful orthogonality relations exist when μ and n are non-negative integers

$$\sum_{\nu=0}^{\infty} e^{-\theta\nu}l_n(\nu)l_m(\nu) = \begin{cases} 0 & n \neq m \\ e^{-n\theta}(1-e^{-\theta})^{-1} & n = m \end{cases} \quad (I.11a)$$

$$\sum_{n=0}^{\infty} e^{n\theta}l_n(\nu)l_n(\mu) = \begin{cases} 0 & \mu \neq \nu \\ e^{\nu\theta}(1-e^{-\theta})^{-1} & \mu = \nu. \end{cases} \quad (I.11b)$$

Gottlieb has also shown that for a fixed complex number x and large n

$$l_n(x) = (-1)^n(1-e^{-\theta})^{-x-1}\binom{x}{n} + O(n^{-Rex-2}). \quad (I.12)$$

[16] M. J. Gottlieb, Am. J. Math. **60**, 455 (1938).
[17] Erdelyi, Magnus, Oberhettinger, and Tricomi, *Higher Transcendental Functions* (McGraw-Hill Book Company, Inc., New York, 1953), Vol. 2, p. 225.

Hence as $n \to \infty$, $l_n(x) \to 0$. This shows that our $l_n(x)$'s satisfy the physical requirement that the population of states characterized by the quantum number n decreases as $n \to \infty$ for θ fixed.

We can now construct a general solution of (1.7) by superposition of terms (I.1),

$$x_n(t) = \sum_{\mu=0}^{\infty} a_\mu l_n(\mu) e^{-\mu \tau} \quad \text{(I.13a)}$$

$$= \sum_{\mu=0}^{\infty} a_\mu e^{\theta(\mu-n)} l_\mu(n) e^{-\mu \tau} \quad \text{(I.13b)}$$

where the constants a_μ are determined in the usual manner through the orthogonality relations. We find

$$a_\mu = (1 - e^{-\theta}) \sum_{n=0}^{\infty} x_n(0) l_\mu(n) \quad \text{(I.14a)}$$

or alternatively

$$a_\mu = (1 - e^{-\theta}) e^{-\mu \theta} \sum_{n=0}^{\infty} e^{n\theta} l_n(\mu) x_n(0). \quad \text{(I.14b)}$$

When $x_n(0)$ is chosen to be a Boltzmann distribution, one can by repeated use of the generating function $F(w,\mu)$ of (I.5) readily obtain Eq. (4.4) for $x_n(t)$.

Rubin and Shuler[6] previously derived the following formula for $x(n,t) = x_n(t)$ which is to be valid when θ is very small; ($L_n(x)$ is the nth Laguerre polynomial)

$$x(n,t) = e^{-n\theta} \sum_{\mu=0}^{\infty} a_\mu L_\mu(n\theta) \exp(-\mu \theta t k_{10}) \quad \text{(I.15a)}$$

where

$$a_\mu = \theta \int_0^{\infty} x(y,0) L_\mu(\theta y) dy. \quad \text{(I.15b)}$$

This result can be obtained from (I.14a) and (I.13b) by letting $\theta \to 0$ in certain terms and retaining it in others. Gottlieb pointed out that

$$\lim_{\theta \to 0} l_n(x/\theta, \theta) = L_n(x). \quad \text{(I.16)}$$

If we write $l_\mu(n) = l_\mu(n\theta/\theta, \theta)$ and suppose $n\theta$ is fixed while $\theta \to 0$, and if we let $1 - \exp(-\theta) \simeq \theta$, (I.14a) becomes

$$a_\mu \simeq \theta \sum_{\nu=0}^{\infty} x(y,0) L_\mu(y\theta) \simeq \theta \int_0^{\infty} x(y,0) L_\mu(\theta y) dy.$$

The conversion of the summation to an integration is valid if $x(y,0)$ and $L_\mu(\theta y)$ are slowly varying functions of y. Equation (I.13b) reduces to (I.15a) if one uses (I.16), sets $\exp \theta \mu = 1$ and writes $\tau = t k_{10}(1 - e^{-\theta}) \simeq k_{10} \theta t$.

APPENDIX II. THE RELAXATION EQUATION FOR VIBRATIONAL-TRANSLATIONAL ENERGY EXCHANGE

We now wish to derive the analog of the relaxation equation (1.6) when the energy exchange is between the vibration of the harmonic oscillators and the translational degrees of freedom of the heat bath molecules. The formulation presented here is in principle analogous to that given by Rubin and Shuler[6] but is more detailed in that it takes explicit account of the matrix elements for the transitions between the translational "energy levels."

Let

$x_n(t) =$ fraction of excited oscillators in vibrational level n

$y_i =$ concentration of heat bath molecules (or atoms) with momentum† p_i

$P_{n,\,n+1;\,i,\,i-1} =$ joint probability per collision for the energy transfer $nh\nu \to (n+1)h\nu$ as $p_i \to p_{i-1}$.

The relaxation equation can now be written as [see (1.2)]

$$\frac{dx_n(t)}{dt} = -Z\Big[x_n\Big(\sum_{i=0}^{\infty} y_i P_{n,\,n+1;\,i,\,i-1}$$

$$+ \sum_{i=0}^{\infty} y_i P_{n,\,n-1;\,i,\,i+1}\Big)$$

$$- x_{n+1} \sum_{i=0}^{\infty} y_i P_{n+1,\,n;\,i,\,i+1}$$

$$- x_{n-1} \sum_{i=0}^{\infty} y_i P_{n-1,\,n;\,i,\,i-1}\Big] \quad \text{(II.1)}$$

where Z is again the collision number. The joint transition probabilities can now be written as

$$P_{n,\,n+1;\,i,\,i-1} = P_{10}(n+1) Q_{i(-)}$$
$$P_{n,\,n-1;\,i,\,i+1} = P_{10} n Q_{i(+)} \quad \text{(II.2)}$$

where $(n+1) P_{10}$ and $n P_{10}$ are the vibrational transition probabilities (see 1.1) and where the Q's are the translational transition probabilities to be determined. Substitution of (II.2) in (II.1) leads to

$$\frac{dx_n(t)}{dt} = Z P_{10}(n+1)$$

$$\times \Big[x_{n+1} \sum_i y_i Q_{i(+)} - x_n \sum_i y_i Q_{i(-)}\Big]$$

$$+ Z P_{10} n \Big[x_{n-1} \sum_i y_i Q_{i(-)} - x_n \sum_i y_i Q_{i(+)}\Big]. \quad \text{(II.3)}$$

At equilibrium, (i.e., as $t \to \infty$), we have

$$\frac{dx_n(\infty)}{dt} = 0$$

$$\frac{x_{n+1}(\infty)}{x_n(\infty)} = \frac{x_n(\infty)}{x_{n-1}(\infty)} = e^{-\theta} \quad \text{(II.4)}$$

† The indices i, $i \pm 1$ do not represent successive translational energy levels but are used to indicate translational levels separated by $h\nu$.

and Eq. (II.3) becomes

$$0 = ZP_{10}(n+1)x_n(\infty)[e^{-\theta}\sum_i y_i Q_{i(+)} - \sum_i y_i Q_{i(-)}]$$
$$- ZP_{10}nx_{n-1}(\infty)[e^{-\theta}\sum_i y_i Q_{i(+)} - \sum_i y_i Q_{i(-)}]. \quad (II.5)$$

Since the heat bath is assumed to remain at its Maxwell-Boltzmann equilibrium distribution at all times t, $y_i \neq f(t)$ and $y_i(\infty) = y_i$.

The first two terms in Eq. (II.5) refer to the rate of the (vibrational) transition $n \rightleftarrows n+1$ and the last two terms to the rate of the transition $n \rightleftarrows n-1$. By the principle of detailed balancing at equilibrium, the net rate of each of these transitions must be independently zero to satisfy (II.5). We thus find that

$$\sum_i y_i Q_{i(-)} = e^{-\theta} \sum_i y_i Q_{i(+)}. \quad (II.5a)$$

Substitution of (II.5a) into (II.3) leads to

$$\frac{dx_n(t)}{dt} = ZP_{10} \sum_i y_i Q_{i(+)} \{ne^{-\theta}x_{n-1}$$
$$- [n+(n+1)e^{-\theta}]x_n + (n+1)x_{n+1}\} \quad (II.6)$$

which is of the same form as (1.6) except for the replacement of $k_{10}(1-e^{-\theta})^{-1}$ by $ZP_{10}\sum_i y_i Q_{i(+)}$. To evaluate the term in front of the braces in (II.6) we write, with Landau and Teller[4]

$$y_i = y_w = 2N(m/2kT)^{\frac{3}{2}}w^3 \exp\left(-\frac{mw^2}{2kT}\right) \quad (II.7)$$

for the number of heat bath molecules with momentum mw undergoing collisions (N is the total number of heat bath molecules) and

$$Q_{i(+)} = Q_{w(+)} = Q_0 e^{-2\pi\nu a/w} \quad (II.8)$$

where Q_0 is a constant, ν is the frequency of the oscillator, and where a is a length characteristic of the interaction forces in the collisions between the oscillators and the heat bath molecules. Replacing the summation over i indicated in (II.6) by integration over the velocities w one obtains

$$\frac{dx_n(t)}{dt} = k_{10}Q_0 \left(\frac{6}{\pi}\frac{kT}{\epsilon}\right)^{-\frac{1}{2}} \exp\left[-\frac{3}{2}\left(\frac{\epsilon}{kT}\right)^{\frac{1}{3}}\right]$$
$$\times \{ne^{-\theta}x_{n-1} - [n+(n+1)e^{-\theta}]x_n$$
$$+ (n+1)x_{n+1}\} \quad (II.9)$$

where $k_{10} = ZNP_{10}$ has been defined in connection with (1.6) and where $\epsilon = m(2\pi\nu a)^2$ with m equal to the effective mass of the collision system. One thus obtains finally

$$\frac{dx_n(t)}{dt} = k_{10}'\{ne^{-\theta}x_{n-1}$$
$$- [n+(n+1)e^{-\theta}]x_n + (n+1)x_{n+1}\} \quad (II.9a)$$

where

$$k_{10}' = k_{10}Q_0\left(\frac{6}{\pi}\frac{kT}{\epsilon}\right)^{-\frac{1}{2}} \exp\left[-\frac{3}{2}\left(\frac{\epsilon}{kT}\right)^{\frac{1}{3}}\right] \quad (II.9b)$$

is the transition probability used in (1.7a) for the vibration-translation energy exchange.

A more accurate evaluation of the joint vibrational-translational transition probabilities in (II.2) could be obtained from the quantum mechanical treatments of Jackson and Mott[18] and Herzfeld and his co-workers.[19]

---

[18] T. M. Jackson and N. F. Mott, Proc. Roy. Soc. (London) **A137**, 703 (1932).
[19] Slawsky, Schwartz, and Herzfeld, J. Chem. Phys. **20**, 1591 (1952); R. N. Schwartz and K. F. Herzfeld, *ibid.*, **22**, 767 (1954); see also K. F. Herzfeld in *Thermodynamics and Physics of Matter* (Princeton University Press, Princeton, 1955), Sec. H.

Random Walks on Lattices. II

ELLIOTT W. MONTROLL

*Institute for Fluid Dynamics and Applied Mathematics,
University of Maryland, College Park, Maryland*

AND

GEORGE H. WEISS

National Institute of Health, Bethesda, Maryland

(Received 10 September 1964)

Formulas are obtained for the mean first passage times (as well as their dispersion) in random walks from the origin to an arbitrary lattice point on a periodic space lattice with periodic boundary conditions. Generally this time is proportional to the number of lattice points.

The number of distinct points visited after n steps on a k-dimensional lattice (with $k \geq 3$) when n is large is $a_1 n + a_2 n^{\frac{1}{2}} + a_3 + a_4 n^{-\frac{1}{2}} + \cdots$. The constants $a_1 - a_4$ have been obtained for walks on a simple cubic lattice when $k = 3$ and a_1 and a_2 are given for simple and face-centered cubic lattices. Formulas have also been obtained for the number of points visited r times in n steps as well as the average number of times a given point has been visited.

The probability $F(c)$ that a walker on a one-dimensional lattice returns to his starting point before being trapped on a lattice of trap concentration c is $F(c) = 1 + [c/(1 - c)] \log c$.

Most of the results in this paper have been derived by the method of Green's functions.

A NUMBER of problems in solid-state physics are directly or indirectly related to various aspects of random walks on periodic space lattices. The theory of such random walks on infinite lattices was first discussed by Polya[1] who was especially concerned with the effect of dimensionality on the probability that a walker starting at a given point eventually returns to that point. Some other types of problems which are of special interest involve the average time required by a walker to go from a given lattice point to another preassigned point for the first time and with the average number of distinct points occupied in a walk of a given number of steps. Results on these topics as well as the effect of a small number of lattice defects on random walks have been discussed in the first paper of this series.[2]

That paper is concerned mainly with random walks which involve jumps to nearest-neighbor lattice points only. Many of the results are generalized here to be applicable to walks which involve steps to more distant neighbors. We also discuss the average number of points occupied k times in an n-step walk as well as the number of times a given point has been occupied in such a walk. The average number of points occupied in an n-step walk was first estimated by Dvoretsky and Erdös,[3] further analysis having been made by Vineyard[4] and one of the authors.[2] Repeated occupancy was first considered by Erdös and Taylor.[5]

Green's function techniques and Tauberian theorems are the main mathematical tools used in this paper. Although emphasis is placed on walks in which steps are taken at regular time intervals, the generalization to those in which the steps are taken at random times is developed in Sec. V.

We also discuss the effect of traps of a given con-

---

[1] G. Polya, Math. Ann. **84**, 149 (1921).
[2] E. W. Montroll, Proc. Symp. Appl. Math. Am. Math. Soc. **16**, 193 (1964).
[3] A. Dvoretzky and E. Erdos, Proc. 2nd Berkeley Sympos. Math. Stat. and Prob., (University of California Press, Berkeley, 1951), p. 33.
[4] G. H. Vineyard, J. Math. Phys. **4**, 1191 (1963).
[5] P. Erdos and S. J. Taylor, Acta Math. Acad. Sci. Hung. **11**, 137 (1960).

centration on the probability of a walker on a one-dimensional lattice returning to his starting point before being trapped.

Since the first draft of this article was completed, a book by Spitzer[6] has appeared which contains a discussion of some of the topics included here.

I. LATTICE GREEN'S FUNCTIONS AND RANDOM-WALK GENERATING FUNCTIONS

We begin by studying discrete random walks on lattices with periodic boundary conditions (i.e., toroidal lattices), and in particular will assume that there exists an integer N such that the lattice points $\mathbf{s} = (s_1, s_2, \cdots, s_k)$ satisfy

$$(s_1 + j_1 N, s_2 + j_2 N, \cdots, s_k + j_k N) = (s_1, s_2, \cdots, s_k)$$

when the j's are integers.

There are N^k distinct lattice points on our k-dimensional lattice. Let $P_n(\mathbf{s})$ be the probability that the random walker is at a point $\mathbf{s}$ after the nth step. In view of the periodic boundary conditions,

$$P_n(s_1 + j_1 N, s_2 + j_2 N, \cdots, s_k + j_k N) = P_n(\mathbf{s}), \quad (I.1)$$

when the j's are integers. The $\{P_n(\mathbf{s})\}$ satisfy the recursion formula

$$P_{n+1}(\mathbf{s}) = \sum_{\mathbf{s}'} p(\mathbf{s} - \mathbf{s}') P_n(\mathbf{s}'), \quad (I.2)$$

if $p(\mathbf{s})$ represents the probability that any step results in a vector displacement $\mathbf{s}$ by a walker. We find the Fourier expansion of $p(\mathbf{s})$

$$\lambda(2\pi\mathbf{r}/N) = \sum_{\mathbf{s}} p(\mathbf{s}) \exp(2\pi i \mathbf{r} \cdot \mathbf{s}/N), \quad (I.3)$$

which we call the structure function of the walk, to be of considerable importance. In particular

$$\sum_{\mathbf{s}} p(\mathbf{s}) = 1 \quad \text{and} \quad \lambda(0) = 1 \quad (I.4)$$

when walkers are conserved; i.e., when walkers are neither created nor destroyed in the walk. The reader can easily verify that

$$\lambda(\boldsymbol{\vartheta}) = \begin{cases} (c_1 + c_2 + \cdots + c_k)/k \\ \quad \text{for } k\text{-D simple cubic lattice} \\ (c_1 c_2 + c_2 c_3 + c_3 c_1)/3 \\ \quad \text{for 3-D face-centered cubic lattice} \\ c_1 c_2 c_3, \\ \quad \text{for 3-D body-centered cubic lattice,} \end{cases} \quad (I.5)$$

where

$$c_i = \cos \vartheta_i \quad \text{and} \quad \vartheta_i = 2\pi r_i/N. \quad (I.5a)$$

---
[6] F. Spitzer, *Principles of Random Walks* (D. Van Nostrand, Inc., Princeton, New Jersey, 1964).

Properties of random walks can be described effectively through the random-walk generating function

$$P(\mathbf{s}, z) = \sum_0^\infty z^n P_n(\mathbf{s}). \quad (I.6)$$

We restrict ourselves now to the initial condition

$$P_0(\mathbf{s}) = \delta_{\mathbf{s},0} \quad (I.7)$$

which corresponds to walks which start from the origin, $\mathbf{s} = 0$. By multiplying (2) by z^n, summing over all n, and applying (7), one finds that $P(\mathbf{s}, z)$ satisfies the Green's function equation

$$P(\mathbf{s}, z) - z \sum_{\mathbf{s}'} p(\mathbf{s} - \mathbf{s}') P(\mathbf{s}', z) = \delta_{\mathbf{s},0}. \quad (I.8)$$

This equation can be solved for our generating function $P(\mathbf{s}, z)$ by considering the function

$$u(z, 2\pi\mathbf{r}/N) = \sum_{\mathbf{s}} P(\mathbf{s}, z) \exp(2\pi i \mathbf{r} \cdot \mathbf{s}/N). \quad (I.9)$$

If we multiply (8) by $\exp(2\pi i \mathbf{s} \cdot \mathbf{r}/N)$, sum over $\mathbf{s}$ and employ (9) and (3) we find

$$u(z, 2\pi\mathbf{r}/N) = \{1 - z\lambda(2\pi\mathbf{r}/N)\}^{-1}. \quad (I.10)$$

Since $P(\mathbf{s}, z)$ is the Fourier inverse of $u(z, 2\pi\mathbf{r}/N)$, we find

$$P(\mathbf{s}, z) = N^{-k} \sum_{\mathbf{r}} \frac{\exp(-2\pi i \mathbf{r} \cdot \mathbf{s}/N)}{1 - z\lambda(2\pi\mathbf{r}/N)}. \quad (I.11)$$

In the case of an infinite lattice, $N \to \infty$ and

$$P(\mathbf{s}, z) = \frac{1}{(2\pi)^k} \int_{-\pi}^{\pi} \cdots \int \frac{\exp(-i\mathbf{s} \cdot \boldsymbol{\vartheta}) d^k \boldsymbol{\vartheta}}{1 - z\lambda(\boldsymbol{\vartheta})}. \quad (I.12)$$

From this it is clear that

$$P_n(\mathbf{s}) = \frac{1}{(2\pi)^k} \int_{-\pi}^{\pi} \cdots \int [\lambda(\boldsymbol{\vartheta})]^n e^{-i\mathbf{s} \cdot \boldsymbol{\vartheta}} d^k \boldsymbol{\vartheta}. \quad (I.13)$$

Also since we assume walkers to be conserved

$$\sum_{\mathbf{s}} P_n(\mathbf{s}) = 1, \quad (I.14a)$$

and

$$\sum_{\mathbf{s}} P(\mathbf{s}, z) = (1 - z)^{-1}. \quad (I.14b)$$

In all the analysis above we assume $|z| \leq 1$.

We will find it expedient to separate out the singular and nonsingular parts of $P(\mathbf{s}, z)$ by writing

$$P(\mathbf{s}, z) = (1 - z)N^{-k} + \varphi(\mathbf{s}, z), \quad (I.15)$$

where

$$\varphi(\mathbf{s}, z) = N^{-k} \sum_{\mathbf{r}}' \frac{\exp(2\pi i \mathbf{r} \cdot \mathbf{s}/N)}{1 - z\lambda(2\pi\mathbf{r}/N)}, \quad (I.16)$$

in which the prime indicates that the term with $r_1 = r_2 = \cdots = r_k = 0$ is to be omitted. In general, when the limit $N \to \infty$ is taken, the sum can be replaced by an integral. Although the resulting integral may be a singular function of z the singularity is weaker than $(1 - z)^{-1}$ as we show later.

We will also be interested in the properties of the first passage time, and for this purpose we define $F_n(\mathbf{s})$ to be the probability that a random walker reaches the point $\mathbf{s}$ for the *first time* at step n. The generating function of the $F_n(\mathbf{s})$ will be denoted by $F(\mathbf{s}, z)$:

$$F(\mathbf{s}, z) = \sum_{n=1}^{\infty} F_n(\mathbf{s}) z^n. \quad (I.17)$$

It is possible to relate the $F_n(\mathbf{s})$ to the $P_n(\mathbf{s})$ since, if the random walker is at step n he must first have reached there at some step j and then returned to $\mathbf{s}$ in $n - j$ steps. Taking account of the initial condition of Eq. (I.3), we find

$$P_n(\mathbf{s}) = \delta_{n,0}\delta_{\mathbf{s},0} + \sum_{j=1}^{n} F_j(\mathbf{s})P_{n-j}(0)$$

The generating functions therefore satisfy

$$F(\mathbf{s}, z) = [P(\mathbf{s}, z) - \delta_{\mathbf{s},0}]/P(0, z). \quad (I.18)$$

The probability that the walker reaches point s at some time is just $F(s, 1)$. For $N < \infty$ the probability of reaching any point on the lattice is one, independent of the dimension. When $N = \infty$ the probability of a return to the origin is $1 - [F(0, 1)]^{-1}$. In one and two dimensions $F(0, 1) = \infty$ and the walker returns to the origin with probability one. In higher dimensions the return to the origin occurs with probability less than one. The same results are true for the first passage to any point $\mathbf{s}$.

Another function that will be useful later is $F_n^{(r)}(\mathbf{s})$, the probability that the random walker reaches $\mathbf{s}$ for the rth time at step n. This function satisfies the recurrence formula

$$F_n^{(r)}(\mathbf{s}) = \sum_{j=1}^{n} F_{n-j}^{(r-1)}(\mathbf{s})F_j(0), \quad (I.19)$$

and its generating function $F^{(r)}(\mathbf{s}, z)$ is therefore given by

$$F^{(r)}(\mathbf{s}, z) = [F(0, z)]^{r-1}F(\mathbf{s}, z) = \sum_{n=1}^{\infty} F_n^{(r)}(\mathbf{s}) z^n. \quad (I.20)$$

II. STATISTICS OF FIRST-PASSAGE TIME

The first results to be given will be those related to first-passage times. Let $\langle n^j(\mathbf{s}) \rangle$ be the jth moment of the first-passage time to reach point $\mathbf{s}$. In terms of $F(\mathbf{s}, z)$, $\langle n^i(\mathbf{s}) \rangle$ can be written

$$\langle n^i(\mathbf{s}) \rangle = (z\, \partial/\partial z)^i F(\mathbf{s}, z)]_{z=1}. \quad (II.1)$$

In particular, if we substitute the representation of Eq. (II.15) for $P(\mathbf{s}, z)$ into (II.18) we find, for the first two moments

$$\langle n(\mathbf{s}) \rangle = \begin{cases} N^k[\varphi(0, 1) - \varphi(\mathbf{s}, 1)], & \mathbf{s} \neq 0, \quad (II.2a) \\ N^k, & \mathbf{s} = 0, \quad (II.2b) \end{cases}$$

$$\langle n^2(\mathbf{s}) \rangle = [2N^k\varphi(0, 1) + 1]\langle n(\mathbf{s}) \rangle$$
$$+ 2N^k\left[\frac{\partial\varphi(0, z)}{\partial z} - \frac{\partial\varphi(\mathbf{s}, z)}{\partial z}\right]_{z=1} \text{ if } \mathbf{s} \neq 0, \quad (II.3a)$$

$$\langle n^2(0) \rangle = 2N^{2k}\varphi(0, 1) + N^k. \quad (II.3b)$$

Notice that *the expected number of steps required to return to the origin is N^k*, the total number of lattice points, independently of the structure of the lattice. The second moment of the expected number of steps required to return to the origin for the first time does depend on lattice structure as is indicated by the function $\varphi(0, 1)$. Moments of the number of steps to reach other points on the lattice for the first time all depend on the structure.

So far we have given formal results valid for any k-dimensional periodic lattice. In the next few paragraphs we shall illustrate the general theory by evaluating some of the relevant functions for particular lattices. In our evaluation we will need some analytic properties of the functions $\lambda(\vartheta)$ and $\varphi(\mathbf{r}, z)$ which appear in many of the formulas derived above. We shall be interested only in symmetric random walks, hence the expansion of $\lambda(\vartheta)$ in a neighborhood of the origin is

$$\lambda(\vartheta) = 1 - \frac{1}{2} \sum_{1}^{k} \sigma_i^2 \vartheta_i^2 + O(\vartheta^4), \quad (II.4)$$

where

$$\sigma_i^2 = \sum_m m_i^2 p(m). \quad (II.5)$$

We will make use of $\varphi(\mathbf{s}, z)$ for an infinite lattice in the limit $z = 1^-$. The expression for $\varphi(\mathbf{s}, z)$ when $N = \infty$ is

$$\varphi(\mathbf{s}, z) = \frac{1}{(2\pi)^k}$$
$$\times \int_{-\pi}^{\pi}\cdots\int \frac{\exp(i\vartheta\cdot\mathbf{s})\, d^k\vartheta}{1 - z\lambda(\vartheta)} = P(\mathbf{s}, z). \quad (II.6)$$

The function $\varphi(\mathbf{s}, 1)$ is singular in one and two dimensions. We can see this by considering the contribution to $\varphi(\mathbf{s}, 1)$ from a neighborhood of the origin $\vartheta = 0$,

$$\int\cdots\int \frac{d^k\vartheta}{\sigma_1^2\vartheta_1^2 + \cdots + \sigma_k^2\vartheta_k^2}. \quad (II.7)$$

If the integrand is transformed to polar coordinates, there arise contributions of the form

$$\int \cdots \int \frac{r^{k-1}}{r^2} d\mathbf{r}, \qquad (\text{II.8})$$

which diverges in one and two dimensions, but remains finite in higher dimensions. We will be interested in the behavior of $\varphi(\mathbf{s}, z)$ for $\mathbf{s} = 0$ and $\mathbf{s} = (s_1^2 + \cdots + s_k^2)^{\frac{1}{2}}$ large but not large enough to violate the condition $\mathbf{s} \ll N^k$. It will be demonstrated that the properties of $\langle n(\mathbf{s}) \rangle / N^k$ can be obtained fairly simply for large distances from the origin.

Let us begin by decomposing the integral defining $\varphi(\mathbf{s}, z)$ into two parts:

$$\varphi(\mathbf{s}, z) = \frac{1}{(2\pi)^k} \int_{-\pi}^{\pi} \cdots \int \frac{\exp(i\boldsymbol{\vartheta} \cdot \mathbf{s}) \, d^k\boldsymbol{\vartheta}}{1 - z + \frac{1}{2}z(\sigma_1^2\vartheta_1^2 + \cdots + \sigma_k^2\vartheta_k^2)}$$

$$+ \frac{1}{(2\pi)^k} \int_{-\pi}^{\pi} \cdots \int e^{i\boldsymbol{\vartheta} \cdot \mathbf{s}} \left\{ \frac{1}{1 - z\lambda(\boldsymbol{\vartheta})} \right.$$

$$\left. - \frac{1}{1 - z + \frac{1}{2}z(\sigma_1^2\vartheta_1^2 + \cdots + \sigma_k^2\vartheta_k^2)} \right\} d^k\boldsymbol{\vartheta}$$

$$= \varphi_1(\mathbf{s}, z) + \varphi_2(\mathbf{s}, z). \qquad (\text{II.9})$$

The singularity in one and two dimensions at $z = 1$ comes from the function $\varphi_1(\mathbf{s}, z)$ since the integral for $\varphi_2(\mathbf{s}, 1)$ has the form

$$\int \cdots \int r^{k-1} dr$$

at the origin of θ space. In higher dimensions both $\varphi_1(\mathbf{s}, z)$ and $\varphi_2(\mathbf{s}, z)$ approach zero as $\mathbf{s} \to \infty$, but

$$\lim_{\mathbf{s} \to \infty} [\varphi_2(\mathbf{s}, 1)/\varphi_1(\mathbf{s}, 1)] = 0. \qquad (\text{II.10})$$

This limit can be established by examining the behavior of the integrands in the neighborhood of $\boldsymbol{\vartheta} = 0$, which gives the principal contribution in the range of large $\mathbf{s}$. A detailed justification is given in Appendix A. We therefore see that the significant analytic properties of $\varphi(\mathbf{s}, z)$ are contained in $\varphi_1(\mathbf{s}, z)$ for large $\mathbf{s}$.

We shall recast the form of this function as a Laplace transform and begin by using the identity

$$u^{-1} = \int_0^\infty e^{-ut} dt$$

to rewrite it as

$$\varphi_1(\mathbf{s}, z) = \int_0^\infty e^{-(1-z)t} dt$$

$$\times \prod_{i=1}^k \left\{ \frac{1}{2\pi} \int_{-\pi}^{\pi} \exp\left(i\vartheta_i s_i - \tfrac{1}{2} zt\sigma_i^2\vartheta_i^2\right) d\vartheta_i \right\}, \qquad (\text{II.11})$$

where the interchange of orders of integration can be justified in detail. Thus $\varphi_1(\mathbf{s}, z)$ can be expressed as a Laplace transform

$$\varphi_1(\mathbf{s}, z) = \int_0^\infty e^{-(1-z)t} F(z, t) \, dt, \qquad (\text{II.12})$$

where $F(z, t)$ is the product of integrals in Eq. (II.11). Since each of the integral factors of $F(z, t)$ is analytic in z at $z = 1$, we may expand $F(z, t)$ in a Taylor series around $z = 1$,

$$F(z, t) = F(1, t) + (z - 1)[\partial F/\partial z]_{z=1} + \cdots . \qquad (\text{II.13})$$

Since we are interested in the behavior of $\varphi_1(\mathbf{s}, z)$ at $z = 1$ we can invoke an Abelian theorem for Laplace transforms[7] which states in the present context that the behavior of $\varphi(\mathbf{s}, z)$ at $z = 1$ is determined by the behavior of $F(z, t)$ at $t = \infty$.

To determine this behavior we note that as $t \to \infty$ the integrand of each of the integrals in $\varphi_1(\mathbf{s}, z)$ is peaked sharply at the origin with negligible contribution coming from values of ϑ_i greater than $2/\sigma_i t^{\frac{1}{2}} z^{\frac{1}{2}}$. Hence the ranges of integration on the ϑ integrals $(-\pi, \pi)$ can, as $t \to \infty$ be replaced by $(-\infty, \infty)$ so that

$$F(z, t) \sim \prod_{i=1}^k (2\sigma_i^2 \pi tz)^{-\frac{1}{2}} \exp(-s_i^2/2zt\sigma_i^2)$$

and

$$F(1, t) \sim [\sigma_1 \cdots \sigma_k (2\pi t)^{k/2}]^{-1} \exp(-\lambda^2/2t), \qquad (\text{II.14a})$$

where

$$\lambda^2 = \sum_{i=1}^k (s_i/\sigma_i)^2. \qquad (\text{II.14b})$$

In one dimension we find

$$\varphi_1(s, z) \sim \frac{1}{\sigma(2\pi)^{\frac{1}{2}}} \int_0^\infty e^{-(1-z)t - (s^2/2t\sigma^2)} t^{-\frac{1}{2}} dt$$

$$= \frac{2^{\frac{1}{2}} s^{\frac{1}{2}}}{\pi^{\frac{1}{2}} \sigma^{\frac{1}{2}}(1-z)^{\frac{1}{4}}} K_{\frac{1}{2}}\!\left(\frac{s}{\sigma}[2(1-z)]^{\frac{1}{2}}\right)$$

$$= \frac{\exp\{-(s/\sigma)[2(1-z)]^{\frac{1}{2}}\}}{\sigma[2(1-z)]^{\frac{1}{2}}}, \qquad (\text{II.15})$$

where $K_{1/2}(x)$ is a Bessel function of the third kind of imaginary argument. The two-dimensional form $\varphi_1(\mathbf{s}, z)$ for $\lambda \neq 0$ is

$$\varphi_1(\mathbf{s}, z) \sim \frac{1}{2\pi\sigma_1\sigma_2} \int_0^\infty t^{-1} \exp\{-(1-z)t - \lambda^2/2t\} dt$$

$$= \frac{1}{\pi\sigma_1\sigma_2} K_0(\lambda[2(1-z)]^{\frac{1}{2}}). \qquad (\text{II.16})$$

---

[7] D. V. Widder, *The Laplace Transform* (Princeton University Press, Princeton, New Jersey, 1941).

When $\lambda = 0$ we may use an Abelian theorem for Laplace transforms[7] to show that it follows from the asymptotic form $F(1, t) \sim (2\pi\sigma_1\sigma_2 t)^{-1}$ that

$$\varphi_1(0, z) \sim -(1/2\pi\sigma_1\sigma_2) \log (1 - z) \quad (II.17)$$

for $z \to 1^-$. In three dimensions and higher $\varphi_1(0, 1)$ is defined by a convergent integral and must be calculated numerically. For large λ^2 an asymptotic expression for $\varphi_1(\mathbf{s}, 1)$ is

$$\varphi_1(\mathbf{s}, 1) \sim \frac{1}{(2\pi)^{k/2}\sigma_1 \cdots \sigma_k} \int_0^\infty e^{-\lambda^2/2t} t^{-k/2} \, dt$$

$$= \lambda^{2-k} \Gamma(\tfrac{1}{2}k - 1)/2\sigma_1 \cdots \sigma_k \pi^{k/2}. \quad (II.18a)$$

The 3-D expression for $P(\mathbf{s}, z)$ as $z \to 1$ is

$$P(\mathbf{s}, z) \sim (2\pi\lambda\sigma_1\sigma_2\sigma_3)^{-1} \exp\{-\lambda[2(1-z)]^{\frac{1}{2}}\}. \quad (II.18b)$$

Let us now consider some results for specific lattices. The simplest case is that of a one-dimensional lattice with jumps to nearest neighbors with probability $\tfrac{1}{2}$. For this case we can calculate an explicit expression for[2] $P(s, z)$:

$$P(s, z) = \frac{1}{N} \sum_{r=0}^{N-1} \frac{\exp(2\pi i r s/N)}{1 - z \cos(2\pi r/N)}$$

$$= \frac{(1-z^2)^{-\frac{1}{2}}(U^s + U^{N-s})}{(1-U^N)}, \quad (II.19a)$$

where

$$U = z^{-1}\{1 - (1-z^2)^{\frac{1}{2}}\}. \quad (II.19b)$$

It is known that the mean recurrence time for return to the origin is infinite for an infinite lattice, even though the return probability[8] is 1. Likewise the expected time to reach any point is infinite although the probability of reaching any point is 1. This difficulty is avoided in the case of a finite lattice. Here, in contrast to the Polya case, return to the origin or to any lattice point occurs with probability one in any number of dimensions. We shall calculate the expected time to reach any point for the first time for nearest-neighbor jumps, and then present the generalization for different one-dimensional random walks, in the limit of large N. For the lattice with jumps to nearest neighbors only, we find[2] by an exact calculation starting from Eqs. (II.2a) and (II.19)

$$\langle n(s) \rangle = s(N - s). \quad (II.20)$$

To treat the case of the general one-dimensional walk for which $N \gg s$, we use Eqs. (II.2a) and (II.15)

---

[8] W. Feller, *An Introduction to Probability Theory and its Applications* (John Wiley & Sons, Inc., New York, 1951).

to find

$$\lim_{N\to\infty} \langle n(\mathbf{s}) \rangle / N = \varphi(0, 1) - \varphi(\mathbf{s}, 1)$$

$$\sim \lim_{z\to 1} [\varphi_1(0, z) - \varphi_1(\mathbf{s}, z)] = s/\sigma^2. \quad (II.21)$$

In the two-dimensional case, since

$$K_0\{\lambda[2(1-z)]^{\frac{1}{2}}\} \sim -\tfrac{1}{2} \log(1-z) - \log\lambda + O(1)$$

for large λ, we have the expression

$$\lim_{N\to\infty} \langle n(\mathbf{s}) \rangle / N^2 \sim \frac{\log \lambda}{\pi \sigma_1 \sigma_2}. \quad (II.22)$$

For the symmetric random walk on a simple square lattice with jumps to nearest neighbors, $\sigma_1 = \sigma_2 = 2^{-\frac{1}{2}}$ and the mean passage time is

$$\lim_{N\to\infty} \frac{\langle n(\mathbf{s}) \rangle}{N^2} = \frac{2 \log s}{\pi}. \quad (II.23)$$

The three-dimensional first-passage time is given by

$$\lim_{N\to\infty} \frac{\langle n(\mathbf{s}) \rangle}{N^3}$$

$$= \varphi(0, 1) - \frac{1}{2\pi\sigma_1\sigma_2\sigma_3\lambda} + \cdots, \quad \mathbf{s} \neq 0. \quad (II.24)$$

It is interesting to note that, in one and two dimensions, the first term in the asymptotic expansion for the mean first-passage time depends only on λ and the σ_i and not on any further detailed description of the lattice. Furthermore, $\langle n(\mathbf{s}) \rangle / N^k$ is an increasing function of λ for large λ in one and two dimensions. In three and higher dimensions the mean first-passage time depends in a detailed way on the lattice [through $\varphi(0, 1)$] and to a first approximation is a constant, independent of λ.

Calculation of the variances of the first passage times is considerably more difficult because, at the very least, the expression for the variance contains $\varphi(0, 1)$. For the one-dimensional random walk with jump probabilities of $\tfrac{1}{2}$ to either nearest neighbor, the detailed expansion of $P(s, z)$ around $z = 1$ is from Eq. (II.19)

$$P(s, z) = 1 - s(N - s)(1 - z) + \tfrac{1}{6}s(N - s)$$

$$\times (N^2 + sN - s^2 - 5)(1 - z)^2 + \cdots. \quad (II.25)$$

From this expression we derive

$$\sigma^2(s) = \langle n^2(s) \rangle - \langle n(s) \rangle^2$$

$$= \tfrac{1}{3}s(N - s)[N^2 - 2s(N - s) - 2], \quad s \neq 0. \quad (II.26)$$

For $s = 0$ we have

$$F(0, z) = 1 - [P(0, z)]^{-1} = 1 - N(1 - z)$$

$$+ \tfrac{1}{6}N(N^2 - 1)(1 - z)^2 - \cdots, \quad (II.27)$$

from which it follows that[2]

$$\sigma^2(0) = \tfrac{1}{3}N(N-1)(N-2). \quad (II.28)$$

It is possible to derive an asymptotic value for $\sigma^2(0)$ for any 1-D transition probabilities by noticing that in the limit $N = \infty$, the principal contributions in

$$\varphi(0,1) = \frac{1}{N}\sum_{j=1}^{N-1}\left\{1 - \lambda\left(\frac{2\pi j}{N}\right)\right\}^{-1}$$

$$= \frac{2}{N}\sum_{j=1}^{[\frac{1}{2}(N-1)]}\left\{1 - \lambda\left(\frac{2\pi j}{N}\right)\right\}^{-1}$$

$$+ \frac{1}{2N}[1 + (-1)^N]\left\{1 - \lambda\left(\frac{2\pi[\frac{1}{2}(N-1)] + 2\pi}{N}\right)\right\}^{-1}$$

come from small j. We therefore expand $\lambda(2\pi j/N)$ according to Eq. (II.4) and find

$$\varphi(0,1) \sim \frac{N}{\pi^2\sigma_1^2}\sum_{j=1}^{[\frac{1}{2}(N-1)]} j^{-2}. \quad (II.29)$$

In the limit of large N the series can be replaced by its sum to infinity $\tfrac{1}{6}\pi^2$, and the asymptotic expression for the variance becomes

$$\sigma^2(0) \sim N^3/3\sigma_1^2 \quad (II.30)$$

in agreement with the special result given in Eq. (II.28). It is also possible to derive an expression for $\sigma^2(s)$ for $s \ll N$ by this technique. A calculation similar to the preceding serves to show that

$$\frac{\partial \varphi(0,1)}{\partial z} - \frac{\partial \varphi(s,1)}{\partial z} = \frac{Ns^2}{6\sigma_1^4}; \quad (II.31)$$

hence the principal contribution to $\sigma^2(s)$ comes from the first term in the expression for $\langle n^2(s)\rangle$. Using the expression for $\varphi(0,1)$ given in Eq. (II.29) we find

$$\sigma^2(s) = (sN^3)/(3\sigma_1^4). \quad (II.32)$$

It is shown in Appendix B that the asymptotic form for $\varphi(0,1)$ in the 2-D case as $N \to \infty$ is

$$\varphi(0,1) \sim (\pi\sigma_1\sigma_2)^{-1}\log N. \quad (II.33)$$

Hence in 2D the variance in the return time to the origin is

$$\sigma^2(0) \sim (2/\pi\sigma_1\sigma_2)N^4 \log N. \quad (II.34)$$

The sum defining $\varphi(0,1)$ converges in three dimensions and greater. As $N \to \infty$, $\varphi(0,1)$ has the integral form (II.6). These integrals have been calculated by Watson[9] for cubic lattices. From the numerical values one obtains the following estimates of $\sigma^2(0)$

[9] G. N. Watson, Quar. J. Math. Oxford, Ser. 10, 266 (1939).

for the cubic lattices:

s.c. $\sigma^2(0)/N^6$

$$\sim \frac{2}{(2\pi)^3}\iiint_{-\pi}^{\pi}\frac{d^3\vartheta}{1 - \tfrac{1}{3}(c_1 + c_2 + c_3)} - 1 = 2.032,$$

f.c. $\sigma^2(0)/N^6$

$$\sim \frac{2}{(2\pi)^3}\iiint_{-\pi}^{\pi}\frac{d^3\vartheta}{1 - \tfrac{1}{3}(c_1c_2 + c_2c_3 + c_3c_1)} - 1 = 1.690,$$

b.c. $\sigma^2(0)/N^6$

$$\sim \frac{2}{(2\pi)^3}\iiint_{-\pi}^{\pi}\frac{d^3\vartheta}{1 - c_1c_2c_3} - 1 = 1.786.$$

It is of incidental interest that the expression for the variance

$$\sigma^2(0) = N^{2k}[2\varphi(0,1) - 1] + N^k \quad (II.35)$$

shows that, for $k \geq 3$,

$$\varphi(0,1) \geq \tfrac{1}{2}, \quad (II.36)$$

a result which seems otherwise difficult to prove.

III. NUMBER OF POINTS VISITED r TIMES IN AN n-STEP WALK

We now turn to the statistics of the number of distinct lattice points visited during an n-step walk. We will be concerned mainly with the large n case, although some results will also be given for any integer n.

Let S_n be the average number of lattice points visited in an n-step walk. Then

$$S_n = 1 + \sum_s' \{F_1(s) + F_2(s) + \cdots + F_n(s)\}, \quad (III.1)$$

where the primed summation proceeds over all lattice points except the origin. The integer 1 represents the fact that the walker was originally at the origin. As before $F_j(s)$ is the probability that the walker arrives at s for the first time after the jth step. Hence the summand of (III.1) represents the probability that the point s has been occupied at least once in the first n steps.

It is convenient to define a quantity Δ_k by

$$\Delta_k = S_k - S_{k-1}, \quad k = 1, 2, \cdots. \quad (III.2)$$

Since $S_0 = 1$ and $S_1 = 2$ we find $\Delta_1 = 1$. Then

$$\Delta_n = \sum_s' F_n(s) = -F_n(0) + \sum_s F_n(s). \quad (III.3)$$

Hence the generating function for Δ_k is

$$\Delta(z) = \sum_{1}^{\infty} z^n \Delta_n$$
$$= -\sum_{1}^{\infty} z^n F_n(0) + \sum_{s} \sum_{n=1}^{\infty} z^n F_n(s), \quad (III.4)$$

so that

$$\Delta(z) = -F(0, z) + \sum_{s} F(s, z). \quad (III.5)$$

However from Eq. (I.18)

$$F(s, z) = \frac{P(s, z) - \delta_{s,0}}{P(0, z)} \quad (III.6)$$

and

$$\Delta(z) = -1 + \sum_{s} \frac{P(s, z)}{P(0, z)}. \quad (III.7)$$

Then Eq. (I.14b) implies

$$\Delta(z) = -1 + \{(1 - z)P(0, z)\}^{-1}. \quad (III.8)$$

The generating function $S(z)$ can be obtained immediately from this expression since

$$S_0 = 1,$$
$$S_1 = 1 + \Delta_1,$$
$$\cdots$$
$$S_n = 1 + \Delta_1 + \Delta_2 + \cdots + \Delta_n \text{ etc.}, \quad (III.9)$$

we find

$$S(z) = \frac{1}{1-z} + \frac{z\Delta_1}{1-z} + \frac{z^2\Delta_2}{1-z} + \cdots$$
$$= \frac{1}{1-z} + \frac{\Delta(z)}{1-z}.$$

Hence from (III.8)

$$S(z) = \{(1-z)^2 P(0, z)\}^{-1}. \quad (III.10)$$

The asymptotic properties of S_n as $n \to \infty$ can be inferred from the analytic behavior of $\Delta(z)$ as $z \to 1$ by employing the following Tauberian theorem[10]:
Let $A(y) = \sum a_n \exp(-ny)$ be convergent for all $y > 0$ and let $a_n > 0$ for all n. If as $y \to 0$

$$A(y) \sim \varphi(y^{-1}), \quad (III.11a)$$

where (i) $\varphi(x) = x^\sigma L(x)$ is a positive increasing function of x for x greater than some x_0, and which increases monotonically to infinity for x sufficiently large; (ii) σ is ≥ 0; and (iii) $L(cx) \sim L(x)$ as $x \to \infty$; then as $n \to \infty$

$$a_1 + a_2 + \cdots + a_n \sim \varphi(n)/\Gamma(\sigma + 1). \quad (III.11b)$$

In our problem we interpret $a_1 + \cdots + a_n$ as $\Delta_1 + \cdots + \Delta_n$ and $A(y)$ as $\Delta(e^{-y})$.

[10] G. H. Hardy, *Divergent Series* (Oxford University Press, New York, 1949).

As $z \to 1$ the asymptotic behavior of $P(0, z)$ in one, two, and three dimensions is as follows[2]:

1D $\quad P(0, z) = (1 - z^2)^{-\frac{1}{2}}, \quad$ (III.12a)

2D $\quad P(0, z) \sim -\pi^{-1} \log(1 - z), \quad$ (III.12b)

3D $\quad P(0, z) \sim P(0, 1)$
$\quad\quad\quad + a(1 - z)^{\frac{1}{2}} + \cdots, \quad$ (III.12c)

where a is a constant which depends on the lattice. Then, if we let $z = \exp(-y)$ and let $y \to 0$,

1D $\quad \Delta(z) \sim (2/y)^{\frac{1}{2}}, \quad$ (III.13a)

2D $\quad \Delta(z) \sim (\pi/y)[1/\log(1/y)], \quad$ (III.13b)

3D $\quad \Delta(z) \sim [yP(0, 1)]^{-1}. \quad$ (III.13c)

The Tauberian theorem given above applies directly to our problem[2] if we choose

1D $\quad \sigma = \frac{1}{2}, \quad L(x) = 2^{\frac{1}{2}}, \quad$ (III.14a)

2D $\quad \sigma = 1, \quad L(x) = \pi/\log x, \quad$ (III.14b)

3D $\quad \sigma = 1, \quad L(x) = 1/P(0, 1). \quad$ (III.14c)

We therefore find for the number of distinct lattice points visited after n steps

1D $\quad S_n \sim (8n/\pi)^{\frac{1}{2}}, \quad$ (III.15a)

2D $\quad S_n \sim \pi n/\log n, \quad$ (III.15b)

3D $\quad S_n \sim n/P(0, 1). \quad$ (III.15c)

These results have been derived by Erdos and Dvoretzky[3] and by Vineyard[4] by somewhat different methods. The values of $P(0, 1)$ are 1.5164 for a simple cubic lattice, 1.3445 for a face-centered and 1.3932 for a body-centered cubic lattice.[2,4]

It is interesting to note that the 2-D S_n/π has the same asymptotic behavior as the number of primes less than n. Perhaps one can find some deep connection between random walks on square lattices and the distribution of primes.

We have shown in Appendix C that the generating function for the average number of lattice points visited at least r times, $S_n^{(r)}$, is

$$S^{(r)}(z) = \left\{1 - \frac{1}{P(0, z)}\right\}^{r-1} \frac{1}{(1-z)^2 P(0, z)}, \quad (III.16)$$

while that of

$$\Delta_n^{(r)} = S_n^{(r)} - S_{n-1}^{(r)}$$

is

$$\Delta^{(r)}(z) = \left\{1 - \frac{1}{P(0, z)}\right\}^{r-1} \frac{1}{(1-z)P(0, z)},$$
$$r \geq 2. \quad (III.17)$$

The average number of lattice points visited *exactly*

r times after n steps, $V_n^{(r)}$ is given by

$$V_n^{(r)} = S_n^{(r)} - S_n^{(r+1)}. \quad (III.18)$$

Its generating function is

$$V^{(r)}(z) = \sum_0^\infty V_n^{(r)} z^r$$

$$= \frac{1}{(1-z)^2 [P(0,z)]^2} \left\{ 1 - \frac{1}{P(0,z)} \right\}^{r-1}. \quad (III.19)$$

By applying the above Tauberian theorem to Eq. (III.17) we can generalize (III.15c) to find $S_n^{(r)}$, the average number of points occupied at least r times in a walk of n steps on a three-dimensional lattice. If we set $z = e^{-y}$ and let $y \to 0$ we find

$$\Delta^{(r)}(e^{-y}) \sim \left\{ 1 - \frac{1}{P(0,1)} \right\}^{r-1} \frac{1}{P(0,1)y}, \quad (III.20)$$

so that in the notation of the Tauberian theorem $\sigma = 1$ and

$$L(x) = \frac{1}{P(0,1)} \left\{ 1 - \frac{1}{P(0,1)} \right\}^{r-1}$$

$$= \text{constant}. \quad (III.21)$$

Hence, since for $r > 1$

$$S_n^{(r)} = \Delta_1^{(r)} + \Delta_2^{(r)} + \cdots + \Delta_n^{(r)},$$

Eq. (III.16) implies that as $n \to \infty$

$$S_n^{(r)} \sim \frac{n}{P(0,1)} \left\{ 1 - \frac{1}{P(0,1)} \right\}^{r-1}. \quad (III.22)$$

Noting that the quantity $f = 1 - [P(0,1)]^{-1}$ is the probability that a random walker who starts from the origin ever returns to the origin, we can write

$$S_n^{(r)} \sim n(1-f)f^{r-1}. \quad (III.23)$$

The values of f for the three cubic lattices are sc 0.34056, bcc 0.28223, and fcc 0.25632.

As $n \to \infty$, the average number of points occupied on a 3-D lattice exactly r times in an n step walk is

$$V_n^{(r)} = S_n^{(r)} - S_n^{(r+1)} \sim n(1-f)^2 f^{r-1}. \quad (III.24)$$

If one wishes to find correction terms to the asymptotic formulas (III.15) for S_n, the number of points visited at least once in an n step walk, he must proceed in a more systematic manner. In the 3-D case it is shown in Appendix D that

$$P(0,z) = u_0 - u_1(1-z)^{\frac{1}{2}} + u_2(1-z)$$

$$- u_3(1-z)^{\frac{3}{2}} + \cdots. \quad (III.25)$$

The numbers u_0 for the various cubic lattices are given in Eq. (D.3) of that appendix.[9] It was also shown that

$$u_1 = \begin{cases} (3/\pi)(\frac{3}{2})^{\frac{1}{2}} = 1.1695454 & \text{sc}, \quad (III.26a) \\ 1/2^{\frac{1}{2}}\pi = 0.2250791 & \text{bcc}, \quad (III.26b) \\ 3^{\frac{1}{2}}/4\pi = 0.4134967 & \text{fcc}. \quad (III.26c) \end{cases}$$

The values of u_2 and u_3 have not been calculated for the bcc and fcc; however, for the sc lattice,[11]

$$u_2 = 1.384761, \quad (III.27a)$$

$$u_3 = \frac{9}{4\pi} \left(\frac{3}{2}\right)^{\frac{1}{2}} = 0.877159. \quad (III.27b)$$

Equation (III.25) can be substituted into the generating function $S(z)$ [see Eq. (III.10)] to obtain

$$S(z) = [u_0(1-z)^2]^{-1}$$
$$+ (u_1/u_0^2)(1-z)^{-\frac{3}{2}} + [(u_1^2 - u_2 u_0)/u_0^3](1-z)^{-1}$$
$$+ [(u_1^3 - 2u_0 u_1 u_2 + u_3 u_0^2)/u_0^4](1-z)^{-\frac{1}{2}} + \cdots. \quad (III.28)$$

Now the coefficient of z^n in the series expansion of $(1-z)^m$ is

for $m = -2$: $(n+1)$, $\quad (III.29a)$

for $m = -\frac{3}{2}$: $(2n+1)!/2^{2n} n! n!$, $\quad (III.29b)$

for $m = -1$: 1, $\quad (III.29c)$

for $m = -\frac{1}{2}$: $(2n-1)!/2^{2n-1} n! (n-1)!$. $\quad (III.29d)$

One can use Stirling's expansion for large n to find

$$\frac{(2n+1)!}{2^{2n} n! n!} \sim 2\left(\frac{n}{\pi}\right)^{\frac{1}{2}}$$

$$\times \left[1 + \frac{3}{8n} - \frac{7}{128n^2} + \cdots \right], \quad (III.30a)$$

$$\frac{(2n-1)!}{2^{2n-1} n! (n-1)!} \sim \frac{1}{(n\pi)^{\frac{1}{2}}}$$

$$\times \left[1 - \frac{1}{8n} + \frac{1}{128n^2} - \cdots \right]. \quad (III.30b)$$

Then

$$S_n \sim \frac{n}{u_0} + \frac{2u_1}{u_0^2} \left(\frac{n}{\pi}\right)^{\frac{1}{2}} + (u_1^2 - u_2 u_0 + u_0^2)/u_0^3$$

$$+ (3u_1 u_0^2 + 4u_1^3 - 8u_0 u_1 u_2$$

$$+ 4u_3 u_0^2)/[4u_0^4(\pi n)^{\frac{1}{2}}] + O(1/n). \quad (III.31)$$

In the case of the bcc lattice

$$S_n \sim \frac{4\pi^2 n}{[\Gamma(\frac{1}{4})]^4} + \frac{16\pi^5}{[\Gamma(\frac{1}{4})]^8} \left(\frac{2n}{\pi}\right)^{\frac{1}{2}} + O(1)$$

$$= 0.71777001n + 0.130846n^{\frac{1}{2}} + O(1). \quad (III.32)$$

---

[11] A. Maradudin, E. Montroll, G. Weiss, R. Herman, and H. Milnes, "Green's Functions for Monatomic Simple Cubic Lattices," Acad. Roy. Belg. Cl. Sci. Mem. Coll. in 4° (2) 14 (1960) No. 7.

In the case of the fcc lattice

$$S_n \sim \frac{2^{11/3} n \pi^4}{9\{\Gamma(\frac{1}{3})\}^6} + \frac{2^{19/3} \pi^7}{9\{\Gamma(\frac{1}{3})\}^{12}} \left(\frac{n}{3\pi}\right)^{1/2} + O(1)$$

$$= 0.74368182n + 0.258048n^{1/2} + O(1), \quad (\text{III}.33)$$

while with the extra information available for the sc lattice one finds in that case

$$S_n \sim 0.65946267n + 0.573921n^{1/2}$$
$$+ 0.449530 + 0.40732n^{-1/2} + \cdots . \quad (\text{III}.34)$$

A similar expression can be obtained for the number of points occupied at least once after n steps on a 1-D lattice walk in which the walker steps only to a nearest-neighbor point on each step (steps in either direction being equally probable). Then from (III.10)

$$S(z) = \frac{(1-z^2)^{\frac{1}{2}}}{(1-z)^2} = \frac{[2-(1-z)]^{\frac{1}{2}}}{(1-z)^{\frac{3}{2}}}$$

$$= 2^{\frac{1}{2}}\{(1-z)^{-\frac{3}{2}} - \tfrac{1}{4}(1-z)^{-\frac{1}{2}}$$
$$- \tfrac{1}{32}(1-z)^{\frac{1}{2}} - \tfrac{1}{128}(1-z)^{\frac{3}{2}} - \cdots\}, \quad (\text{III}.35)$$

so that

$$S_n \sim \frac{2^{\frac{1}{2}}(2n+1)!}{2^{2n} n! \, n!} \left\{ 1 - \frac{1}{4(2n+1)} \right.$$
$$\left. - \frac{1}{32(4n^2-1)} - \cdots \right\}. \quad (\text{III}.36)$$

If n is chosen to be as small as 4 this yields 3.347 as compared with the exact value 3.375 given in Table II. By using Stirling's approximation [see Eq. (III.30)] for the factorials we find the somewhat simplified expression

$$S_n \sim \left(\frac{8n}{\pi}\right)^{\frac{1}{2}} \left\{ 1 + \frac{1}{4n} - \frac{3}{64n^2} + \cdots \right\}. \quad (\text{III}.37)$$

The generating function for the number of points which are occupied exactly once in an n step 1-D

TABLE I. Values of $P(\mathbf{s}, 1)$ for a simple cubic lattice when $s^2 = s_1^2 + s_2^2 + s_3^2 < 15$. These numbers correspond to the symmetrical case with $P(s_1 s_2 s_3, 1) = P(s_2 s_1 s_3, 1) = \cdots$, etc. This function is the lattice Green's function defined by (II.6) and (I.5) when $z = 1$.

(s_1, s_2, s_3)	$P(\mathbf{s}, 1)$	(s_1, s_2, s_3)	$P(\mathbf{s}, 1)$
001	0.516387	023	0.132451
002	0.257336	111	0.261470
003	0.165271	112	0.191792
011	0.331149	113	0.144196
012	0.215590	122	0.156953
013	0.153139	123	0.126946
022	0.168331	222	0.135908

TABLE II. $S_n^{(r)}$ = Average number of points occupied at least r times in a 1-D walk of n steps.

n/r	1	2	2	4	5
0	1	0	0	0	0
1	2	0	0	0	0
2	5/2	1/2	0	0	0
3	3	1	0	0	0
4	27/8	11/8	1/4	0	0
5	15/4	7/4	1/2	0	0
6	65/16	33/16	3/4	1/8	0
7	35/8	19/8	1	1/4	0

walk is, from (III.19) and (III.12a)

$$V^{(1)}(z) = [(1-z)P(z,0)]^{-2}$$
$$= (1-z^2)/(1-z)^2 = (1+z)/(1-z)$$
$$= 1 + 2z + 2z^2 + 2z^3 + \cdots . \quad (\text{III}.38)$$

Hence

$$V_1^{(1)} = 1 \quad \text{and} \quad V_n^{(1)} = 2 \quad \text{for} \quad n > 1. \quad (\text{III}.39)$$

The asymptotic expression for $V_n^{(1)}$ for large n on a 2-D square lattice can be obtained by finding the generating function for

$$D_n^{(1)} = V_n^{(1)} - V_{n-1}^{(1)},$$
$$D^{(1)}(z) = -1 + (1-z)^{-1}[P(z,0)]^2 \quad (\text{III}.40)$$

(here $D_n^{(1)}$ is analogous to Δ_n in the calculation of S_n). By employing (III.40) and our Tauberian theorem we find

$$V_n^{(1)} \sim n\pi^2/(\log n)^2 \quad (\text{III}.41)$$

to be the asymptotic number of points occupied exactly once in the 2-D case. The result has also been derived by Erdos and Taylor[5] by a different method.

The 1-D generating function for $S_n^{(2)}$, the number of points occupied at least twice in an n-step walk is

$$S^{(2)}(z) \equiv \{1 - (1-z^2)^{\frac{1}{2}}\} \frac{(1-z^2)^{\frac{1}{2}}}{(1-z)^2}$$

$$= S^{(1)}(z) - \left(\frac{1+z}{1-z}\right). \quad (\text{III}.42a)$$

Hence when $n \geq 2$

$$S_n^{(2)} = S_n^{(1)} - 2, \quad (\text{III}.42b)$$

which can be verified in Table II. Similarly,

$$S^{(3)}(z)$$
$$= [1 - 2(1-z^2)^{\frac{1}{2}} + (1-z^2)](1-z^2)^{\frac{1}{2}}/(1-z)^2$$
$$= (2-z^2)S^{(1)}(z) - 2(1+z)/(1-z)$$

Hence

$$S_n^{(3)} = 2S_n^{(1)} - S_{n-2}^{(1)} - 4 \quad \text{if} \quad n \geq 4. \quad (\text{III}.42c)$$

TABLE III. $V_n^{(r)}$ = Average number of points occupied exactly r times in a 1-D walk of n steps.

n/r	1	2	3	4	5
0	1	0	0	0	0
1	2	0	0	0	0
2	2	1/2	0	0	0
3	2	1	0	0	0
4	2	9/8	1/4	0	0
5	2	5/4	1/2	0	0
6	2	21/16	5/8	1/8	0
7	2	11/8	3/4	1/4	0

Similarly

$$S_n^{(4)} = 4S_n^{(1)} - 3S_{n-2}^{(1)} - 6 \quad \text{if} \quad n \geq 6, \quad \text{(III.42d)}$$

etc.

This scheme can be continued further and when these formulas are combined with (III.36) and (III.37), very accurate asymptotic expansions can be found for $S_n^{(r)}$ for 1-D walks, $r \ll n$.

IV. THE NUMBER OF VISITS TO A GIVEN LATTICE POINT DURING A WALK OF n STEPS

The probability that a point $\mathbf{s}$ is visited at least r times in an n-step walk is

$$\sum_{j=1}^{n} F_j^{(r)}(\mathbf{s}) \quad \text{if} \quad \mathbf{s} \neq 0,$$

$$\sum_{j=1}^{n} F_j^{(r-1)}(0) \quad \text{if} \quad \mathbf{s} = 0,$$

so that the probability that $\mathbf{s}$ is visited *exactly* r times is

$$\beta_n^{(r)}(\mathbf{s}) = \begin{cases} \sum_{j=1}^{n} [F_j^{(r)}(\mathbf{s}) - F_j^{(r+1)}(\mathbf{s})], \\ \qquad\qquad\qquad\qquad \text{if} \quad \mathbf{s} \neq 0, \\ \sum_{j=1}^{n} [F_j^{(r-1)}(0) - F_j^{(r)}(0)], \\ \qquad\qquad\qquad\qquad \text{if} \quad \mathbf{s} = 0. \end{cases} \quad \text{(IV.1)}$$

The formulas for $\beta_n^{(r)}(0)$ are distinctive because the walker starts at the origin.

The generating function for $\beta^{(r)}(\mathbf{s}, z)$,

$$\beta^{(r)}(\mathbf{s}, z) \equiv \sum_{1}^{n} z^n \beta_n^{(r)}, \quad \text{(IV.2)}$$

is easily seen from (I.20) to be

$$\beta^{(r)}(\mathbf{s}, z) = \begin{cases} (1-z)^{-1} F(\mathbf{s}, z)[1 - F(0, z)] \\ \qquad \times [F(0, z)]^{r-1}, \quad \mathbf{s} \neq 0, \\ (1-z)[F(0, z)]^{r-1} \\ \qquad \times [1 - F(0, z)], \quad \mathbf{s} = 0. \end{cases} \quad \text{(IV.3)}$$

The mean number of times the point $\mathbf{s}$ has been visited after n steps is

$$M_n(\mathbf{s}) = \sum r \beta_n^{(r)}(\mathbf{s}). \quad \text{(IV.4)}$$

This has the generating function

$$M(\mathbf{s}, z) = \sum_{r=1}^{\infty} \sum_{n=1}^{\infty} r \beta_n^{(r)}(\mathbf{s}) z^n$$

$$= \sum_{r=1}^{\infty} r \beta^{(r)}(\mathbf{s}, z)$$

$$= \frac{F(\mathbf{s}, z)}{(1-z)[1 - F(0, z)]} \quad \text{if} \quad \mathbf{s} \neq 0$$

$$= (1-z)^{-1} P(\mathbf{s}, z). \quad \text{(IV.5)}$$

If $\mathbf{s} = 0$,

$$M(0, z) = \{(1-z)[1 - F(0, z)]\}^{-1}$$

$$= (1-z)^{-1} P(0, z). \quad \text{(IV.6)}$$

Hence (IV.5) is valid for all $\mathbf{s}$ including $\mathbf{s} = 0$.

The asymptotic form for $M_n(0)$ for 3-D lattices can be obtained by using the expression for $P(0, z)$ given in Appendix F. There it is shown that

$$P(0, z) \sim u_0 - [2(1-z)]^{\frac{1}{2}} / \pi \sigma_1 \sigma_2 \sigma_3 + \cdots, \quad \text{(IV.7)}$$

where the u_0 and σ's are defined generally and evaluated for walks on cubic lattices where only steps to nearest-neighbor points (all with equal probability) are taken. By combining Eqs. (IV.6), (IV.7), and (III.29) we obtain

$$M_n(0) \sim u_0 - (2/\pi n)^{\frac{1}{2}} / \pi \sigma_1 \sigma_2 \sigma_3 + O(1/n). \quad \text{(IV.8)}$$

The numerical results are

$$M_n(0) \sim \begin{cases} 1.51639 - 1.31969 n^{-\frac{1}{2}} + \cdots & \text{sc}, \\ 1.39320 - 0.25397 n^{-\frac{1}{2}} + \cdots & \text{fcc}, \\ 1.34466 - 0.46658 n^{-\frac{1}{2}} + \cdots & \text{bcc}. \end{cases} \quad \text{(IV.9)}$$

As $n \to \infty$

$$M_n(\mathbf{s}) \to P(\mathbf{s}, 1). \quad \text{(IV.10)}$$

These functions have been tabulated for[11] sc lattices when $s^2 < 25$. Some values are given in Table I.

When $\mathbf{s}$ is large and $z \to 1$ we have from (I.18b) [where $\lambda^2 = \sum (s_i/\sigma_i)^2$]

$$P(\mathbf{s}, z) \sim \frac{\exp\{-\lambda[2(1-z)]^{\frac{1}{2}}\}}{\lambda \sigma_1 \sigma_2 \sigma_3 (2\pi)^2}$$

$$\sim \frac{1}{\lambda \sigma_1 \sigma_2 \sigma_3 (2\pi)^2} \{1 - \lambda[2(1-z)]^{\frac{1}{2}} + \cdots\}. \quad \text{(IV.11)}$$

Hence, from (IV.5), (IV.7), and (III.29), when s

and n are both large, but still with $s \ll n^{\frac{1}{2}}$,

$$M_n(\mathbf{s}) \sim \frac{1}{\lambda \sigma_1 \sigma_2 \sigma_3 (2\pi)^2}$$
$$\times \left\{ 1 - \lambda \left(\frac{2}{\pi n} \right)^{\frac{1}{2}} + \cdots \right\}. \quad (\text{IV.12})$$

Employing the σ values given in Appendix D for the walks involving only steps to nearest-neighbor lattice points on cubic lattices we find

$$M_n(\mathbf{s}) \sim \begin{cases} \frac{3}{4s\pi^2} \left[1 - s\left(\frac{6}{\pi n} \right)^{\frac{1}{2}} + \cdots \right] \text{sc} \\ \frac{1}{4s\pi^2} \left[1 - s\left(\frac{2}{\pi n} \right)^{\frac{1}{2}} + \cdots \right] \text{bcc} \\ \frac{3}{8s\pi^2} \left[1 - s\left(\frac{3}{\pi n} \right)^{\frac{1}{2}} + \cdots \right] \text{fcc}. \end{cases} \quad (\text{IV.13})$$

A word of caution should be given concerning these results. One would expect that $M_n(\mathbf{s})$ should appear with some ordering with respect to nearest neighbors. However the bcc results are not between the sc and fcc. This is because all lattices were obtained by restricting walks on a fundamental sc lattice. If the unit cells of each of the lattices were made the same size and s reexpressed as a length the results would fall properly in order.

V. LATTICE WALKS FOR CONTINUOUS TIME VARIABLE

The preceding results can be used as a basis for the analysis of continuous time random walks on discrete lattices. In this theory we shall be interested in functions like $\bar{P}(\mathbf{s}, t)$ and $\bar{F}(\mathbf{s}, t)$ (the probability of being at $\mathbf{s}$ at time t) and the probability density for reaching $\mathbf{s}$ for the first time at time t, respectively. We shall assume that jumps are made at random times $t_1, t_2, t_3, \cdots$ where the random variables

$$T_1 = t_1, \quad T_2 = t_2 - t_1, \cdots ,$$
$$T_n = t_n - t_{n-1}, \cdots \quad (\text{V.1})$$

have a common density $\psi(t)$. It will be convenient to define a further class of probability densities $\{\psi_n(t)\}$ by

$$\psi_0(t) = \delta(t), \quad \text{V.2}$$

$$\psi_n(t) = \int_0^t \psi(\tau) \psi_{n-1}(t - \tau) \, d\tau,$$
$$n = 1, 2, 3, \cdots . \quad (\text{V.3})$$

These are the probability densities for the occurrence of the nth step at time t. The most significant property of the $\psi_n(t)$ is the fact that their Laplace transforms are

$$\int_0^\infty e^{-ut} \psi_n(t) \, dt = [\psi^*(u)]^n, \quad (\text{V.4})$$

where

$$\psi^*(u) = \int_0^\infty e^{-ut} \psi(t) \, dt. \quad (\text{V.5})$$

In terms of the $F_n(\mathbf{s})$ defined in Sec. 1, $\bar{F}(\mathbf{s}, t)$ is given by

$$\bar{F}(\mathbf{s}, t) = \sum_{n=0}^\infty F_n(\mathbf{s}) \psi_n(t) \quad (\text{V.6})$$

and its Laplace transform is

$$\bar{F}^*(\mathbf{s}, u) = \int_0^\infty \bar{F}(\mathbf{s}, t) e^{-ut} \, dt$$
$$= \sum_{n=0}^\infty F_n(\mathbf{s}) [\psi^*(u)]^n$$
$$= F[\mathbf{s}, \psi^*(u)], \quad (\text{V.7})$$

where $F(\mathbf{s}, z)$ is the generating function of Eq. (I.17).

The function $\bar{P}(\mathbf{s}, t)$ is almost as simply related to the generating function $P(\mathbf{s}, z)$. Let $Q(\mathbf{s}, t)$ be the probability density for the random walk to reach $\mathbf{s}$ at time t (not necessarily for the first time) and let

$\Psi(t)$ = probability that walker remains fixed in time interval $(0, t)$

$$= 1 - \int_0^t \psi(x) \, dx = \int_t^\infty \psi(x) \, dx. \quad (\text{V.8})$$

Then

$$\bar{P}(\mathbf{s}, t) = \int_0^t Q(\mathbf{s}, \tau) \Psi(t - \tau) \, d\tau, \quad (\text{V.9})$$

or, in terms of Laplace transforms,

$$\bar{P}^*(\mathbf{s}, u) = Q^*(\mathbf{s}, u)[1 - \psi^*(u)]/u. \quad (\text{V.10})$$

But $Q(s, t)$ is given by

$$Q(\mathbf{s}, t) = \sum_{n=0}^\infty P_n(\mathbf{s}) \psi_n(t) \quad (\text{V.11})$$

or

$$Q^*(\mathbf{s}, u) = \sum_{n=0}^\infty P_n(\mathbf{s}) [\psi^*(u)]^n$$
$$= P[\mathbf{s}, \psi^*(u)], \quad (\text{V.12})$$

so that only the generating functions already discussed need be calculated.

Moments for various quantities of interest are easily derived from the formulas above. For example the first moment and variance of the first-passage time to s are

$$\bar{t} = -\frac{\partial F[\mathbf{s}, \psi^*(u)]}{\partial u}\bigg]_{u=0^+} = \langle n(\mathbf{s})\rangle \bar{T}, \quad (V.13a)$$

$$\overline{t^2} - \bar{t}^2 = [\langle n^2(\mathbf{s})\rangle - \langle n(\mathbf{s})\rangle^2]\bar{T}^2 + \langle n(\mathbf{s})\rangle[\overline{T^2} - \bar{T}^2], \quad (V.13b)$$

where $\overline{T^n}$ is the nth moment of the time between steps and $\langle n(s)\rangle$ and $\langle n^2(s)\rangle$ are given by (II.2) and (II.3).

Continuous analogues of other discrete results are obtained in the same manner. For example, the probability density for the random walker to reach s for the rth time is

$$\overline{F^{(r)}}(\mathbf{s}, t) = \sum_{n=0}^{\infty} F_n^{(r)}(\mathbf{s})\psi_n(t) \quad (V.14a)$$

or

$$\overline{F^{(r)}}*(\mathbf{s}, u) = F^{(r)}[\mathbf{s}, \psi^*(u)], \quad (V.14b)$$

where $F^{(r)}(\mathbf{s}, z)$ is given by (I.20).

We can also consider the statistics of the number of distinct steps visited after a time t. Let $S(t)$ be the average number of lattice points visited at least once in time t. Then

$$S(t) = \sum_{\mathbf{s}} \int_0^t F^{(r)}(\mathbf{s}, \tau)\, d\tau. \quad (V.15a)$$

Hence the Laplace transform of $S(t)$ is

$$\mathcal{L}\{S(t)\} = \frac{\psi^*(u)}{u[1 - \psi^*(u)]P(0, \psi^*(u))}. \quad (V.15b)$$

To find the large t behavior of $S(t)$ it is necessary to use the expansion

$$\psi^*(u) = 1 - u\bar{T} + o(u) \quad (V.16)$$

in Eq. (V.5), together with the asymptotic forms of Eq. (III.12) for the behavior of $P(0, z)$ in the neighborhood of $z = 1$. In this way, we find that in one dimension

$$\mathcal{L}\{S(t)\} = (2/\bar{T})^{\frac{1}{2}} u^{-\frac{3}{2}} + O(u^{-1}) \quad (V.17)$$

in the neighborhood of $u = 0$. But, by a Tauberian theorem[7] this implies that

$$S(t) = (8t/\pi\bar{T})^{\frac{1}{2}} + O(1). \quad (V.18)$$

In three dimensions the result is

$$S(t) = (t/\bar{T})/P(0, 1) + O(1). \quad (V.19)$$

The results are in agreement with (III.15a) and (III.15b) since the number of steps n is just $t/\bar{T}$ in the case of steps at regular time intervals.

VI. EFFECT OF TRAPS ON PROBABILITY OF RETURN TO THE ORIGIN ON A 1-D LATTICE

Another type of random walk problem is concerned with effect of traps on the probability of a walker eventually returning to the origin. We shall limit ourselves here to a discussion of the 1-D case while an analysis of the 2-D and 3-D problems, which are much more difficult, will be given elsewhere.

It has been shown[2] that in the presence of one trap at l_1 and another at l_2 with $l_2 < 0 < l_1$ the probability that a walker initially at the origin is trapped before return to the origin is

$$(l_1 - l_2)/2l_1(-l_2) = \tfrac{1}{2}(l_1^{-1} - l_2^{-1}).$$

This probability is not changed by the addition of any number of new traps *which are not located in the interval* $l_2 < 0 < l_1$.

Let c be the concentration of independently located traps. Then, if it is known that the origin is not a trap, the probability that a trap exists at l_1 and at l_2 and none in between is

$$c(1 - c)^{-l_2-1}(1 - c)^{l_1-1}c.$$

Hence the probability of our walker being trapped before returning to the origin is

$$\sum_{l_1=1}^{\infty} \sum_{l_2=-1}^{-\infty} c^2(1 - c)^{l_1-l_2-2}(\tfrac{1}{2})(l_1^{-1} - l_2^{-1})$$

$$= c^2(1 - c)^{-1}\left\{\sum_{l=1}^{\infty}(1 - c)^{l-1}\right\}\left\{\sum_{l_1=1}^{\infty}(1 - c)^{l_1}/l_1\right\}$$

$$= -[c/(1 - c)] \log c.$$

Then as a function of concentration of traps, the probability of a walker returning to the origin before being trapped is

$$F(c) = 1 + [c/(1 - c)] \log c.$$

APPENDIX A. ASYMPTOTIC FORM OF $\varphi(s, z)$ AS $s \to \infty$

The Green's function

$$\varphi(\mathbf{s}, z) = \frac{1}{(2\pi)^k} \int_{-\pi}^{\pi}\cdots\int \frac{\exp i\mathbf{s}\cdot\boldsymbol{\vartheta}\, d^k\boldsymbol{\vartheta}}{1 - z\lambda(\boldsymbol{\vartheta})} \quad (A.1)$$

can be expressed as

$$\varphi(\mathbf{s}, z) = \frac{1}{(2\pi)^k} \int_0^{\infty} e^{-\alpha}\, d\alpha \int_{-\pi}^{\pi}\cdots\int e^{i\mathbf{s}\cdot\boldsymbol{\vartheta}}$$
$$\times e^{\alpha z\lambda(\boldsymbol{\vartheta})}\, d^k\boldsymbol{\vartheta}. \quad (A.2)$$

When s is very large the main contribution to the $\boldsymbol{\vartheta}$ integration comes from small values of $|\boldsymbol{\vartheta}|$. In this range in a symmetrical random walk (see I.4)

$$\lambda(\boldsymbol{\vartheta}) = 1 - \tfrac{1}{2} \sum \sigma_i^2 \vartheta_i^2 + \tfrac{1}{4} \sum_{ij} \mu_{ij} \vartheta_i^2 \vartheta_j^2 - \cdots . \quad \text{(A.3)}$$

If we let $s_i \vartheta_i = \varphi_i$ and $\lambda_i = s_i/\sigma_i$, then $\varphi(s, z)$ becomes

$$\varphi(s, z) = \frac{1}{(2\pi)^k} \int_0^\infty e^{-\alpha(1-z)} \, d\alpha$$

$$\times \int_{-\pi s_1}^{\pi s_1} \cdots \int_{-\pi s_k}^{\pi s_k} e^{i(\varphi_1 + \varphi_2 + \cdots)} e^{-\tfrac{1}{2}\alpha z \sum \lambda_i^{-2} \varphi_i^2}$$

$$\times \{1 + \tfrac{1}{4}\alpha z \sum (\mu_{ij}/s_i^2 s_j^2)\varphi_i^2 \varphi_j^2 + \cdots\} \, d^k \varphi / s_1 \cdots s_k. \quad \text{(A.4)}$$

As all $s_i \to \infty$ the limits on the φ integration can be extended to $\pm\infty$ with errors of only $O[\exp(-cs_i^2)]$ appearing.

If we let

$$R_n(a) = \int_{-\infty}^\infty e^{i\varphi} e^{-\tfrac{1}{2}a\varphi^2} \varphi^n \, d\varphi, \quad \text{(A.5)}$$

then

$$\varphi(s, z) \sim (2\pi)^{-k} \int_0^\infty e^{-\alpha(1-z)} \, d\alpha \Big\{ \prod_{\nu=1}^k R_0(\alpha z \lambda_\nu^{-2}) \Big\}$$

$$\times \Big\{ 1 + \tfrac{1}{4}\alpha z \sum_{i=1}^k (\mu_{ii}/s_i^4)(R_4/R_0)_{\alpha z \lambda_i^{-2}}$$

$$+ \tfrac{1}{4}\alpha z \sum{}' \frac{\mu_{ij}}{s_i^2 s_j^2} \Big(\frac{R_2}{R_0}\Big)_{\alpha z \lambda_i^{-2}} \Big(\frac{R_2}{R_0}\Big)_{\alpha z \lambda_j^{-2}} + \cdots \Big\},$$

where as usual the prime in the summation indicates that the terms with $i = j$ are to be omitted.

From standard integral tables one finds

$$R_0(a) = (2\pi/a)^{\tfrac{1}{2}} \exp(-1/2a),$$
$$(R_2/R_0)_a = a^{-1}(1 - a^{-1}),$$
$$(R_4/R_0)_a = a^{-2}(3 - 6a^{-1} + a^{-2}).$$

Then, if we let

$$S_n(z) = \int_0^\infty \alpha^{-\tfrac{1}{2}n} e^{-\alpha(1-z)} \exp\Big(-\frac{1}{2\alpha z} \sum \lambda_\nu^2\Big) d\alpha,$$

we find

$$\varphi(s, z) \sim \frac{(2\pi z)^{-\tfrac{1}{2}k}}{\sigma_1 \cdots \sigma_k} \Big[S_k + \sum_{i=1}^k \frac{\mu_{ii}}{4z\sigma_i^4}$$

$$\times (3S_{k+2} - 6z^{-1}\lambda_i^2 S_{k+4} + z^{-2}\lambda_i^4 S_{k+6})$$

$$+ \tfrac{1}{4} \sum{}' \frac{\mu_{ij}}{z\sigma_i^2 \sigma_j^2} [S_{k+2} - z^{-1}(\lambda_i^2 + \lambda_j^2)S_{k+4}$$

$$+ z^{-2}\lambda_i^2 \lambda_j^2 S_{k+6}] + \cdots \Big].$$

In the special case $z = 1$, we see that

$$S_n(1) = (2/\lambda^2)^{\tfrac{1}{2}(n-2)} \Gamma(\tfrac{1}{2}n - 1)$$

where

$$\lambda^2 = \sum \lambda_\nu^2 = \sum_{\nu=1}^k s_\nu^2/\sigma_\nu^2.$$

Generally,

$$S_n(z) = 2\Big(\frac{2}{\lambda}\Big)^{\tfrac{1}{2}n-1} [z(1-z)]^{\tfrac{1}{2}(n-2)} K_{\tfrac{1}{2}n-1}\Big(\Big[\frac{\lambda^2}{z}(1-z)\Big]^{\tfrac{1}{2}}\Big)$$

where K_ν is the νth modified Bessel Function of the second kind. When $z = 1$

$$\varphi(\mathbf{s}, 1) = \frac{\Gamma(\tfrac{1}{2}k - 1)}{2\sigma_1 \cdots \sigma_k \pi^{\tfrac{1}{2}k} \lambda^{k-2}} \Big\{ 1 - \frac{(1 - \tfrac{1}{2}k)}{2\lambda^2} \sum_{i=1}^k \frac{\mu_{ii}}{\sigma_i^2}$$

$$\times \Big[3 - 6k\Big(\frac{\lambda_i}{\lambda}\Big)^2 + k(k+2)\Big(\frac{\lambda_i}{\lambda}\Big)^4 \Big]$$

$$- (1/2\lambda^2)(1 - \tfrac{1}{2}k) \sum{}' \frac{\mu_{ij}}{\sigma_i^2 \sigma_j^2} [1 - k(\lambda_i^2 + \lambda_j^2)/2\lambda^2$$

$$+ k(k+2)\lambda_i^2 \lambda_j^2/\lambda^4] + O(\lambda^{-4}) \Big\}.$$

If $\varphi_1(\mathbf{s}, z)$ is defined [see Eq. (I.9)] as

$$\varphi_1(\mathbf{s}, z)$$

$$= \frac{1}{(2\pi)^k} \int_{-\pi}^{\pi} \cdots \int \frac{\exp(i\boldsymbol{\vartheta} \cdot \mathbf{s}) \, d^k \boldsymbol{\vartheta}}{1 - z + \tfrac{1}{2}z(\sigma_1^2 \vartheta_1^2 + \cdots + \sigma_k^2 \vartheta_k^2)},$$

then as $\mathbf{s} \to \infty$ and $z \to 1$ when $k \geq 3$, then [see Eq. (I.186)]

$$\varphi_1(\mathbf{s}, z) \sim \frac{\Gamma(\tfrac{1}{2}k - 1)}{2\sigma_1 \cdots \sigma_k \pi^{\tfrac{1}{2}k} \lambda^{k-2}},$$

which is the leading term in $\varphi(\mathbf{s}, 1)$. Hence if we let

$$\varphi(\mathbf{s}, z) = \varphi_1(\mathbf{s}, z) + \varphi_2(\mathbf{s}, z)$$

where $\varphi_2(s, z)$ is defined as $\varphi(s, z) - \varphi_1(s, z)$, we see that when $k \geq 3$

$$\lim_{\mathbf{s} \to \infty} \frac{\varphi_1(\mathbf{s}, 1)}{\varphi_2(\mathbf{s}, 1)} = 0.$$

APPENDIX B. CALCULATION OF 2-D $\varphi(0, 1)$ FOR $N \times N$ LATTICE AS $N \to \infty$

The expression for $\varphi(0, 1)$ in a finite lattice is (Eq. I.16)

$$\varphi(0, 1) = N^{-2} \sum_{r_1=1}^{N-1} \sum_{r_2=1}^{N-1} \Big\{ 1 - \lambda\Big(\frac{2\pi \mathbf{r}}{N}\Big) \Big\}^{-1}$$

$$= 4N^{-2} \sum_{1}^{[(N-1)/2]} \sum_{1}^{[(N-1)/2]} \{1 - \lambda\}^{-1}$$

$$+ O(1/N). \quad \text{(B.1)}$$

As $N \to \infty$ this sum approaches a divergent integral, the divergence being related to the smallness of $1 - \lambda(2\pi \mathbf{r}/N)$ as $(2\pi \mathbf{r}/N) \to 0$. Hence one would expect the main contribution to $(0, 1)$ to result from small integral values of r_1 and r_2. In this range one can approximate λ by [see Eq. (I.4)]

$$\lambda(2\pi r/N) \sim 1 - (2\pi^2/N^2)(\sigma_1^2 r_1^2 + \sigma_2^2 r_2^2) + \cdots. \quad (B.2)$$

We now restrict ourselves to $\sigma_1 = \sigma_2$, the more general case being amenable to a similar analysis.

The range of summation in (B.1) is divided into two parts; the first part containing those lattice points (r_1, r_2) such that $(r_1^2 + r_2^2)^{\frac{1}{2}} < \alpha N$ where α is small enough so that (B.2) is a good approximation of λ for all these points, and the second containing the remainder of the lattice points. It can be shown that the contribution of the second set to $\varphi(0, 1)$ remains bounded as $N \to \infty$. The contribution of the first set is

$$4N^{-2} \sum\sum_{1<(r_1^2+r_2^2)^{\frac{1}{2}}<\alpha N} (N^2/2\pi^2 \sigma_1^2)(r_1^2 + r_2^2)^{-1}.$$

When N is sufficiently large, the sum is well approximated by the corresponding integral, which we express in polar coordinates

$$\sum_{1<(r_1^2+r_2^2)^{\frac{1}{2}}<\alpha N} (r_1^2 + r_2^2)^{-1} \sim \int_1^{\alpha N} \frac{2\pi j\, dj}{j^2} = \frac{\pi}{2} \log \alpha N.$$

Hence, as $N \to \infty$ for fixed α,

$$\varphi(0, 1) \sim (1/\pi\sigma_1^2) \log N.$$

In the unsymmetric case $\sigma_1 \neq \sigma_2$, one finds

$$\varphi(0, 1) \sim (1/\pi\sigma_1\sigma_2) \log N.$$

The above ideas can, with a little effort, be made completely rigorous.

APPENDIX C. GENERATING FUNCTION FOR AVERAGE NUMBER OF POINTS VISITED AT LEAST r TIMES IN AN n-STEP WALK

Let $S_n^{(r)}$ be the average number of lattice points visited at least r times in an n-step walk. Then

$$S_n^{(r)} = F_1^{(r-1)}(0) + \cdots + F_n^{(r-1)}(0)$$
$$+ \sum_{s}{}' \{F_1^{(r)}(\mathbf{s}) + F_2^{(r)}(\mathbf{s}) + \cdots + F_n^{(r)}(\mathbf{s})\}, \quad (C.1)$$

where the primed summation proceeds over all lattice points except the origin. As usual $F_j^{(r)}(\mathbf{s})$ is the probability that the walker arrives at $\mathbf{s}$ for the rth time on the jth step. The sum

$$F_1^{(r)}(\mathbf{s}) + F_2^{(r)}(\mathbf{s}) + \cdots + F_n^{(r)}(\mathbf{s})$$

represents the probability that the point $\mathbf{s}$ has been occupied *at least* r times in n steps. The reason

$$F_1^{(r-1)}(0) + \cdots + F_n^{(r-1)}(0)$$

is chosen to represent $(r - 1)$ returns to the origin instead of r is that the walker started at the origin, so visiting the origin r times means *returning* to it $r - 1$ times.

It is convenient to define a quantity

$$\Delta_k^{(r)} = S_k^{(r)} - S_{k-1}^{(r)}. \quad (C.2)$$

Since $S_0^{(1)} = 1$ and $S_1^{(1)} = 2$ while $S_0^{(r)} = S_1^{(r)} = 0$ for $r > 1$,

$$\Delta_1^{(1)} = 1 \quad \text{and} \quad \Delta_1^{(r)} = 0 \quad \text{if} \quad r > 1. \quad (C.3)$$

Also

$$S_n^{(r)} = \delta_{r,1} + \Delta_1^{(r)} + \Delta_2^{(r)} + \cdots + \Delta_n^{(r)}. \quad (C.4)$$

Through the use of an appropriate Tauberian Theorem we will be able to find the asymptotic properties of $S_n^{(r)}$ in terms of the properties of the generating function

$$\Delta^{(r)}(z) = \sum_{n=1}^{\infty} z^n \Delta_n^{(r)}. \quad (C.5)$$

Note that

$$\Delta_n^{(r)} = F_n^{(r-1)}(0) + \sum_{s}{}' F_n^{(r)}(\mathbf{s})$$
$$= [F_n^{(r-1)}(0) - F_n^{(r)}(0)] + \sum_s F_n^{(r)}(\mathbf{s}). \quad (C.6)$$

Hence if we multiply this equation by z^n and sum from $n = 1$ to ∞ we find

$$\Delta^{(r)}(z) = \{F^{(r-1)}(0, z) - F^{(r)}(0, z)\} + \sum_s F^{(r)}(\mathbf{s}, z).$$

From Eq. (I.20) we obtain

$$\Delta^{(r)}(z) = \{F(0, z)\}^{r-1}\{1 - F(0, z) + \sum_s F(\mathbf{s}, z)\},$$

while Eq. (I.18) implies

$$\Delta^{(r)}(z) = \left\{1 - \frac{1}{P(0, z)}\right\}^{r-1}$$
$$\times \left\{\frac{1}{P(0, z)} + \sum_s \left[\frac{P(\mathbf{s}, z) - \delta_{s,0}}{P(0, z)}\right]\right\}.$$

Finally from (I.14b)

$$\Delta^{(r)}(z) = \left\{1 - \frac{1}{P(0, z)}\right\}^{r-1} \{(1 - z)P(0, z)\}^{-1}.$$

From this expression and (C.4) one finds

$$S^{(r)}(z) = \left\{1 - \frac{1}{P(0, z)}\right\}^{r-1} \{(1 - z)^2 P(0, z)\}^{-1}. \quad (C.7)$$

APPENDIX D. THE ASYMPTOTIC FORM OF $P(0, z)$ AS $z \to 1$ FOR 3-D LATTICES

The generating function

$$P(0, z) = \frac{1}{(2\pi)^3} \iiint_{-\pi}^{\pi} \frac{d^3\varphi}{1 - z\lambda(\varphi)} \quad (D.1)$$

can be expressed as

$$\frac{1}{(2\pi)^3}\iiint_{-\pi}^{\pi}\frac{d^3\varphi}{1-\lambda(\varphi)}$$

$$-\frac{(1-z)}{(2\pi)^3}\iiint_{-\pi}^{\pi}\frac{\lambda(\varphi)\,d^3\varphi}{[1-\lambda(\varphi)][1-z\lambda(\varphi)]}=u_0-\delta. \quad (D.2)$$

The first part, u_0 has been found by G. N. Watson[9] for simple, body-centered, and face-centered cubic lattices. His results are

sc 1.5163860591,

bcc $(4\pi^3)^{-1}[\Gamma(\tfrac{1}{4})]^4 = 1.3932039297,$ (D.3)

fcc $9\{\Gamma(\tfrac{1}{3})\}^6 2^{-11/3}\pi^{-4} = 1.3446610732.$

We shall be concerned with the determination of δ as $z \to 1$.

The main contribution to δ as $z \to 1$ comes from values of φ close to the origin. We can write

$$\lambda(\varphi) = 1 - \tfrac{1}{2}(\sigma_1^2\varphi_1^2 + \sigma_2^2\varphi_2^2 + \sigma_3^2\varphi_3^2) + O(\varphi^4). \quad (D.4)$$

For example in the case of steps to the nearest-neighbor lattice points only on cubic lattices one finds from (I.5) that

sc $\sigma_1 = \sigma_2 = \sigma_3 = (\tfrac{1}{3})^{\frac{1}{2}},$ (D.5a)

bcc $\sigma_1 = \sigma_2 = \sigma_3 = 1,$ (D.5b)

fcc $\sigma_1 = \sigma_2 = \sigma_3 = (\tfrac{2}{3})^{\frac{1}{2}}.$ (D.5c)

As $\varphi \to 0$ and $z \to 1$ the integrand of δ becomes $2/\{(\sigma_1^2\varphi_1^2 + \sigma_2^2\varphi_2^2 + \sigma_3^2\varphi_3^2)$
$\times[(1-z) + \tfrac{1}{2}(\varphi_1^2\sigma_1^2 + \varphi_2^2\sigma_2^2 + \varphi_3^2\sigma_3^2) + \cdots]\}.$

It can be shown that as $z \to 1$ the range of integration can be made infinite in δ if one is concerned only with terms first order in $(z-1)$. Then, if we let $x_i = \sigma_i\varphi_i$ and calculate δ using polar coordinates with $r^2 = x_1^2 + x_2^2 + x_3^2$ we find

$$\delta \sim \frac{1-z}{\sigma_1\sigma_2\sigma_3\pi^2}\int_0^\infty\frac{dr}{(1-z)+\tfrac{1}{2}r^2} = \frac{[\tfrac{1}{2}(1-z)]^{\frac{1}{2}}}{\sigma_1\sigma_2\sigma_3\pi},$$

so that

$$P(0,z) \sim u_0 - [2(1-z)]^{\frac{1}{2}}/\sigma_1\sigma_2\sigma_3\pi + O(1-z). \quad (D.6)$$

It is much harder to calculate the term of $O(1-z)$. It has only been done for the simple cubic lattice. Since

$$1/\sigma_1\sigma_2\sigma_3 = \begin{cases} 3^{\frac{1}{2}} & \text{sc,} \\ 1 & \text{bcc,} \\ (\tfrac{3}{2})^{\frac{1}{2}} & \text{fcc.} \end{cases} \quad (D.7)$$

we find that for bcc

$$P(0,z) \sim \frac{1}{4\pi^3}\{\Gamma(\tfrac{1}{4})\}^4 - \frac{1}{\pi}[\tfrac{1}{2}(1-z)]^{\frac{1}{2}} + \cdots; \quad (D.8a)$$

for fcc

$$P(0,z) \sim \frac{9\{\Gamma(\tfrac{1}{3})\}^6}{2^{11/3}\pi^4} - \frac{3^{\frac{1}{2}}}{4\pi}(1-z)^{\frac{1}{2}} + \cdots. \quad (D.8b)$$

More terms have been obtained for the sc lattice[11]:

$$P(0,z) \sim 1.516386 - \frac{3}{\pi}\left(\frac{3}{2}\right)^{\frac{1}{2}}(1-z)^{\frac{1}{2}}$$
$$+ 1.384761(1-z) - \frac{9}{4\pi}\left(\frac{3}{2}\right)^{\frac{1}{2}}(1-z)^{\frac{3}{2}} + \cdots. \quad (D.9)$$

Generalized Master Equations for Continuous-Time Random Walks[1]

V. M. Kenkre,[2] E. W. Montroll,[2] and M. F. Shlesinger[2]

Received April 17, 1973

> An equivalence is established between generalized master equations and continuous-time random walks by means of an explicit relationship between $\psi(t)$, which is the pausing time distribution in the theory of continuous-time random walks, and $\phi(t)$, which represents the memory in the kernel of a generalized master equation. The result of Bedeaux, Lakatos-Lindenburg, and Shuler concerning the equivalence of the Markovian master equation and a continuous-time random walk with an exponential distribution for $\psi(t)$ is recovered immediately. Some explicit examples of $\phi(t)$ and $\psi(t)$ are also presented, including one which leads to the equation of telegraphy.
>
> **KEY WORDS:** Generalized master equations; random walks; statistical mechanics; transport theory.

1. INTRODUCTION

A standard starting point for the discussion of various random walks and other transport processes is the master equation

$$d\tilde{P}(l, t)/dt = -\alpha \tilde{P}(l, t) + \alpha \sum_{l' \neq l} p(l, l') \, \tilde{P}(l', t) \tag{1}$$

This study was partially supported by ARPA and monitored by ONR Contract No. (N00014-17-C-0308).

[1] For continuity, the reader is directed to the article entitled "Random Walks on Lattices. IV. Continuous Time Walks and Influence of Absorbing Boundaries," by E. W. Montroll and H. Scher, which will appear in Volume 9, Number 2, of this journal, and which should precede the following article. Regrettably, the two articles were inadvertently switched during processing.

[2] Institute for Fundamental Studies, Department of Physics and Astronomy, University of Rochester, Rochester, New York.

where $\tilde{P}(l, t)$ is the probability that a system of interest is in state l at time t and $\alpha p(l, l')$ is the probability per unit time of a transition from l' to l. Equation (1) is of course equivalent to the "gain–loss" form of the master equation

$$d\tilde{P}(l, t)/dt = \alpha \sum_{l'} [p(l, l')\, \tilde{P}(l', t) - p(l', l)\, \tilde{P}(l, t)] \tag{2a}$$

since transition probabilities have the normalization

$$\sum_{l'} p(l', l) = 1 \tag{2b}$$

We now interpret the states $\{l\}$ to be lattice points on a periodic space lattice and the system to be a random walker on the lattice. This intepretation is not necessary but it is made to give a direct contact with the results of Ref. 1. It was emphasized there that certain interesting random walks cannot be described by (1).

The basic quantity employed in the preceding paper is the pausing time distribution function $\psi(t)$ (the probability density function for the time t between the arrival of a walker at a given lattice point and the initiation of the next step to another site). All lattice points were postulated to be equivalent (periodic boundary conditions being used) so that $\psi(t)$ can be taken to be universal for all points. The methods of the preceding paper involve the random walk generating function which satisfies the Green's function equation [with $\tilde{P}(l, 0) = \delta_{l,0}$]

$$G(l, z) - z \sum_{l'} p(l - l')\, G(l', z) = \delta_{l,0} \tag{3a}$$

The form for $G(l, z)$ on a d-dimensional periodic lattice with $N \times N \times N \times \cdots$ lattice points in each direction (with periodic boundary conditions) is

$$G(l, z) = N^{-d} \sum_{\{s_j=1\}}^{N} \cdots \sum\ e^{ik \cdot l}/[1 - z\lambda(k)] \tag{3b}$$

where $k_j = 2\pi s_j/N$ and $\lambda(k)$ is the so-called structure function

$$\lambda(k) = \sum_l p(l)\, e^{ik \cdot l} \tag{3c}$$

The quantity $\tilde{P}(l, t)$ was shown to be related to $G(l, z)$ through the inverse Laplace transform formula[1,2]

$$P(l, t) = \frac{1}{2\pi i} \int_{c-i\infty}^{c+i\infty} e^{ut}\, du\, \{[1 - \psi^*(u)]/u\}\, G(l, \psi^*(u)) \tag{4}$$

where $\psi^*(u)$ is the Laplace transform of $\psi(t)$.

It was emphasized in Ref. 1 that Eq. (4) can be used to analyze processes which do not lie within the reach of the master equation (1); furthermore it has been shown by Bedeaux et al.[3] that when one chooses $\psi(t)$ to be an exponential distribution, expression (4) is identical to the solution of (1) that corresponds to the initial condition $\tilde{P}(l, 0) = \delta_{l,0}$. For no other form of $\psi(t)$ are the two expressions for $\tilde{P}(l, t)$ identical for all $t > 0$.

Generalized master equations have appeared naturally for non-Markovian processes associated with nonequilibrium phenomena.[4] They have the form

$$d\tilde{P}(l, t)/dt = \int_0^t d\tau \sum_j [K_{lj}(t - \tau) \tilde{P}(j, \tau) - K_{jl}(t - \tau) \tilde{P}(l, \tau)] \qquad (5)$$

In the theory of nonequilibrium statistical mechanics the kernels $\{K_{lj}(t)\}$ are derived from dynamics. We shall not be concerned with those results but merely consider Eq. (5) to characterize certain stochastic processes.

The main result of this note is to establish an equivalence between generalized master equations such as (5) and continuous-time random walks characterized by (4).

2. THE EQUIVALENCE

Since in the special lattice walk with which we are concerned all lattice points are equivalent, the form of (5) appropriate for our walk would have kernels of the following type:

$$K_{lj}(t) \equiv \phi(t) \, p(l - j) \qquad (6)$$

where $p(l)$ is just the transition probability which appears in (3a). Then (5) becomes

$$d\tilde{P}(l, t)/dt = \int_0^t \phi(t - \tau)[-\tilde{P}(l, \tau) + \sum_{l'}{}' p(l - l') \tilde{P}(l', \tau)] \, d\tau \qquad (7)$$

where we have used the conservation of probability equation

$$\sum_{l'}{}' p(l' - l) = 1$$

If initially the walker is at the origin so that $\tilde{P}_l(0) = \delta_{l,0}$, it is easily shown by taking Laplace transforms of (7) and comparing the resulting equation with (3) that, $\phi^*(u)$ being the Laplace transform of $\phi(t)$,

$$\tilde{P}(l, t) = (1/2\pi i) \int_{c-i\infty}^{c+i\infty} e^{ut} \, du \, \{[u + \phi^*(u)]^{-1} \, G(l, \phi^*/[u + \phi^*])\} \qquad (8)$$

By comparing (8) and (4), it is evident that Eq. (7) is an appropriate characterization of the random walks described in Ref. 1 provided that one sets

$$\phi^*(u) = u\psi^*(u)/[1 - \psi^*(u)] \qquad (9a)$$

which is equivalent to

$$\psi^*(u) = \phi^*(u)/[u + \phi^*(u)] \qquad (9b)$$

If one sets $u = i\omega$, $\phi^*(u)$ is closely related to the frequency-dependent diffusion constant.[5] The reciprocal of $\phi^*(u)$ also appears naturally in the application of linear response theory to random walk transport.[6]

Note that $\psi(t)$ and $\phi(t)$ are related by the integrodifferential equation

$$d\psi/dt + 2\,\delta(t)\,\psi(0) = \phi(t) - \int_0^t \phi(\tau)\,\psi(t-\tau)\,d\tau \qquad (10)$$

Note that the random walks described in Ref. 1 can also be characterized through the following equation provided $l \neq 0$:

$$\tilde{P}(l, t) = \int_0^t d\tau\, \psi(t - \tau) \sum_{l'} p(l - l')\, \tilde{P}(l', \tau) \qquad (11)$$

This form of the equation can be easily established with the help of the preceding analysis and it shows how the introduction of the pausing time distribution function $\psi(t)$ brings about a generalization of the Chapman–Kolmogorov equation to non-Markovian situations.

3. SOME SPECIAL EXAMPLES

We now consider several special examples of the pausing time distribution function and find the differential equations appropriate to those $\psi(t)$. The first example will be chosen to yield the Markovian master equation (1). Let

$$\psi(t) = \alpha e^{-\alpha t} \qquad (12a)$$

Then by taking Laplace transforms and applying (9), we find

$$\psi^*(u) = \alpha/(\alpha + u) \quad \text{and} \quad \phi^*(u) = \alpha \qquad (12b)$$

so that

$$\phi(t) = 2\alpha\delta(t) \qquad (12c)$$

When this expression is substituted into (7) the simple master equation is obtained. In light of the relation between the $\psi(t)$ and $\phi(t)$ established in this paper, the analysis of Ref. 3 which leads to this conclusion is seen to be exactly equivalent to the remark that (7) leads to (1) when $\phi(t)$ is a delta function.

Bedeaux et al.[3] have also shown that when the moments

$$\mu_n \equiv \int_0^\infty t^n \psi(t)\, dt \tag{13}$$

are all finite, the master equation is appropriate for a description of a random walk at times which are large compared with

$$t^* = \sup[\mu_n/n!]^{1/n} \tag{14}$$

Our second example is chosen to derive an equation which at early times after the walk has started yields results which are quite different from those that would follow from (1) and yet which at large times become equivalent to them.

Let us consider (with $\lambda^2 > 4a$)

$$\psi(t) = 2a(\lambda^2 - 4a)^{-1/2}\, e^{-\lambda t/2} \sinh[\tfrac{1}{2} t(\lambda^2 - 4a)^{1/2}] \tag{15}$$

which is equivalent to the difference between two exponentials. The corresponding expression for $\phi(t)$ is

$$\phi(t) = a e^{-\lambda t} \tag{16}$$

When (7) is differentiated with respect to t and (16) is substituted into the resulting expression the ensuing equation is

$$P_{tt}(l, t) + \lambda P_t(l, t) = a\left[-P(l, t) + \sum_{l'} p(l - l')\, P(l', t)\right] \tag{17}$$

If steps are taken to nearest-neighbor points only and if the lattice spacings are made very small, then by proceeding to the continuum limit this equation takes the form of the telegrapher's equation (with $P_t \equiv \partial P/\partial t$, etc.)

$$a^{-1} P_{tt} + (\lambda/a)\, P_t = D P_{xx} + k P_x \tag{18}$$

the D and λ being appropriately defined. It is known that at early times an initial pulse propagates as a wave, while at later times it propagates as a diffusion packet. This phenomenon was observed in the early days of telegraphy.[7] Signal diffusion reduced the data rate in long cables such as the early Atlantic cable. More recent applications of (17) or (18) have been to propagation of impulses in nerves and to exciton transport in photosynthetic units.[8]

The concept of the pausing time associated with a nondelta $\phi(t)$ or a nonexponential $\psi(t)$ is particularly physical in the problem of exciton transport in photosynthetic units. (An exciton hops from site to site but pauses at each site. Lattice vibrations provide a relaxation mechanism which yields a pausing time of the order of 10^{-12} sec.) The traditional theory of exciton transport postulates a Markovian random walk which corresponds to an exponential $\psi(t)$ and therefore a deltalike $\phi(t)$. However, this represents an instantaneous relaxation of the exciton at every site. A general theory which takes into account the actual non-Markovian nature of the process and the finite magnitude (10^{-12} sec) of the relaxation time has been developed recently.[8] Equations like (7) and (17) have been used[9] to analyze the oscillatory approach to equilibrium observed in a study of certain models in nonequilibrium statistical mechanics.

Since it was pointed out in Ref. 1 that there is some evidence in transient photoconductivity experiments that $\psi(t)$ may not have any finite moments, we close our discussion with a consideration of one of the forms of $\psi(t)$ without moments which was presented there. We choose

$$\psi(t) = 4a^2[\exp(ta^2)] \, i^2 \, \mathrm{erfc}(at^{1/2}) \tag{19a}$$

Since

$$\psi^*(u) = [1 + (u^{1/2}/a)]^{-2} \tag{19b}$$

$$\phi^*(u) = a^2\{1 - [1 + (u^{1/2}/2a)]^{-1}\} \tag{19c}$$

whose Laplace transform is

$$\phi(t) = 2a^2\delta(t) - 2a^3(\pi t)^{-1/2} + 4a^4[\exp(4ta^2)] \, \mathrm{erfc}(2at^{1/2}) \tag{20}$$

REFERENCES

1. E. W. Montroll and H. Scher, *J. Stat. Phys.* **9**(2) (1973).
2. E. W. Montroll and G. H. Weiss, *J. Math. Phys.* **6**:167 (1965).
3. D. Bedeaux, K. Lakatos-Lindenberg, and K. E. Shuler, *J. Math. Phys.* **12**:2116 (1971).
4. L. Van Hove, *Physica* **23**:441 (1957); I. Prigogine and P. Resibois, *Physica* **27**:629 (1961); R. W. Zwanzig, *Physica* **30**:1109 (1964); E. W. Montroll, in *Fundamental Problems in Statistical Mechanics*, E. G. D. Cohen, ed., North-Holland, Amsterdam (1962).
5. H. Scher and M. Lax, *J. Non-Cryst. Solids* **8**:497 (1972).
6. K. Lakatos-Lindenburg and D. Bedeaux, *Physica* **57**:157 (1972).
7. O. Heaviside, *Phil. Mag.* **II**:135 (1876); W. Thomson (Lord Kelvin), *Proc. Roy. Soc.* **VII**:382 (1855); G. Kirchhoff, *Ann. d. Phys.* C **193**:25 (1857).
8. V. M. Kenkre, submitted to *J. Chem. Phys.*
9. V. M. Kenkre, submitted to *Phys. Rev. A*.

Printed by the St Catherine Press Ltd., Tempelhof 37, Bruges, Belgium.

Anomalous transit-time dispersion in amorphous solids

Harvey Scher
Xerox Webster Research Center, 800 Phillips Road, Webster, New York 14580

Elliott W. Montroll
Institute for Fundamental Studies,* Department of Physics and Astronomy, University of Rochester, Rochester, New York 14627
(Received 13 January 1975)

Measurements of the transient photocurrent $I(t)$ in an increasing number of inorganic and organic amorphous materials display anomalous transport properties. The long tail of $I(t)$ indicates a dispersion of carrier transit times. However, the shape invariance of $I(t)$ to electric field and sample thickness (designated as universality for the classes of materials here considered) is incompatible with traditional concepts of statistical spreading, i.e., a Gaussian carrier packet. We have developed a stochastic transport model for $I(t)$ which describes the dynamics of a carrier packet executing a time-dependent random walk in the presence of a field-dependent spatial bias and an absorbing barrier at the sample surface. The time dependence of the random walk is governed by hopping time distribution $\psi(t)$. A packet, generated with a $\psi(t)$ characteristic of hopping in a disordered system [e.g., $\psi(t) \sim t^{-(1+\alpha)}$, $0 < \alpha < 1$], is shown to propagate with a number of anomalous non-Gaussian properties. The calculated $I(t)$ associated with this packet not only obeys the property of universality but can account quantitatively for a large variety of experiments. The new method of data analysis advanced by the theory allows one to directly extract the transit time even for a featureless current trace. In particular, we shall analyze both an inorganic ($a\text{-}As_2Se_3$) and an organic (trinitrofluorenone-polyvinylcarbazole) system. Our function $\psi(t)$ is related to a first-principles calculation. It is to be emphasized that these $\psi(t)$'s characterize a realization of a non-Markoffian transport process. Moreover, the theory shows the limitations of the concept of a mobility in this dispersive type of transport.

I. INTRODUCTION

The development of modern photocopying machines has motivated experimental work on amorphous materials, some of which display anomalous transport properties. In such a machine the surface of a film of amorphous material is charged by a corona and excited by a light pulse which creates pairs of charge carriers as exhibited in Fig. 1(a). The electric field due to the deposited charge transports one of the carrier components, say the positive charge, away from the optically excited surface, leaving the negative charge behind. A powder of negatively charged particles (toner) is then attached to the remaining positive spots (the latent image) on the film. The toner is finally fixed on a paper which passes over the charged surface. It is natural that in the investigation of amorphous materials used in such a technology, one should have measured the mobility of the positive carrier (holes) in films of the materials. From these measurements, experimenters were surprised to find that carrier mobilities in some of these amorphous materials depend on the thickness of the material instead of being an intrinsic property of the material. It is this observation, as well as others, which we describe as "anomalous transport properties."

Two of the materials used commercially are As_2Se_3, and an organic charge-transfer complex of trinitrofluorenone and polyvinylcarbazole (TNF-PVK).

Measurements of the transient current in insulators, due to injected charge, have often been the only means to probe the transport properties of these materials. Earlier investigations of sulfur[1] and amorphous selenium (a-Se)[2] have yielded detailed information on the drift mobility of both electrons and holes and on deep-trapping lifetimes. Charge injection with strongly absorbed light has, in addition, enabled one to separate the transport of charge from the photogenerated supply efficiency.[3] Recent drift-mobility observations in[4] a-Si have provided the only substantial experimental evidence for the often-discussed "mobility edge" in amorphous semiconductors.[5]

Let us now proceed to summarize the experimental background to our theory and the *main ideas* of our theoretical model. This model will be analyzed in detail in Secs. II and III.

A. Experiments

The basic measurement technique is illustrated in Fig. 1(b). The sample is sandwiched between two planar contacts, one or both of which are semi-transparent. A short, strongly absorbed, light flash causes a conduction current to flow in the sample, which is held at a constant voltage V if the circuit RC time is much less than the time scale of the experiment. The measured current $I(t)$, under these conditions, is simply the space-averaged conduction current

$$I(t) = \frac{1}{L}\int_0^L j_c(x,t)\,dx,\qquad(1)$$

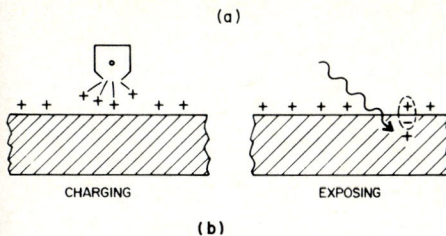

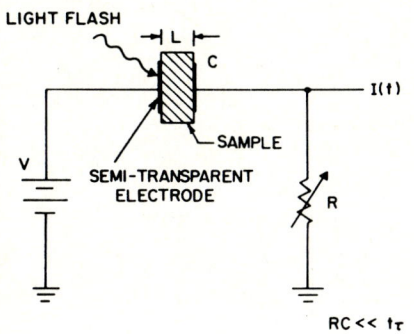

FIG. 1. (a) Initial steps in the xerographic copying process: (1) A high-resistivity photoconductor is charged by a corotron. (2) An absorbed photon creates an electron-hole pair. The electron neutralizes the positive ion on the surface while the hole moves through the photoconductor to neutralize the counterpart on the electrode. The remaining positive ions constitute the latent image. (b) Schematic diagram for a transient photoconductivity measurement. A light flash of duration much shorter than the transit time t_τ is absorbed in a depth much less than the sample thickness L. Carriers of one sign move across the sample, inducing a time-dependent current $I(t)$ in the external circuit.

where L is the sample thickness. An idealized $I(t)$ following the light flash is shown in Fig. 2. It depicts a sheet of charge moving with constant velocity and departing from the sample at the transit time t_τ. An unambiguous drift mobility μ can be determined by the relation

$$\mu = L^2/t_\tau V . \tag{2}$$

Departures from the idealized current shape in Fig. 2 might be the result of several influences: (a) RC response time, (b) loss of carriers due to deep trapping, (c) variation in drift mobility as a function of x due to sample inhomogeneity, (d) local electric-field variation due to trapped space charge, and (e) spreading of the sheet of charge to a width comparable to L. (We shall restrict ourselves to the "small-signal" case, i.e., we will not consider spreading due to the mutual Coulomb repulsion of the trap-free-space-charge-limited or "large signal" case.[6])

The first two of the above influences usually determine the practical limitations of the transient experiment. One must have

$$RC \ll t_\tau \ll \tau_D , \tag{3}$$

where τ_D is the deep-trapping lifetime. The next two influences can cause a variation in the current level. The carriers can still move across the sample as a sheet of charge. However, the changes in drift velocity as a function of x will cause a variation in the step height of the current in Fig. 2 but not necessarily change the sharp drop at the transit time. Recently, Pai[7] has used small incremental steps in the current to map a profile of charged impurities in a-Se.

The influences so far discussed are characterized by definite mechanisms. The last item, which involves a discernible spreading of the sheet of charge, is really a generic symptom of a number of possible causes rather than any one specific departure from the ideal behavior shown in Fig. 2. The spreading can arise from statistical fluctuations associated with a variety of processes, including multiple trapping, hopping, or dispersion due to material inhomogeneity (e.g., granular effects, nonuniform thickness). This type of spreading can result in a decreasing current level and a considerable smearing of the "transit edge."

In recent experiments on an increasingly wider class of amorphous materials, As_2Se_3,[8,9] PVK,[10] and TNF-PVK,[11] one typically obtains the current trace shown in Fig. 3 instead of the current shape shown in Fig. 2, which is realized in S and in a-Se at room temperature. The exhibited trace is derived from measurements[18] on As_2Se_3; however, except for scaling factors, it is identical in appearance to those obtained in PVK,[10] TNF-PVK.[11] Examining the features of Fig. 3, we observe that the current spike, immediately after the onset of the short light pulse, is followed by a soft "plateau" in

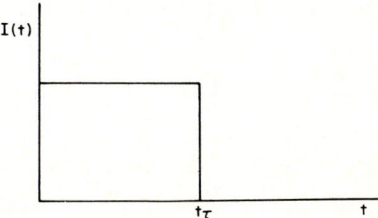

FIG. 2. Idealized transient-current trace measured with the technique shown in Fig. 1(b). The trace depicts a sheet of charge moving with constant velocity until it leaves the sample at time t_τ.

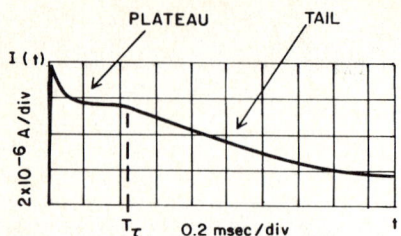

FIG. 3. Highly dispersive transient photocurrent trace $I(t)$ measured on As_2Se_3 by Scharfe.

the current level, then a "shoulder" or transition region, and finally a ubiquitous long "tail." These distortions of the idealized response, indicated in Fig. 2, immediately suggest some statistical process causing a spread in the transit times (or a distribution in surface release times[12]) as discussed above. The tail represents the dribble of the "slow" carriers. The shoulder region or onset of the tail has been chosen as the transit time; however, as will be shown, one often does not even see a shoulder.

A very revealing treatment of the $I(t)$ data advanced by Scharfe[8] is shown by the plot in Fig. 4. The current is normalized to $I(t)_\tau$ and the time is measured in relative units of t_τ. Scharfe has chosen t_τ to be the time of the onset of the tail. The $I(t)$ data, corresponding to a wide range of transit time (obtained by changing E or L), is shown in Fig. 4 to collapse into one curve in this type of plot. We designate this feature as "universality." This universality is the clue, as shown below, that an unusual statistical process is taking place, a process in which the ratio of the mean position $\langle l \rangle$ of the propagating packet of carriers to the rms spread σ is independent of time, i.e.,

$$\langle l \rangle / \sigma = \text{const} \qquad (4)$$

for t of the order of t_τ, i.e., $t = O(t_\tau)$.

B. Theory

The universality is most clearly exhibited on a $\log I$-$\log t$ plot. The nature of the carrier transport is also easier to interpret on these plots. In Fig. 5, we have a plot of this kind for an inorganic film and in Fig. 6, for an organic. A schematic form of these current curves is displayed in Fig. 7.

These curves will be analyzed in the main body of this paper through a hopping-time distribution function $\psi(t)$. In an amorphous material there is a dispersion in the separation distances between nearest-neighbor localized sites available for hopping carriers and a dispersion in the potential barriers between these sites. Both of these variables strongly affect the hopping time, the time between a carrier arrival on successive sites. Hence, the distribution of these hopping times, $\psi(t)$, would have a long tail. We will propose tails of the form

$$\psi(t) \sim \text{const} \times t^{-(1+\alpha)}, \quad 0 < \alpha < 1 \qquad (5)$$

indicating an extremely large hopping-time dispersion. Classical Gaussian propagation as exhibited in Fig. 8 is associated with an asymptotic exponential tail with $\psi(t) \sim e^{-\lambda t}$. The $\psi(t)$ in Eq. (5) will be related to a first-principles calculation of the distribution function in a disordered system in Sec. III. The main point to emphasize here is the contrast between Eq. (5) and a $\psi(t) \sim e^{-\lambda t}$, which is characteristic of a system with a single transition rate λ.

When the inverse-power tail (5) exists, a considerable fraction of the carriers remain at the point of their formation for a long time. Those carriers whose local environment at their initial point permits their immediate motion (fast hops) will sooner

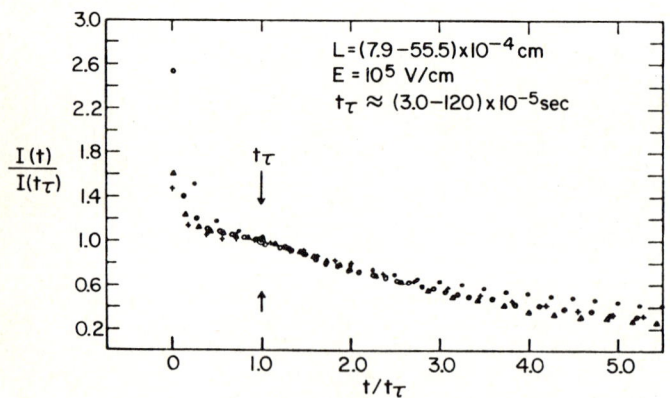

FIG. 4. Superposition of normalized current traces, for a range of transit time t_τ, as a function of t/t_τ. The invariance of the shape of $I(t)$ to t_τ is designated as universality. E is the electric field and L is the sample thickness.

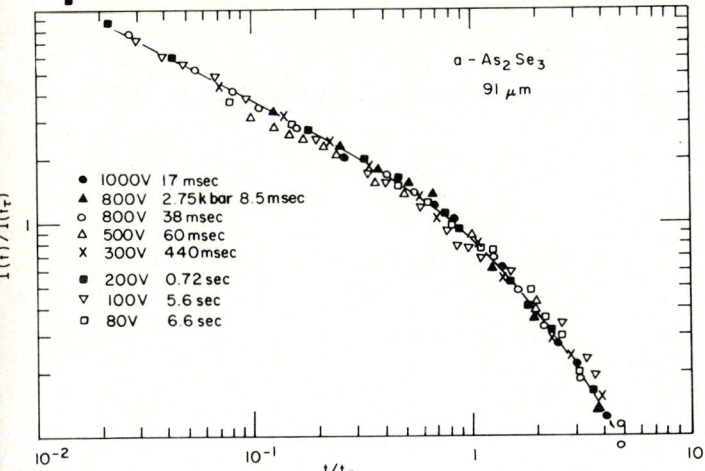

FIG. 5. A log$I$-log$t$ plot for a-As$_2$Se$_3$ for the range of transit time listed in the figure. The measurements by Pfister also include a data set at high pressure (2.75 kbar). The solid line is the theoretical curve.

or later find themselves immobilized at some site (long hops), thus reducing their contribution to the current, until they escape. The propagating packet associated with Eq. (5) has the form exhibited in Fig. 9. While the peak of the Gaussian packet of Fig. 8 and the mean carrier of the packet are located at the same position and move with the same velocity, the mean carrier of the packet of Fig. 9 propagates with a velocity which decreases with time as it separates from the peak which remains nearly fixed at the point of origin of the carriers.

The type of current trace expected for a moving Gaussian packet is shown in Fig. 10. As the mean moves with constant velocity, the current is constant until the mean carrier encounters the absorbing barrier at the sample surface; then, the current level drops to zero. The rounding of the "transit edge" reflects the spread in the transit times of the carriers in the packet. As observed in Fig. 8, for a Gaussian packet, the ratio $\sigma/\langle l \rangle$ decreases with time ($\sigma/\langle l \rangle \sim t^{-1/2}$). Hence, in the plot of $I(t)/I(t_\tau)$ vs t/t_τ shown in Fig. 10, there will be a relative sharpening of the transit edge for the longer transit times as indicated by curve (1), while the current trace for shorter t_τ [curve (2)] will show more relative dispersion. *Universality of the current is incompatible with a propagating Gaussian packet!* In Fig. 9, the packets spread in the same manner as the moving mean; $\sigma/\langle l \rangle$ is independent of time, and the current traces associated with these packets are universal.

In an infinite sample, the decrease in velocity of the mean carrier in these packets (Fig. 9) implies a continuous reduction of the current. It will be shown in Sec. II that the current variation derivable from Eq. (5) is

$$I(t) \sim \text{const} \times t^{-(1-\alpha)} \ . \qquad (6)$$

This is the form plotted in Fig. 7 when t/t_τ is less than 1.0. In a sample of finite thickness, an additional reduction of current results from the loss of carriers into the absorbing barrier when a group of the fastest carriers reaches the barrier. The time t_τ is to be identified with this change in the time dependence of $I(t)$. For $t \gg t_\tau$ we will show

$$I(t) \sim \text{const} \times t^{-(1+\alpha)} \ . \qquad (7)$$

Note that a check of the model would be the observation that, in a plot of log$I$ as a function of log$t$, the sum of the slopes at times $t/t_\tau < 1$ and at times $t/t_\tau \gg 1$ would be $-[(1+\alpha)+(1-\alpha)] = -2$. The data

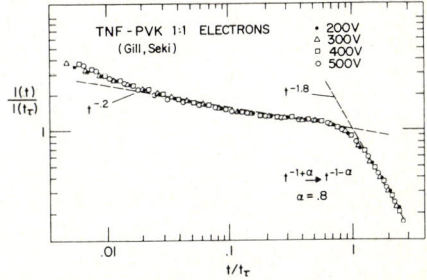

FIG. 6. A log$I$-log$t$ plot for 1:1 TNF-PVK measured by Gill and taken from a paper by Seki. The slopes of the dashed lines are -0.2, and -1.8, respectively.

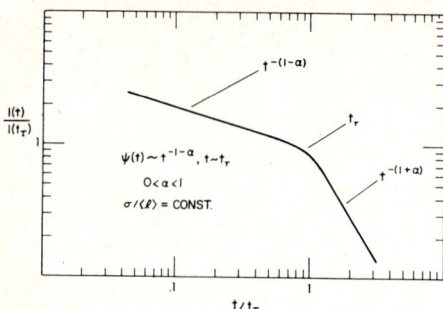

FIG. 7. A log$I$-log$t$ plot indicating the current $I(t)$ associated with a packet of carriers moving, in an electric field, with a hopping-time distribution function $\psi(t) \sim t^{-1-\alpha}$, $0 < \alpha < 1$, towards an absorbing barrier at the sample surface.

obtained for a number of As$_2$Se$_3$ samples exhibited in Fig. 5 indicate that $\alpha = 0.45$, while Fig. 6 shows that for amorphous TNF-PVK, $\alpha \simeq 0.80$.

Traditional transport theory, which is characterized by the motion of Gaussian packets, is generally described through a Markoffian transport mas-

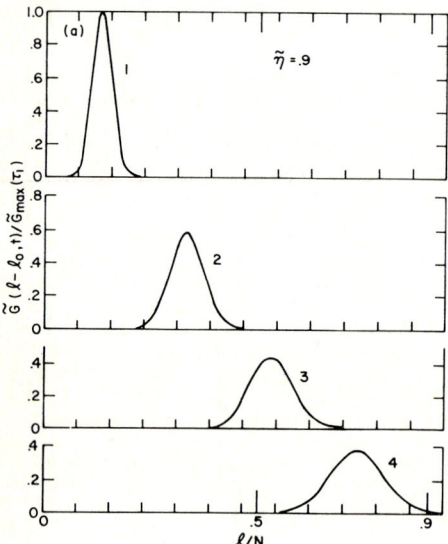

FIG. 8. Propagator for a carrier packet $\bar{G}(l-l_0, t)$ vs l/N for a range of time $t/t_\tau = (2n-1)/10$, $n = 1, 2, 3, 4$. The "transit time" is defined by $\langle l(t_\tau)\rangle/N = 1$. The random walk is based on $\psi(t)/\lambda = e^{-\lambda t}$. The plot is scaled by the peak value of $\bar{G}(l-l_0, t)$ at the earliest time. The spatial bias factor $\tilde{\eta} = 0.9$.

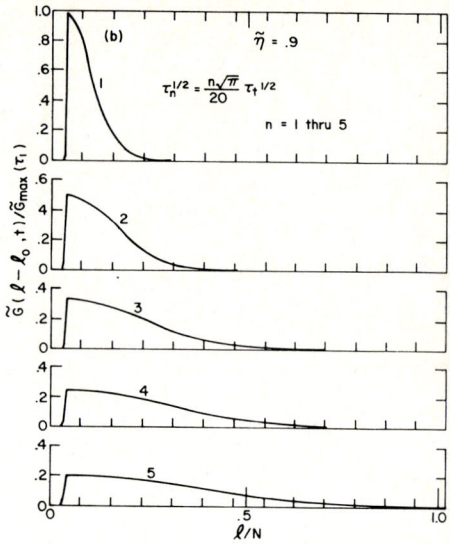

FIG. 9. Propagator for a carrier packet $\bar{G}(l-l_0, t)$ vs l/N for a range of time $\tau_n^{1/2} = (n\sqrt{\pi}/20)\tau_t^{1/2}$ $n = 1-5$, and τ_t defined in the text. The random walk is based on $\psi_2(t) \sim \pi^{-1/2} t^{-3/2}$. The plot is scaled by the peak value of $\bar{G}(l-l_0, t)$ at the earliest time. The spatial bias factor $\tilde{\eta} = 0.9$.

ter equation which might be written as

$$d\tilde{P}(l,t)/dt = \lambda \sum_{l'} [p(l-l')\tilde{P}(l',t) - p(l'-l)\tilde{P}(l,t)]. \tag{8}$$

Here l represents a cell number in a system which is broken into equal-sized cells, and $\tilde{P}(l,t)$ is the

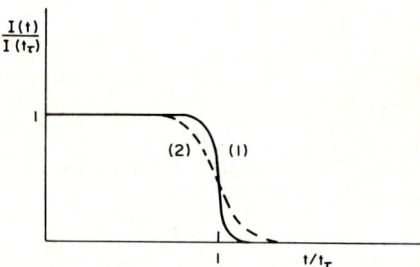

FIG. 10. Normalized current trace $I(t)/I(t_\tau)$ vs t/t_τ expected for a propagating Gaussian packet. Curve (1) corresponds to the longer t_τ and curve (2) corresponds to the shorter t_τ. This figure illustrates the incompatibility of a Gaussian with the universality of $I(t)$.

probability that a cell l is occupied by a carrier at time t. The first term on the right-hand side represents the flow rate into cell l from transitions from l', while the second represents flow out of l into l'; λ is a proportionality constant. The general non-Gaussian, non-Markoffian transport process which we proposed above can be described by a generalized master equation[13] with a relaxation function $\phi(t)$:

$$d\tilde{P}(l,t)/dt = \int_0^t \phi(t-x) \sum_{l'} [\, p(l-l')\tilde{P}(l',x) - p(l'-l)P(l,x)\,]dx \,. \quad (9)$$

The relaxation function $\phi(t)$ and the hopping-distribution functions are related through their Laplace transforms $\phi^*(u)$ and $\psi^*(u)$ with

$$\phi^*(u) = u\psi^*(u)/[1-\psi^*(u)] \,. \quad (10)$$

An exponential $\psi(t) = \lambda e^{-\lambda t}$ yields $\phi(t) = 2\lambda\delta(t)$, so that Eq. (9) reduces to Eq. (8) in that case.

It is now apparent that one must go beyond the traditional transport concepts embodied in Eq. (8), in order to generate propagating carrier packets that can account for all the experimental evidence.

In Sec. II we will define the model, present the mathematical problem generated by the application of the model, and review the essential features of the analytic solution of the problem. In Sec. III we will exhibit the main theoretical results of our model, discuss in some detail their conceptual implications, and we will compare these results with the available experimental data on a wide class of amorphous insulators. The very satisfactory agreement with theory (as already shown in Figs. 4–7) will be advanced as strong evidence that the same phenomenon is indeed occurring in each of the different materials. We will further indicate the direction of future work on this unusual phenomenon.

II. MODEL

An amorphous insulating material might be considered a network of localized sites for electrons or holes. Electron or hole transport in such a material could be characterized by a succession of hops from one site to another. The distances between various neighboring sites have some variation about a mean value, and the effective intersite transition rates, which sensitively depend on these distances, will suffer a wide statistical dispersion. This in turn yields a broad distribution of hopping times.

In our model we postulate our material to be divided into a regular lattice of equivalent cells, with each cell containing many randomly distributed localized sites available for hopping carriers.

Carrier transport is a succession of carrier hops from one localized site to another and finally from one cell to another. We define the hopping time to be the time interval between the moment of arrival of a carrier into one cell and the moment of arrival into the next cell into which it lands. The random distribution of sites and hence the disorder of an amorphous material is incorporated into a hopping-time *distribution function* $\psi(t)$. Then $\psi(t)dt$ is the probability that after a carrier arrives in a given cell, it will arrive in its next cell in the time interval $(t, t+dt)$. We identify the cells by dimensionless lattice vectors with components l_1, l_2, l_3, where $l_i = 1, 2, \ldots, N_i$. The set $\{N_i\}$ defines the size of the lattice in units of the individual cell dimensions. At first, when we consider our lattice to be periodic, the numbers N_1, N_2, and N_3 will represent the number of cells in various directions.

In the practical computation of $\psi(t)$, as in Ref. 14, one could have a single site per cell. In that case, $\psi(t)$ is the distribution of intersite hopping times. One can consider hopping on a lattice because the main effects of intersite spatial variations are to produce enormous variations in the hopping times. These effects are incorporated into a calculation of $\psi(t)$ for a random medium as discussed in Ref. 14 and in Sec. III.

Transient photoconductivity, according to this general transport model, would be characterized by a group of carriers (Fig. 11) executing a three-dimensional time-dependent random walk biased by an electric field E. The directional bias will be characterized by a dimensionless asymmetry factor $\bar{\eta}(E)$. A group of carriers is pictured in Fig. 11 to be initially near the (+) electrode of a planar capacitor. They drift a distance L to the right along the x axis toward the (−) electrode, where they become absorbed.

The analytic description of an asymmetric non-Markoffian continuous-time random walk in the presence of planar absorbing barriers has been given by the authors (Ref. 15). We now review the main features of this process.

The basic quantity of our theory is $\tilde{G}(l,t)$, the probability that a walker is found at l at time t if at time $t=0$ it was at the origin. The function $\tilde{G}(l,t)$ completely specifies the propagation of the carrier packet (in the absence of the absorbing boundary). This quantity is related to two functions, $\psi^*(u)$ and $G(l,z)$, through the inverse-Laplace-transform contour integral[15]

$$\tilde{G}(l,t) = \frac{1}{2\pi i}\int_{c-i\infty}^{c+i\infty} (du/u)\,e^{ut}\,[1-\psi^*(u)]\,G(l,\psi^*(u)) \,. \quad (11)$$

The function $\psi^*(u)$ is the Laplace transform of the hopping-time distribution $\psi(t)$,

$$\psi^*(u) = \int_0^\infty e^{-ut}\psi(t)\,dt \,, \quad (12)$$

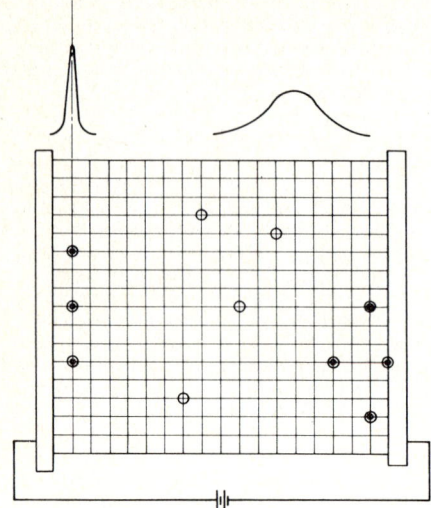

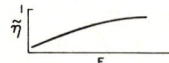

FIG. 11. Schematic diagram for the random-walk model of transient photoconductivity. The diagram is a composite in time. The carriers (●) are injected as a narrow distribution at $t=0$ in the localized sites (○). The packet spreads and propagates to the right, $t>0$, with a spatial bias $\bar{\eta}$, towards an absorbing barrier.

and $G(l, z)$ is a random-walk generating function, which for periodic boundary conditions has the form

$$G(l, z) = \frac{1}{N_1 N_2 N_3} \sum_{s_1=1}^{N_1} \sum_{s_2=1}^{N_2} \sum_{s_3=1}^{N_3} \frac{e^{i\vec{l}\cdot\vec{k}}}{1 - z\lambda(k)}, \quad (13)$$

with $k_j = 2\pi s_j/N_j$. The structure function $\lambda(k)$ is the Fourier representation of the transition probability of hops from cell to cell:

$$\lambda(k) = \sum_l p(l) e^{i\vec{l}\cdot\vec{k}}. \quad (14)$$

Here $p(l)$ is the probability of a hop from a given cell to one displaced by a vector l from it, with

$$\sum_l p(l) = 1. \quad (15)$$

Our hopping process is then characterized by the two functions $\psi(t)$ and $p(l)$, $\psi(t)$ being the hopping-time distribution at any cell and $p(l)$ being the transition probability of a hop by a vector distance l. All cells are defined to be equivalent, so that $\psi(t)$ and $p(l)$ are universal functions for all cells. All the disorder is included in the fact that the functions are *distributions*, and these distributions are determined from an ensemble of random systems (cf. Sec. III).

The carrier current $I(t)$ in the x direction (l_1) in an assembly of hopping charges is proportional to the time rate of change of the mean of the carrier packet

$$I(t) \propto d\langle l_1\rangle/dt, \quad (16)$$

where

$$\langle l_1 \rangle = \sum_{l_1=1}^{N_1} \sum_{l_2=1}^{N_2} \sum_{l_3=1}^{N_3} l_1 \bar{G}(l,t). \quad (17)$$

The field is in the x direction; hence $\langle l \rangle = \langle l_1 \rangle$.

We shall now compute Eqs. (11) and (16) for a se of prototype forms of $\psi(t)$, illustrating the dramatic change in the carrier-packet propagation with characteristically different $\psi(t)$. Throughout our discussion we assume that carrier displacements are between nearest-neighbor cells with probabilities (with $\bar{\eta} > \frac{1}{2}$)

$$p(0, \pm 1, 0) = p(0, 0, \pm 1) = q, \quad (18)$$

$$p(1, 0, 0) = 2\bar{\eta}p, \ p(-1, 0, 0) = 2(1-\bar{\eta})p \quad (19)$$

[all other $p(l_1, l_2, l_3)$ vanishing]. These transition probabilities correspond to nearest-neighbor hopping on a simple cubic lattice with a bias for hops in the $+x$ direction over that in the $-x$ direction, due to the electric field, i.e., $\bar{\eta} \equiv \bar{\eta}(E)$. The adoption of Eqs. (18) and (19) is consistent with the assertion that the spatial fluctuations are trivial compared to the hopping-time fluctuations.[14] The asymmetry which is introduced by E in the basic hopping-transition probabilities of the system is reflected in the phenomenological function $\bar{\eta}(E)$. Discussion and generalization of this approach is included at the end of the next section.

The two types of $\psi(t)$ we shall consider are

$$\psi_1(t) = We^{-\tau} \quad (20)$$

and

$$\psi_2(t) = 4W_M e^\tau i^2 \, \text{erfc}(\tau^{1/2}), \quad (21)$$

where W and W_M are rate constants, τ is a dimensionless time (Wt or $W_M t$), and $i^2 \text{erfc} z$ is the second repeated integral of the complementary error function, the nth repeated integral being

$$i^n \text{erfc} z \equiv (2/\pi^{1/2} n!) \int_z^\infty (t-z)^n e^{-t^2} dt. \quad (22)$$

The distribution $\psi_1(t)$ is a rapidly decaying function of τ. It generates a random walk with only *one* intersite-hopping transition rate, W. It decays rapidly because in such a process the probability of hopping is in sharp contrast for the two time regimes, $t < W^{-1}$ or $t > W^{-1}$. The choice $\psi_1(t)$ implies that

$$\langle l \rangle = \bar{l}\tau , \quad (23a)$$

where

$$\bar{l} = 2p(2\bar{\eta} - 1) , \quad (23b)$$

$\bar{l}$ being the mean displacement for a single hop and computed with the model characterized by Eqs. (18) and (19); hence,

$$d\langle l \rangle/dt = \bar{l}W . \quad (23c)$$

The dispersion is

$$\sigma^2 = \langle (l_1 - \langle l \rangle)^2 \rangle = 4p\tau , \quad (24)$$

and

$$\sigma/\langle l \rangle \sim \tau^{-1/2} . \quad (25)$$

In Fig. 8 we have plotted $\bar{G}(l, t)$ for this case. As can be judged from the figure, the curves become Gaussian after a δ-function pulse initially near the origin has propagated a small fraction of N. Note that in Eq. (23c) the propagation rate is constant at all time. Also note that for a Gaussian, $\sigma/\langle l \rangle$ decreases with increasing τ. The results summarized above are valid not only for an exponential $\psi(t)$, but remain appropriate for $\psi(t)$ generally when at least the first two moments of $\psi(t)$ exist, and $t > \bar{t}$, defined by

$$\bar{t} = \int_0^\infty t\psi(t)\, dt . \quad (26)$$

All moments of the exponential distribution $\psi_1(t)$ exist.

When $\psi(t)$ has a long tail [e.g., $\psi(t) \sim t^{-\beta}$, $\beta \leq 3$, as $t \to \infty$], the above results are no longer applicable. The function $\psi_2(t)$, defined by Eq. (21), is of this type with the asymptotic tail

$$\psi_2(t)/W_M \sim \pi^{-1/2} \tau^{-3/2} , \quad \tau \gg 1 , \quad (27)$$

and has been investigated in considerable detail by the authors.[15] We will show later that experimental results always correspond to $\tau \gg 1$. The fact that $\psi_2(t)$ has no finite positive integral moments, not even the first, leads to the unusual character of the propagated pulse exhibited in Fig. 9. The mean position of the propagating packet is, if $\tau \gg 1$,

$$\langle l \rangle \sim \bar{l}(\tau/\pi)^{1/2} . \quad (28a)$$

The peak in the distribution, however, remains near the initial position. Hence the current $I(t)$, which is proportional to

$$d\langle l \rangle/dt \sim \tfrac{1}{2} \bar{l}(\pi\tau)^{-1/2} , \quad (28b)$$

decreases with increasing time until it finally vanishes, even in the absence of an absorbing boundary!

This observation that the mean position of a carrier in the packet is a fractional power of time will be the basis of our discussion of anomalously propagating packets. The behavior of Eq. (28a) is a consequence of the long tail of $\psi_2(t)$ as exhibited in Eq. (27). The slowly varying time dependence of $\psi_2(t)$ corresponds to a large dispersion in hopping times. One can physically understand the decaying current in Eq. (28b) on this basis. At early times in the transport process, most carriers are moving with the relatively more probable short hopping times. However, with increasing t all carriers eventually encounter at least one long hopping time. Since long hopping times are analogous to a deep trap, the carrier is immobilized temporarily. Hence, as $t \to \infty$, most carriers become so immobilized that the current becomes very small.

Figure 9 shows that the peak in the packet remains near its starting position while the mean advances according to Eq. (28a). Thus, the dispersion grows in the same manner as the mean carrier displacement, $\langle l \rangle$, so that

$$\sigma^2 \sim \bar{l}^2 \tau(\tfrac{1}{2}\pi - 1)/\pi . \quad (29)$$

Hence

$$\sigma/\langle l \rangle \sim (\tfrac{1}{2}\pi - 1)^{1/2} , \quad (30)$$

a constant independent of t. This is to be compared with the classical Gaussian result Eq. (25). Equation (30) is the basis for our understanding of the universality of $I(t)$ provided in Eq. (4). It is now clear that a distribution of the type $\psi_2(t)$ can generate carrier packets displaying the appropriate qualitative transport behavior.

An immediate mathematical generalization of $\psi_2(t)$ is the class of hopping-time distribution functions with long tails which decay asymptotically as $t \to \infty$ in the following manner:

$$\psi(t) \sim [At^{1+\alpha}\,\Gamma(1 - \alpha)]^{-1} , \quad 0 < \alpha < 1. \quad (31)$$

A detailed physical justification for the application of members of this class to our anomalous transport phenomenon is given in the beginning of the next section. Now we might ask if results such as Eqs. (28) and (30) are a consequence of the general behavior of Eq. (31). Shlesinger[16] has shown by applying certain Tauberian theorems in a manner presented in Appendix A of this paper, that Eq. (31) implies that

$$\langle l \rangle \sim \bar{l}At^\alpha/\Gamma(\alpha + 1) \quad (32)$$

and

$$\frac{\sigma}{\langle l \rangle} \sim \{[2\Gamma^2(1 + \alpha)/\Gamma(1 + 2\alpha)] - 1\}^{1/2} , \quad (33)$$

with $\bar{l} \neq 0$. A $\psi(t)$ of the form (31) thus yields a fractional-power time dependence of $\langle l \rangle$ and a ratio of dispersion to mean independent of time (33).

We now complete the mathematical development of our model by calculating the important influence of the absorbing plane at $l_1 = N \equiv 0$ (N being the number of cells in the x direction). When $\langle l \rangle \sim N$, one switches from the small-slope portion of $I(t)$ (Fig. 7) to the large-slope portion, since carriers disappear in large numbers into the boundary after that time. We have shown[15] that for an injected carrier initially at the plane $l_1 = l_0$, that the free-space propagator $\tilde{G}(l - l_0, t)$ is to be replaced by

$$\tilde{P}(l, t) = \tilde{G}(l - l_0, t) - \int_0^t \tilde{G}(l, t - x) \tilde{F}(N - l_0, x) \, dx, \tag{34}$$

where $\tilde{F}(l, t)$ is the first passage-time distribution function for the transition $l_0 \to N$ ($N \equiv 0$). The probability $\tilde{P}(l, t)$ for a carrier, starting at l_0 at $t = 0$, to be found at l at time t, is equal to the unperturbed propagator for the transition $l_0 \to l$ minus the contribution of all paths that have crossed the boundary at N. These are represented by the integral in Eq. (34). All the paths that cross the boundary can be specified by first grouping them into a probability per unit time of a carrier reaching N from l_0 for the first time, $\tilde{F}(N - l_0, x)$,

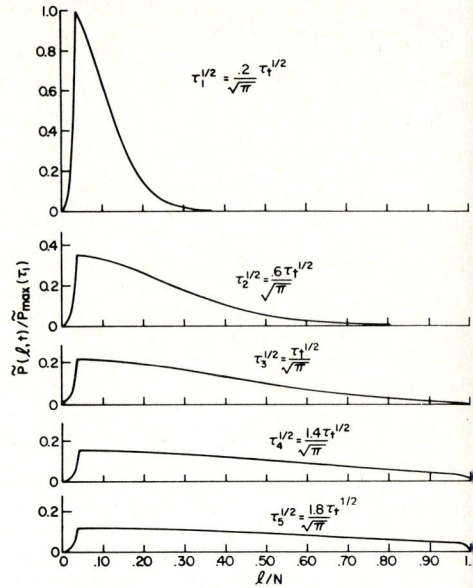

FIG. 13. Same as Fig. 12 with $\tilde{\eta} = 0.6$.

at some earlier time x, and then returning to l in the remaining time $(t - x)$ without concern for whether or not it had again encountered the boundary. Note that $\tilde{G}[-(N - l), t - x] \equiv \tilde{G}(l, t - x)$. The integration in Eq. (34) represents a sum over all times $x < t$ when the first crossing might have been made.

The method we use to compute $\tilde{P}(l, t)$ is the same as the one used to construct the graphs of $\tilde{G}(l, t)$ shown in Figs. 8 and 9. The first step is to construct the spatial Fourier transform (FT) of $\tilde{P}(l, t)$,

$$\Gamma(k, \tau) = \sum_{l=1}^{N} e^{-ikl} \tilde{P}(l, t), \tag{35}$$

analytically from a chosen $\psi(t)$. Then this FT is inverted numerically. This procedure is applied in Appendix B for the interesting function $\psi_2(t)$ to obtain the formula (B19). In Figs. 12–14 we plot the time series of $\tilde{P}(l, t)$ for various values of the asymmetry factor $\tilde{\eta}$. Note that the shape of the propagating packet differs from the free-carrier form (in Fig. 9) of $\tilde{G}(l, t)$ only in the truncation region in which the packet overlaps the absorbing boundaries at the ends of the sample. We are primarily interested in the effect of the boundary on $\langle l \rangle$ and, thus, on $I(t)$. For $\psi_2(t)$ we can compute $\langle l \rangle$ analytically from

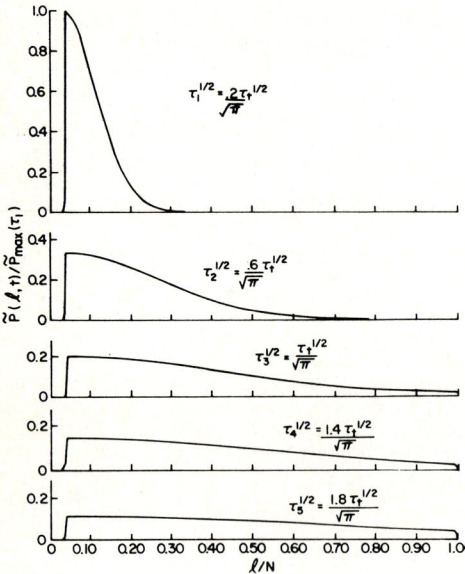

FIG. 12. Propagator for a carrier packet $\tilde{P}(l, t)$ in the presence of absorbing barriers at $l/N = 0, 1$, for a range of time $\tau_n^{1/2} = 0.2(2n-1)\tau_t^{1/2}/\sqrt{\pi}$, $n = 1, 5$. The transit time τ_t is defined in the text. The random walk is based on $\psi_2(t) \sim \pi^{1/2} \tau^{-3/2}$. The spatial bias factor $\tilde{\eta} = 0.9$.

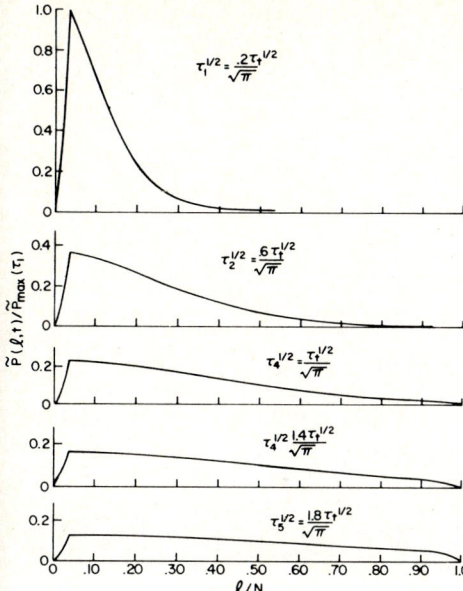

FIG. 14. Same as Fig. 12 with $\bar{\eta} = 0.53$.

and summing, one has

$$\langle l(\tau) \rangle \equiv \sum_{l=1}^{N} l \tilde{P}(l, t) = l_0 + \langle l(\tau) \rangle_0$$

$$- \int_0^\tau d\tau' \tilde{F}(N - l_0, \tau') \langle l(\tau - \tau') \rangle_0 \qquad (40a)$$

$$= l_0 + \langle l(\tau) \rangle_0 - \frac{\beta l}{\pi} \int_0^\tau d\tau' \tau'^{-3/2}$$

$$\times e^{-\beta^2/\tau'} (\tau - \tau')^{1/2}. \qquad (40b)$$

The integral in Eq. (40b) can be transformed to

$$\frac{\beta \bar{l}}{\pi} e^{-\beta^2/\tau} \int_0^\infty \frac{dy\, y^{1/2}}{y+1} e^{-(\beta^2/\tau)y}$$

$$= (\beta \bar{l}/\pi^{1/2}) e^{-\beta^2/\tau} [\beta^{-1} \tau^{1/2} - \pi^{1/2} e^{\beta^2/\tau} \operatorname{erfc}(\beta/\tau^{1/2})], \qquad (41)$$

where we have used Formula 2.1.2 of Ref. 17. One now has

$$\langle l \rangle = l_0 + \bar{l}(\tau/\pi)^{1/2}(1 - e^{-\beta^2/\tau}) + \beta \bar{l} \operatorname{erfc}(\beta/\tau^{1/2}), \qquad (42)$$

which at early and late times has the asymptotic expressions

$$\langle l \rangle \sim \begin{cases} l_0 + \bar{l}(\tau/\pi)^{1/2} + \cdots, & \beta/\tau^{1/2} > 1, \\ N - \bar{l}\beta^2(\pi\tau)^{-1/2} + \cdots, & \beta/\tau^{1/2} < 1. \end{cases} \qquad (43)$$

In the early-time regime, $\langle l \rangle \simeq \langle l(\tau) \rangle_0$, while as $\tau \to \infty$, $\langle l \rangle \to N$, the thickness of the sample. It is more revealing to examine the current

$$I(t) \propto \frac{d\langle l \rangle}{d\tau} = \tfrac{1}{2} \bar{l} (\pi\tau)^{-1/2}(1 - e^{-\beta^2/\tau}). \qquad (44)$$

At early times, when $\beta/\tau^{1/2} \gg 1$, the current is just the unperturbed value (28b) of an infinite medium. For $\beta/\tau^{1/2} \ll 1$, the expansion of the exponential yields

$$I(t) \propto (\bar{l}\beta^2/2\pi^{1/2})\tau^{-3/2}. \qquad (45)$$

Thus the dominant influence of the absorbing boundary on the current is to decrease its magnitude sharply after some characteristic time. The general behavior of the current is most clearly evident in a $\log I$-$\log \tau$ plot of the expression in Eq. (44) shown in Fig. 15. A characteristic time t_τ can easily be obtained from the change in slope in Fig. 15, with

$$\tau_t \simeq [(N - l_0)/\bar{l}]^2 \qquad (46)$$

$$\langle l \rangle = \left. \frac{\partial \Gamma(k, \tau)}{\partial k} \right|_{k=0}, \qquad (36)$$

using Eq. (B19) of Appendix B. An alternative calculation of $\langle l \rangle$ can also be made directly from Eq. (34) by inserting the asymptotic value of the mean computed with $\tilde{G}(l, t)$ in Eq. (28a) and the asymptotic value of $\tilde{F}(N - l_0, t)$, both valid for $\tau \gg 1$, into that formula. When $\tau \gg 1$, the hopping-time distribution $\psi_2(t)$ is known [Eqs. (B5)–(B9), cf. also Ref. 15] to yield the asymptotic formula

$$\tilde{F}(N - l_0, \tau)/W_M \sim (\beta/\pi^{1/2}\tau^{3/2})e^{-\beta^2/\tau}, \quad \tau \gg 1, \qquad (37)$$

where $\beta \equiv (N - l_0)/\bar{l}$. The expression (37) is vanishingly small until

$$[(N - l_0)/\bar{l}]^2/\tau \sim 1 \qquad (38)$$

or, from Eq. (28a)

$$\langle l(\tau) \rangle_0 \propto \bar{l}\tau^{1/2} \sim N - l_0, \qquad (39)$$

i.e., until the mean distance of travel is of the order of the distance from the source of carrier to the distant boundary of the sample. Hence, the integral in Eq. (34) makes little contribution until $\tau \sim \beta^2$.

After inserting Eq. (37) into (34), multiplying by l

($\tau_t \equiv W_M t_\tau$). The change of slope is also shown in Fig. 7, and it is to be emphasized that the t_τ defined in this manner is not the same as the one associated with the plateau, discussed in Sec. I. The fraction of carriers which have survived until time τ, $S(\tau)$, can be determined

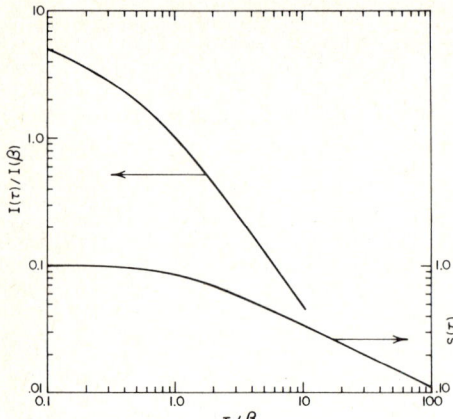

FIG. 15. Plot of Eq. (44) for the current $I(\tau)$ derived from an analytic solution to the random walk governed by $\psi_2(t)$ in the presence of an absorbing boundary. The lower curve is the fraction of surviving carriers $S(\tau)$ as a function of time in units of the transit time $\beta \equiv (N-l_0)/\bar{l}$.

by summing (24) from 1 to N. If one uses the identity

$$\sum_{l=1}^{N} G(l - l_0, t) = 1 \qquad (47)$$

to obtain

$$S(\tau) \equiv \sum_{l=1}^{N} \tilde{P}(l, \tau) = 1 - \int_0^\tau d\tau' \, \tilde{F}(N - l_0, \tau')/W_M$$

$$= 1 - \beta \pi^{-1/2} \int_0^\tau dx\, x^{-3/2} e^{-\beta^2/x} = \mathrm{erf}(\beta/\tau^{1/2}). \qquad (48)$$

This expression could also have been obtained from Eq. (35) by setting $k = 0$. The explicit form for this appears in Eq. (B14), with the term proportional to κ^{-l_0} omitted. At the transit time τ_t, the fraction of survivors is

$$S(\tau_t) = \mathrm{erf}(1) = 0.84. \qquad (49)$$

The time τ_x at which the fraction has declined to 0.5 is $\tau_x/\tau_t = 4.3$. We exhibit a log-log plot of $S(\tau)$ in Fig. 15.

With the choice $\psi_2(t)$, we have demonstrated the basic behavior of a packet of carriers drifting towards an absorbing boundary under the influence of a field. Now $\psi_2(t)$ is a representative of the class (31) that can be expected to prevail for the time regime $\tau \sim \tau_t$ of the duration of transport experiments of disordered material (cf. Sec. 3).

In the general case (31), a simple closed-form expression for $\langle l \rangle$ does not seem to exist, except for some specific values such as $\alpha = \frac{1}{2}$, $\frac{1}{3}$, etc. However, its asymptotic behavior can be derived from that of the first passage-time distribution function $\tilde{F}(N - l_0, \tau)$. We have studied this function through its Laplace transform in Appendix C and found, for $0 < \alpha < 1$,

$$\tilde{F}(N - l_0, \tau)/W_M \sim \frac{\exp\{-[(1-\alpha)/\alpha](\alpha b/\tau^\alpha)^{1/(1-\alpha)}\} h(\tau)}{\tau[2\pi(1-\alpha)(\tau^\alpha/\alpha b)^{1/(1-\alpha)}]^{1/2}}, \qquad (50)$$

where

$$h(\tau) \sim \begin{cases} 1, & b/\tau^\alpha \gg 1 \\ \tau^{-(1-\alpha/2)/(1-\alpha)}, & b/\tau^\alpha \ll 1 \end{cases} \qquad (51)$$

The calculation for $\langle l(\tau) \rangle$ based on Eqs. (40a), (32), (50), and an interpolation formula for $h(\tau)$, will be carried out elsewhere. However, in the long-time limit $\beta/\tau^\alpha \ll 1$, Shlesinger[16] has carried out the computation with the use of Tauberian theorems; a review of his work is included in Appendix A. We shall illustrate the content of these theorems by considering a contour-integral expression of $I(t)$. The procedure is as follows: one represents $d[\sum l \tilde{P}(l, t)]/d\tau$ in terms of the Laplace transform of Eq. (34) [cf. Eqs. (152-162) of Ref. 15],

$$I(t) = \frac{1}{2\pi i} \int_{c - i\infty}^{c + i\infty} d\tau\, e^{s\tau} \tilde{I}(s), \qquad (52a)$$

$$\tilde{I}(s) \propto [1 - \psi^*(s)] \left[\sum l G(l - l_0, \psi^*) - \left(\frac{G(N - l_0, \psi^*)}{G(0, \psi^*)} \right) \sum l G(l, \psi^*) \right], \qquad (52b)$$

where s is the dimensionless Laplace-transform variable, $s \equiv u/W_M$. The ratio of the two G functions in Eq. (52b) is just the Laplace transform of $\tilde{F}(N - l_0, \tau)$. For $\tau \gg 1$, one must consider the $s \ll 1$ limit of $\tilde{I}(s)$ in Eq. (52b). Using Eqs. (B6), (B7), and

$$\epsilon \equiv [1 - \psi^*(s)]/\bar{l} \xrightarrow[s \ll 1]{} c s^\alpha/\bar{l} \qquad (53)$$

[from Eq. (155) of Ref. 15], one can obtain

$$\tilde{I}(s) \sim (1 - e^{-(N-l_0)\epsilon})/\epsilon, \qquad (54)$$

where we have retained the leading singular terms of Eq. (52b). Inserting Eq. (54) into (52a), we proceed in the same manner as we did in Appendix C for the asymptotic development of $\tilde{F}(N - l_0, \tau)$ for $\beta/\tau^\alpha \ll 1$. The contour of Eq. (52a) is displaced, as in Fig. 16, to C, encircling the branch point of $\tilde{I}(s)$. Corresponding to the two time regimes discussed above, one has

$$I(t) \propto \begin{cases} \dfrac{\bar{l}}{2\pi i} \displaystyle\int_C \dfrac{ds\, e^{st}}{cs^\alpha} = \dfrac{\bar{l}}{c\Gamma(\alpha)\tau^{1-\alpha}}, & (N-l_0)\epsilon \gg 1 & (55a) \\[2ex] \dfrac{\bar{l}}{2\pi i} \displaystyle\int_C ds\, e^{s\tau}\left[\dfrac{N-l_0}{\bar{l}} - \dfrac{1}{2}\left(\dfrac{N-l_0}{\bar{l}}\right)^2 cs^\alpha + \cdots\right] = \dfrac{1}{2}\bar{l}\left(\dfrac{N-l_0}{\bar{l}}\right)^2 \dfrac{c}{[-\Gamma(-\alpha)\tau^{1+\alpha}]}, & (N-l_0)\epsilon \ll 1, & (55b) \end{cases}$$

where only terms containing branch points contribute to the integral evaluated on C. Thus, from a mathematical point of view, the effect of the absorbing boundary is to cause a "time-dependent" transition in the branch point ($s^{-\alpha} \to s^\alpha$) in the Laplace transform of $I(t)$. Equations (55a) and (55b) were first obtained by Shlesinger using Tauberian theorems (cf. Appendix A). The time dependence of $I(t)$ for a $\psi(t) \propto t^{-(1+\alpha)}$ is shown schematically in Fig. 7. One observes that it is a generalization of the behavior shown in Fig. 15 for the special case $\alpha = \frac{1}{2}$. It is schematic in the sense that one does not have a closed expression like Eq. (44) but knows that $I(t)$ undergoes a transition, as indicated in Eqs. (55a) and (55b), from $\tau^{-(1-\alpha)} \to \tau^{-(1+\alpha)}$, as β/τ^α goes through unity with increasing τ. One notes that the slopes of the log$I$-logτ plot sum to -2. This can be considered a crucial test of our model when experimental data are being analyzed.[18] Furthermore, the fact that the transit time τ_t varies as $\beta^{1/\alpha}$ will be shown to have considerable significance.

III. DISCUSSION OF RESULTS AND CONCLUSIONS

The traditional macroscopic description of carrier transport is associated with the central-limit theorem of probability. The total displacement of a carrier in a biasing field is considered to be a succession of independent displacements such that the time intervals between the initiation of successive displacements, which we call hopping times, have a narrow distribution function, as do the lengths of the individual displacements. The small dispersions in these two distributions imply that an initially narrow carrier packet remains narrow as it traverses the sample. The spatial distribution of the packet becomes Gaussian with increasing time, while spreading becomes negligible compared with the distance traversed; the velocity of propagation of the packet becomes constant. Thus, a clearly defined transit time proportional to the thickness of the sample appears naturally in this traditional description. As noted earlier, however, observations on charge-carrier mobility in some insulators seem to violate this description.

We have introduced in Sec. II a new stochastic model in which we relaxed the hypothesis of a small dispersion in the hopping-time distribution function $\psi(t)$. A small dispersion in hopping times would imply that the duration of the experiment t_τ is large compared to a characteristic time for an individual hop, e.g., the mean hopping time $\bar{t}$. For a carrier hopping across an amorphous insulating film, this does not seem to be the case. The introduction of a long tail in $\psi(t)$ such that

$$\psi(t) \sim \text{const} \times t^{-(1+\alpha)}, \quad t = O(t_\tau) \tag{56}$$

with $0 < \alpha < 1$, yields an anomalous non-Gaussian transport which, has been seen, has properties consistent with experimental measurements on two diverse systems, As_2Se_3,[8,9] and TNF-PVK.[11] If $\psi(t)$ satisfies Eq. (56) for all $t_0 < t < \infty$ (where t_0 is some arbitrarily chosen time), then $\bar{t} = \infty$. However, as emphasized in Eq. (56), the algebraic tail of $\psi(t)$ is operative for the restricted time range of the experiment. We shall discuss this point more completely, below. By using the distribution Eq. (56) to calculate the packet propagator $P(l, t)$, we have shown that the mean position of a carrier in the packet is given at time t by

$$\langle l \rangle \propto t^\alpha, \quad 0 < \alpha < 1, \tag{57}$$

and that the ratio of the dispersion of the packet to its mean position is independent of the time,

$$\sigma/\langle l \rangle = \text{const.} \tag{58}$$

The current $I(t)$ associated with such a moving packet exhibits the time dependence of Fig. 7, with

$$I(t) \propto \begin{cases} t^{-(1-\alpha)}, & \langle l \rangle \ll L \\ t^{-(1+\alpha)}, & \langle l \rangle \gtrsim L \end{cases} \tag{59}$$

We now discuss experiments reported in Refs. 8, 9, and 11, on the basis of Eqs. (57)–(59), and present a justification of the employment of Eq. (56).

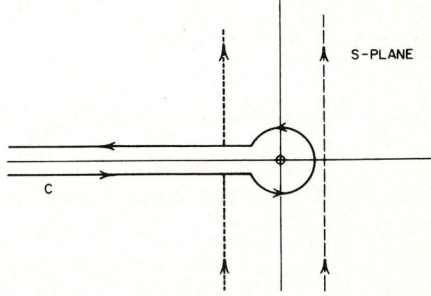

FIG. 16. Integration contours in the complex s-plane used for the computation of the current $I(t)$ (cf. text).

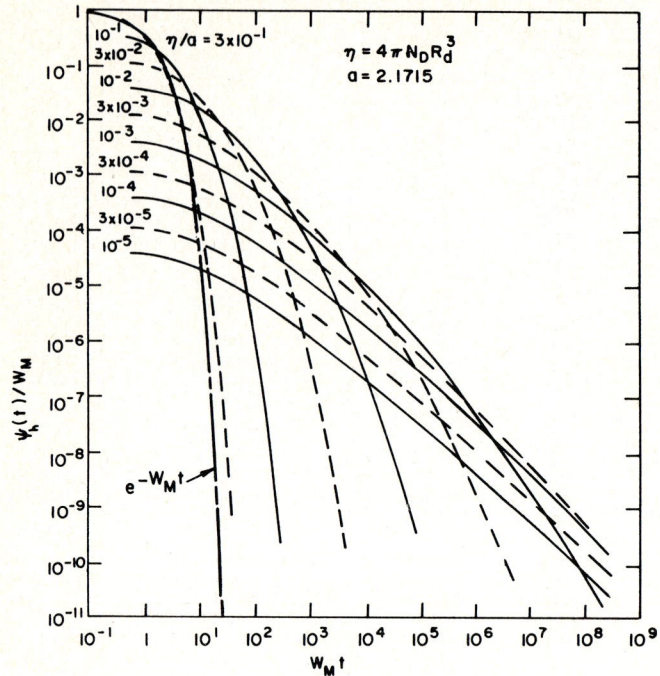

FIG. 17. Hopping-time distribution function $\psi_h(t)$ calculated for a random medium (Ref. 14). The parameter η is a dimensionless measure of the localized site density. The broken line curve (———— ————) is $\psi_h(t)/W_M = \exp(-W_M t)$.

A first-principles calculation of $\psi(t)$ for hopping through a random medium has been considered in Ref. 14. We outline the procedure in Appendix D and reproduce the results in Fig. 17. We designate that choice of $\psi(t)$ by $\psi_h(t)$. The reduced $\psi_h(t)/W_M$ is a function of two dimensionless variables, $\tau \equiv W_M t$, $\eta \equiv 4\pi N_D R_d^3$, where N_D is the density of localized sites for hopping carriers, and R_d is the charge-localization radius (one-half the effective Bohr radius). The function $\psi_h(t)$ is related to an ensemble average of the probability of hopping between sites. It is a direct measure of the degree of fluctuation of this probability. In the computation of $\psi_h(t)$, only positional disorder of the localized sites has been considered. The temperature dependence is contained in $W_M \propto e^{-\Delta/kT}$, where Δ is an average hopping activation energy. The transition rate $W(\vec{r}) = W_M e^{-r/R_d}$. If one includes fluctuations in Δ, then $\psi_h(t)$ would depend on an additional variable,[19] Δ_0/kT, where Δ_0 is a measure of the dispersion of Δ. If there is no disorder, only one site-to-site transition rate $W(\vec{r}) = W$ exists, and $\psi_h(t)/W = e^{-Wt}$. As η increases, the dispersion in the transition rate decreases, and $\psi_h(t)$ approaches a limit, the exponential function. Conversely, as η decreases, the dispersion in the transition rate increases, and $\psi_h(t)$ approaches a function which for large t has an inverse-power decline. The long tail of $\psi(t)$ can be understood as follows: With a wide range of $W(\vec{r})$, one can always find a pair of sites corresponding to a specified hopping time τ; with a single W, a cutoff W^{-1} exists; τ is either less than or greater than the hopping time W^{-1}.

More precisely, in Fig. 17,

$$\psi_h(t)/W_M = \eta(\ln\tau)^2/\tau^{1+(\eta/3)(\ln\tau)^2} . \tag{60}$$

In calculating an ac conductivity $\sigma(\omega)$ for hopping motion when the frequency ω is varied over many decades, one must use the expression for $\psi_h(t)$ in Eq. (60), especially when one encompasses the $\sigma(\omega)$ range from the ω^s region to the dc limit.[14] However, for a t_τ measurement, one can approximate $\psi_h(t)$ by Eq. (56) where

$$\alpha \simeq \tfrac{1}{3}\eta(\ln\tau)^2, \quad \tau = O(\tau_t), \quad \tau_t \gg 1. \tag{61}$$

This approximation increases the mathematical tractability of our model.

Thus, the physical justification for the use of Eq. (56) is that the time dependence of $\psi_h(t)$, calculated for a spectrum of $W(\vec{r})$ appropriate for a random medium, is simulated by $t^{-(1+\alpha)}$ for $t \simeq t_\tau$. Here,

α is a function of η, W_M, and t_τ. However, in contrast to Eq. (56), $\psi_h(t)$ has a finite first moment,

$$\overline{\tau} = (\pi \eta^{-1/2})^{1/2} \exp(\tfrac{2}{3}\eta^{-1/2}). \tag{62}$$

If one used $\psi_h(t)$ in the computation of $\tilde{P}(l, t)$, *then the carrier packet would approximate a Gaussian shape in the range* $\tau_t \gg \overline{\tau}$. This fact emphasizes our point that the unusual results [Eqs. (57)–(59)] derived from the stochastic (hopping) process are not so much due to a difference in genre, as they are to a crucial interaction between the *conditions of the experiment determining* τ_t (i.e., E, L) and the *parameters of the system determining* $\overline{\tau}$ (i.e., N_D, R_d).

To stress that the non-Gaussian results are not due merely to a sampling of a small number of events (a few hops per transit), one must have $\tau_t > \tau_{1/2}$. The (median) time $\tau_{1/2}$ is defined to be the time (after arrival) at which the probability of a carrier remaining on the site of arrival is equal to $\tfrac{1}{2}$. One summarizes the above discussion with the constraint

$$\tau_{1/2} \ll \tau_t \lesssim \overline{\tau}. \tag{63}$$

For $\psi_h(t)$,

$$\tau_{1/2} = \exp(3\ln 2/\eta)^{1/3}, \tag{64}$$

and one has, for a value $\eta = 10^{-3}$,

$$\overline{\tau} = 1.42 \times 10^{10},$$
$$\tau_{1/2} = 3 \times 10^5, \tag{65}$$

a difference between these characteristic times of almost five orders of magnitude!

As discussed in Sec. I the surprising element in the results of the transient photocurrent measurement [Fig. 1(b)] in an increasing number of amorphous insulators is not the long tail (Fig. 3), which indicates some sort of dispersion in transit times, but the universality in the shape of $I(t)$ over a wide range of t_τ (Fig. 4). This universal current trace $I(t)/I(t_\tau)$ vs t/t_τ was first quantitatively demonstrated by Scharfe[8] in his measurements on a-As$_2$Se$_3$. Another remarkable aspect of this dispersed transport, analyzed by Scharfe, can be seen in a log-log plot of t_τ^{-1} vs E/L. All the inverse transit times for a considerable range of both field and *sample thickness* lie on one straight line, with a slope $\simeq 2$, i.e., $t_\tau^{-1} \propto (E/L)^2$. By relating t_τ to the effective drift mobility in the conventional way, with the use of Eq. (2), one would obtain

$$\mu \propto E/L. \tag{66}$$

A field-dependent μ has become a familiar object in the phenomenology of insulating materials, but a carrier drift mobility that depends upon sample thickness is most unusual!

The actual choice of t_τ has depended on discerning some plateau in the trace of $I(t)$—a residue of the idealized current shape shown in Fig. 2. However, quite a number of current traces on As$_2$Se$_3$ and some polymeric films display a completely featureless time decay. Moreover, the presence or absence of a small plateau is often a function of the immediate sample history, e.g., the number and frequency of incident light flashes and the degree of dark resting of the sample. A more revealing analysis of these featureless traces is obtained[18] with a log-log plot $I(t)/I(t_\tau)$ vs t/t_τ on As$_2$Se$_3$ made by Pfister[9] in connection with his study of the pressure dependence of t_τ (Fig. 5). The data (including one set at high pressure) exhibit a universality in the current shape over three-orders-of-magnitude range in t_τ. The data clearly show two slopes, each persisting for over an order of magnitude in t. The time dependence of $I(t)$ corresponds to $t^{-0.55}$ with a transition to $t^{-1.45}$. This behavior is in complete conformity with the theoretical results [Eq. (57) and Fig. 7] for $\alpha = 0.45$. The solid line in Fig. 5 is derived from the analytic result for $\alpha = \tfrac{1}{2}$ (Fig. 15, slightly "tilted" to accomodate the small change in α, computed in Sec. II and Appendix B). For general α, one can determine the two slopes [cf. Eq. (57)] and estimate from Eq. (C12) the time dependence of the transition region. However, in the case of $\alpha = \tfrac{1}{2}$, one can derive an analytic expression Eq. (44) for $I(t)$ covering the entire range of t.

The solid line is in excellent agreement with the experimental data. As detailed in Sec. II, the change in slope for the log-log plot of $I(t)$ corresponds to the carriers encountering the oppositely charged electrode (the absorbing barrier). The transit time is associated with this change, i.e., the step change in the quantity $-d[\ln I(t)]/d(\ln t)$.

In Fig. 18 we show traces of both a continuous recording of an electronically produced $\log I$-$\log t$ and a photograph of the linear $I(t)$ vs t displayed on a cathode-ray tube. The time designated as t_τ is marked with an arrow on both traces. The transit time is defined by the $\log I$-$\log t$ plot. The corresponding time on the $I(t)$ plot is an undistinguished point on a completely featureless curve. A convenient choice for t_τ is the time defined by the intersection of the extrapolated lines of constant slope (dashed lines in Fig. 18). For $\alpha = \tfrac{1}{2}$, this point is

$$\tau_t^{1/2} = 0.92(N - l_0)/\overline{l} \simeq 0.92 L/\overline{l}\rho, \tag{67}$$

where ρ is the average intersite separation. The theoretical curve for $\alpha = 0.45$ is but slightly modified, so we take

$$\tau_t^{0.45} = 0.95 L/\overline{l}\rho. \tag{68}$$

There are As$_2$Se$_3$ samples that exhibit another value of α, and Pfister encountered materials problems connected with thin samples ($L < 30$ μm); these

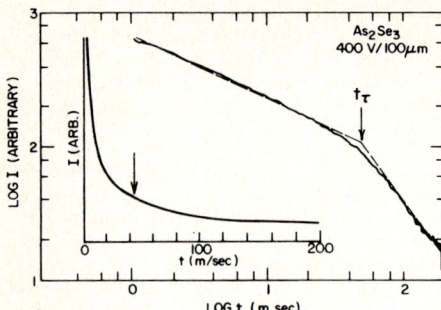

FIG. 18. Trace of an electronically produced log I - log t plot for As_2Se_3 measured by Pfister. The dashed lines have slopes of -0.55 and -1.45, respectively. The intersection of the lines is denoted as t_τ. The same t_τ is indicated on the trace of the oscillogram of I vs. t in the inset.

will be discussed in another work.

The dependence on the electric field E is contained in $\bar{l}$, which is equal to $2p[2\bar{\eta}(E)-1]$. The parameter $\bar{\eta}$ is a measure of the spatial asymmetry introduced by E in the transition rates between the sites. Moreover, $\bar{\eta}(E)$ can be a general (nonlinear) function of E. The exact form of $\bar{\eta}$ for any specific system must be determined by a first-principles calculation of the transition rates, which in turn depends on a knowledge of the effect of an electric field on the site wave functions and site energy levels or intersite barrier.

One can make a prediction of the sample thickness dependence of τ_t independently of $\bar{\eta}(E)$,

$$\tau_t = cL^{1/\alpha}/(\bar{l}\rho)^{1/\alpha}, \qquad (69)$$

where c is a numerical constant of order unity. In the case of As_2Se_3, from Eq. (68), one has $1/\alpha = 2.2$, $c = 0.89$, and

$$\tau_t \propto L^{2.2}. \qquad (70)$$

Hence, *for fixed E*, a current trace of the form in Figs. 5 and 18, *must* correspond to the L dependence in Eq. (70). In Fig. 19 we exhibit a log-log plot of t_τ vs L for a range of L varying from 31 to 100 μm. The values of the transit time have all been determined by the intersection of the lines with slopes -0.55 and -1.45 and at a field of 10^5 V/cm. The slope of the solid line through the data points is equal to 2.2, in excellent agreement with Eq. (70). Thus, the anomaly of a thickness-dependent mobility, as discussed in Eq. (66), is resolved. The apparent L dependence of μ is related to the propagation characteristics of the carrier packet.

The conventional relation Eq. (2) is based on the assumption that the mean position $\langle l \rangle$ of the carriers changes linearly in time. Expression (57) for $\langle l \rangle$ requires a *time-dependent mobility* to interpret the transit-time measurements in the usual way with Eq. (2). Therefore, the insistence on the use of a mobility μ to describe the packet motion implies that μ depends on any parameters, such as L and E, that change the time scale of the measurement.

Let us consider the nature of the field dependence. First, we must estimate $\bar{\eta}(E)$. Since we have little information of the nature of the localized states in As_2Se_3, we shall make the simplest assumptions possible. By postulating a field-independent mobility, we write

$$2\bar{\eta}(E) - 1 \propto E, \qquad (71)$$

the asymmetry being linear in E. Inserting Eq. (71) into (69), one has

$$\tau_t \propto (L/E)^{2.2}. \qquad (72)$$

In Fig. 20 we show a log-log plot of t_τ^{-1} vs E/L. The range of transit time covers four orders of magnitude and is a composite of varying both L and E. The slope of the solid line is equal to 2.2, in agreement with Eq. (72). Thus the apparent

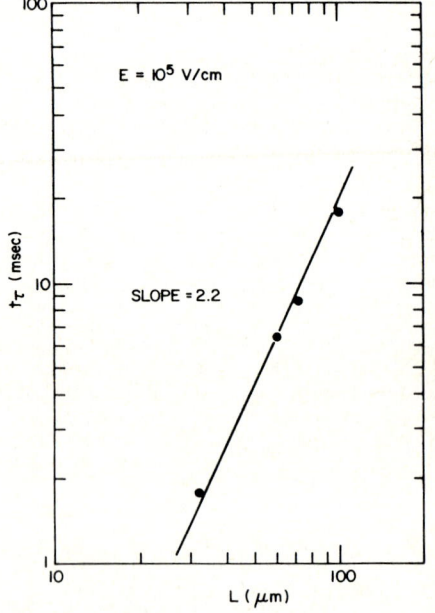

FIG. 19. Log-log plot of the transit time t_τ vs L. The solid line of slope 2.2 is the theoretical curve.

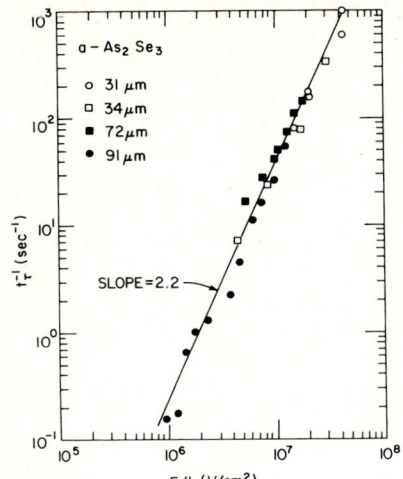

FIG. 20. Log-log plot of the inverse transit time t_T^{-1} vs E/L. The solid line of slope 2.2 is the theoretical curve.

field dependence of the mobility appears on the same basis as the thickness dependence, the field being another parameter that changes the time scale of the experiment. It will be shown below that field-dependent mobilities in other systems cannot be explained entirely on this basis.

To determine the various physical parameters in As_2Se_3 we proceed in a self-consistent manner. With the lowest field that Pfister used in his measurements, we set

$$2\tilde{\eta}(E) - 1 \lesssim 0.1. \tag{73}$$

To obtain τ_t we choose a value of ρ in Eq. (68) and determine if the magnitude of τ_t is then consistent with this ρ via the constraint imposed by the relation in Eq. (61) $[\eta \equiv \frac{3}{8}(a_B/\rho)^3]$. Let $N_D = 1.5 \times 10^{17}$ cm^{-3}, corresponding to $\rho = 1.17 \times 10^{-6}$ cm, and, with Eq. (68), we have $\tau_t = 5.7 \times 10^{11}$ ($L = 91$ μm). Inserting this value for τ in Eq. (61), we determine $\eta = 1.8 \times 10^{-3}$ for $\alpha = 0.45$. Therefore, for this low field ($t_\tau = 6.6$ sec), we have the following (self-consistent) parameters for As_2Se_3:

$$N_D = 1.5 \times 10^{17} \text{ cm}^{-3}, \ a_B = 19.7 \text{ Å}, \ W_M \lesssim 5 \times 10^{10} \text{ sec}^{-1}. \tag{74}$$

The values of N_D and a_B in Eq. (74) are very reasonable. They are close to the values $N_{1oc} = 10^{17}$ cm^{-3}, and $\frac{1}{2}a_B = 10$ Å, determined for a-Ge.[20]

One can view the present theory as a one-parameter model. We choose the value of α to fit the tail of $I(t)$. This choice then predicts *two* independent aspects of the transport, the initial time dependence of $I(t)$ [$I(t) \propto t^{-(1-\alpha)}$] and the dependence of τ_t on L and E, as in Eq. (69). The determination in Eq. (74) of a reasonable set of physical parameters in As_2Se_3, based on $\psi_h(t)$ as the physical justification for the use of a $\psi(t) \propto t^{-(1+\alpha)}$, adds to the hopping model of the transport. The parameters specified in Eq. (74) imply that the lower bound for the average number of hops required to transverse the 91-μm sample is $L/\rho \simeq 8 \times 10^3$. This large number hardly corresponds to a case of "small statistics."

There are systems, TNF-PVK mixtures, in which one has independent knowledge of a hopping transport and of some of the pertinent parameters such as N_D and a_B. We shall now analyze the measurements of Gill[11] on those systems.

By varying the relative composition of the TNF monomer in the polymer PVK, Gill was able to demonstrate that the electron mobility μ_e has an exponential dependence on the average separation of the TNF molecules,

$$\mu_e \propto R_{TNF}^2 \exp(-R_{TNF}/R_0), \tag{75}$$

where

$$R_0 = 0.9 \text{ Å}. \tag{76}$$

Gill also found a similar dependence of the hole mobility on the separation of the uncomplexed vinylcarbazole units. Relation (75) is good evidence for a hopping mechanism. Moreover, by varying the composition, Gill established the TNF monomer as the hopping site for electrons. By equating R_d and R_0, one now has knowledge of two important parameters N_D and R_d ($\equiv \frac{1}{2}a_B$). In Fig. 6 we reproduce a log-log plot of Gill's data for $I(t)$, for 1:1 molar ratio of TNF-PVK, as presented in a recent paper by Seki.[21] One notes again the universal shape of $I(t)$. The transit-time range in Fig. 6, however, extends over only one order of magnitude. The sum of the slopes of the solid lines through data points in Fig. 6, −0.2 and −1.8, again equal to −2.0, in excellent agreement with the theoretical prediction (59) for $\alpha = 0.8$. For this higher value of α, one observes a more rapid transition between the asymptotic slope lines (as compared to Fig. 5 for $\alpha = 0.45$), in agreement with the time dependence in (C12). Now, $\alpha = 0.8$ corresponds to

$$\tau_t \propto L^{1.25}, \tag{77}$$

which is a nearly linear dependence on thickness, in agreement with reported results.[11]

In contrast to As_2Se_3, one can calculate $\eta(\equiv 4\pi N_D R_d^3)$, a priori, for TNF-PVK. One obtains $\eta \simeq 10^{-2}$ for the 1:1 molar ratio with $N_D = N_{TNF} \simeq 10^{21}$ cm^{-3} and $R_d \simeq 0.9$ Å, where N_{TNF} is the concentration of complexed TNF.[22] Using this value of η in Eq. (61) and $\alpha = 0.8$, one determines that the range of τ for the transit time is

1:1 TNF-PVK, $\tau_t \simeq 5 \times 10^6$. (78)

The magnitude of τ_t in Eq. (78) will now be shown to be consistent with that obtained with the use of Eq. (69). For $L = 5$ μm and $\rho = 6.2$ Å (corresponding to $N_{TNF} = 10^{21}$) one has

$$\tau_t = \left(\frac{2.4 \times 10^4}{2\bar{\eta}(E) - 1}\right)^{1.25}. \quad (79)$$

With a nominal value of $2\bar{\eta}(E) - 1 \approx 0.1$ in Eq. (79), one obtains $\tau_t \simeq 5 \times 10^6$, in good agreement with the value in Eq. (78)

To rephrase the above numerical argument, we have for TNF-PVK only one unknown physical parameter of the hopping model, namely, W_M, ($\equiv W_0 e^{-\Delta/\kappa T}$), the prefactor of the intersite transition rate. Inserting the experimental values for N_D and R_d, one must satisfy two independent constraints, Eqs. (61) and (69), for the value of W_M. We have obtained a value for W_M that does satisfy both! The transit time measured by Gill for $L = 5$ μm and $E = 5 \times 10^5$ V/cm [$2\bar{\eta}(E) - 1 \simeq 0.1$ used above is assumed to correspond to this field value] is $t_\tau = 2.0$ msec. Hence for $\tau_t \simeq 4 \times 10^6$, one has

$W_M = 2.5 \times 10^9$ sec^{-1}. (80)

The smaller value of W_M for TNF-PVK in Eq. (80) compared to W_M for As$_2$Se$_3$ in Eqs. (74) is consistent with the larger activation energy for TNF,

$\Delta_{TNF} \simeq 0.65$ eV, $\Delta_{As_2Se_3} \simeq 0.5$ eV. (81)

In contrast to As$_2$Se$_3$, the effective mobility in TNF-PVK appears to be field-dependent. Gill has used a field dependence of the form

$\mu = \mu_0 \exp[-(\Delta - \beta E^{1/2})/\kappa T_{eff}]$. (82)

The strongest evidence for Eq. (82) is contained in Fig. 7 of Ref. 11, a plot of the activation energy of μ, E_a vs $E^{1/2}$. We stress this point because Eq. (82) in general would indicate a very rapid field dependence for $\mu(E)$. However, over the available limited mobility range shown in Fig. 4 of Ref. 11, the field dependence of $\mu(E)$ is quite mild and could be fit with an algebraic dependence $\mu(E) \propto E^{1.3}$. We are not seriously suggesting that this is the case. However, we wish to emphasize that, although the form of $\mu(E)$ in Eq. (82) does imply a very rapid field dependence for μ, this has not been experimentally established, whereby μ changes by orders of magnitude over some field range. Nonetheless, with $\alpha = 0.8$, one would predict an effective $\mu \propto E^{0.25}$ if one assumed no intrinsic $\mu(E)$ (i.e., the asymmetry linear in E as for As$_2$Se$_3$). This is clearly not the case and one needs, at least, $2\bar{\eta}(E) - 1 \propto E^{1.8}$ over the experimental range of E.

To consider the field dependence from a more fundamental approach, one would have to calculate the effect of the field on the transition rates [$W(\vec{r})$] and then incorporate these effects into a calculation of $\psi(s, t)$,[14,19] the probability rate of hopping a distance $\vec{s}$ in a time t. We reserve this approach for another investigation. A few remarks may, however, be in order. There is some difficulty in understanding the magnitude of the activation energies in (81) purely on the basis of the idea that Δ represents the width of the energy dispersion of the localized levels[21] (the disorder energy). One can conjecture that Δ consists of two parts, a self-trapping or small-polaron part Δ_p and the disorder-energy part Δ_d. In single-crystal sulfur,[4] there is transport evidence that the electron propagates as a small polaron, but the Δ_p is independent of electrical field. Thus, a reasonable hypothesis, which will be explored in future work, is that the E dependence is entirely in Δ_d.

In summary, the stochastic transport model has enabled us to understand the nature of the transient photoconductivity in at least two different amorphous systems, one inorganic (As$_2$Se$_3$) and the other organic (TNF-PVK).

First, one has been able to understand a current-trace shape as shown in Fig. 18. There is no "flat part," characteristic of the idealized current response shown in Fig. 2, to specify a transit time. $I(t)$ traces, such as obtained in Fig. 18, were formerly "unanalyzable." By transposing such an $I(t)$ to a log-log plot, the transit time is indicated by the region where $-d[\ln I(t)]/d(\ln t)$ experiences a step change. Furthermore, with the log-log display of the current, in both As$_2$Se$_3$ and TNF-PVK, the theory has accounted for the overall general shape and, in fact, for the detailed time dependence with the appropriate choice of one parameter α. Whatever the choice of α ($0 < \alpha < 1$), the theory predicts that the slopes of the lines, on the log-log plot, should sum to -2.0. There is excellent agreement with this firm prediction in Figs. 5, 6, and 18, for both As$_2$Se$_3$ and TNF-PVK. Thus, with a value of α chosen to fit the tail of $I(t)$, there is an independent prediction of the initial time decay. Moreover, this choice of α must further correlate the shape of $I(t)$ with the thickness dependence of the transit time. For As$_2$Se$_3$ with $\alpha = 0.45$, $\tau_t \propto L^{2.2}$, and TNF-PVK with $\alpha = 0.8$, $\tau_t \propto L^{1.25}$, a nearly linear dependence. Both of these correlations have been established experimentally. In the case of As$_2$Se$_3$, with an assumption of no *intrinsic* field-dependent mobility, the nonlinear E dependence of τ_t has been understood on the same basis as the L dependence. In the TNF-PVK system, there is now evidence for an intrinsic field dependence of the effective electron (hole) mobility, and the origin of this behavior will be pursued in a further study.

On a more microscopic level, a set of parameters N_D, R_d, and W_M has been self-consistently determined for As_2Se_3, in conformity with the value of α, using $\psi_h(t)$ as the physical justification for using a $\psi(t) \propto t^{-(1+\alpha)}$. Knowing N_D and R_d for TNF-PVK, one obtained a value of W_M that satisfied two independent constraints (61) and (69) for $\alpha = 0.8$. In addition, the increase of α in 1:1 TNF-PVK relative to As_2Se_3 (i.e., less dispersive transport) can be ascribed to the increase of $N_D R_d^3$ between the two systems.

The next step is to calculate the basic transition rates in the two materials and account for the magnitudes of the parameter W_M.

On a conceptual level, the theory establishes the reason for the universality of $I(t)$ and has shown the limitations of the notion of a mobility in this dispersive type of transport. If one insists on the conventional relation (2) to define an effective μ, then, besides the possibility of an apparent field dependence, one would have to rationalize a thickness dependence of μ! Unless the mean $\langle l \rangle \propto t$, the idea of a μ depending solely on the properties of the material breaks down. For example, if one has $\langle l \rangle \propto t^\alpha$, $0 < \alpha < 1$, one could write

$$\langle l \rangle = (V_0/t^{1-\alpha}) t, \tag{83}$$

and the effective μ would be *time dependent*. Thus, for this type of transport, the present theory relates the transit time to intrinsic-rate processes in the material, but the simple notion of a mobility, field dependent or otherwise, is very limited.

Dispersive transit-time experiments in disordered solids are perhaps only one of a number of physical phenomena in which this very different kind of statistical process, i.e., one that does not obey the central limit theorem, plays a central role.

ACKNOWLEDGMENTS

We wish to thank both M. E. Scharfe and G. Pfister for enlightening discussions about the experiments on As_2Se_3 and the use of unpublished data. We are also indebted to M. Shlesinger for informative discussions on asymptotic properties of random walks and to L. C. Hebel, E. M. Conwell, G. Pfister, R. Zallen, and J. Mort for a critical reading of the manuscript.

APPENDIX A: ASYMPTOTIC DEVELOPMENT OF THE MEAN AND DISPERSION WITH TAUBERIAN THEOREMS

While the asymptotic formulas (29)-(32) were obtained for the specific model characterized by Eqs. (19) and (28), they can immediately be generalized, as recently shown by Shlesinger,[16] through the application of certain Tauberian theorems. It is convenient to characterize the random walk of interest by two functions, the Laplace transform, $\psi^*(u)$, of the waiting-time distribution function (12), and the Fourier component of the probability $\bar{G}(l, t)$,

$$\gamma(k, t) = \sum_l \bar{G}(l, t) e^{-i\vec{k} \cdot \vec{l}} . \tag{A1}$$

Then

$$\gamma(k, t) = \mathcal{L}^{-1}\left([1 - \psi^*(u)]\{u[1 - \gamma(k)\psi^*(u)]\}^{-1}\right). \tag{A2}$$

Also,

$$\langle l_1(t) \rangle \equiv \sum_l l_1 \bar{P}(l, t) = \bar{l}\left(\frac{\partial \gamma(k, t)}{\partial \lambda}\right)_{\lambda=1}$$

$$= \bar{l}\mathcal{L}^{-1}\{\psi^*(u)/u[1 - \psi^*(u)]\}, \tag{A3}$$

where

$$\bar{l} = \sum_l l p(l) . \tag{A4}$$

Similarly,

$$\sigma^2(t) = \langle l^2 \rangle_{\text{av}} \left(\frac{\partial \gamma}{\partial t}\right)_{\lambda=1} + \bar{l}^2 \left[\frac{\partial^2 \gamma}{\partial \lambda^2} - \left(\frac{\partial \gamma}{\partial \lambda}\right)^2\right]_{\lambda=1}. \tag{A5}$$

When the first two integral moments $\bar{t}$ and $\langle t^2 \rangle_{\text{av}}$ are finite, with

$$\langle t^n \rangle_{\text{av}} \equiv \int_0^\infty t^n \psi(t) dt , \quad \langle l^2 \rangle_{\text{av}} \equiv \sum l^2 p(l) , \tag{A6}$$

$$\psi^*(u) = 1 - \bar{t}u + \tfrac{1}{2} u^2 \langle t^2 \rangle_{\text{av}} + \int_0^\infty (e^{-ut} - 1 + tu - \tfrac{1}{2} t^2 u^2) \psi(t) dt . \tag{A7}$$

The integral contribution can be shown to be $o(u^2)$. Hence

$$\psi^*(u) \sim 1 - \bar{t}u + \tfrac{1}{2} u^2 \langle t^2 \rangle_{\text{av}} . \tag{A8}$$

On the other hand, if $\bar{t}$ is infinite and, as $t \to \infty$,

$$\psi(t) \sim [At^{1+\alpha} \Gamma(1-\alpha)]^{-1} \text{ for } 0 < \alpha < 1 , \tag{A9}$$

Feller[23] has shown that

$$\psi^*(u) \sim 1 - u^\alpha / A . \tag{A10}$$

An asymptotic expression for $\langle l(t) \rangle$ is obtained from Eqs. (A3), (A8), and (A10), by applying the following Tauberian theorem of Hardy and Littlewood, and Karamata:

If $f(u) \sim Au^{-k}$ as $u \to 0$ with $k > 0$,

then $g(t) \sim At^{k-1}/\Gamma(k)$ as $t \to \infty$ (A11)

when $\mathcal{L}[g(t)] = f(u)$, A being a constant.

Shlesinger has also used the generalization of this theorem in which A is a slowly varying function of u.

When Eq. (A8) is applicable, we have, from Eqs. (A3), (A8), and (A11), if we set

$$f(u) \equiv \psi^*(u)/u[1 - \psi^*(u)] \sim (\bar{t}u^2)^{-1} , \tag{A12}$$

$$g(t) \simeq \langle l(t) \rangle / l \sim t/\bar{t} ,\qquad (A13)$$

which is the classical random-walk result, $\langle l(t) \rangle \sim (\bar{l}/\bar{t})t$.

When Eq. (A9) and (A10) are applicable, the Tauberian theorem implies that

$$f(u) \sim A/u^{\alpha+1} \qquad (A14)$$

and

$$\langle l(t) \rangle \sim \bar{l} A t^{\alpha}/\Gamma(\alpha+1) . \qquad (A15)$$

With an absorbing boundary let us consider the Fourier component of $\tilde{P}(l,t)$, $\Gamma(k,t)$, which is related to $\tilde{P}(l,t)$ in the same manner that $\gamma(k,l)$ is related to $\tilde{G}(l,t)$ as given by Eq. (A1). Then

$$\Gamma(k,t) = \gamma(k,t) e^{-ikl_0} - \int_0^t dx\, \gamma(k,t-x) \tilde{F}(N-l_0, x) , \qquad (A16)$$

where it is known that

$$\tilde{F}(N-l_0, \tau) = \mathcal{L}^{-1} G(N-l_0, \tilde{\psi}(s))/G(0, \psi(s)) . \quad (A17)$$

It was shown in Eqs. (158), (138a), and (138b) of Ref. 15 that

$$\frac{G(N-l_0, z)}{G(0, z)} = \frac{\alpha_1^{N-l_0} - (\alpha_1/\alpha_2)^N \alpha_1^{-l_0} + (1-\alpha_1^N)\alpha_2^{-l_0}}{1 - (\alpha_1/\alpha_2)^N}$$

$$\sim \alpha_1^{N-l_0} + (1-\alpha_1^N)\alpha_2^{-l_0} , \qquad (A18)$$

where $|\alpha_2| > 1 > |\alpha_1|$ and

$$\alpha_1 = 1 - \epsilon , \quad \alpha_2 = [\bar{\eta}/(1-\bar{\eta})](1+\alpha_1) , \qquad (A19)$$

the parameter $\bar{\eta}$ being related to the bias in the walk as defined by Eq. (28b). Also,

$$\epsilon(z) = (1-z)/2p(2\bar{\eta}-1)$$
$$+ (1-z)^2 [2p(2\bar{\eta}-1)^2 - \eta]/4p^2(2\bar{\eta}-1)^3 + O(1-z)^3 . \qquad (A20)$$

Let us again consider the case

$$\psi^*(u) \sim 1 - u^{\alpha}/A_1 + u^{2\alpha}/A_2 - \cdots , \qquad (A21)$$

for which Eq. (A15) was derived. Then

$$\alpha_1 \sim (u^{\alpha}/A_1 - u^{2\alpha}/A_2 + \cdots)/2p(2\bar{\eta}-1) , \qquad (A22)$$

$$\alpha_2 \sim [\bar{\eta}/(1-\bar{\eta})][1 + u^{\alpha}/pA_1(2\bar{\eta}-1) + \cdots] . \qquad (A23)$$

The Laplace transform of $\Gamma(k,t)$ is, applying the convolution theorem,

$$\mathcal{L}[\Gamma(k,t)] = \mathcal{L}[\gamma(k,t)][e^{-ikl_0}$$
$$- G(N-l_0, \tilde{\psi}(u))/G(0, \tilde{\psi}(u))] . \qquad (A24)$$

Since $\langle l \rangle = i \lim_{k \to 0} (\partial \Gamma/\partial k)$, we find

$$\langle l \rangle = l_0 + \bar{l} \mathcal{L}^{-1} \{\psi^*(u)/u[1 - \psi^*(u)]\}$$
$$\times \{1 - [G(N-l_0, \tilde{\psi}(u))/G(0, \tilde{\psi}(u))]\} . \qquad (A25)$$

The Tauberian theorems discussed above can be used to find the long-time behavior of $\langle l \rangle$ from the asymptotic properties as $u \to 0$ of the function to the right of the $\mathcal{L}^{-1}$ operator in Eq. (A25). From Eq. (A18) and (A20), we find that as $u \to 0$,

$$1 - G(N-l_0, \tilde{\psi}(u))/G(0, \tilde{\psi}(u)) \sim 1 - \alpha_1^{N-l_0} - (1-\alpha_1^N)\alpha_2^{-l_0}$$
$$= \epsilon(N-l_0)[1 - \tfrac{1}{2}\epsilon(N-l_0-1)] + \epsilon N[1 - \tfrac{1}{2}\epsilon(N-1)]$$
$$\times [2\eta/(1-\eta)]^{-l_0}(1 + \tfrac{1}{2}\epsilon l_0) + O(\epsilon^3)$$
$$= c_1 u^{\alpha} + c_2 u^{2\alpha} + \cdots , \qquad (A26)$$

where $c_1, c_2, \ldots$ are numbers which depend on $N, l_0, \eta,$ and p. We also have

$$\psi^*(u)/u[1 - \psi^*(u)] \sim u^{-1-\alpha} A_1 [1 + u^{\alpha}(A_1^2 - A_2)/A_1 A_2$$
$$+ O(u^{2\alpha})] . \qquad (A27)$$

The product of Eqs. (A26) and (A27), whose Laplace inverse is required in (A25), is

$$u^{-1} c_1 A_1 (1 - u^{\alpha} B/c_1 A_1 + \cdots) , \qquad (A28)$$

with

$$B = -[(A_1^2 - A_2) c_1 + c_2 A_1 A_2]/A_2 . \qquad (A29)$$

Hence, upon application of the Tauberian theorem (A11) we find, as $t \to \infty$,

$$\langle l \rangle \sim l_0 + \bar{l} A_1 c_1 - \bar{l} B/t^{\alpha} \Gamma(\alpha+1) + \cdots , \qquad (A30)$$

and

$$\frac{d\langle l \rangle}{dt} \sim B\bar{l}/t^{\alpha+1} \Gamma(\alpha) . \qquad (A31)$$

The significance of results (A15) and (A31) are thoroughly discussed in the text.

APPENDIX B: CHARACTERISTIC FUNCTION $\Gamma(k,t)$ FOR $\psi_2(t)$ WITH ABSORBING BOUNDARY

The solution for an absorbing-plane boundary for the asymmetric continuous-time random walk (CTRW) is described in the text [Eq. (34)]. The final form for the propagation is

$$\tilde{P}(l, \tau) = \tilde{G}(l - l_0, \tau) - \int_0^{\tau} d\tau' \tilde{G}(l, \tau - \tau')$$
$$\times \tilde{F}(N - l_0, \tau')/W_M , \qquad (B1)$$

where $\tau \equiv W_M t$. The characteristic functions are defined as

$$\Gamma(k, \tau) = \sum_{l=1}^{N} e^{-ikl} \tilde{P}(l, \tau) , \qquad (B2)$$

$$\gamma(k, \tau) = \sum_{l=1}^{N} e^{-ikl} \tilde{G}(l, \tau) , \qquad (B3)$$

where $k \equiv 2\pi r/N$, r = integer. We substitute the expression for $\tilde{P}(l, \tau)$ in Eq. (B1) into Eq. (B2), and obtain for the characteristic function

$$\Gamma(k, \tau) = \gamma(k, \tau) e^{-il_0 k} - \int_0^{\tau} d\tau' \gamma(k, \tau - \tau')$$
$$\times \tilde{F}(N - l_0, \tau')/W_M . \qquad (B4)$$

For the region of τ of interest in the present paper

($\tau \gg 1$), we can determine $\tilde{F}(N - l_0, \tau)$ from the asymptotic behavior of its Laplace transform

$$\frac{\tilde{F}(N - l_0, \tau)}{W_M} = \frac{1}{2\pi i} \int_{c-i\infty}^{c+i\infty} ds\, e^{s\tau} \frac{G(N - l_0, \psi^*(s))}{G(0, \psi^*(s))}, \quad \text{(B5)}$$

$$\frac{G(N - l_0, \psi^*(s))}{G(0, \psi^*(s))} \simeq \alpha_1^{N-l_0} + (1 - \alpha_1^N) \alpha_2^{-l_0}, \quad \text{(B6)}$$

where relative contributions of order $(\alpha_1/\alpha_2)^N \simeq \kappa^{-N}$ in Eq. (B6) have been neglected [cf. Eq. (A18)]. [$\kappa \equiv \eta/(1-\eta)$]. The complete expression for $\alpha_{1,2}$ is defined in Eqs. (130) and (128), of Ref. 15 (with $z = \psi^*(s)$). For $\tau \gg 1$ we only need the form of $\alpha_{1,2}$ for small s,

$$\alpha_1 \simeq e^{-\epsilon}, \quad \alpha_2 \simeq \kappa e^{+\epsilon}, \quad \text{(B7)}$$

$$\epsilon \equiv [1 - \tilde{\psi}_2(s)]/\bar{l} \simeq 2 s^{1/2}/\bar{l}. \quad \text{(B8)}$$

Inserting Eqs. (B7) and (B8) into Eq. (B6), we obtain the inverse Laplace transform (using formula 3.2, 14 of Ref. 17)

$$\tilde{F}(N - l_0, \tau)/W_M = [(N - l_0) e^{-(N-l_0)^2/\bar{l}^2 \tau} + \kappa^{-l_0} l_0 e^{-l_0^2/\bar{l}^2 \tau} - \kappa^{-l_0}(N + l_0) e^{-(N+l_0)^2/\bar{l}^2 \tau}]/\bar{l} \pi^{1/2} \tau^{3/2}. \quad \text{(B9)}$$

The first passage-time distribution $\tilde{F}(N - l_0, \tau)$ consists of a sum of terms of the form $\beta \tau^{-3/2} e^{-\beta^2/t}$. The characteristic function for $\tilde{G}(l, \tau)$ computed with $\psi_2(t)$ has been given previously,[15]

$$\gamma(k, \tau) = \tfrac{1}{2} \lambda^{-1/2}(k) \sum_{\pm}(\pm) c_\pm e^{\tau c_\pm^2} \text{erfc}(\tau^{1/2} c_\mp), \quad \text{(B10)}$$

$$c_\pm \equiv 1 \pm \lambda^{1/2}(k), \quad \text{(B11)}$$

$$\lambda(k) = 1 - 2p + 2p\bar{\eta} e^{-ik} + 2p(1 - \bar{\eta}) e^{ik}. \quad \text{(B12)}$$

Thus, inserting Eqs. (B10) and (B9) into Eq. (B4), one must calculate integrals of the form

$$\frac{\beta}{\pi^{1/2}} \int_0^\infty d\tau' \frac{e^{-\beta^2/\tau'}}{\tau'^{3/2}} e^{c^2(\tau - \tau')} \text{erfc}[c(\tau - \tau')^{1/2}], \quad \text{(B13)}$$

in order to determine $\Gamma(k, \tau)$. We introduce a new integration variable $t = (\tau/\tau') - 1$ and change Eq. (B13) to

$$\frac{\beta e^{-\beta^2/\tau}}{(\pi\tau)^{1/2}} \int_0^\infty \frac{dt\, e^{-(\beta^2/\tau) t}}{(t+1)^{1/2}}$$

$$\times e^{-c^2 \tau t/(t+1)} \text{erfc}\{c\tau^{1/2} [t/(t+1)]^{1/2}\}. \quad \text{(B14)}$$

The integral in Eq. (B14) now has the form of a Laplace transform. We take advantage of this by inserting into Eq. (B14) the series expression

$$e^{z^2} \text{erfc}\, z = \sum_{n=0}^\infty \frac{(-z)^n}{\Gamma(\tfrac{1}{2} n + 1)} \quad \text{(B15)}$$

and use formula 2.9 of Ref. 17

$$\int_0^\infty \frac{e^{-pt} t^{\nu-1} dt}{(t+a)^{\nu-1/2}} = 2^{\nu-1/2} \Gamma(\nu) p^{-1/2} e^{ap/2} D_{1-2\nu}[(2ap)^{1/2}], \quad \text{(B16)}$$

where D_ν is the parabolic cylinder function. We obtain for Eq. (B14)

$$e^{-\beta^2/\tau} \sum_{n=0}^\infty (-2c\tau^{1/2})^n e^{\beta^2/\tau} i^n \text{erfc}(\beta/\tau^{1/2}), \quad \text{(B17)}$$

where i^n erfc is the nth repeated integral of the error function. We now insert the integral representation of $e^{z^2} i^n$ erfc z into Eq. (B17) and sum the series:

$$\frac{2e^{-\beta^2/\tau}}{\pi^{1/2}} \int_0^\infty ds \sum_{n=0}^\infty \frac{(-2c\tau^{1/2} s)^n}{n!} e^{-s^2 - 2s\beta/\tau^{1/2}}$$

$$= \frac{2e^{-\beta^2/\tau}}{\pi^{1/2}} \int_0^\infty ds \exp[-s^2 - 2(\beta\tau^{-1/2} + c\tau^{1/2}) s]$$

$$= e^{-\beta^2/\tau} \exp[(\beta\tau^{-1/2} + c\tau^{1/2})^2]$$

$$\times \text{erfc}(\beta\tau^{-1/2} + c\tau^{1/2}) \equiv f(\beta, c; \tau). \quad \text{(B18)}$$

Thus the final expression for the characteristic function is

$$\Gamma(k, \tau) = \gamma(k, \tau) e^{-il_0 k} - \tfrac{1}{2} \lambda^{-1/2}(k) \sum_\pm (\pm) c_\pm$$

$$\times \left[f\!\left(\frac{N - l_0}{\bar{l}}, c_\mp; \tau\right) + \kappa^{-l_0} f\!\left(\frac{l_0}{\bar{l}}, c_\mp; \tau\right) \right.$$

$$\left. - \kappa^{-l_0} f\!\left(\frac{N + l_0}{\bar{l}}, c_\mp; \tau\right) \right]. \quad \text{(B19)}$$

This expression is used in the numerical inversion of Eq. (B2) to obtain the $\tilde{P}(l, \tau)$ shown in Figs. 12–14.

APPENDIX C: GENERAL ASYMPTOTIC EVALUATION OF THE FIRST PASSAGE-TIME DISTRIBUTION $F(N-l_0\tau)$

As shown by Eqs. (34) and (37), and discussed in Appendix B, the solution for the propagator in the presence of absorbing boundaries involves the evaluation of the first passage-time distribution $\tilde{F}(N - l_0, \tau)$. In the region of interest, $\tau \gg 1$, $\tilde{F}(N - l_0, \tau)$ can be determined from the small-s behavior of its Laplace transform. Following Eqs. (B5)-(B7), one needs to consider integrals of the form

$$f(\tau) = \frac{1}{2\pi i} \int_{c-i\infty}^{c+i\infty} ds\, e^{s\tau} e^{-bs^\alpha}, \quad \text{(C1)}$$

corresponding to

$$1 - \tilde{\psi}(s) \propto s^\alpha. \quad \text{(C2)}$$

The function $f(\tau)$ is essentially a Lévy distribution.[24] The final expression for $\tilde{F}(N - l_0, \tau)$ contains a sum of terms, each proportional to Eq. (C1) for various values of the constant b. By a simple change in the integration variable, one can group the parameters of the integral in Eq. (C1) to obtain

$$\tau f(\tau) = \frac{1}{2\pi i} \int_{\bar{c}-i\infty}^{\bar{c}+i\infty} dz \exp[z - (b/\tau^\alpha) z^\alpha]. \quad \text{(C3)}$$

We can now evaluate the integral in Eq. (C3) in two time regimes: (a) $b/\tau^\alpha \ll 1$, and (b) $b/\tau^\alpha \gg 1$. The constant b is typically a large number ($\sim N/\overline{l}$), so both regimes correspond to $\tau \gg 1$ [therefore justifying the use of Eq. (C1)]. The transition regime $b/\tau^\alpha \gtrsim 1$ corresponds to the transit time. In regime (a) one can develop an asymptotic series for $\tau f(\tau)$ by expanding $\exp(-bz^\alpha/t^\alpha)$ and integrating term by term,

$$\tau f(\tau) = \sum_{l=0}^{\infty} \frac{1}{l!} \left(-\frac{b}{\tau^\alpha}\right)^l \frac{1}{2\pi i} \int_C dz\, z^{l\alpha} e^z\,. \quad (C4)$$

The contour C in Eq. (C4) is shown in Fig. 16. The original contour in Eq. (C3) is displaced, to the left in the z plane, to one consisting of a broken circle around the branch point at the origin and line segments parallel to the imaginary axis (dotted line). The latter part of contour makes an exponentially small contribution and thus, the countour can be further changed to C. The integral in Eq. (C4) is simply the Hankel representation of the inverse Γ function,[25] thus

$$\tau f(\tau) = \sum_{l=0}^{\infty} \frac{(-b/\tau^\alpha)^l}{\Gamma(l+1)\Gamma(-l\alpha)}\,. \quad (C5)$$

Now, $1/\Gamma(-l\alpha) = 0$ for $l\alpha = 0, 1, 2, 3, \ldots$. For such values of $l\alpha$, there is no branch point, and hence the integral evaluated on C vanishes. The above derivation of Eq. (C5) is essentially a proof of the Tauberian theorem.

Now, using the familiar reflection formula for the γ function

$$\Gamma(z)\Gamma(1-z) = \pi \csc \pi z\,, \quad (C6)$$

one can rewrite Eq. (C5) as

$$\tau f(\tau) = -\frac{1}{\pi} \sum_{l=0}^{\infty} \left(\frac{-b}{\tau^\alpha}\right)^l \sin \pi l \alpha\, \frac{\Gamma(l\alpha+1)}{\Gamma(l+1)}\,. \quad (C7)$$

Using the well-known asymptotic properties of the gamma function, one can see that the function defined by the sum in Eq. (C7) is an entire function for $0 < \alpha < 1$. For some values of α, where $\Gamma(l\alpha+1)$ is a factor of $\Gamma(l+1)$, one can analytically sum the series:

$$\alpha = \tfrac{1}{2},\quad \tau f(\tau) = \frac{b}{2\pi^{1/2}\tau^{1/2}} e^{-b^2/4\tau}\,, \quad (C8)$$

$$\alpha = \tfrac{1}{3},\quad \tau f(\tau) = \frac{\sin(\pi/3)}{\pi}\, x K_{1/3}(x)\,,$$

$$x \equiv 2\left(\frac{b}{3\tau^{1/3}}\right)^{3/2}, \quad (C9)$$

where $K_\nu(x)$ is the modified Bessel function of order ν. More generally, for any rational value of α, $\tau f(\tau)$ can be represented as a finite sum of generalized hypergeometric functions ${}_iF_k(a_1, \ldots, a_i; c_1, \ldots, c_k; x)$, e.g.,

$$\alpha = \tfrac{3}{4},\quad \tau f(\tau) = -\left(\frac{8}{3\pi}\right)^{1/2} \sum_{n=1}^{3} \sin\left(\frac{3\pi n}{4}\right) z^n$$

$$\times {}_2F_2\left(\frac{1}{3}+\frac{n}{4}, \frac{2}{3}+\frac{n}{4}; \frac{1}{2}, \frac{n(n-1)}{8},\right.$$

$$\left.\frac{n(7-n)}{8};\, -z^4\right),\quad z \equiv -\frac{b}{4}\left(\frac{\tau}{3}\right)^{-3/4}. \quad (C10)$$

In regime (b) one can evaluate the integral in Eq. (C3) with the saddle-point method.[26] One seeks the stationary points of $z - (b/\tau^\alpha)z^\alpha$, i.e.,

$$z_0 = (ab/\tau^\alpha)^{1/(1-\alpha)}\,. \quad (C11)$$

Expanding the exponent in (C3) about z_0 one finds that the path of steepest descent through z_0 is parallel to the imaginary axis. Hence,[27]

$$\tau f(\tau) \simeq \frac{\exp\{-[(1-\alpha)/\alpha](\alpha b/\tau^\alpha)^{1/(1-\alpha)}\}}{[2\pi(1-\alpha)(\tau^\alpha/\alpha b)^{1/(1-\alpha)}]^{1/2}},\quad \frac{b}{\tau^\alpha} \gg 1\,. \quad (C12)$$

We can now compare the general asymptotic form to the analytic results in Eqs. (C8) and (C9). For $\alpha = \tfrac{1}{2}$ the expression in Eq. (C12) is exact, and is equal to Eq. (C8). For $\alpha = \tfrac{1}{3}$, we must first obtain the asymptotic limit of Eq. (C9). For $x \to \infty$,

$$x K_{1/3}(x) \to (\pi x/2)^{1/2} e^{-x}$$

and

$$\tau f(\tau) \simeq (3/4\pi)^{1/2} (b/3\tau^{1/3})^{3/4} \exp[-2(b/3\tau^{1/3})^{3/2}]\,. \quad (C13)$$

The expression agrees exactly with the one in Eq. (C12) for $\alpha = \tfrac{1}{3}$. Hence, we have been able to establish a representation of the function $\tau f(\tau)$ in both time regimes of pertinence to our transport study.

The function in Eq. (C1) is, as mentioned above, the Lévy distribution, and therefore of interest in other areas of stochastic processes. The representation of $f(\tau)$ we have established will be further developed in another paper.

APPENDIX D: CALCULATION OF $\psi(t)$

We outline the main steps in the computation of $\psi(t)$; the details can be found in Ref. 14. We consider an arbitrary site as the origin in a random medium and define $Q(t)$ to be equal to the probability that a carrier remains on the site for a time interval t after arrival. This probability can decrease in time via all the *parallel decay channels* for it to transfer to surrounding sites:

$$\frac{dQ}{dt} = -Q \sum_j W(\vec{r}_j)\,, \quad (D1)$$

where $W(\vec{r})$ is the transition rate to a site located at $\vec{r}$. One now solves (D1) for the fixed (random) configuration $\{r_j\}$, computes the configuration average,

$$\Psi(t) \equiv \langle Q(t) \rangle = \left\langle \exp\left[-t \sum W(\vec{r}_j)\right] \right\rangle$$

$$= \exp\left(-\int d^3r\, p(\vec{r})\{1 - \exp[-W(\vec{r})t]\}\right), \quad \text{(D2)}$$

and determines $\psi(t)$ with the relation

$$\psi(t) = -\frac{d\Psi(t)}{dt}. \quad \text{(D3)}$$

In Eq. (D2), $p(\vec{r})d^3r$ is the probability a site is located in a volume d^3r centered about $\vec{r}$. An excellent approximation to the integrand in Eq. (D2) in the limit of large τ (cf. Appendix A of Ref. 14) is a unit step function,

$$1 - \exp[-W(\vec{r})t] = \begin{cases} 1, & W(\vec{r})t > 1 \\ 0, & W(\vec{r})t < 1 \end{cases}. \quad \text{(D4)}$$

Due to the general exponential dependence of $W(\vec{r})$ on r, the transition from 1 to 0 in Eq. (D4) is very rapid as a function of $\vec{r}$. With

$$W(\vec{r}) = W_M e^{-r/R_d}, \quad \text{(D5)}$$

the assumption of a totally random distribution

$$p(r) = N_D, \quad \text{(D6)}$$

and the use of Eq. (D4), one obtains

$$\ln \Psi(t) = -4\pi N_D \int_0^{r_\tau} r^2 dr = -4\pi N_D r_\tau^3/3, \quad \text{(D7)}$$

where

$$W(r_\tau)t \equiv \tau e^{-r_\tau/R_d} = e^{-c} \simeq 1, \quad \text{(D8)}$$

or

$$r_\tau = R_d \ln e^{c\tau}. \quad \text{(D9)}$$

We now absorb the constant e^c into the definition of W_M and, inserting Eqs. (D7) and (D9) into (D3), one has

$$\frac{\psi(t)}{W_M} = \frac{-d\Psi}{d\tau} = \frac{\eta(\ln \tau)^2}{\tau^{1+(\eta/3)(\ln\tau)^2}}, \quad \tau \gg 1, \quad \text{(D10)}$$

which is the result we quoted for $\psi_h(t)$ in Eq. (60). The long tail in $\psi(t)$ is related to the absence of a truncation of the transition rate spectrum, i.e., at an arbitrary t one can "find," with a finite probability, a $W(\vec{r})$, such that $W(\vec{r})t \simeq 1$. The "smoothness" of the tail is related to the random assumption (D6). If instead of Eq. (D5) one used

$$W(\vec{r}) = W_M \exp[-(r/R_D)^\gamma], \quad \text{(D11)}$$

with $\gamma \geq 1$, then

$$r_\tau = R_d (\ln \tau)^{1/\gamma} \quad \text{(D12)}$$

and

$$\Psi(t) = \exp[-(\eta/3)(\ln \tau)^{3/\gamma}]. \quad \text{(D13)}$$

Hence for $\gamma > 1$, the time dependence of $\psi(t)$ would be slower than Eq. (D10) for a given value of η. *The more rapid the dependence of $W(r)$ on r, the slower the time decay of $\psi(t)$!* If one can change $W(r)$ significantly by a small variation in r, then for any arbitrary t one can locate a nearest-neighbor site separation that satisfies $W(r) \simeq 1$. There is direct experimental verification of these concepts in the study of radiative recombination in semiconductors.[14,28]

---

*Partially supported by ARPA and monitored by ONR (N00014-17-0308).

[1] A. R. Adams and W. E. Spear, J. Phys. Chem. Solids 25, 1113 (1964); D. J. Gibbons and W. E. Spear, J. Phys. Chem. Solids 27, 1917 (1966).

[2] W. E. Spear, Proc. Phys. Soc. Lond. B 70, 669 (1957).

[3] M. D. Tabak and P. J. Warter, Phys. Rev. 173, 899 (1968); D. M. Pai and S. W. Ing, Phys. Rev. 173, 7: (1968).

[4] P. G. LeComber and W. E. Spear, Phys. Rev. Lett. 25, 509 (1970).

[5] N. F. Mott, J. Non-Cryst. Solids 1, 1 (1968).

[6] A. Many and G. Rakavy, Phys. Rev. 126, 1980 (1962).

[7] D. M. Pai (private communication).

[8] M. E. Sharfe, Phys. Rev. B 2, 5025 (1970); D. M. Pai and M. E. Scharfe, J. Non-Cryst. Solids 8–10, 752 (1972); M. E. Scharfe, Bull. Am. Phys. Soc. 18, 454 (1973); M. E. Scharfe (private communication).

[9] G. Pfister, Phys. Rev. Lett. 33, 1474 (1974); and private communication.

[10] J. Mort and A. I. Lakatos, J. Non-Cryst. Solids 4, 117 (1970). The first mention of the universal shape of $I(t)$ is contained in this reference.

[11] W. D. Gill, J. Appl. Phys. 43, 5033 (1972).

[12] M. Silver, K. S. Dy, and D. L. Huang, J. Non-Cryst. Solids 8–10, 773 (1972).

[13] V. M. Kenkre, E. W. Montroll, and M. F. Shlesinger, J. Stat. Phys. 9, 45 (1973).

[14] H. Scher and M. Lax, Phys. Rev. B 7, 4491 (1973); 7, 4502 (1973).

[15] E. W. Montroll and H. Scher, J. Stat. Phys. 9, 101 (1973).

[16] M. Shlesinger, J. Stat. Phys. 10, 421 (1974).

[17] G. E. Roberts and H. Kaufman, *Tables of Laplace Transforms* (Saunders, Philadelphia, 1966).

[18] A preliminary account of this analysis has appeared in H. Scher, *Amorphous and Liquid Semiconductors*, edited by J. Stuke and W. Brenig (Taylor and Francis, London, 1974), p. 135.

[19] S. C. Maitra and H. Scher (unpublished).

[20] M. L. Knotek, M. Pollak, and T. M. Donovan, in Ref. 18, p. 225.

[21] H. Seki, in Ref. 18, p. 1015.

[22] G. Weiser, J. Appl. Phys. 43, 5028 (1972).

[23] W. Feller, *An Introduction to Probability Theory and its Applications*, 2nd ed. (Wiley, New York, 1971), Vol II.

[24] P. Levy, *Processus Stochastiques et Mouvement Brownien*, 2nd ed. (Gauthier-Villars, Paris, 1965).

[25] Formula No. 6.1.4 in *Handbook of Mathematical Func-*

tions, edited by M. Abramowitz and I. A. Stegun (U. S. GPO, Washington, D. C. 1964).

[26] E. T. Copson, *Asymptotic Expansions* (Cambridge U.P., Cambridge, England, 1965).

[27] We have discovered that similar functions have been discussed in E. W. Barnes, Phil. Trans. Roy. Soc. A **206**, 249 (1906).

[28] D. G. Thomas, J. J. Hopfield, and W. M. Augustyniak, Phys. Rev. **140**, A202 (1965); R. C. Enck and A. Honig, Phys. Rev. **177**, 1182 (1969).

TRAFFIC DYNAMICS: STUDIES IN CAR FOLLOWING

Robert E. Chandler, Robert Herman, and Elliott W. Montroll*

Research Staff, General Motors Corporation, Detroit, Michigan

(Received November 8, 1957)

> The manner in which vehicles follow each other on a highway (without passing) and the propagation disturbances down a line of vehicles has been investigated. Experimental data is presented which indicates that the acceleration at time t of a car which is attempting to follow a leader is proportional to the difference in velocity of the two cars at a time $(t-\Delta)$, Δ being about 1.5 sec and the proportionality constant being about 0.37 sec^{-1}. It is shown theoretically that the motion of a long line of vehicles becomes unstable when the product of the lag time and the proportionality constant exceeds one-half. The experimental data implies that driving is done on the verge of instability. A variety of other laws of following is analyzed theoretically.

THE VITAL DEPENDENCE of our daily activities on the efficient and safe flow of vehicular traffic has stimulated the accumulation of enormous amounts of relevant empirical data by traffic engineers.[1] These data and the parallel research in road construction have been the basis of the development of our modern highways. However, it is only recently that serious thought has been devoted to the analysis of the fundamental mechanisms which operate to control the movement of traffic.

Several interesting theoretical approaches to the characterization of these mechanisms have been proposed. A review of these, as well as an extensive bibliography, has been given by GERLOUGH AND MATHEWSON.[2]

PIPES[3] has studied the dynamics of a linear array of vehicles whose

---

* The last-named author is consultant to the Research Staff, General Motors Corporation. His permanent address is The Institute of Fluid Dynamics and Applied Mathematics, University of Maryland, College Park, Maryland.

motion is characterized by rules given in the California Motor Vehicle Code Summary, namely, "a good rule for following another vehicle at a safe distance is to allow yourself the length of a car (about 15 feet) for every ten miles an hour you are traveling." He showed how lines of cars stop and start and perform other following operations on the assumption that responses are immediate and that no inertial effects exist in the vehicles or response lags in the operators. He also discusses several other mechanisms of following. Similar analyses have also been made by REUSCHEL.[4]

LIGHTHILL AND WHITHAM[5] and RICHARDS[6] have postulated the density of traffic on a long highway to be a continuous function of position along the highway and of time. The traffic is then treated as a fluid flowing along the highway. The mathematical methods of fluid dynamics have been applied to a discussion of various highway phenomena, such as the development of shock waves when sudden stops and starts are made. PRAGER[7] has made a two-dimensional continuum model of the flow of traffic in large areas, such as cities.

NEWELL[8] has stressed the analogy between the motion of vehicles on a sparsely populated highway and the behavior of molecules in rarified gases. The motion of both is a 'free flow' except during occasional encounters with other elements. When a fast car overtakes a slow one, the encounter usually results in a loss of time, namely, that required for the passing operation, or an equivalent reduction in the mean velocity of the fast car. Occasionally the opposite effect occurs when a driver on a low density highway speeds up in preparation for and during the passing operation.

Considerable interest exists in the simulation of traffic with high speed computers. For example, GERLOUGH AND MATHEWSON[2] and GOODE[9] have been simulating the behavior of vehicles at road intersections.

Although the fluid flow approach mentioned above shows considerable promise of providing a framework for a general theory of traffic, we feel that it is worthwhile to investigate the possible application of another highly developed branch of modern applied mathematics, namely, the theory of servomechanisms and network analysis. In its most general form this theory is merely that of the analysis of the propagation of assorted signals through 'black boxes' arranged in various topological configurations. In traffic analysis we might consider individual vehicles or certain sets of vehicles as the signals and the highway as the network.

An important 'black box' in a traffic network is an intersection with or without a traffic light. The four outputs, the traffic leaving the intersection in four directions at time t, are related to the four inputs, vehicles approaching the intersection during some time interval $t-\tau$. The dependence of the outputs on the inputs characterizes the intersection 'black box.'

The manner in which a given length of intersectionless highway fits into the black box pattern can be seen by considering a two-lane highway. Suppose two types of vehicles are using the highway—low speed trucks and high speed passenger cars. First consider the case of traffic flowing in opposite directions in the two lanes with the occasional passing of low by high speed vehicles. At low traffic densities only a small amount of time is lost in passing so that the output of fast vehicles in one lane is simply related to the input of both fast and slow vehicles of the same lane at some previous time interval. As the traffic density increases an interaction develops between the flow in the two lanes—opportunities for passing become rarer and the output of a given lane is related to the input of both lanes (and perhaps also to the output of the other lane since a jam in the second lane prevents passing in the first). Finally, as the density becomes very high no passing can occur. In the case of both lanes of traffic proceeding in the same direction, the output of fast vehicles from a given length of highway depends on the input of all types. The resistance to flow of fast vehicles depends increasingly on the number of slow ones as the over-all density increases, since a passenger car trapped behind a truck in the slow lane has difficulty in escaping when other passenger cars are whizzing by in the fast lane. The detailed relations between inputs and outputs in a stretch of highway gives the characteristics of a schematic black box that might be used in a network analysis.

Once the characteristics of the elements of the traffic network are understood, we can expect to be able to employ some analogies between traffic and communication theory, since one of the main problems of a communications engineer is to pass as much information on a given circuit per unit time as possible while the traffic engineer attempts to pass as many vehicles as possible. As in communication theory, various sources of noise exist in traffic theory, e.g., pedestrians.

Instabilities of two types exist in traffic—traffic jams and accidents. Of the two kinds of accidents the spontaneous (caused by such driver failure as falling asleep or committing errors in judgment, and such mechanical failure as blowouts) and the inherent (which results from the accumulation of small effects over which nobody has complete control and leads to systems instability), only the second is amenable to some theoretical analysis (the first being statistical in nature).

A driver programs his driving operations in various ways. In the absence of other interfering vehicles, he attempts to keep his speed fairly constant at a set point determined by a compromise between the urge to minimize trip duration and maximize safety. When following other vehicles whose speed is of the order of his set point speed the driver introduces a new set point, the inter-car spacing whose value depends on his speed.

The servomechanism approach is especially useful in clarifying the role and interaction of the three components of the traffic system—the road topology (number of lanes, nature of intersection, signals, warning signs, etc.), the vehicle characteristics (speed, acceleration and deceleration qualities, signaling mechanisms, vision, etc.) and the operator's behavior (range of perception, lags between perception and response, etc.). This approach gives one the opportunity of making the study of traffic an experimental as well as an observational science.

One can set up artificial traffic situations to correspond to various elements or 'black boxes' in the traffic network and by controlling the nature of the inputs the dependence of outputs on inputs might be established with greater dispatch than is possible by a detailed analysis of traffic on real highways. We are optimistic enough to believe that the dynamics of real traffic can be synthesized from results of experiment and theory. One of the results of this type of investigation is that quantitative information might be obtained on the effect of the introduction of new signaling devices on cars and roads and of the behavior of abnormal drivers (tired, drunk, etc.) on the elements of the traffic network. Finally if the vehicle of the future is to be automatic as well as automobile, its design can only follow an understanding of the traffic system as a servomechanism.

This paper is our first discussion of a traffic element treated as a servomechanism. We consider the theory of the manner in which one car follows another, and are especially interested in determining the conditions required for stable following. We shall propose various models of the game of 'Follow that car!' and compare such models with experimental data on how cars are actually followed. There is some merit in studying models which do not correspond to general practice since some of these may be more stable (and safer) and might be put into use by installing appropriate signaling devices on cars. The theory discussed here is not limited to automobile traffic but might be applied to other 'follow the leader situations.' We hope in future publications to discuss a variety of traffic network elements and to make remarks about complete traffic systems.

THEORY OF FOLLOW THE LEADER

ACCIDENTS caused by improper following can occur in two ways. If a driver follows the car in front so closely that he cannot avoid an accident caused by a sudden perturbation, he has merely been using bad judgment and no mathematical analysis is required. However, accidents frequently occur in collisions which involve cars considerably behind the car that initiated some fluctuation. It is such accidents that may result in the multiple car pile-ups which are sometimes observed on congested superhighways, especially at high speeds. It is this latter case that results from an amplification of the original perturbation as it is transmitted down the line of traffic.

Let us consider a line of identical vehicles that are attempting to follow each other in a steady or stable manner. We assume that if such a state could be achieved, the separation distance between vehicles plus the car length would have a constant value* a and each vehicle would have the same velocity v. The spacing a would in general depend on v. We let $u_n(t)$, the deviation of the velocity of the nth vehicle from the velocity v, be given by

$$u_n(t) = dx_n/dt - v, \qquad (1)$$

where x measures distance and $y_n(t)$ the spacing of vehicles given by

$$y_n(t) = x_{n-1}(t) - x_n(t). \qquad (2)$$

As the operator of the nth vehicle observes variations in $u_n(t)$ or $y_n(t)$, he applies either his accelerator or brakes to keep from lagging or closing in on his leader. Two factors prevent this operator from immediately reproducing the leader motions. His delayed response and that of the mechanisms which transmit brake and acceleration signals to the vehicle contribute a lag in the follow-the-leader process as does the inertia of the vehicle itself.

The accelerating force (other than the force required to maintain the steady motion) applied to the nth vehicle at time t can be expected to depend on its instantaneous velocity deviation $u_n(t)$ and on some functional of the difference in velocities of the $(n-1)$st and nth vehicles $u_{n-1}(\tau) - u_n(\tau)$ (for some range of τ with $\tau \leq t$) as well as on a functional of the spacing $y_n(\tau)$. The equations of motion of the individual vehicles assumed to have the same mass, M, are then given by

$$M\, du_n(t)/dt = F\{u_n(t);\; f_1[u_{n-1}(\tau) - u_n(\tau)];\; f_2[y_n(\tau)]\}, \qquad (3)$$

where $u_0(t)$ refers to the velocity pattern of the lead vehicle.

In the past, vehicle operating data has not been analyzed with a view to determining the precise form of the functional F. We shall discuss the results of preliminary experiments carried out for this purpose later in this paper. The purpose of the present section is to investigate the stability characteristics of various choices of the functional F. Even though some of these forms may not be generally prevalent in automobile operation today, some knowledge of their consequences may be of interest in that they indicate dangerous types of behavior and might suggest new

---

* When the mean separation distance is very large each driver tends to behave independently and the theory developed is no longer applicable. We are concerned primarily with the high traffic density situation in which no passing is allowed. We hope to develop a phenomenological theory of passing at a later time.

forms of signaling devices for the improvement of responses. The development of the automatic automobile of the future will require an understanding of the follow-the-leader process. The mathematical models given below are linear. As will be pointed out later linear equations appear to give surprisingly good agreement with an experiment that corresponds to the high density follow-the-leader case. The introduction of a nonlinear functional causes no fundamental difficulty in solving the equations of motion. This is so because the equation of motion for a particular vehicle depends only on the behavior of its predecessor so that the equations can be solved successively. Complications would arise if the influence of vehicles other than nearest neighbors were included.

Proportionate Control

As a first example we postulate that the applied force is proportional to the instantaneous difference in the velocity of a given vehicle and its predecessor, or the case of 'proportional control' in the language of servomechanism theory. The equations of motion of a line of N identical vehicles each of mass M is

$$M \, du_n/dt = \lambda \, (u_{n-1} - u_n), \qquad (n = 1, 2, \cdots, N) \quad (4)$$

where λ is the sensitivity of the control mechanism. At instants in which a lead car is going faster than the following car, the follower applies an accelerating force and vice versa. We assume in equation (4) and throughout this paper that the sensitivities for acceleration and deceleration are identical. Although this is a reasonable approximation in a properly functioning car at low speed, it is certainly not the case at high speed or when for example either the brakes are poor or an engine is not well tuned. The solution of these equations depends on the velocity pattern, $u_0(t)$, of the lead vehicle. The stability of a line of traffic depends on whether a local fluctuation in velocity is damped out or amplified as it propagates down the line of cars. There are two types of instability, local and asymptotic instability. We are concerned with the latter. It should be noted that even asymptotic stability conditions depend on the equilibrium spacing and velocity. If the equilibrium spacing is small, then one does not have to go back far in the line of vehicles behind the initial perturbation to find the occurrence of a collision. Although from our solutions of the equations of motion one can determine where down the line an accident occurs we are primarily interested in the criteria for the growth or decay of a disturbance.

Since the system now under consideration is linear, this stability question can be investigated in terms of the Fourier components of the driving function $u_0(t)$. Let us assume that the driving function is monochromatic with the frequency ω so that

$$u_0(t) = e^{i\omega t}. \tag{5}$$

Of course an arbitrary driving function can be expressed as a linear combination of monochromatic components by the usual Fourier analysis. By substituting

$$u_n(t) = f_n\, e^{i\omega t}, \qquad f_0 = 1, \tag{6}$$

into equation (4) we find

$$(i\omega M/\lambda)\, f_n = f_{n-1} - f_n, \tag{7}$$

so that

$$f_n = (1 + i\omega M/\lambda)^{-n}\, f_0, \tag{8}$$

and

$$u_n(t) = (1 + \omega^2 M^2/\lambda^2)^{-n/2} \exp\{i\,[\omega t - n\cos^{-1}(1+\omega^2 M^2/\lambda^2)^{-1/2}]\}. \tag{9}$$

The amplitude of the velocity deviation decreases with increasing n for all frequencies, masses, and sensitivities. Hence instantaneous proportional control is stable under all circumstances. The phase velocity of a signal of frequency ω, in terms of car spacings per second is

$$\frac{dn}{dt} = \omega \cos^{-1}(1+\omega^2 M^2/\lambda^2)^{-1/2} \cong \begin{cases} \omega^2 M/\lambda & \text{for } \omega \to 0 \\ \omega\pi/2 & \text{for } \omega \to \infty. \end{cases} \tag{10}$$

The spacing between the $(n+1)$st and nth vehicle is

$$\begin{aligned} y_n(t) &= y_n(t_0) + \int_{t_0}^{t} [u_{n-1}(t) - u_n(t)]\, dt, \\ &= y_n(t_0) + (M/\lambda)(1+i\omega M/\lambda)^{-n}\, (e^{i\omega t} - e^{i\omega t_0}). \end{aligned} \tag{11}$$

Even though the decay of y_n with n implies asymptotic stability the amplitude of say y_1 might be sufficiently large to cause local instability. Suppose one chooses t_0 to be a time at which the spacing $y_n(t_0)$ has the normal value a. Then the greater the sensitivity λ, the more stable the spacing for all t and n. In principle one would like to make λ as large as possible. However, we shall see below that time lags in control systems limit the sensitivity λ for stable driving. Qualitatively the limitation results from the fact that if both the lag and λ are large, then large corrective measures are taken for observed variations whose effects might die out more quickly than the time required for the responses to make themselves felt.

Response Lag

Equation (4) can be generalized to include the lag in the response of the operator through the introduction of a weight function $\sigma(t)$. Then the following relation

$$M \frac{du_n}{dt} = \int_0^\infty [u_{n-1}(t-\tau) - u_n(t-\tau)]\, d\sigma(\tau), \qquad (12)$$

indicates that the total force applied at a given time t depends on a weighted average of all earlier differences in u_{n-1} and u_n. The choice

$$\sigma(\tau) = \lambda\, H(\tau - \Delta) = \begin{cases} 0, & (\tau < \Delta) \\ \lambda, & (\tau > \Delta) \end{cases} \qquad (13)$$

where H is the Heaviside step function, or

$$\sigma'(\tau) = \lambda\, \delta(\tau - \Delta), \qquad (13a)$$

$\delta(x)$ being the Dirac delta function, corresponds to a time lag Δ between the observation of a velocity difference and the application of a correcting force. Equation (12) then becomes

$$M\, du_n(t)/dt = \lambda\, [u_{n-1}(t-\Delta) - u_n(t-\Delta)]. \qquad (14)$$

As before we let $\qquad u_0(t) = e^{i\omega t},$

and substitute equation (6) into equation (14). We then find

$$(i\omega M/\lambda)\, e^{i\Delta\omega} f_n = f_{n-1} - f_n, \qquad (15)$$

or

$$2\, i\mu\omega\, e^{i\Delta\omega} f_n = f_{n-1} - f_n, \qquad (15a)$$

where $\qquad \mu = M/(2\,\lambda),$

so that $\qquad f_n = (1 + 2\, i\mu\omega\, e^{i\Delta\omega})^{-n} f_0, \qquad (16)$

and

$$u_n(t) = (1 + 4\,\mu^2\omega^2 - 4\,\mu\omega\, \sin\Delta\omega)^{-n/2}$$
$$\times \exp\{i\,[\omega t - n\, \cos^{-1}(1 + 4\,\mu^2\omega^2 - 4\,\mu\omega\, \sin\Delta\omega)^{-1/2}]\}. \qquad (17)$$

The amplitude factor decreases with increasing n if

$$1 + 4\,\mu^2\omega^2 - 4\,\mu\omega\, \sin\Delta\omega > 1,$$

i.e., if $\qquad 4\,\mu^2\omega > 4\,\mu\, \sin\omega\Delta. \qquad (18)$

Low frequencies give the greatest limitations on sensitivities. As $\omega \to 0$, λ must satisfy the inequality

$$\lambda < M/(2\,\Delta), \qquad (19)$$

or $\qquad \Delta < \mu. \qquad (19a)$

Hence, for a given lag Δ, a stable operation results as long as the inequality is satisfied.

As in the previous case the spacing between the $(n-1)$st and nth vehicle is given by

$$y_n(t) = y_n(t_0) + 2\ \mu[e^{i\omega(t+\Delta)} - e^{i\omega(t_0+\Delta)}]/(1 + 2\ i\mu\omega\ e^{i\Delta\omega})^n. \qquad (20)$$

A more realistic response function $\sigma(t)$ is one with a deal period lag followed by a continuous response

$$\sigma(\tau) = \begin{cases} \lambda(1 - e^{-(\tau-\Delta)/\delta}), & (\tau > \Delta) \\ 0 & (\tau < \Delta) \end{cases} \qquad (21)$$

Then our fundamental equation becomes

$$M\frac{du_n(t)}{dt} = \frac{\lambda}{\delta} \int_\Delta^\infty [u_{n-1}(t-\tau) - u_n(t-\tau)]\ e^{-(\tau-\Delta)/\delta}\ d\tau. \qquad (22)$$

Differentiation of equation (22) with respect to t yields

$$M\frac{d^2 u_n(t)}{dt^2} = -\frac{\lambda}{\delta} \int_\Delta^\infty e^{-(\tau-\Delta)/\delta}\ \frac{d}{d\tau} [u_{n-1}(t-\tau) - u_n(t-\tau)]\ d\tau,$$

so that after integration by parts we find

$$M\frac{d^2 u_n(t)}{dt^2} + \frac{M}{\delta} \frac{du_n(t)}{dt} = \frac{\lambda}{\delta} [u_{n-1}(t-\Delta) - u_n(t-\Delta)]. \qquad (23)$$

We again set $u_n(t) = f_n\ e^{i\omega t}$ and find

$$u_n(t) = \left\{ \frac{\lambda\ e^{-i\omega\Delta}}{\lambda\ e^{-i\omega\Delta} + i\omega M - \delta M \omega^2} \right\}^n e^{i\omega t}. \qquad (24)$$

It is easy to show by the methods discussed above that fluctuations in our line of traffic will be damped out rather than amplified if

$$M\omega\ (1+\delta^2\omega^2) > 2\ \lambda\ [\sin\omega\Delta + \omega\delta\ \cos\omega\delta].$$

As before, the most restrictive condition on time lags and relaxation times exists at low frequencies. Stability exists at all frequencies if

$$\lambda < M/[2(\delta+\Delta)]. \qquad (25)$$

Notice that the time lag Δ and the relaxation time δ are additive in determining stability conditions.

Constant Spacing

A mode of driving that is unstable even without control-response lags is that in which an operator attempts to keep the distance between vehicles constant and applies a force proportional to the deviation of this distance from the required spacing when fluctuations occur. We introduce a moving coordinate system which progresses with the mean velocity of the lead car and has its origin at the position the lead car would have if it always moved with this velocity. Then, if a is the required spacing, we let $x_n(t)$ be the deviation of the position of the nth vehicle from the point $-an$ in the moving coordinate system.

The equations of motion of a line of vehicles that employs this mode of control are

$$M\, d^2 x_n/dt^2 = K\,(x_{n-1} - x_n). \tag{26}$$

Again suppose
$$x_0(t) = e^{i\omega t}$$
and
$$x_n(t) = f_n e^{i\omega t};$$
then
$$-M\omega^2 f_n = K\,(f_{n-1} - f_n),$$
so that
$$f_n = (1 - MK^{-1}\omega^2)^{-n} f_0 \tag{27}$$
or
$$x_n(t) = (1 - MK^{-1}\omega^2)^{-n} e^{i\omega t}. \tag{27a}$$

Note that for any value of ω a resonance condition exists when

$$\omega = (K/M)^{1/2}, \tag{27b}$$

so that fluctuations in separation distance would be amplified. This situation is of importance when a group of cars follows one another at very small velocity independent distances such as occurs frequently on our super highways during rush hours. Then a fluctuation in position of one car amplifies down the line of cars and can cause an accident if the line of cars is sufficiently long.

California Code

A control scheme whose stability is rather insensitive to lags can be devised by following a rule suggested in the California Vehicle Code Summary:[3] "A good rule for following another vehicle at a safe distance is to allow yourself the length of a car (about fifteen feet) for every ten miles per hour you are traveling." This rule implies that

$$x_{n-1} = x_n + b + T\,v_n + L_{n-1},$$

where b is the standard distance between vehicles at rest, L_n is the length of the nth vehicle, and T is the time constant inferred by the California Code $[T \cong 15\text{ ft}/(14.67\text{ ft/sec}) \cong 1\text{ sec}]$. We assume L_n to be a constant, $c-b$, for all vehicles and write

$$x_{n-1} = x_n + c + T\,v_n. \tag{28}$$

Fluctuations in lead car performance would, as a result of various response lags, cause equation (28) to be violated in spite of the best intentions of followers. If

$$\delta_n(t) = x_{n-1}(t) - x_n(t) - c - T\,v_n(t) > 0, \tag{29}$$

the nth driver would accelerate in order to recover the equality in equation (28) and vice versa. Let us suppose that at any time t a force proportional to $\delta_n(t-\Delta)$ is applied to the nth car. Then the equations of motion of our line of vehicles are

$$M\, d^2x_n/dt^2 = K\,[x_{n-1}(t-\Delta) - x_n(t-\Delta) - c - T\, dx_n(t-\Delta)/dt]. \quad (30)$$

The constant c can be eliminated by letting

$$x_n = x_n' - ctT^{-1}.$$

Then x_n' satisfies the equation

$$M\, d^2x_n'(t)/dt^2 = K\,[x'_{n-1}(t-\Delta) - x_n'(t-\Delta) - T\, dx_n'(t-\Delta)/dt]. \quad (31)$$

As usual, we investigate stability by letting $x_n'(t) = f_n e^{i\omega t}$. Then

$$f_n = (1 + i\omega T - MK^{-1}\omega^2\, e^{i\omega\Delta})^{-n},$$

and our stability criterion is

$$T^2 + (MK^{-1}\omega)^2 > 2\, MK^{-1}\,[\cos\omega\Delta + \omega T\sin\omega\Delta]. \quad (32)$$

The low frequency condition, $\omega \to 0$,

$$T^2 > 2\, M/K \quad (33)$$

is sufficient to insure stability at all frequencies independently of the lag, Δ. Note that if this condition is not satisfied, resonances might occur. Suppose the lag Δ is very small then equation (32) becomes

$$T^2 + (MK^{-1}\omega)^2 > 2\, MK^{-1}\,[1 - \tfrac{1}{2}\omega^2\Delta^2 + \omega^2 T\Delta]$$

or
$$\omega^2\,[M^2K^{-2} + MK^{-1}\Delta^2 - T\Delta] > 2\, MK^{-1} - T^2. \quad (34)$$

Hence, if equation (33) is not satisfied, resonances occur at frequencies w for which the absolute value of the denominator in equation (31a) vanishes:

$$1 + \omega^2 T^2 + M^2 K^{-2}\omega^4 = 2\, MK^{-1}\omega^2\,(\cos\omega\Delta + T\omega\sin\omega\Delta), \quad (35)$$

which reduces to equation (27b) when $T = \Delta = 0$.

The inequality $T^2 > 2\, MK^{-1}$ is to be interpreted as meaning that stability exists for any sensitivity K provided that the time constant T is made sufficiently large. Remember that a large value of T implies a conservative or greater spacing between cars.

Propagation of a Perturbation

One can follow the details of the propagation of a perturbation down a line of cars through the use of the Laplace transform. As an example let us suppose that the dynamical equations are those given in equation (12) and that no disturbance in velocity occurs for $t<0$. Then

$$M \frac{du_n}{dt} = \int_0^t [u_{n-1}(t-\tau) - u_n(t-\tau)] \, d\sigma(\tau), \qquad (n=1, 2, \cdots) \quad (12)$$

Furthermore we assume that the velocity variation of the first car from the average velocity $\bar{v}$ is given by

$$u_0(t) = f(t), \tag{36}$$

where $f(t) = 0$ if $t<0$. We define the Laplace transforms of $u_n(t)$, $\sigma(t)$, and $f(t)$, respectively, to be

$$U_n(s) = \int_0^\infty u_n(t) \, e^{-st} \, dt, \tag{37}$$

$S_1(s)$ and $F(s)$. Then it can be shown that

$$U_n(s) = \left[\frac{S_1(s)}{sM + S_1(s)}\right]^n f(s). \tag{38}$$

The standard Laplace transform inversion formula yields

$$u_n(t) = \frac{1}{2\pi i} \int_{c-i\infty}^{c+i\infty} f(s) \left[\frac{S_1(s)}{sM + S_1(s)}\right]^n e^{st} \, ds, \tag{39}$$

while the spacing between the $(n-1)$st and nth car is given by

$$y_n(t) = a + \frac{n}{2\pi i} \int_{c-i\infty}^{c+i\infty} f(s) \left[\frac{S_1(s)}{sM + S_1(s)}\right]^n [S_1(s)]^{-1} (e^{-st} - 1) \, ds, \tag{40}$$

where a is the normal spacing.

As an example of the application of equation (39) we consider the propagation of disturbances in a system with a dead period lag Δ. Then using equation (13a) and

$$S_1(s) = \lambda \, e^{s\Delta},$$

we have

$$u_n(t) = \frac{1}{2\pi i} \int_{c-i\infty}^{c+i\infty} f(s) \left[\frac{1}{2 \, \mu s \, e^{-s\Delta} + 1}\right]^n e^{st} \, ds. \tag{41}$$

Let us assume that no singularities exist in the integrand in the right half plane Re $s \geq 0$. Then we can set $c=0$ and $s=i\omega$ to obtain

$$u_n(t) = \frac{1}{2\pi} \int_{-\infty}^{\infty} f(i\omega) \left[1 + 2 \, i\mu\omega e^{-i\omega\Delta}\right]^{-n} e^{i\omega t} \, d\omega. \tag{42}$$

One can show that if the stability condition $\lambda < M/2\Delta$ is satisfied the quantity in square brackets in equation (42) achieves its maximum absolute value when $\omega=0$. Hence when n is large we can expect values of ω near 0 to give the main contribution to $u_n(t)$. In this region

$$[1+2\,i\mu\omega\,e^{-i\omega\Delta}]^{-1}=\exp\{-2\,i\mu\omega-2\,\mu\,(\mu-\Delta)\,\omega^2+O(\omega^3)\}, \quad (43)$$

so that

$$u_n(t)=\frac{1}{2\pi}\int_{-\infty}^{\infty}[f(0)+i\omega\,f'(0)+\cdots]\exp[i\omega(t-2\,\mu n)] \quad (44)$$

$$\times\exp[-2\,\mu\,(\mu-\Delta)\,n\omega^2+O(n\omega^3)]\,d\omega.$$

If we let
$$z=[2\,\mu n\,(\mu-\Delta)]^{1/2}\,\omega, \quad (45)$$

then, as $n\to\infty$,

$$u_n(t)\cong f(0)\,[8\,\pi^2\mu n\,(\mu-\Delta)]^{-1/2}\int_{-\infty}^{\infty}e^{-z^2}\exp\left\{i\,\frac{t-2\,\mu n}{[2\,\mu n\,(\mu-\Delta)]^{1/2}}\right\}y\,dy, \quad (46)$$

and finally

$$u_n(t)\cong f(0)\,[8\,\pi\mu n\,(\mu-\Delta)]^{-1/2}\exp\left\{-\frac{(t-2\,\mu n)^2}{8\,\mu n\,(\mu-\Delta)}\right\}. \quad (47)$$

This shows that under stable conditions the low frequency component of a disturbance is transmitted over the greatest distance. The velocity of propagation, in number of car separations per unit time, is

$$n/t=\lambda/M=1/(2\,\mu). \quad (48)$$

As a result of experiments that will be described later the quantity λ/M is of the order of 0.4 sec^{-1} for a typical modern vehicle used in the experiment. The width of a time pulse is of the order of

$$[2\,\mu n\,(\mu-\Delta)]^{1/2} \quad (49)$$

Notice that as the lag Δ increases (i.e., as $\Delta\to\mu$) the amplitude of $u_n(t)$ grows until instability is reached when $\Delta=\mu$. When $\Delta>\mu$ in the unstable range, equation (47) is no longer valid because the denominator of the integrand of equation (46) has a pole to the right of the imaginary axis. When one wishes to follow the details of the development of an instability in a line of cars separate integrations of equation (41) must be made for each value of n.

Velocity-Dependent Sensitivity

It should be pointed out that it would be surprising if the sensitivity λ were velocity independent. Suppose as a rough correction we assume that

$$\lambda=\lambda_0\,(1+\alpha v). \quad (50)$$

Then in a range of small velocity variations about the average $\bar{v}$ the stability condition in equation (19) becomes

$$\lambda_0 (1+\alpha\bar{v}) < M/(2\,\Delta). \tag{51}$$

Hence if the velocity coefficient α were positive λ_0 would have to be reduced with increasing velocity to preserve stability.

Emergency Control

When two cars become closer than some critical distance, X (whose value might depend on the velocity of the second car), an emergency decelerating force is applied by the second car to prevent a collision. The operator would slam on his breaks to give the maximum deceleration mechanically feasible. The law of following might then be approximated by

$$\ddot{x}_n(t) = \alpha\,[\dot{x}_{n-1}(t-\Delta) - \dot{x}_n(t-\Delta)][1 - H(z_n)] - \beta\,H(z_n)$$

where H is the Heaviside step function defined in equation (13) and

$$z_n \equiv X - x_{n-1}(t-\Delta') + x_n(t-\Delta').$$

The differential equations become nonlinear and although they can be solved analytically, their solution is clumsy. We have therefore programmed them for machine solution. The new parameters β, X, and Δ' must be determined experimentally.

EXPERIMENTS AND THEIR INTERPRETATION

IN ORDER to obtain statistical estimates of certain functions and parameters for a preliminary evaluation of the mathematical models previously mentioned, it was necessary to design and conduct an experimental study to collect quantitative information regarding driver-car performance in a two-lane highway in which one car cannot pass another owing to the high traffic density in the opposite direction.

We now give a brief discussion of the experimental apparatus employed in the experiment and consider the process of one car following another without passing. Let $x_l(t)$ and $x_f(t)$ be the positions of the lead and following car at a time t so that the spacing between the cars is $x_l - x_f$. Also let the velocities of the respective cars be represented by v_l and v_f so that the relative velocity of the two cars is $v_l - v_f$.

To measure the spacing and the relative velocity of the two cars, a car follower, which is shown in Fig. 1, was designed and installed in a test car. The car follower consists essentially of a reel and a power unit mounted on a small platform which was fastened on the front bumper of the test car. Several hundred feet of fine wire were wound on the reel, and the end of the wire was fastened on the rear bumper of a lead car. A constant wire tension was maintained by means of a slipping friction clutch.

Inasmuch as the power unit kept the wire very taut at all times, $x_l - x_f$ was measured by the position of the reel at any particular instant, which depends on the amount of wire stretched between the two cars. This measurement was made by using a multiple turn potentiometer geared to

Fig. 1. Photograph of car follower showing wire reel and power unit.

a reel shaft. A direct current generator tachometer operating off the same shaft gave a measure of the rate at which the wire was wound or unwound, which is proportional to $v_l - v_f$. A fifth wheel attached to the test car measured v_f, while an accelerometer mounted in the car indicated the car's longitudinal acceleration which is designated by a_f.

The totality of this information, i.e., $x_l - x_f$, $v_l - v_f$, v_f, and a_f, was recorded simultaneously by an oscillograph installed in the back seat of the test car.

Eight male drivers participated in the study. These people, all employees of the Research Staff of the General Motors Technical Center, ranged in age from 24 to 38 years. Prior to testing each subject drove the test car, a 1957 Oldsmobile, until he indicated that he was sufficiently familiar with the car's response, controls, etc., to operate the car safely in congested traffic. Each driver then operated the car behind a lead car in an actual experimental run on the test track at the General Motors Technical Center. Testing time was approximately 20 to 30 minutes per driver.

The directions given to the drivers were simply, "Follow the lead car at what you consider to be a minimum safe distance at all times."

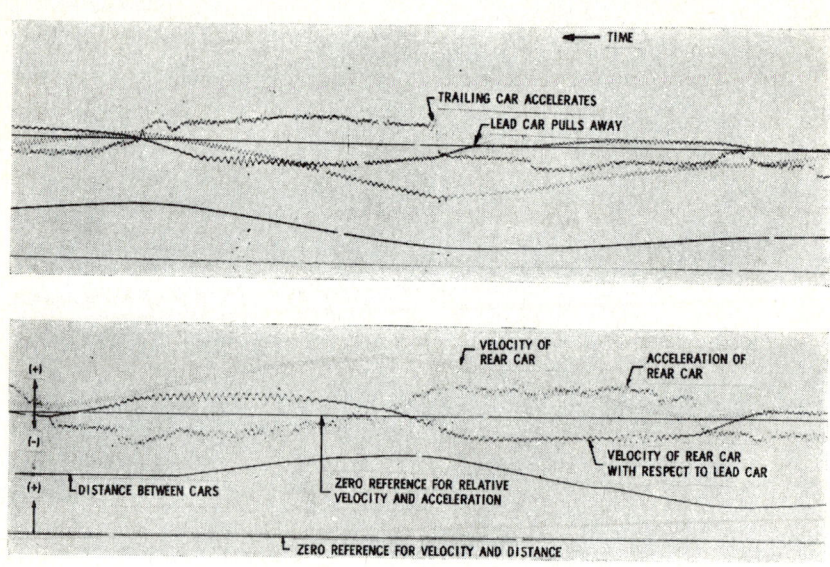

Fig. 2. The oscillograph recording shown below identifies the various curves recorded in the car-following experiments. The top strip is a typical recording from such an experiment.

These directions were employed in an attempt to produce a driving situation that would evoke driver behavior similar to that which might be observed as a person drives in dense traffic. The driver of the lead car, in all cases, pursued no prescribed program or driving pattern, but randomly varied his speed within the range of 10 to 80 mph and included several braking actions.

The information recorded on the oscillograph was of the type shown in Fig. 2. The records were inspected to identify a continuous section in each record where the test conditions were more or less dynamic. In other words, sections of the records in which spacing, $x_l - x_f$, and speed, v_f, are constant, are trivial and of no interest in the present study.

The aim of our data analysis was to obtain a relation between the acceleration, a_f, the relative velocity, v_l-v_f, and the spacing, x_l-x_f, of the form

$$a_f = f_1(v_l-v_f) + f_2(x_l-x_f). \qquad (52)$$

(Note that we recorded v_f-v_l on the tracings shown in Fig. 2 for ease of measurement.) The analysis was made by reading points equally spaced in time from the relevant parts of the three curves on the experimental records. The functions f_1 and f_2 were first assumed to be linear and by the method of least squares a multiple correlation coefficient was derived from the record of each driver.

It was discovered that the space dependent function $f_2(x_l-x_f)$ did not contribute significantly to the correlation. Consequently, this function was dropped from equation (52). Since the choice of a linear form of $f_1(v_l-v_f)$ with the omission of f_2 led to relatively high correlation coefficients in the neighborhood of 0.80–0.90, and in view of the preliminary character of our experiment, it was deemed unnecessary to examine non-linear forms for the f's.

An appreciation of the physical factors involved in the experimentation dictates that the best linear correlation would be achieved through the introduction of a time lag Δ. Hence our statistical problem was to determine the values of the constants b and Δ, which yield the best least squares fit to the equation

$$a_f(t) = b\,[v_l(t-\Delta) - v_f(t-\Delta)]. \qquad (53)$$

Correlation procedures for this type of analysis have been recently reviewed by MERRILL AND BENNETT.[10] The relation in equation (53) is exactly that given in equation (14) and the constant b is identified as λ/M.

The lag constant Δ is the sum of three more elementary lags. We note that the (v_l-v_f) can be regarded as stimuli input to the driver, i.e., the information which tells him to effect a change in his car's acceleration. After an acceleration change is made by the driver of the lead car, the response of the trailing vehicle depends upon its driver's perception time, t_1, his response time, t_2, and the time of the response of the vehicle, t_3. Inasmuch as each driver-car combination has its own parameters, Δ and λ/M, we readily discern the necessity for limiting our present discussion to the particular eight drivers and the test car used in this experiment.

Since we do not know the individual t_i's, we can let Δ take on various values. Then by plotting Δ versus the correlation coefficient, r, we can identify an optimum for each driver. The constants b and Δ for a given driver are those associated with the maximum of his r versus Δ curve and are given in Table I.

The fact that the mean value $(2\lambda\Delta/M)_{\text{Av}} = 1.12$ is so close to unity shows that the model of follow-the-leader given by equation (14) is a fairly accurate description of the dynamics of a line of cars. Although the stability condition $\Delta/\mu < 1$ is violated slightly, the degree of violation is within the experimental error. Two extra stabilizing influences exist in actual highway traffic. A given driver generally notices the behavior of the vehicle two ahead of him as well as that which follows him (through a rear view mirror or horn signals by his follower).

It would be interesting to extend the experiments here described to more extended lines of cars to evaluate the degree of coupling of a car with rear and second nearest front neighbors and to introduce these interactions into the dynamical equations. Of course new stability conditions would

TABLE I
PARAMETERS OF EQUATION (53)

Driver	Δ (r=max)	$b = \lambda/M$	r	$2\lambda\Delta/M$
1	1.4 sec	0.74 sec^{-1}	0.87	2.08
2	1.0	0.44	0.90	0.88
3	1.5	0.34	0.86	1.03
4	1.5	0.32	0.49	0.97
5	1.7	0.38	0.74	1.29
6	1.1	0.17	0.86	0.37
7	2.2	0.32	0.82	1.43
8	2.0	0.23	0.85	0.93
Average	1.55	0.368		1.12

result. Anyone who has done considerable driving notices that the margin between stable and unstable operation is very narrow. In practice one would expect that even the added stabilizing influences would yield values of the appropriate parameters in the dynamical equations, which would make driving conditions merely a shade on the stable side.

A few conservative drivers interspersed in a chain of cars add tremendously to the stability because they effectively cut the chain by leaving such large gaps that disturbances that might have grown earlier in the chain have time to damp out. In dense traffic such gaps are however soon filled by their more impatient brethren so that their good influence is frequently nullified.

It is to be emphasized that a phenomenological theory of traffic dynamics lumps together a large number of mechanical and human attributes that can only with great difficulty be handled individually. This, however, is what makes the use of phenomenological models so powerful in the unravelling of so complicated a set of events.

TOPICS FOR FUTURE INVESTIGATION IN THE THEORY OF TRAFFIC FLOW ON THE UNLIMITED HIGHWAY

THE PREVIOUS sections of this paper have been concerned with dense traffic situations in which no passing is possible. We close with a few remarks on passing, bunching, and acceleration noise, topics that we hope to discuss both theoretically and experimentally in future publications.

In the high density limit, passing can be treated as a queuing problem. Suppose a fixed obstacle exists on a two-lane highway. Cars that accumulate behind the obstacle are only able to go around it when appearance time gaps larger than a certain critical value exist between successive vehicles in the opposite lane. If large gaps (which allow two or more cars to go around the obstacle per gap) are rare the rate of growth of the line behind the obstacle and the reduction in traffic flow current caused by the obstacle can be discussed by standard queuing theory. The distribution of service times of the queue is the distribution of time intervals between the required long gaps. The distribution of appearance times is of course that of the time intervals between the appearance of successive cars at the obstacle. A slow driver is a moving obstacle and can be treated in the same manner as a fixed one through the use of a moving coordinate system. When very large gaps are common so that two or more cars may occasionally pass the obstacle together the queuing theory becomes more complicated. Queuing theory is also applicable to the analysis of the effect of a bad curve or very steep grade on traffic flow.

An alternative approach to the passing problem can be made by setting up continuum flow equations for each lane, including cross terms which characterize the interactions between the lanes.

Another effect caused by the existence of a speed distribution in medium density traffic conditions is bunching. Everyone has seen clusters of cars form and evaporate. It would be interesting to observe the distribution of cluster sizes as a function of mean speed and density and to find the gel point at which clusters congeal to form a jammed traffic situation.

The estimation of the state of the traffic on an open road is a highly personal matter. The driver who is satisfied in maintaining a speed of 35 mph while his fellow travelers are racing along at 70 mph considers them to be lunatics. The speedier drivers consider our snail to be a menace. A resistance to flow caused by speed dispersion can be defined in a different subjective way for each driver. The mean resistance averaged over all drivers might serve as a useful parameter of the traffic stream. A quantity sensitive to the resistance to flow is the acceleration noise experienced by a given vehicle. We define this noise as the dispersion in the acceleration distribution function. The only measurement we have of this quantity at the moment is that obtained from the records of the follow-the-leader experiments discussed earlier. The acceleration distributions are essentially Gaussian with mean zero and dispersion of the order of $\sim 0.15\ g$. We plan to make more extensive measurements of this quantity under real and well-specified highway conditions.

Finally, a car moving with the average speed of the stream would have a very narrow acceleration distribution pattern, while one that moves faster than the average stream speed would be expected to have a broadened acceleration distribution that would increase with the speed differential. One might try to relate the resistance of the stream to the acceleration noise of its component cars. A car moving with a speed lower than the stream average will cause a reduction of the stream velocity, the magnitude of the reduction increasing with the density.

ACKNOWLEDGMENTS

THE AUTHORS take this opportunity to express their thanks to MR. R. T. BUNDORF and MR. LESTER V. OSTRANDER, JR. of the Research Staff, General Motors Corporation, for the design and installation of the experimental apparatus. One of us, E. W. M., wishes to thank DR. R. W. HART for several interesting discussions. We also thank MR. J. B. BIDWELL for his co-operation and assistance in some of the experiments, for his interest, and for comments concerning this paper. Finally we express our appreciation to MR. JOHN M. CAMPBELL and DR. L. R. HAFSTAD for their continued interest and encouragement.

REFERENCES

1. T. M. MATSON, W. S. SMITH, AND F. W. HURD, *Traffic Engineering*, McGraw-Hill, New York, 1955.
2. D. L. GERLOUGH AND T. H. MATHEWSON, "Approaches to Operational Problems in Street and Highway Traffic—A Review," *Opns. Res.* **4**, 32 (1956).
3. L. A. PIPES, *J. Ap. Phys.* **24**, 274 (1953); Inst. of Transportation and Traffic Engineering, U. of Col. report, "A Proposed Dynamic Analogy of Traffic" (1951).
4. A. REUSCHEL, *Zeits. d. Oesterreich. Ing. u. Arch. Vereines* **95**, 59 and 73 (1950); *Oesterreich. Ing. Archiv.* **4**, 193 (1950).
5. M. J. LIGHTHILL AND G. B. WHITHAM, *Proc. Royal Soc.* **229**, 317 (1955).
6. P. I. RICHARDS, "Shock Waves on the Highway," *Opns. Res.* **4**, 42 (1956).
7. W. PRAGER, "Problems of Traffic and Transportation," Brown University, Mathematics Division (1954).
8. G. F. NEWELL, "Mathematical Models for Freely-Flowing Highway Traffic," *Opns. Res.* **3**, 176 (1955).
9. H. H. GOODE, C. H. POLLMAR, AND J. B. WRIGHT, "The Use of a Digital Computer to Model a Signalized Intersection," IP-146, University of Michigan (1956).
10. W. J. MERRILL, JR. AND C. A. BENNETT, "The Application of Temporal Correlation Techniques in Psychology," *J. Appl. Psychol.* **40**, 272–280 (1956).

Reprinted from

Proc. Natl. Acad. Sci. USA
Vol. 75, No. 10, pp. 4633-4637, October 1978
Applied Mathematical Sciences

Social dynamics and the quantifying of social forces

(logistic curve/replacement and evolution/transportation)

ELLIOTT W. MONTROLL

Institute for Fundamental Studies, Department of Physics, University of Rochester, Rochester, New York 14627

Contributed by Elliott W. Montroll, July 10, 1978

ABSTRACT Social and industrial evolutionary processes are considered to be a sequence of replacements or substitutions: new ideas for old, new labor patterns for old, new technologies for old. The logistic equation has often been used to describe population growth processes and replacement processes. It sometimes suffers from contradicting observational data. It is shown here that the deviations are often associated with unusual intermittent events—wars, strikes, economic panics, etc.—and that in many cases a few years after the event it can be abstracted as an instantaneous δ function impulse. After the event, the evolutionary process continues along its normal course. A formula is derived to use the observational data to determine the strength of the impulse modeling an event.

During any period of history, the middle-aged and elderly lament upon the changes "in the world" during their lifetime, complaining that "things" are no longer what they were. The things that have changed the most vary from one generation to another. In this century the automobile has replaced the horse; the light bulb, the gas mantle; the movie theatre, home entertainment (finally, television, the movie theatre); the Balkan States of the Soviet Bloc, the old Austro-Hungarian Empire; the meaningful relationship, formal marriage; to name just a few. The continuing change in our habits and loyalties is bound to affect economic and social structures. On this basis, social science models should include the evolution of social processes.

The aim of this paper is to present an elementary point of view for the development of a style for the description of social dynamics. A physical scientist will immediately observe the influence of Newton's laws of particle dynamics upon the structure of this style. Following in the Newtonian tradition, I propose three laws of social dynamics (1).

THREE "LAWS" OF SOCIAL DYNAMICS

Newton's first law of dynamics is the postulate that, in the absence of an external force, every body in a state of motion will remain in that state of motion—i.e., it will continue to move in a straight line with a constant velocity. Of course, this situation never prevails in earthbound experiments, but it is still a good starting point for the construction of mathematical models of dynamical systems.

The first law of social dynamics is here chosen to have two similarly stated parts

(*a*) In the absence of any social, economic, or ecological force, the rate of change of the logarithm of a population, $N(t)$, of an "organism" is constant,

$$d\log N(t)/dt = \text{constant}. \qquad [1a]$$

Without the prescribed forces this equation is also postulated to be valid for the variation of the population of objects of production (automobiles, radios, etc.), and for the change in a population of a new social group.

(*b*) In the absence of any social, economic, or ecological force, the rate of change of the logarithm of the price of maintenance $P(t)$ (per unit time) of an "organism" is also constant

$$d\log P(t)/dt = \text{constant}. \qquad [1b]$$

In the case of objects of production, $P(t)$ is to be interpreted as a unit cost.

A discussion of part *b* of the first law and of the influence of prices on part *a* will be given elsewhere.

The population of inanimate objects is included so that population growth models might be applied to production of and competition between manufactured items.

Eq. 1a is, of course, nothing but the Malthusian law of exponentiation of populations (2), and Eq. 1b is a statement of the accountants' "discounting" principle and the housewives' observation that things are always getting more expensive. The constant in Eq. 1a might be negative as well as positive because interest in some items just dies away.

It might be claimed that the first law of social dynamics is more often applicable to the real world than is Newton's first law of mechanics to real dynamical systems, since social reformers as well as conservative politicians frequently seek means of violating Eqs. 1a and 1b. Numerous observations exhibiting the first law are in refs. 3 and 4.

Newton was very astute in his use of the second law as the definition of a force. Who can go wrong by making definitions if he does not make too many of them? The second law is just the statement that a force is that which causes the first law to be violated. I will not attempt to outdo the master on this point.

The second "law" of social dynamics is the postulate:

Eq. 1a or 1b or both are violated when a social, economic, or ecological force is applied. How is the force to be chosen or measured? By observing the manner in which the first law is violated!

Several examples of applied forces will follow the statement of:

The third "law." Evolution is the result of a sequence of replacements. Newton never tried to derive his laws of force from first principles. The restoring force for a displaced mass in a harmonic oscillator was merely the force of greatest mathematical simplicity and one found to be useful in describing many physical phenomena. The postulation of the inverse square law of the gravitation force was natural for a genius whose geometric intuition told him it was just what was required to produce Kepler's empirical observations on elliptical planetary orbits.

One of the simplest mathematical forms for a force that might replace the constant on the right-hand side of Eq. 1a is a linear force:

$$F\{N(t)\} \equiv k - \alpha N(t) \qquad [2]$$

which represents a deterence to population growth. If one sets

4633

$\alpha = k/\theta$, then Eq. 1a becomes the well-known Verhulst equation (with population saturation at $N = \theta$) (3, 5)

$$d\log\{N(t)/\theta\}/dt = k\{1 - (N(t)/\theta)\} \quad [3]$$

whose solution yields a remarkably accurate representation of the population growth in many countries. The growth to saturation according to the solution of Eq. 3 is often called the logistic curve. The characterization of deviations will be discussed in the next section.

If one sets $x \equiv N/\theta$ and $y = 1/x$, then

$$-d(y - 1)/dt = k(y - 1), \text{ and} \quad [4]$$

$$y(t) - 1 = [y(0) - 1] \exp(-kt) \quad [5]$$

so that

$$\log[x/(1 - x)] = \log[x(0)/[1 - x(0)]] + kt. \quad [6]$$

A more general (two-parameter) force law (3) is

$$F\{N(t)\} = k\{1 - [N(t)/\theta]^\nu\}/\nu \equiv G_\nu(N/\theta) \quad [7]$$

which yields Verhulst's equation when $\nu = 1$ and the Gompertz equation (6) when $\nu = 0$:

$$dN/dt = -Nk \log(N/\theta). \quad [8]$$

The simplest interaction between two species is obtained by choosing the second species to apply a force linear in its population to the first species and vice versa so that

$$d\log N_1/dt = k_1 + C_{12}N_2$$

$$d\log N_2/dt = k_2 + C_{21}N_1.$$

The case $k_1 = \alpha_1 > 0$, $k_2 = \alpha_2 < 0$, $C_{12} = \lambda_1 < 0$, and $C_{21} = \lambda_2 > 0$ yields the well-known Lotka–Volterra equations for competing species (7–9)

$$dN_1/dt = \alpha_1 N_1 - \lambda_1 N_1 N_2 \quad [9a]$$

$$dN_2/dt = -\alpha_2 N_2 + \lambda_2 N_1 N_2 \quad [9b]$$

such that the predator species 2 preys upon species 1 in such a manner that species 2 would disappear without species 1 and species 1 would propagate in a Malthusian manner without species 2. These equations have periodic solutions (8, 9).

Finally, in an assembly of many interacting species, one might add the influence of other species to the right-hand side of Eq. 3 as a random force, $F(t)$, to yield the equation

$$d\log[N(t)/\theta]/dt = kG_\nu(N/\theta) + F(t). \quad [10]$$

The special case $G_1(x) = 1 - x$ was first considered by Leigh (10) and the general case was analyzed in ref. 9. If one assumes $F(t)$ to be generated by a Gaussian random process with

$$\langle F(t_1)F(t_2)\rangle_{av} = \sigma^2 \delta(t_2 - t_1), \quad [11]$$

then it can be shown in the standard manner that the probability distribution of N at time t, $P[N(t)]$, satisfies the Fokker–Planck equation:

$$\frac{\partial P}{\partial t} = -\frac{\partial}{\partial v} k\{PG(\exp v)\} + \frac{1}{2}\frac{\partial^2 P}{\partial v^2}\sigma^2 \quad [12a]$$

in which

$$v = \log(N/\theta). \quad [12b]$$

The steady-state solution of this equation with $\dot{P} = 0$ is easily verified (9) to be

$$P(v,t) = P_0 \exp\left[2\sigma^{-2}\int_0^v G(\exp v)dv\right] \quad [13]$$

in which P_0 is a normalization constant. In the Gompertz and Verhulst cases, the steady-state distribution functions are, respectively,

Gompertz: $P = P_0 \exp(-kv^2/\sigma^2)$, $v = \log(N/\theta)$ [14]

Verhulst: $P = P_0 \exp 2k[v - \exp v]/\sigma^2$
$$= P_0(N/\theta)^{2k/\sigma^2}\exp(-2kN/\theta\sigma^2). \quad [15]$$

Eq. 15 was first derived by Leigh and Eq. 14 was first given in ref. 9. A noisy Gompertz growth process can be shown to be equivalent to the Ornstein–Uhlenbeck process for the theory of Brownian motion in the variable v (11).

Of course, many more models have been discussed in the literature. My main interest here is to discuss the third law, the law of evolution, as a sequence of replacements. I choose the canonical form for the evolutionary force to be the right-hand side of the Verhulst (or logistic) equation (Eq. 3). This was exploited by Fisher and Pry (12) as the appropriate model for the dynamics of industrial replacement. I discuss the replacement as an evolutionary process in the next section where I shall also analyze cases in which the canonical form is violated.

DYNAMICS OF REPLACEMENT PROCESS AND MEASUREMENT OF FORCES ACCELERATING AND DETERRING REPLACEMENT OR EVOLUTIONARY PROCESS

A set of Fisher and Pry's typical evolutionary curves is plotted in Fig. 1. They represent the replacement of one industrial process (or product) by another. They can also be interpreted as market penetration curves. Here x is the fraction of the market captured by the new process (or product) and $(1 - x)$ is that remaining for the old. Fisher and Pry (12) have found over twenty examples of their industrial replacement mechanism and other authors have supplemented the list (13–15).

Herman and I (16) have noticed that the industrial revolution in both the United States and Sweden has also evolved in the logistic pattern, as exhibited in Fig. 2, with $(1 - x)$ there being the fraction of the work force performing agricultural tasks and x being the fraction performing nonagricultural tasks. The data

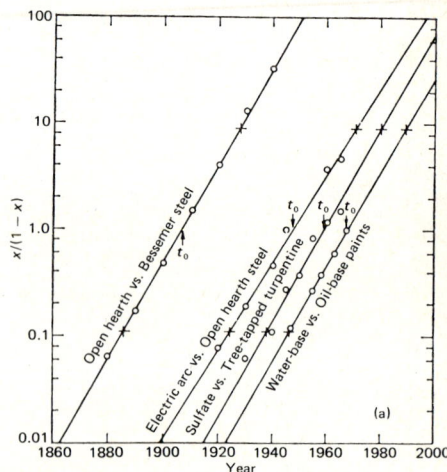

FIG. 1. Replacement dynamics of four technologies, following the model and data of Fisher and Pry (12).

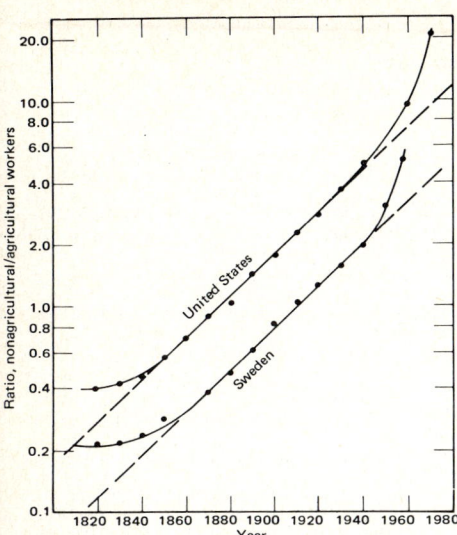

FIG. 2. Variation of the ratio of nonagricultural workers to agricultural workers in the American and Swedish labor force (16).

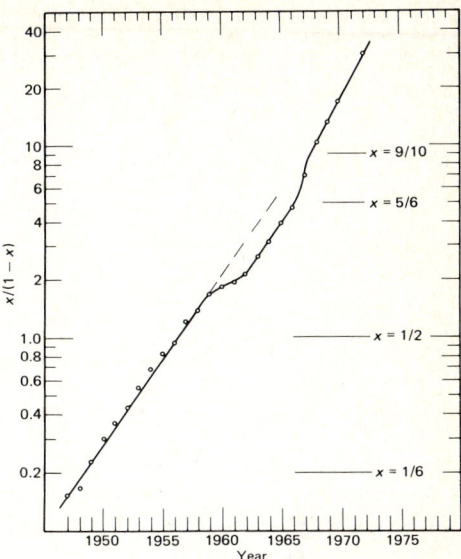

FIG. 3. Manner in which intercity passenger travel by rail has been replaced by air travel.†

follow the solution of the Verhulst equation for over 100 years. A new social force emerged in 1940 through a flood of telegrams from the White House, starting with the word "Greetings . . ." and received by young farmers the country over. These telegrams accelerated the decline in agricultural work force beyond the expected rate of the canonical Verhulst process.

An example of the influence of a deterring force is evident in Fig. 3, a record of the replacement of rail service by air service in intercity passenger travel.† There,

$$x = \frac{\text{annual air passenger miles}}{\text{annual air \& rail passenger miles}} \qquad [16]$$

Notice that in the period 1947–1959 the data are in accord with Eq. **6b**. The deterrence in replacement evolution during 1960–1961 seems to have been due to several unusually long strikes by airline workers and to the public's response to a series of serious airplane accidents. By 1962, the system recovered and the replacement curve continued with its old slope until the late 1960s. Then, accelerated replacement to the end of the decade occurred at the time the largest rail passenger carrier, the Penn-Central line, was suffering through its prebankruptcy and bankruptcy condition. The management of passenger rail service was reorganized in the 1970s by Amtrak.

The role of intercity buses was omitted from the above discussion because, during the period investigated, the fraction of the total passenger traffic maintained by them changed very little.

Other examples of intermittent accelerating and deterring forces are easily found. Deviations from the logistic national population curves during and after wars and during economic depressions are common. These will be discussed elsewhere. R. Bolt (personal communication) has shown that the United States has been experiencing an educational evolution since 1900 in

† Montroll, E. W. (1978) *Proceedings of Environmental Protection Agency Conference on Mathematical Modeling*, in press.

that the fraction of persons receiving university degrees has been following a logistic pattern with intermittent forces appearing during and after World War I and World War II. The replacement of the sailboat by the steamboat is discussed at the end of this section. I now present a systematic procedure for measuring the magnitude of intermittent forces.

Let the solid straight line in Fig. 4 represent the canonical evolutionary curve for a replacement process and suppose that an accelerating force was applied at time τ and withdrawn at a later time. After withdrawal, the evolutionary curve will parallel the unaccelerated one and will reach a level A at a time t^*, somewhat earlier than the time t that would have been required to reach the same level in the normal fashion. If $F(t)$ is the accelerating force, the accelerating process is characterized by

$$d\log x/dt = k(1 - x) + F(t). \qquad [17]$$

In the time regime after the withdrawal of an intermittent force, the continuing evolutionary curve parallels the unaccelerated one. The time gained,

$$\Delta t = t - t^* \qquad [18]$$

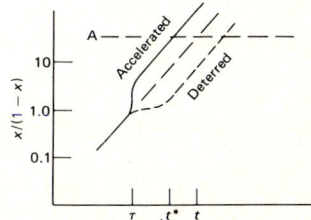

FIG. 4. Schematic of the manner in which an intermittent accelerating or deterring force affects replacement dynamics.

in reaching the level A could have been achieved by the introduction of an equivalent δ function impulse. The object of the remainder of this section is to derive an expression for the amplitude α of the δ function impulse

$$F(t) = \alpha \delta(t - \tau) \qquad [19]$$

necessary to yield the observed gained time Δt. We then consider α to be the measure of the magnitude of the originally applied accelerating force, whatever its detailed nature. The form used for the δ function is

$$F(t) = \begin{cases} 0 \text{ if } t < \tau \text{ or } \tau + \delta < t \\ \alpha/\delta \text{ if } \tau \leq t \leq \tau + \delta \end{cases}. \qquad [20]$$

Our final required equations will be obtained by taking the limit $\delta \to 0$. The solution of a canonical Verhulst equation (Eq. 3) was shown to be Eqs. 5 and 6.

Upon introduction of the δ function force, the form of Eq. 17 is

$$dx/dt = kx(1-x) + x\alpha/\delta$$
$$\text{with } \tau < t < \tau + \delta \equiv T, \text{ or} \qquad [21a]$$

$$-dy/dt = y\{k + (\alpha/\delta)\} - k = k(y-1) + (y\alpha/\delta),$$
$$\text{if } y = 1/x. \qquad [21b]$$

In the regime $t > T$, Eq. 3 is again applicable with its solution being

$$\log \{x(t^*)/[1 - x(t^*)]\} = -\log[y(T) - 1] + k(t^* - T) \qquad [22]$$

at the time t^* when the accelerated evolution reaches level A in Fig. 4. Then from Eqs. 22 and 6, because the function on the left side of both equations corresponds to the level A,

$$k\Delta t \equiv k(t - t^*) = -\log\{[y(T) - 1]/[y(0) - 1]\} - kT. \qquad [23]$$

The quantity $y(T)$ is found by solving the accelerated evolution equation (Eq. 21b). Before doing so, we note that

$$\log \left\{\frac{y(T) - 1}{y(0) - 1}\right\} = \log \left\{\frac{y(\tau) - 1}{y(0) - 1}\right\} \cdot \left\{\frac{y(T) - 1}{y(\tau) - 1}\right\}$$

$$= -k\tau + \log\{[y(T) - 1]/(y(\tau) - 1)\} \qquad [24]$$

$$k(t - t^*) = -\log \left\{\frac{y(T) - 1}{y(\tau) - 1}\right\} - k\delta,$$

since

$$T - \tau = \delta. \qquad [25]$$

Since δ is small

$$y(T) - 1 = y(\tau) - 1 + \delta dy/dt + \ldots.$$

By using Eq. 21b, as $\delta \to 0$

$$\log \left\{\frac{y(T) - 1}{y(\tau) - 1}\right\} = \log \left\{1 + \frac{\delta \, dy/dt}{y(\tau) - 1}\right\} \sim \delta(dy/dt)/[y(\tau) - 1]$$
$$\to \alpha y(\tau)/[1 - y(\tau)]. \qquad [26]$$

Then, since $y(\tau) = 1/x(\tau)$, we have, from Eq. 25, as $\delta \to 0$

$$k\Delta t \equiv k(t - t^*) = \alpha/[1 - x(\tau)] \qquad [27a]$$

or

$$\alpha = k(\Delta t)[1 - x(\tau)]. \qquad [27b]$$

Incidentally, the general solution of Eq. 17 is

$$x(t) = x(0) \left\{\exp \int_0^t [k + F(t)]dt\right\}$$
$$\times \left\{1 + kx(0) \int_0^t d\tau \exp \int_0^\tau [k + F(t')]dt'\right\}^{-1}$$

Eq. 27b can be derived by incorporating Eq. 19 in this expression. However, the algebra is more tedious than that given above.

Let us calculate the value of α appropriate for the deterring intermittent force applied in the period 1960–1961 to airline evolution as indicated in Fig. 3. The Δt value at the level $x = \frac{5}{6}$ is -1.8 years; $x/(1-x) = 1.7$ or $x = 0.63$ at the time of the application of the force. To estimate k, note that 11 years were required for $x/(1-x)$ to be multiplied by a factor 10 in evolving from 0.1 to 1.0 so that, from Eq. 6b,

$$k = (1/11 \text{ year}) \log_e 10 = 0.209/\text{year}.$$

Hence, from Eq. 27b

$$\alpha = -(0.209)(1.8)(0.37) = -0.139 \simeq -0.14.$$

The replacement pattern of sail by steam in the U.S. Merchant Marine during the period 1830–1950 was rather like that of rail by air travel in the middle 1900s. Data from tables Q417–432 of ref. 17 are summarized in Fig. 5. The launching of steam ships in significant numbers dates from 1825. During the two decades 1830–1850, the logarithm of the fraction of tonnage in steamers to that in sailing ships followed the canonical straight line, as it did again in the interval 1870–1915. At first, steamboats appeared in river traffic and then in coastal traffic; later on they began transatlantic runs, but only after the Civil War could they successfully compete with the clipper ships on the longer Pacific passages and on the voyages around the Horn connecting the East Coast with San Francisco.

The first fast clipper ships appeared in the 1830s, at about the same time steamboats were becoming practical. They were

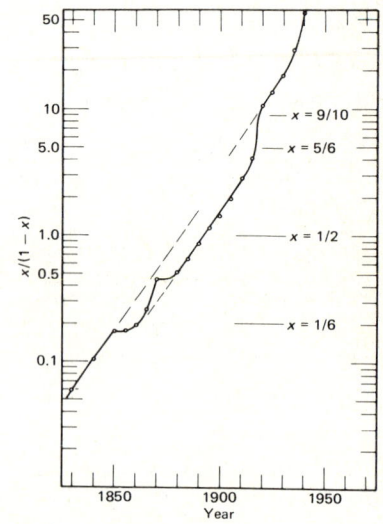

FIG. 5. Manner of replacement of sailing ships by steamships in the U.S. merchant fleet: data from ref. 17.

Table 1. Gross tonnage, according to type, in United States merchant fleet*

Year	Gross tonnage, tons × 10⁻³	
	Steam	Sail
1840	202	1978
1845	326	2091
1850	526	3010
1855	770	4442
1860	868	4480

* Data from ref. 17.

built in large numbers during the decade 1845–55 (peaking between 1850 and 1853). Two important events of 1849 stimulated their production, perturbing the take-over by steam: (*i*) discovery of gold in California; and (*ii*) repeal of the British Navigation Acts and the breaking of the China trade monopoly long enjoyed by the British merchant marine. The expansion of trade to the West Coast, the Orient, and Australia by adventurous American skippers created an enormous demand for the speedy clippers as indicated in Table 1.

"Among the most famous builders of Clipper ships was Donald McKay of Boston who launched the Sovereign of the Sea of 2421 tons registered in 1852 . . . . this Clipper achieved a speed of 411 miles in one day. But it was [his] Lightening (1854) . . . that established the best record for a single days run, 436 miles, a record not surpassed by a steam ship for many years . . . . Construction of the extreme Clipper capable of making up to 18 knots generally stopped after 1854 because of a financial slump" (18). During the tight money episode of the Panic of 1857, all types of construction were curtailed.

Steamboat construction was favored over sail for the increased local transport required by the Civil War. "In order to keep up with the times [even] Donald McKay, during the Civil War, changed his yard over so that he could build iron ships, marine engines, etc." (19). The considerably improved steamboat models dominated naval construction in the postwar reconstruction period 1865–1873 only to be abated by the Panic of 1873 whose effects persisted for several years. With the return to normalcy, the steamboat replacement curve followed its canonical course until 1915 when the shipping requirements of World War I stimulated an accelerated naval construction program. By that time, no one considered new sailing vessels as suitable for commercial shipping.

Now, let us estimate the magnitude of an instantaneous impulsive force that would, in the long run, have the deterrent effect observed from the combination of two panics (1857 and 1873), the Civil War, and Mr. McKay's bold attempt to demonstrate the superiority of sail to steam. The deterrence time of these combined influences, measured at $x = \frac{1}{2}$ is $t - t^* = -11.25$ years. The value of k appropriate for the rate constant in a Verhulst equation to reproduce observed $x/(1 - x)$ values in 1830 and 1850 is $k = 0.052/\text{year}$. The $x(t)$ to be identified in Fig. 5 with 1850 (the year of the application of the deterring force) is 0.145. Hence, from Eq. **27b**,

$$\alpha = (0.052)(-11.25)[1 - 0.145] = -0.50$$

A number of α values for other processes will be presented elsewhere.

NATURE OF CONTINUING PROGRAM

A further elaboration of the above ideas is possible through the following data analysis program. The values of the rate constants, k, appropriate to numerous industrial (12–15), scientific (20), and social (16–18) evolutionary processes as well as those of population growth (3, 21) can be assembled. α values can be derived from those evolutionary curves that indicate the existence of intermittent forces. It is hoped that, as the collection of k and α values grows, one will develop some intuition on the evolutionary rate constants associated with various classes of processes and the influence of various types of accelerating and deterring forces. In cases exhibiting a significant random component to a force law (cf. Eq. **10**) the statistical parameters can be estimated.

This study was partially supported by the General Electric Foundation and General Electric Corporate Research and Development.

1. Montroll, E. W. (1978) *Nonlinear Equations in Abstract Spaces*, ed. Lakshmikantham, V. (Academic, New York), pp. 161–216.
2. Malthus, T. R. (1798) *An Essay on the Principle of Population as It Affects the Future Improvement of Society*.
3. Montroll, E. W. & Badger, W. W. (1974) *Introduction to Quantitative Aspects of Social Phenomena* (Gordon and Breach, New York).
4. Lapp, R. E. (1973) *The Logarithmic Century* (Prentice-Hall, Englewood Cliffs, NJ).
5. Verhulst, P. F. (1844) *Mem. Acad. R. Bruxelles*, **28**, 1.
6. Gompertz, R. (1825) *Philos. Trans. R. Soc. London*, **115**, 513–585.
7. Lotka, A. J. (1956) *Elements of Mathematical Biology* (Dover Reprints, New York).
8. Volterra, V. (1931) *Lecons sur la Theorie mathematique de la lutte pour la vie* (Gauthier-Villars, Paris).
9. Goel, N., Maitra, S. & Montroll, E. W. (1971) *Rev. Mod. Phys.* **43**, 231–276.
10. Leigh, E. G. (1969) *Some Mathematical Problems in Biology* (American Mathematical Society, Providence, RI), Vol. 1.
11. Montroll, E. W. & West, B. J. (1979) *Studies in Statistical Mechanics* (North-Holland, Amsterdam, Netherlands), in press.
12. Fisher, J. C. & Pry, R. H. (1971) *Tech. Forecast. Soc. Change* **3**, 75–88.
13. Blackman, A. W., Jr. (1974) *Tech. Forecast. Soc. Change* **6**, 41–63.
14. Marchetti, C. (1975) *Second Status Report on the IIASA Project on Energy Systems*, ed. Häfele, W. (International Institute for Applied Systems Analysis, Luxenburg, Austria), pp. 203–217.
15. Stern, M. O., Ayres, R. W. & Shapanka, A. (1975) *Tech. Forecast. Soc. Change*, **7**, 57–79.
16. Herman, R. & Montroll, E. W. (1972) *Proc. Natl. Acad. Sci. USA* **69**, 3019–3023.
17. U.S. Census Bureau (1976) *Historical Statistics of the United States, Colonial Times to 1970* (U.S. Government Printing Office, Washington, DC).
18. Imlay, M. H. in *Encyclopedia Britannica* (1967 Edition), (Wm. Benton, Chicago), Vol. 5, p. 930.
19. McKay, R. C. (1928) *Some Famous Sailing Ships and Their Builder, Donald McKay* (Putnam, New York).
20. DeSola Price, D. J. (1967) *Little Science, Big Science* (Columbia Univ. Press, New York).
21. Pearl, R. (1924) *Studies in Human Biology* (Williams and Wilkins, Baltimore, MD).

Contents of Volumes I–XII

Vols. I–V, under the editorship of J. de Boer and G.E. Uhlenbeck; out of print.

Vols. VI–XI, under the editorship of E.W. Montroll and J.L. Lebowitz

VI. *The kind of motion we call heat*, by S.G. Brush

1.	The kinetic theory in the history of physics	3
2.	Herapath	107
3.	Waterston	134
4.	Clausius	160
5.	Maxwell	183
6.	Boltzmann	231
7.	Van der Waals	249
8.	Mach	274
9.	The wave theory of heat	303
10.	Foundations of statistical mechanics 1845–1915	335
11.	Interatomic forces and the equation of state	386
12.	Viscosity and the Maxwell–Boltzmann transport theory	422
13.	Heat conduction and the Stefan–Boltzmann law	469
14.	Randomness and irreversibility	543
15.	Brownian movement	655
16.	The literature of kinetic theory	705
Index		xv

VII. Fluctuation phenomena, eds. E.W. Montroll and J.L. Lebowitz

1. Fluctuations, *M. Kac and J. Logan*	1
2. On an enriched collection of stochastic processes, *E.W. Montroll and B.J. West*	61
3. Stochastic geometry: aspects of amorphous solids, *R. Zallen*	177
4. Statistical mechanical theory of the kinetics of phase transitions, *H. Metiu, K. Kitahara and J. Ross*	229
5. Towards a rigorous theory of metastability, *O. Penrose and J.L. Lebowitz*	293
Author index	341
Subject index	347

VIII. The liquid state of matter: Fluids, simple and complex, eds. E.W. Montroll and J.L. Lebowitz

On the equilibrium theory of fluids: An introductory overview, *J.L. Lebowitz and E.M. Waisman*	1
Non-uniform fluids, *J.K. Percus*	31
Dense conducting liquids, *N.W. Ashcroft*	141
The equilibrium statistical mechanics of simple ionic liquids, *B. Hafskjold and G. Stell*	175
Equilibrium theory of polyatomic fluids, *D. Chandler*	275
Low frequency dielectric properties of liquid and solid water, *F.H. Stillinger*	341
Subject index	433
Name index	435

IX. Perspectives in statistical physics, ed. H.J. Raveché

1. The Lorentz gas, *B.J. Alder and W.E. Alley*	3
2. Some recent developments in the kinetic theory of gases, *J.R. Dorfman*	23
3. Non-equilibrium fluctuations and the hierarchy, *M.H. Ernst and E.G.D. Cohen*	59
4. Green's contributions to non-equilibrium statistical mechanics revisited, *L.S. Garcia-Colin and J.L. Del Rio*	75
5. Stochastic description of many-body systems, *N.G. Van Kampen*	89
6. H-theorems for Markoffian processes, *R. Kubo*	101
7. Energy flow and thermal conductivity in one-dimensional, harmonic, isotopically disordered crystals, *R.J. Rubin*	111

8. Where do we go from here?, *R. Zwanzig* 123
9. A new model Hamiltonian for a correlated electron system within the general framework of critical phenomena and phase transitions, *C. Di Castro* 137
10. Membrane flux: conditions for limit cycle oscillations, *A.G. De Rocco and G.L. Clark* 155
11. Critical phenomena—A model illustration of scientific method, *C. Domb* 173
12. Coarse-grained Helmholtz free energy functional, *K. Kawasaki, T. Imaeda and J.D. Gunton* 201
13. Exact renormalization in two dimensional ising systems, *J.M.J. van Leeuwen* 225
14. How close is "close to the critical point"?, *J.M.H. Levelt Sengers and J.V. Sengers* 239
15. The interfaces between fluid phases, *B. Widom* 273
16. On higher order WKB approximations for the calculation of energy levels, *F.T. Hioe, E.W. Montroll and M. Yamawaki* 295
17. Equilibrium density matrix for fluids, *J.E. Mayer* 323
18. The mechanisms of stochasticity in classical dynamical systems, *A.S. Wightman* 343
Subject index 365

X. *Nonequilibrium phenomena I: The Boltzmann equation, eds. E.W. Montroll and J.L. Lebowitz*

1. On a derivation of the Boltzmann equation, *O.E. Lanford III* 1
2. Global existence proofs for the Boltzmann equation, *W. Greenberg, J. Polewczak and P.F. Zweifel* 19
3. Exact solutions of the nonlinear Boltzmann equation and related kinetic equations, *M.H. Ernst* 51
4. Solution of the Boltzmann equation, *C. Cercignani* 121
5. Fluid dynamics and the Boltzmann equation, *R.E. Caflisch* 193
6. Fluctuation theory for the Boltzmann equation, *H. Spohn* 225
Subject index 253

XI. *Nonequilibrium phenomena II: From stochastics to hydrodynamics, eds. J.L. Lebowitz and E.W. Montroll*

1. On the wonderful world of random walks, *E.W. Montroll and M.F. Shlesinger* 1
2. A survey of the hydrodynamical behavior of many-particle systems, *A. De Masi, N. Ianiro, A. Pellegrinotti and E. Presutti* 123
Subject index 295

Vol. XII, under the editorship of J.L. Lebowitz

XII. The wonderful world of stochastics. A tribute to Elliott W. Montroll, eds. M.F. Shlesinger and G.H. Weiss

I. Elliott Waters Montroll, *M.F. Shlesinger and G.H. Weiss*	1
1. Dielectric relaxation via the Montroll–Weiss random walk of defects, *J.T. Bendler and M.F. Shlesinger*	31
2. The fascination of old texts, *C. Domb*	47
3. Some statistical and dynamical problems in quantum electronics, *F.T. Hioe*	75
4. Theory of diffusion via an interstitial and vacancy mechanism, *P.H.E. Meijer*	99
5. Mathieu's difference equation, *R.B. Potts*	111
6. Mayer–Montroll equations (and some variants) through history for fun and profit, *G. Stell*	127
7. Illumination in a random medium, *N.G. Van Kampen*	157
8. Random walks in crystallography, *G.H. Weiss and J.E. Kiefer*	169
9. On the quantum Langevin equation: the linear oscillator, *B.J. West and K. Lindenberg*	189
10. Some inequalities for anisotropic rotators, *J. Bricmont, J.L. Lebowitz and C.E. Pfister*	205

Cumulative Index, Volumes VI–XII

Volumes I–V, under the editorship of J. de Boer and G.E. Uhlenbeck; out of print.

Monographs

Brush, S.G., The kind of motion we call heat	VI

Contributed volumes

Alder, B.J., and W.E. Alley, The Lorentz gas	IX,	3
Alley, W.E., see Alder, B.J.	IX,	3
Ashcroft, N.W., Dense conducting liquids	VIII,	141
Bendler, J.T., and M.F. Shlesinger, Dielectric relaxation via the Montroll–Weiss random walk of defects	XII,	31
Bricmont, J., J.L. Lebowitz and C.E. Pfister, Some inequalities for anisotropic rotators	XII,	205
Caflisch, R.E., Fluid dynamics and the Boltzmann equation	X,	193
Cercignani, C., Solution of the Boltzmann equation	X,	121
Chandler, D., Equilibrium theory of polyatomic fluids	VIII,	275
Clark, G.L., see De Rocco, A.G.	IX,	155
Cohen, E.G.D., see Ernst, M.H.	IX,	59

De Masi, A., N. Ianiro, A. Pellegrinotti and E. Presutti, A survey of the hydrodynamical behavior of many-particle systems — XI, 123
De Rocco, A.G., and G.L. Clark, Membrane flux: conditions for limit cycle oscillations — IX, 155
Del Rio, J.L., see Garcia-Colin, L.S. — IX, 75
Di Castro, C., A new model Hamiltonian for a correlated electron system within the general framework of critical phenomena and phase transitions — IX, 137
Domb, C., Critical phenomena — A model illustration of scientific method — IX, 173
Domb, C., The fascination of old texts — XII, 47
Dorfman, J.R., Some recent developments in the kinetic theory of gases — IX, 23

Ernst, M.H., and E.G.D. Cohen, Non-equilibrium fluctuations and the hierarchy — IX, 59
Ernst, M.H., Exact solutions of the nonlinear Boltzmann equation and related kinetic equations — X, 51

Garcia-Colin, L.S., and J.L. Del Rio, Green's contributions to non-equilibrium statistical mechanics revisited — IX, 75
Greenberg, W., J. Polewczak and P.F. Zweifel, Global existence proofs for the Boltzmann equation — X, 19
Gunton, J.D., see Kawasaki, K. — IX, 201

Hafskjold, B., and G. Stell, The equilibrium statistical mechanics of simple ionic liquids — VIII, 175
Hioe, F.T., E.W. Montroll and M. Yamawaki, On higher order WKB approximations for the calculation of energy levels — IX, 295
Hioe, F.T., Some statistical and dynamical problems in quantum electronics — XII, 75

Ianiro, N., see De Masi, A. — XI, 123
Imaeda, T., see Kawasaki, K. — IX, 201

Kac, M., and J. Logan, Fluctuations — VII, 1
Kawasaki, K., T. Imaeda and J.D. Gunton, Coarse-grained Helmholtz free energy functional — IX, 201
Kiefer, J.E., see Weiss, G.H. — XII, 169
Kitahara, K., see Metiu, H. — VII, 229
Kubo, R., H-theorems for Markoffian processes — IX, 101

Lanford III, O.E., On a derivation of the Boltzmann equation	X, 1
Lebowitz, J.L., see Penrose, O.	VII, 293
Lebowitz, J.L., and E.M. Waisman, On the equilibrium theory of fluids: An introductory overview	VIII, 1
Lebowitz, J.L., see Bricmont, J.	XII, 205
Levelt Sengers, J.M.H., and J.V. Sengers, How close is "close to the critical point"?	IX, 239
Lindenberg, K., see West, B.J.	XII, 189
Logan, J., see Kac, M.	VII, 1
Mayer, J.E., Equilibrium density matrix for fluids	IX, 323
Meijer, P.H.E., Theory of diffusion via an interstitial and vacancy mechanism	XII, 99
Metiu, H., K. Kitahara and J. Ross, Statistical mechanical theory of the kinetics of phase transitions	VII, 229
Montroll, E.W., and B.J. West, On an enriched collection of stochastic processes	VII, 61
Montroll, E.W., see Hioe, F.T.	IX, 295
Montroll, E.W., and M.F. Shlesinger, On the wonderful world of random walks	XI, 1
Pellegrinotti, A., see De Masi, A.	XI, 123
Penrose, O., and J.L. Lebowitz, Towards a rigorous theory of metastability	VII, 293
Percus, J.K., Non-uniform fluids	VIII, 31
Pfister, C.E., see Bricmont, J.	XII, 205
Polewczak, J., see Greenberg, W.	X, 19
Potts, R.B., Mathieu's difference equation	XII, 111
Presutti, E., see De Masi, A.	XI, 123
Ross, J., see Metiu, H.	VII, 229
Rubin, R.J., Energy flow and thermal conductivity in one-dimensional, harmonic, isotopically disordered crystals	IX, 111
Sengers, J.V., see Levelt Sengers, J.M.H.	IX, 239
Shlesinger, M.F., see Montroll, E.W.	XI, 1
Shlesinger, M.F., and G.H. Weiss, Elliott Waters Montroll	XII, 1
Shlesinger, M.F., see Bendler, J.T.	XII, 31
Spohn, H., Fluctuation theory for the Boltzmann equation	X, 225
Stell, G., see Hafskjold, B.	VIII, 175

Stell, G., Mayer–Montroll equations (and some variants) through history for fun and profit   XII, 127
Stillinger, F.H., Low frequency dielectric properties of liquid and solid water   VIII, 341

Van Kampen, N.G., Stochastic description of many-body systems   IX, 89
Van Kampen, N.G., Illumination in a random medium   XII, 157
van Leeuwen, J.M.J., Exact renormalization in two dimensional ising systems   IX, 225

Waisman, E.M., see Lebowitz, J.L.   VIII, 1
Weiss, G.H., see Shlesinger, M.F.   XII, 1
Weiss, G.H., and J.E. Kiefer, Random walks in crystallography   XII, 169
West, B.J., see Montroll, E.W.   VII, 61
West, B.J., and K. Lindenberg, On the quantum Langevin equation: the linear oscillator   XII, 189
Widom, B., The interfaces between fluid phases   IX, 273
Wightman, A.S., The mechanisms of stochasticity in classical dynamical systems   IX, 343

Yamawaki, M., see Hioe, F.T.   IX, 295

Zallen, R., Stochastic geometry: aspects of amorphous solids   VII, 177
Zwanzig, R., Where do we go from here?   IX, 123
Zweifel, P.F., see Greenberg, W.   X, 19